MATHEMATICAL SOFTWARE

Martin Peters

18th August, 2002

Proceedings of the First International Congress of Mathematical Software

MATHEMATICAL SOFTWARE

Beijing, China 17–19 August 2002

Editors

Arjeh M. Cohen

Eindhoven University of Technology, The Netherlands

Xiao-Shan Gao

Chinese Academy of Sciences, China

Nobuki Takayama

Kobe University, Japan

World Scientific

New Jersey • London • Singapore • Hong Kong

Published by

World Scientific Publishing Co. Pte. Ltd.

P O Box 128, Farrer Road, Singapore 912805

USA office: Suite 1B, 1060 Main Street, River Edge, NJ 07661

UK office: 57 Shelton Street, Covent Garden, London WC2H 9HE

British Library Cataloguing-in-Publication Data
A catalogue record for this book is available from the British Library.

Cover Design: Lin WANG
Cover Picture: Thomas Banchoff and Davide Cervone

MATHEMATICAL SOFTWARE
Proceedings of the First International Congress of Mathematical Software

ISBN 981-238-048-5

Printed by FuIsland Offset Printing (S) Pte Ltd, Singapore

PREFACE

This proceedings consists of invited papers and contributed papers presented at the First International Congress of Mathematical Software (ICMS 2002) held at Beijing, China. The program of this conference can be found at http://www.mathsoftware.org.

Mathematics has a wide variety of branches. Despite this, we in the first International Congress of Mathematical Software believe that mathematics is fundamentally just one thing. Algebra, geometry and analysis are examples of separate specialities within mathematics, and of course we consider each of them valuable and hence study them. But we take the view that any branch of mathematics can borrow freely from these specialities. It is often worth while delving into them for ideas that might lead to fundamental new discoveries. The same can be said of mathematical software systems: the study of mathematical software is a coherent whole.

We believe that the appearance of mathematical software is one of the most important events in mathematics. Mathematical software systems are used to construct examples, to prove theorems, and to find new mathematical phenomena. Conversely, mathematical research often motivates developments of new algorithms and new systems. Beyond mathematics, mathematical software systems are becoming indispensable tools in many branches of science and technology.

The development of mathematical software systems relies on the cooperation of mathematicians, algorithm designers, programmers, and the feedback from users. The main audiences of this conference are mathematical software developers and programming mathematicians, but we also intend to provide an opportunity to discuss these topics with mathematicians and users from application areas. Topics for the conference include but are not limited to:

1. Software engineering problems for mathematical software
2. Mathematics and media (including user interfaces)
3. Mathematics related to mathematical software (experiments, algorithms)
4. High performance computing
5. Applications of mathematical software
6. Presentation of mathematical software

At ICM 1998 in Berlin, there was a mathematical software session. We have evolved it into a satellite conference of ICM 2002, and this is the first

International Congress of Mathematical Software. Many people have contributed to the organization of ICMS 2002. The idea of this conference came from an e-mail chat between Nobuki Takayama and Bernd Sturmfels. Kazuhiro Yokoyama mentioned this idea to Dongming Wang at Paris, and Dongming Wang introduced Nobuki Takayama to people at MMRC, Academy of Mathematical and System Sciences whose dedicated work has made the conference a great success. The efforts of the program committee members and external referees are essential to ensure the quality of the conference. We thank all of them for their great contribution, help, and cooperation. Especially, many many thanks go to our host MMRC and the people there.

We hope that ICMS 2002 will be a landmark conference in the development of mathematical software systems.

Arjeh M. Cohen, Xiao-Shan Gao, Nobuki Takayama
ICMS 2002, Program Committee Co-chairs

Invited Speakers

Plenary speakers

Henk Barendregt	University of Nijmegen
Jonathan Borwein	Simon Fraser University
John Cannon	University of Sydney
Henri Cohen	University of Bordeaux
Gert-Martin Greuel	University of Kaiserslautern
Michael Joswig	Technische Universitat Berlin

Invited speakers in organized sessions

Manual Bronstein	INRIA
Andreas Enge	Ecole polytechnique
Hans-Christian Hege	ZIB
Ulrich Kortenkamp	Freie Universität Berlin
Masakazu Kojima	Tokyo Institute of Technology
Lorenzo Robbiano	Universita di Genova
Stephen M. Watt	The University of Western Ontario
Fritz Schwarz	GMD
Doron Zeilberger	Rutgers University

Conference Organization

General Advisory Board

David Eisenbud	MSRI (USA)
Bernd Sturmfels	UCB (USA)

Program Committee

Co-Chairs:

Arjeh M. Cohen	Eindhoven (Netherlands)
Xiao-Shan Gao	Chinese Academy of Sciences (China)
Nobuki Takayama	Kobe (Japan)

Members:

Johannes Buchmann	Darmstadt (Germany)
Arjeh M. Cohen	Eindhoven (Netherlands)
Jack Dongarra	Tennessee (USA)
Komei Fukuda	ETH (Switzerland)
Xiao-Shan Gao	Chinese Academy of Sciences (China)
Steve Linton	St. Andrews (Scotland)
Michael Pohst	TU Berlin (Germany)
Konrad Polthier	TU Berlin (Germany)
Mike Stillman	Cornell (USA)
Nobuki Takayama	Kobe (Japan)
Dongming Wang	Paris 6 (France)
Paul S. Wang	Kent (USA)
Kazuhiro Yokoyama	Kyushu (Japan)
Ya-Xiang Yuan	Chinese Academy of Sciences (China)

Poster Session Chair

Masayuki Noro	Kobe (Japan)

Organization committee

Co-Chairs

Zhuojun Liu	Beijing (China)
Nobuki Takayama	Kobe (Japan)

Members:

Arjeh M. Cohen	Eindhoven (Netherlands)
Lian Li	Lanzhou (China)
Ken Nakamula	Tokyo (Japan)
Dingkang Wang	Beijing (China)
Lu Yang	Chengdu (China)

CONTENTS

THE EXPERIMENTAL MATHEMATICIAN: A COMPUTATIONAL GUIDE TO THE MATHEMATICAL UNKNOWN

JONATHAN MICHAEL BORWEIN

FRSC Canada Research Chair in IT and Gordon M. Shrum Professor of Science, Director, Centre for Experimental and Constructive Mathematics, Simon Fraser University, Burnaby, V5A 1S6, BC

I intend to speak primarily from the perspective of an informed consumer, and only secondarily as a producer of, mathematical software (`www.mathresources.com`).

The emergence of powerful mathematical computing environments, the growing availability of correspondingly powerful (multi-processor) computers and the pervasive presence of the internet allow for research mathematicians, students and teachers, to proceed heuristically and 'quasi-inductively'. We may increasingly use symbolic and numeric computation visualization tools, simulation and data mining.

Many of the benefits of computation are accessible through low-end 'electronic blackboard' versions of experimental mathematics [1, 7]. This also permits livelier classes, more realistic examples, and more collaborative learning. Moreover, the distinction between computing (HPC) and communicating (HPN) is increasingly moot.

The unique features of our discipline make this both more problematic and more challenging. For example, there is still no truly satisfactory way of displaying mathematical notation on the web; and we care more about the reliability of our literature than does any other science. The traditional role of proof in mathematics is arguably under siege. Limned by examples, I intend to pose questions ([8]) such as:

- What constitutes secure mathematical knowledge?
- When is computation convincing? Are humans less fallible?
- What tools are available? What methodologies?
- What about the 'law of the small numbers'?
- How is mathematics actually done? How should it be?
- Who cares for certainty? What is the role of proof?

And I shall offer some personal conclusions.

Many of the more sophisticated examples originate in the boundary between mathematical physics and number theory and involve the ζ-function, $\zeta(n) = \sum_{k=1}^{\infty} \frac{1}{k^n}$, and its friends [2]. They often rely on the sophisticated use of *Integer Relations Algorithms* — recently ranked among the 'top ten' algorithms of the century [5, 6]. (See [3,4] and `www.cecm.sfu.ca/projects/IntegerRelations/`.)

As time permits, I shall also describe *WestGrid*, the new Western Canadian computer grid (`www.westgrid.ca`), and my own advanced collaboration facility, *CoLab* (`www.colab.sfu.ca`).

References

1. J.M. Borwein, P.B. Borwein, R. Girgensohn and S. Parnes, "Making Sense of Experimental Mathematics," *Mathematical Intelligencer*, **18**, Number 4 (Fall 1996), 12–18. [CECM 95:032][a]

2. J.M. Borwein and D.M. Bradley, "Empirically Determined Apéry–like Formulae for Zeta(4n+3)," *Experimental Mathematics*, **6** (1997), 181–194. [CECM 96:069]

3. Jonathan M. Borwein and Robert Corless, "Emerging Tools for Experimental Mathematics," *American Mathematical Monthly*, **106** (1999), 889–909. [CECM 98:110]

4. D.H. Bailey and J.M. Borwein, "Experimental Mathematics: Recent Developments and Future Outlook," pp, 51-66 in Volume I of *Mathematics Unlimited — 2001 and Beyond,* B. Engquist and W. Schmid (Eds.), Springer-Verlag, 2000. [CECM Preprint 99:143]

5. J. Dongarra, F. Sullivan, "The top 10 algorithms," *Computing in Science & Engineering,* **2** (2000), 22–23. (See `www.cecm.sfu.ca/personal/jborwein/algorithms.html`.)

6. J.M. Borwein and P.B. Borwein, "Challenges for Mathematical Computing," *Computing in Science & Engineering*, **3** (2001), 48–53. [CECM 00:160].

7. J.M. Borwein, "The Experimental Mathematician: The Pleasure of Discovery and the Role of Proof," submitted *International Journal of Computers for Mathematical Learning*, February 2002. [CECM Preprint 02:178]

8. D.H. Bailey, J.M. Borwein and K. Devlin, *The Experimental Mathematician: A Computational Guide to the Mathematical Unknown,* A.K. Peters Ltd (in preparation, 2002) ISBN: 1-56881-136-5.

[a] All references are available at `http://www.cecm.sfu.ca/preprints/`.

COMPUTER ALGEBRA AND FINITE GROUPS

GERT-MARTIN GREUEL AND
GERHARD PFISTER

Department of Mathematics, University of Kaiserslautern, D-67663 Kaise rslautern

E-mail: greuel@mathematatik.uni-kl.de and pfister@mathematik.uni-kl.de

1 The Problem

The problem we address in this article can simply be stated as: *characterise the class of finite solvable groups by 2–variable identities.*

To explain the problem, note that a group G is abelian if and only if $[x, y] = 1$ for all $x, y \in G$, where $[x, y] = xyx^{-1}y^{-1}$ is the commutator. Recall also that a finite group G is nilpotent if and only if it satisfies the Engel identity $[x, y]_n$ for some n and all $x, y \in G$ (here $[x, y]_1 = [x, y]$ and $[x, y]_{n+1} = \big[[x, y]_n, y\big]$). It has been conjectured that some Engel–like identities should characterise also the class of finite solvable groups. The following, more precise, conjecture is due to B. Plotkin (cf. [9], [10]).

For a given word w in the letters X, Y, X^{-1}, Y^{-1}, consider the sequence $U_n = U_n^w$ of words defined inductively by

$$U_1 = w$$
$$U_{n+1} = [XU_nX^{-1}, YU_nY^{-1}].$$

Conjecture 1.1 (B. Plotkin). There exists a word w such that a finite group G is solvable if and only if there exists an $n \geq 1$ satisfying $U_n^w(x, y) = 1$ for all $x, y \in G$.

Recall that the minimal finite non–solvable groups G (that is, all subgroups of G are solvable) were classified by J. Thompson [11] in 1968:

(1) PSL$(2, p)$, p a prime number, $p = 5$ or $p > 5$ and $p = \pm 2 \bmod 5$.

(2) PSL$(2, 2^n)$, $n \geq 2$, a prime number.

(3) PSL$(2, 3^n)$, n odd, a prime number.

(4) PSL$(3, 3)$.

(5) $Sz(2^n)$, n odd.

In view of this result, Conjecture 1.1 is implied by

Conjecture 1.2. Let G be one of the groups in Thompson's list (1) – (5). Then there exists a word w in X, Y, X^{-1}, Y^{-1}, independent of G, such that for all n the equation $U_n(x, y) = 1$ does not have a solution in G.
Conjecture 1.2, for all cases except for the Suzuki groups in (5), has been proved by G.-M. Greuel, F. Grunewald, B. Kunyavskii, G. Pfister, E. Plotkin, Y. Segev, cf. [8].

Theorem 1.3 ([8]). *Conjecture 1.2 holds for the word $w = X^{-1}YXY^{-1}X$ and the groups (1) – (4) in Thompson's list.*
The proof, in the spirit of [2], proceeds as follows: for any of the groups in (1) – (3), consider its standard linear representation, that is, consider the matrices

$$x = \begin{pmatrix} 0 & -1 \\ 1 & t \end{pmatrix}, \qquad y = \begin{pmatrix} 1 & b \\ c & 1 + bc \end{pmatrix}$$

as elements of $\mathrm{PSL}(2, p^n)$, p any prime number. (The case (4), $\mathrm{PSL}(3, 3)$, is easy.) Then it is easy to see that $U_1(x, y) \neq 1$ for all x, y, as above.

Consider the matrix equation $U_1(x, y) = U_2(x, y)$. If this has a solution for some x, y, as above, then, by construction of the sequence U_n, $U_n(x, y) = U_1(x, y) \neq 1$ for all $n \geq 1$. Hence, it suffices to show that $U_1(x, y) = U_2(x, y)$ has a solution. We shall show this for $\mathrm{PSL}(2, p)$, $p \geq 5$, $\mathrm{PSL}(2, 2^n)$, $n \geq 2$ and $\mathrm{PSL}(2, 3^n)$, $n \geq 2$. Note that the groups $\mathrm{PSL}(2, 2)$ and $\mathrm{PSL}(2, 3)$ are solvable, hence $U_1(x, y) = U_2(x, y)$ cannot have a solution over the prime field in characteristic 2 and 3. Therefore, the groups in (2) and (3) have to be treated separately (by the same method as the groups in (1) for large characteristic).

Consider the ideal I generated by the entries of the matrix $U_1(x, y) - U_2(x, y)$, which is an ideal in $\mathbb{Z}[b, c, t]$.

We are going to show that the variety $V(I)$ has a rational point $(b, c, t) \in (\mathbb{F}_{p^n})^3$ for all (p, n), $p \geq 5$ and $n = 1$ or $p = 2, 3$ and $n \geq 2$, which is equivalent to $U_1(x, y) = U_2(x, y)$ having a solution $x, y \in \mathrm{PSL}(2, p^n)$.

To show that $V(I)$ has a rational point, we use:

- the Hasse–Weil bound for the number of rational points on an absolutely irreducible curve and its generalisation to singular curves,
- simple facts from algebraic geometry and singularity theory,
- simple generalisations of basic results from the theory of standard bases,
- extensive computations, using SINGULAR.

We should mention that there are other words satisfying the statement of Theorem 1.3 (we checked this for $w = x^{-2}y^{-1}x$). In this article we concentrate on those steps of the proof which require computer algebra. Full details, containing the group theoretical background as well as a description of special strategies which helped to speed up the computation, will appear in [8].

2 $\mathbb{F}_{p^n}$–rational Points

Let $w = x^{-1}yxy^{-1}x$ and define $U_1 = w$, $U_2 = [xU_1x^{-1}, yU_1y^{-1}]$ (with $[x,y] = xyx^{-1}y^{-1}$).

Theorem 2.1. *The equation* $U_1 = U_2$ *has a solution* $x = \left(\begin{smallmatrix} 0 & -1 \\ 1 & t \end{smallmatrix}\right)$, $y = \left(\begin{smallmatrix} 1 & b \\ c & 1+bc \end{smallmatrix}\right) \in \mathrm{PSL}(2,p)$ *such that* $U_1(x,y) \neq \left(\begin{smallmatrix} 1 & 0 \\ 0 & 1 \end{smallmatrix}\right)$ *for all prime numbers* $p \geq 5$.

Before we start with the proof of Theorem 2.1, let us mention the following lemma, which is easily proved by looking at the equations.

Lemma 2.2. *The equation* $U_1(x,y) = \left(\begin{smallmatrix} 1 & 0 \\ 0 & 1 \end{smallmatrix}\right)$ *has no solution.*

Let $I \subseteq \mathbb{Z}[b,c,t]$ be the ideal generated by the four polynomials defining $U_1 = U_2$. Denote by $I^{(p)} \subseteq \mathbb{F}_p[b,c,t]$ the ideal induced by I.

Theorem 1.3 for $\mathrm{PSL}(2,p)$, $p \geq 5$, is then a consequence o f the following theorem:

Theorem 2.3. *For all prime numbers* $p \geq 5$ *the variety* $V(I^{(p)}) \subset \mathbb{F}_p^3$ *is not empty.*

For all prime numbers $5 \leq p \leq 2530$ this has been checked by a computer in the most trivial way, by substituting the points of $\mathbb{F}_p^3$ in the polynomial generating $I^{(p)}$ and stopping when a solution was found.

For the other prime numbers, we are going to apply the theorem of Hasse–Weil as generalised by Aubry and Perret to singular curves (cf. [1]) and using the fact that the affine curve C has, at most, $\deg(\overline{C})$ rational points less than the projective closure $\overline{C}$.

Theorem 2.4. *Let* $C \subseteq \mathbb{A}^n$ *be an absolutely irreducible affine curve defined over the finite field* $\mathbb{F}_q$ *and* $\overline{C} \subset \mathbb{P}^n$ *the projective closure, then the number of* $\mathbb{F}_q$*–rational points of* C *is at least* $q + 1 - 2p_a\sqrt{q} - d$ *with* $d =$ *the degree and* p_a *the arithmetic genus of* $\overline{C}$.

Note that the Hilbert function of $\overline{C}$, $H(t) = dt - p_a + 1$, can be computed from the homogeneous ideal I_h of $\overline{C}$, hence we can compute d and p_a without any knowledge about the singularities of $\overline{C}$.

Now we use the following notations. Let $I_h \subset \mathbb{Z}[b,c,t,h]$ denote the homogenisation of I with respect to the homogenising variable h and $I_h^{(p)}$ the induced ideal in $\mathbb{F}_p[b,c,t,h]$ for a prime number p.

Let $X^{(p)}$ be the projective curve in $\mathbb{P}^3_{\mathbb{F}_p}$, defin ed by $I_h^{(p)}$.

Proposition 2.5.

(1) $X^{(p)}$ *is absolutely irreducible for all prime numbers* $p \geq 2530$.

(2) $X^{(2)}$ *and* $X^{(3)}$ *are absolutely irreducible.*

Before proving the proposition, we apply it to prove Theorem 2.3. Since $X^{(p)}$ is absolutely irreducible, we can apply Theorem 2.4 and get

$$\#X^{(p)} \geq p + 1 - 2p_a\sqrt{p} - d\,.$$

We compute the Hilbert function of $X^{(p)}$ (using SINGULAR) as $16t - -24$. This implies $d = 16$ and $p_a = 25$.

We need $p + 1 - 50\sqrt{p} - 16 > 0$, which holds if $p \geq 2530$.

It remains to prove Proposition 2.5. Here it is sufficient to prove that a suitable projection of $X^{(p)}$ to a plane curve is absolutely irreducible. Elimination of the variable c (using SINGULAR) results in a polynomial f of degree $d = 12$, which defines the projection of $X^{(p)}$ (d drops from 16 to 12 since we project from a point of $X^{(p)}$).

To prove that f is absolutely irreducible, we write $f = f_1^{(n)} f_2^{(d-n)}$, f_1 a homogeneous polynomial of degree n in b,t,h with variables $a_1^{(n)}, \ldots, a_{s(n)}^{(n)}$ as coefficients and f_2 a homogeneous polynomial of degree $d-n$ in b,t,h with variables $b_1^{(n)}, \ldots, b_{\ell(n)}^{(n)}$ as coefficients, $d = 12$ and $n = 1, \ldots, 6$.

This gives six systems of equations defined by polynomials in $\mathbb{Z}[\{a_i^{(n)}\}, \{b_j^{(n)}\}]$. Let $I_n \subset \mathbb{Z}[\{a_i^{(n)}\}, \{b_j^{(n)}\}]$ be the corresponding ideal $n = 1, \ldots, 6$. We have to prove that over $\overline{\mathbb{F}}_p$ (the algebraic closure of $\mathbb{F}_p$) none of these ideals has a common zero. This is equivalent to the fact that 1 is contained in the ideals $I_n(\bmod p)$.

For the proof we use the following easy splitting lemma (cf. [9]).

Lemma 2.6. *Let* $I \subset \mathbb{Z}[x_1, \ldots, x_n]$ *be an ideal and assume that* $f \cdot g \in I$ *for some polynomials* $f, g \in \mathbb{Z}[x_1, \ldots, x_n]$. *Then* $\sqrt{I} = \sqrt{\langle I, f\rangle} \cap \sqrt{\langle I, g\rangle}$.

Now we want to apply Corollary 3.4 and Lemma 2.6 to prove that $1 \in I_k\mathbb{F}_p[\{a_\nu^{(k)}\}, \{b_\nu^{(k)}\}]$.

For a finite, not too large, set of primes this can be checked by computing a Gröbner basis of $I_k\mathbb{F}_p[\{a_\nu^{(k)}\}\{b_\nu^{(k)}\}]$. However, we are faced with the problem of checking this for infinitely many points. The idea is to compute a Gröbner basis over $\mathbb{Q}$, taking care of the coefficients.

To be specific, let $S^{(k)} \subseteq \mathbb{Z}[\{a_\nu^{(k)}\},\{b_\nu^{(k)}\}]$ be a Gröbner basis of $I_k\mathbb{Q}[\{a_\nu^{(k)}\},\{b_\nu^{(k)}\}]$. If $e \in \mathbb{Z}$ is in $S^{(k)}$ then, using Corollary 3.4 (2), we obtain $I_k\mathbb{F}_p[\{a_\nu^{(k)}\},\{b_\nu^{(k)}\}] = \mathbb{F}_p[\{a_\nu^{(k)}\},\{b_\nu^{(k)}\}]$ for all p not dividing e.

Unfortunately, this does not work straightforwardly, since e becomes too large. However, if $e \in S^{(k)}$ is too large, we can apply Lemma 2.6 to split the ideal and obtain for the components smaller integers in the standard basis. The splitting tool is used in the following way:

- We factorise the generators of the ideal and split the ideal if one of the generators has a nontrivial factor.
- If all generators are irreducible in $\mathbb{Q}[\{a_\nu^{(k)}\},\{b_\nu^{(k)}\}]$, we compute a Gröbner basis for suitable sub-ideals to obtain new generators which could have a nontrivial factorisation. In our application "suitable" is a mixture of the first 30 generators and the sub-ideal generated by the generators with few monomials.

Finally, we obtained the following list of prime numbers ≥ 2530 for which we checked directly that $1 \in I_n\mathbb{F}_p\left[\{a_\nu^{(n)}\},\{b_\nu^{(n)}\}\right]$ by computing a Gröbner basis separately:
n=2: 3463,7489
n=3: 7129,9041,10343,18413,20857
n=4: 2671,2693,3299,3373,3391,3517,3593,3701,3779,4111,4409,4423,4561, 4657,4793,5273,5399,5659,5987,6949,7487,8011,8243,8887, 9769,10159, 10177,12007,26347
n=6: 6151,6343,7823

It remains to study the equation $U_1 = U_2$ for $\mathrm{PSL}(2,2^n)$, $n \geq 2$, and $\mathrm{PSL}(2,3^n)$, $n \geq 3$ and odd. A direct computation of the Gröbner basis over $\mathbb{F}_2$ (resp. $\mathbb{F}_3$) of the corresponding ideals $I_k\mathbb{F}_p[\{a_\nu^{(k)}\},\{b_\nu^{(k)}\}]$, $p = 2,3$, shows that the curves $X^{(2)}$ and $X^{(3)}$ are also absolutely irreducible. Theorem 2.4 implies that $U_1 = U_2$ has a solution in $\mathrm{PSL}(2,2^n)$ (resp. $\mathrm{PSL}(2,3^n)$) if $n \geq 12$ (resp. $n \geq 8$).

For the remaining cases we checked directly by a computer, trying all matrices in $\mathrm{PSL}(2,2^n)$, resp. $\mathrm{PSL}(2,3^n)$, in order to show that $U_1 = U_2$ has a solution.

The computations were carried out on an AMD Athlon 700 with 256 MB, running Linux and using the computer algebra system SINGULAR [4]. Finding rational points in Theorem 2.3 for $5 \leq p \leq 2530$ (the trivial way) took roughly six hours, showing absolute irreducibility in Proposition 2.5 (1) (except for the primes tested above) took about one day, while the computations for the primes listed above lasted about half a day. Finally, the computation for $\mathrm{PSL}(2,2^n)$ and $\mathrm{PSL}(3,3^n)$ lasted together about half a day.

3 Standard Bases and Specialisation

We recall and generalise some notions and results from the theory of standard bases which are used in the proof of Proposition 2.5.

A *monomial ordering* is a total ordering $>$ on the set of monomials $\mathrm{Mon}(x_1,\ldots,x_n) = \{x^\alpha = x_1^{\alpha_1}\cdot\ldots\cdot x_n^{\alpha_n} \mid \alpha = (\alpha_1,\ldots,\alpha_n) \in \mathbb{Z}^n_{\geq 0}\}$, satisfying: $x^\alpha > x^\beta \Rightarrow x^\gamma x^\alpha > x^\gamma x^\beta$, for all α,β,γ.

We do not assume $>$ to be a well–ordering, in particular, we allow $x_i < 1$ for all i, which is used when we compute invariants of singularities in local rings.

Let R be a Noetherian integral domain and let $f \in R[x] = R[x_1,\ldots,x_n]$ be a nonzero polynomial. With respect to a fixed monomial ordering, write f as a sum of nonzero terms

$$f = a_\alpha x^\alpha + a_\beta x^\beta + \cdots + a_\gamma x^\gamma, \quad x^\alpha > x^\beta > \cdots > x^\gamma\,,$$

with $a_\alpha, a_\beta,\ldots,a_\gamma \in R$. We call

$\mathrm{LM}(f) := x^\alpha$ the *leading monomial* of f,
$\mathrm{LC}(f) := a_\alpha$ the *leading coefficient* of f,
$\mathrm{LT}(f) := a_\alpha x^\alpha$ the *leading term* of f,

and assume always that $f \neq 0$, if we use these notations.

Consider the multiplicatively closed set $S_> := \{f \in R[x] \mid \mathrm{LT}(f) = 1\}$, and define the localisation of $R[x]$ with respect to $S_>$ as

$$R[x]_> := S_>^{-1}(R[x]) = \left\{ \frac{f}{g} \;\middle|\; f,g \in R[x],\ \mathrm{LT}(g) = 1 \right\}.$$

If $>$ is a global ordering (that is, $x_i > 1$ for all i), then $R[x]_> = R[x]$. If $>$ is a local ordering ($x_i < 1$ for all i), then $R[x]_> = R[x]_{\langle x_1,\ldots,x_n\rangle}$, the localisation of $R[x]$ at the prime ideal $\langle x_1,\ldots,x_n\rangle$.

For any ideal $I \subset R[x_1,\ldots,x_n]$, we set

$$L(I) := \langle \mathrm{LT}(f) \mid f \in I \rangle\,,$$

the ideal in $R[x_1,\ldots,x_n]$ generated by all leading terms of elements of I. Recall that a finite set $G \subset I$ is called a *standard basis* of I (over the ring R) if $L(I) = \langle \mathrm{LT}(g) \mid g \in G\rangle$. If R is a field, we can use the leading monomials $\mathrm{LM}(g)$ instead of the leading terms in the above definition. If $>$ is a well–ordering, a standard basis is usually called a *Gröbner basis*.

We do not need standard bases over rings but something weaker, which we call pseudo standard bases and which we explain now.

Definition 3.1. Let $\mathcal{G}$ denote the set of finite ordered sets (or lists) $G \subset R[x] = R[x_1, \ldots, x_n]$. A function

$$\mathrm{NF} : R[x] \times \mathcal{G} \to R[x], \ \ (f, G) \mapsto \mathrm{NF}(f|G) \ ,$$

is called a *pseudo normal form* on $R[x]$ if the following holds:

(1) $\mathrm{NF}(f|G) \neq 0 \Rightarrow \mathrm{LM}\big(\mathrm{NF}(f|G)\big)$ is not divisible by $\mathrm{LM}(g)$ for all $g \in G$.

(2) For all $f \in R[x]$ and $G = \{g_1, \ldots, g_s\} \in \mathcal{G}$ there exists a $u \in R[x]$ with $\mathrm{LT}(u)$ a product of some $\mathrm{LC}(g_i)$, such that $r := uf - \mathrm{NF}(f|G)$ has a *standard representation* with respect to G. This means either $r = 0$, or

$$r = \sum_{i=1}^{s} a_i g_i, \ a_i \in R[x] \ ,$$

such that $\mathrm{LM}(r) \geq \mathrm{LM}(a_i g_i)$ for all i with $a_i g_i \neq 0$.

If in (2) we have $u = 1$, respectively u a unit in $R[x]_>$, then NF is called a normal form, respectively a weak normal form, (cf. [6], [7]).

We define the *s-polynomial* of two polynomials $f, g \in R[x] \smallsetminus \{0\}$, $\mathrm{LT}(f) = ax^\alpha$ with $\mathrm{LT}(g) = by^\beta$ as

$$\mathrm{spoly}(f, g) := bx^{\gamma - \alpha} f - ax^{\gamma - \beta} g \ ,$$

where $\gamma = (\gamma_1, \ldots, \gamma_n)$, $\gamma_i = \max\{\alpha_i, \beta_i\}$, $i = 1, \ldots, n$. If $\mathrm{LM}(g) \mid \mathrm{LM}(f)$, then $\gamma = \alpha$ and we have

$$\mathrm{spoly}(f, g) = bf - ax^{\alpha - \beta} g \ .$$

In our application, we need the following result, which is the reason why we introduced pseudo normal forms (notations from above).

Proposition 3.2. *Let $I \subset R[x]_>$ be an ideal, NF a pseudo normal form on $R[x]_>$ and $G = \{g_1, \ldots, g_s\} \subset I$ satisfying*

(1) G generates $IR[x]_>$;

(2) $\mathrm{NF}\big(\mathrm{spoly}(g_i, g_j) \mid G\big) = 0$ *for all $1 \leq i < j \leq s$.*

Then, for any maximal ideal $\mathfrak{m} \subset R$ such that $\mathrm{LC}(g_i) \notin \mathfrak{m}$ *for all i, the set $\{\bar{g}_1, \ldots, \bar{g}_s\}$ is a standard basis of $I \cdot (R/\mathfrak{m})[x]_>$. Here $\bar{g}_i$ denotes the residue class of g_i in $(R/\mathfrak{m})[x]$.*

The proof is the same as in [6] or [7].

If R is a field and $\mathfrak{m} = \langle 0 \rangle$, this is just Buchberger's criterion for a standard basis (cf. [3]).

Definition 3.3. We call G, satisfying (1) and (2) of Proposition 3.2, a *pseudo standard basis* of I (note that this is *not* a standard basis of I over the ring R).

Corollary 3.4.

(1) Let $R = K[t]$, $t = (t_1, \dots, t_r)$, K a field and let $\{g_1, \dots, g_s\} \subset (K[t])[x]$, $x = (x_1, \dots, x_n)$, be a pseudo standard basis of I. Set $h_i = \mathrm{LC}(g_i)$ and $h = h_1 \cdot \ldots \cdot h_s \in K[t]$. Then, for any $t_0 \in K^s$ such that $h(t_0) \neq 0$, $\{g_1(t_0, x), \dots, g_s(t_0, x)\}$ is a standard basis of $I|_{t=t_0} \subset K[x]$.

(2) Let $R = \mathbb{Z}$, $\{g_1, \dots, g_s\} \subset \mathbb{Z}[x]$ a pseudo standard basis of I, let $h_i = \mathrm{LC}(g_i)$ and set $h = h_i \cdot \ldots \cdot h_s \in \mathbb{Z}$. Then, for any prime p so that $p \nmid h$, $\{\bar{g}_1, \dots, \bar{g}_s\}$ is a standard basis of $\overline{I}$ in $\mathbb{F}_p[x]$.

Hence, we obtain an effective way to determine parameters t_0, respectively prime numbers p, for which $\{g_1, \dots, g_s\}$ induces a standard basis of $I|_{t=t_0}$ (t substituted by t_0), respectively of $\overline{I}$ (the reduction of I mod p). However, these conditions are only sufficient, not necessary (and depend on the strategy of pair selection in the Buchberger algorithm). Indeed, the strategy in Section 2 was chosen to keep $\mathrm{LC}(g_i)$ small.

To see that nontrivial pseudo normal forms exist, we generalise `NFBuchberger` and `NFMora` from [5], [7] to the present situation. As before, R denotes a Noetherian integral domain.

Algorithm 3.5 (NFBuchberger). Assume $>$ to be a global monomial ordering.

Input: $f \in R[x_1, \dots, x_n]$, $G = \{g_1, \dots, g_s\} \in \mathcal{G}$.

Output: $h \in R[x_1, \dots, x_n]$, a pseudo normal form of f with respect to G, having a standard representation

$$cf - h = \sum_{i=1}^{s} a_i g_i$$

with c a product of leading coefficients of elements of G.

`NFBuchberger` is essentially the same as the following algorithm `NFMora` applied to global orderings.

Algorithm 3.6 (NFMora). Let $>$ be any monomial ordering.

Input: $f \subset R[x_1, \ldots, x_n]$, $G = \{g_1, \ldots, g_s\} \in \mathcal{G}$.

Output: $h \in R[x_1, \ldots, x_n]$, a pseudo normal form of f with respect to G, having a standard representation

$$uf - h = \sum_{i=1}^{s} a_i g_i$$

with $\mathrm{LM}(u) = 1$ and $\mathrm{LC}(u)$ a product of leading coefficients of elements of G.

- $h = f$;
- $T = G$;
- while $h \neq 0$ and $T_h = \{g \in T \mid \mathrm{LM}(g) \mid \mathrm{LM}(h)\} \neq \emptyset$

  choose $g \in T_h$ with $\mathrm{ecart}(g)$ minimal;

  if $(\mathrm{ecart}(g) > \mathrm{ecart}(h))$

  $\quad T = T \cup \{h\}$,

  $h = \mathrm{spoly}(h, g)$;
- return(h);

Proof. Termination of `NFMora` can be seen as in [7, Algorithm 1.7.6].

To see correctness, consider the ν–th turn in the while loop. There we have $(h_0 := f)$

$$\begin{aligned} h_\nu &= \mathrm{spoly}(h_{\nu-1}, g'_\nu) = c'_\nu h_{\nu-1} - m_\nu g'_\nu \ , \\ \mathrm{LM}(h_{\nu-1}) &= \mathrm{LM}(m_\nu)\,\mathrm{LM}(g'_\nu) \ , \\ \mathrm{LM}(h_{\nu-1}) &> \mathrm{LM}(h_\nu) \ , \end{aligned}$$

where $c'_\nu = \mathrm{LC}(g'_\nu)$, m_ν is a term and for g'_ν we have two possibilities:

(a) $g'_\nu = g_j \in \{g_1, \ldots, g_s\} = G$ for some j, or

(b) $g'_\nu = h_j \in T \smallsetminus G = \{h_0, \ldots, h_{\nu-3}, h_{\nu-2}\}$.

Suppose, by induction, that at the end of step $\nu - 1$ we have constructed a standard representation $(u_0 := 1, a_i^{(0)} := 0)$

$$u_{\nu-1} f = \sum_{i=1}^{s} a_i^{(\nu-1)} g_i + h_{\nu-1}, \ a_i^{(\nu-1)} \in R[x_1, \ldots, x_n] \ , \qquad (*)$$

with $\mathrm{LM}(u_{\nu-1}) = 1$ and $\mathrm{LC}(u_{\nu-1})$ a product of leading coefficients of elements of G.

In case (a), multiply both sides of $(*)$ with $c_j = \mathrm{LC}(g_j)$ and replace $c_j h_{\nu-1}$ with $h_\nu + m_\nu g_j$. We obtain an expression $u_\nu f = \sum a_i^{(\nu)} g_i + h_\nu$ with $u_\nu = c_j u_{\nu-1}$ and $a_i^{(\nu)} \in R[x_1, \ldots, x_n]$, hence, $\mathrm{LM}(u_\nu) = 1$, $\mathrm{LC}(u_\nu)$ a product of leading coefficients of elements of G.

In case (b), multiply both sides of $(*)$ with $c'_\nu = \mathrm{LC}(h_j)$ and replace $c'_\nu h_{\nu-1}$ with $h_\nu + m_\nu h_j$, where $j < \nu - 1$.

By induction,

$$- - h_j = \sum_{j=1}^{s} a_i^{(j)} g_j - u_j f, \ \ \mathrm{LM}(u_j) = 1$$

with $\mathrm{LM}(u_j) = 1$, $\mathrm{LC}(u_j)$ a product of leading coefficients of elements of G. Hence, $(c'_\nu u_{\nu-1} - m_\nu u_j) f = \sum_{i=1}^{s} a_i^{(\nu)} g_i + h_\nu$, $a_i^{(\nu)} \in K[x_1, \ldots, x_n]$.

Since $\mathrm{LM}(m_\nu)\,\mathrm{LM}(g'_\nu) = \mathrm{LM}(h_{\nu-1}) < \mathrm{LM}(h_j)$ (since $j < \nu - 1$), we obtain $\mathrm{LM}(m_\nu) < 1$ and, hence, $\mathrm{LM}(m_\nu u_j) < 1$.

By induction, $\mathrm{LC}(u_{\nu-1})$ is a product of leading coefficients of G and $\mathrm{LM}(u_{\nu-1}) = 1$, hence $\mathrm{LM}(c'_\nu u_{\nu-1}) = 1$ and setting $u_\nu = c'_\nu u_{\nu-1} - m_\nu u_j$, we obtain the result. □

Acknowledgements

We should like to thank B. Kunyavskii for communcating the problem to us and for continuous discussions. Moreover, we are grateful to the Deutsche Forschungsgemeinschaft, the German-Israeli Foundation for Scientific Research and Development and the Stiftung Rheinland–Pfalz für Innovation for financial support. The first author was supported by DFG Grant, GR 640/9-2, GIF grant G-616-5.C/1999 and the Stiftung Rheinland–Pfalz für Innovation; the second author was supported by the Stiftung Rheinland–Pfalz für Innovation.

References

1. Aubry, Y.; Perret, M.: *A Weil Theorem for Singular Curves.* Arithmetic, Geometry and Coding Theory. Eds.: R. Pellikaan, M. Perret and S.G. Vladut, Walter de Gruyter, Berlin-New York, 1–7 (1996).
2. Bombierei, E.: *Thompson's problem* $\sigma^2 = 3$. Invent. Math. **58**, 77–100 (1980).

3. Buchberger, B.: *Ein Algorithmus zum Auffinden der Basiselemente des Restklassenringes nach einem nulldimensionalen Polynomideal.* PhD Thesis, University of Innsbruck, Austria (1965).
4. Greuel, G.-M.; Pfister, G.; Schönemann, H.: SINGULAR 2.0. *A Computer Algebra System for Polynomial Computations.* Centre for Computer Algebra, University of Kaiserslautern (2001). `http://www.singular.uni-kl.de`.
5. Greuel, G.-M.; Pfister, G.: *Advances and Improvements in the Theory of Standard Bases and Syzygies.* Arch. Math. **66**, 163–176 (1996).
6. Greuel, G.-M.; Pfister, G.: *Gröbner Bases and Algebraic Geometry.* In: B. Buchberger and F. Winkler (ed.): Gröbner Bases and Applications. LNS 251, 109–143, CUP 1998.
7. Greuel, G.-M.; Pfister, G.: *A* SINGULAR *Introduction to Commutative Algebra.* To appear, Springer (2002).
8. Greuel, G.-M.; Grunewald, F.; Kunyavskii, B.; Pfister, G.; Plotkin, E.; Segev, Y.: *On Equations in Finite Linear Groups.* To appear.
9. Grunewald, F.; Kunyavskii, B.; Nikolova, D.; Plotkin, E.: *Two-variable Identities in Groups and Lie Algebras.* Zap. Nauch. Semin. POMI **272**, 161–176 (2000).
10. Plotkin, B.; Plotkin, E.; Tsurkov, A.: *Geometrical Equivalence of Groups.* Comm. Algebra **27**, 4015–4025 (1999).
11. Thompson, J.: *Non-solvable Finite Groups all of whose Local Subgroups are Solvable.* Bull. Amer. Math. Soc. **74**, 383–437 (1968).

SOFTWARE INTEGRATION AND COMPUTER PROOFS

MICHAEL JOSWIG

Institut für Mathematik, MA 6-2, Technische Universität, Str. 17. Juni 136
D-10623 Berlin, Germany, E-mail: joswig@math.tu-berlin.de

The `polymake` system is introduced as a model for the integration of mathematical software components. While distributed computation can contribute to the verification of computer generated results, other concepts are necessary in addition. It is discussed how web based archives like Electronic Geometry Models fit in.

1 Introduction

The Mathematical Software Session at ICM98 in Berlin, and even more so this First International Congress on Mathematical Software, are visible landmarks of the fact that computer based mathematics has reached the mathematical mainstream. This paper wants to outline some challenges to this developing field. The topics discussed are not restricted to a specific area but common to machine supported mathematics in general. Examples to support the author's views have deliberately been chosen from discrete geometry and combinatorial topology, but the phenomena described should be applicable to most other mathematical subjects.

Despite the great success of the general purpose computer algebra systems including `Macsyma`, `Reduce`, `Scratchpad`, and, later, `Maple` and `Mathematica`, mathematical software of today is multi-centric. There is a large number of specialized systems such as `GAP`[1], `LAPACK`[2], and `CPLEX`[3] which are indispensable tools for researchers in group theory, numerical linear algebra, and optimization, respectively. Even the largest general purpose computer algebra systems cannot possibly keep up with the rapid development in all areas simultaneously. On the other hand, each specialized system occupies a small region only. So the need arises to combine the available systems.

Essentially there are two concepts to distinguish: One is to build communication models such as OpenMath[4] (or proprietary solutions like MathLink[5] for Mathematica and Maple Powertools[6] for Maple). Each software component which wants to be part of the system must be adapted, the advantage being that of direct communication between the components. The other concept is integration. A core system sets up the rules, and, typically, communication is not direct but between the core and an individual component only. The advantage of the latter, more centralistic, approach is that, if the core offers enough flexibility, existing components can be added to the system without redesign.

In Section 2 polymake is introduced as one realization of this integrational approach. For a discussion of related concepts see Solomon[7]. While polymake has proved to be useful for computations in the theory of convex polytopes, this paper describes the new TOPAZ module for combinatorial topology. As an application it is shown how TOPAZ can be used to recognize the 16 vertex triangulation of the $K3$ surface by Casella and Kühnel[8].

Although computer mathematics have become very successful, a certain reluctance towards the use of computers for proving theorems remains. Psychological aspects may play a role, but some of the criticism raised is legitimate. This is largely centered around the question: How does one verify a computer generated proof? Some of the ramifications of this question are explored in Section 3. One path leads back to the integration of mathematical software. However, taking the criticism seriously requires to go beyond. To this end the author suggests Electronic Geometry Models as an approach to establish a platform for validating computer generated mathematical results, see Section 4.

The problems touched upon here are most certainly too complicated to be solved a direct way. The concepts introduced are meant as one contribution to a necessary discussion.

2 polymake

The polymake system was originally designed as a framework for the investigation of combinatorial and geometric properties of convex polytopes; for an introduction to polytope theory see Ziegler[9]. Since polytope theory is close to algorithmic geometry it does not come as a surprise that there are particularly many specialized software packages available. polymake offers several kinds of interfaces to many of these systems, as will be outlined below. For more detailed information about the system see the web-site[10] and the reports[11 12].

More recently, polymake was expanded in such a way that its TOPAZ module now also handles finite simplicial complexes. An example application is given in Section 2.3.

2.1 System Overview

The overall system structure uses a client/server type model. Essentially the user is restricted to issue queries about polytope objects (represented by files) to the polymake server, which happens to be a Perl program. Such queries can be quite complicated such as:

- What is the volume of the convex hull of a given finite set of points?
- Are all facets of the bounded intersection of finitely many given affine halfspaces combinatorially equivalent to cubes?

As a key feature `polymake` is a *rule based* system which automatically chooses a sequence of operations to satisfy the user's queries. For instance, for the first query above, `polymake` would decide to use a triangulation producing convex hull algorithm; the volume can easily be read off such a triangulation. Whereas for the second query, `polymake` would start a Fourier-Motzkin elimination type dual convex hull code to obtain a combinatorial description of the input polytope P in terms of vertex facet incidences; then it would check whether the vertex-edge graph of P is bipartite, a necessary condition for a cubical polytope; if this is satisfied, then the polytope is cubical if and only if the flag vector of P satisfies certain conditions which can be checked easily.[11] [13]

Intermediate results and answers to the user's queries are stored into the file representing the polytope. This way asking for the same thing for the second time does not result in a re-computation of already known properties.

The `polymake` server does not know anything about polytopes. The rule basis defines the set of atomic operations known to the system, e. g., a convex hull computation. These rules rely on client programs, which are usually written in C++. Besides these client based rules there are also *interface rules* which incorporate features from external programs into the `polymake` system. Typically, these external programs are developed entirely independently of `polymake`, and the data transfer is organized by means of (ASCII) file exchange. This is certainly slow, but it has the advantage that almost all kinds of programs can be included into the system. For programs aware of `polymake` there is a more powerful C++ interface. Here the communication is based on UNIX sockets; however, this is currently changed to a, much more efficient, shared memory model.

The interface rules are realized as fragments of Perl code which take `polymake` data, convert it into the input format of the external program, run the program, and finally translate the program's output into `polymake` data. This kind of data processing is Perl's strength. It is fairly easy to add further interfaces to other program to the `polymake` system. This way, by relying on simple file exchange, one can make available a great variety of implementations.

At the core of the polymake system there is a C++ template library which provides several container classes which are STL compatible. Moreover, this `polymake` Template Library (PTL) offers C++ wrappers for exact integer and rational arithmetic of the GNU Multiprecision Library[14] (GMP) as well as C++ wrappers for the library versions of the convex hull codes by Avis (**lrs**[15]) and Fukuda (**cdd**[16]).

Besides this functionality which resembles an expert system, **polymake** provides clients for about fifty different constructions of polytopes. Some of these produce standard examples from the literature, others transform given polytopes into new ones.

Below we show an example interface implementation and, in Section 2.3, how **polymake** is used.

2.2 JavaView Interface

polymake does not have means of its own for interactive visualization. Therefore, we offer interfaces to several visualization tools, including **Geomview**[17] and **POVRAY**[18]. The most advanced interface, however, is to **JavaView**[19].

This section serves two purposes: It is sketched how **polymake** integrates programs such as **JavaView**, and an example application is given.

The interface is written in Perl. We list a slightly idealized version of the interface for visualizing 3-dimensional polytopes. This is part of **polymake**'s rule basis. Therefore the actual Perl code is preceded by a header of two lines which specifies which information is required in order to be able to apply this rule.

```
javaview, VISUAL : AMBIENT_DIM, BOUNDED, VERTICES \
                   VIF_CYCLIC_NORMAL, NEIGHBOR_FACETS_CYCLIC_NORMAL
my $jvx = JavaView::file->new();
my @points = give_dehomogenized("VERTICES");
my @facets = map { /\{(.*)\}/ } give("VIF_CYCLIC_NORMAL");
my @neighbors = map { /\{(.*)\}/ } give("NEIGHBOR_FACETS_CYCLIC_NORMAL");
my %parameters = (points => \@points, pointSet_dim => 3,
                  faces => \@facets, neighbors => \@neighbors);
my $polytope = JavaView::solid->new(%parameters);
$jvx->append($polytope);
$jvx->print_it("temporary.jvx");
system("javaview temporary.jvx);
unlink "temporary.jvx";
```

The visualization of a polytope (represented by the file **my.poly**) is triggered by specifying the *pseudo-target* **VISUAL** in the command line call below. **VISUAL** is specified in the rule's header to the left of the colon '**:**'. This means that this rule is applicable provided the information corresponding to the *sources* on the right of the colon is available.

```
> polymake my.poly VISUAL
```

In the example the communication is via file exchange. **JavaView**'s XML based file format is mapped to a set of Perl classes: **JavaView::file** corresponds to a whole `jvx` file, and `JavaView::solid` corresponds to one of possibly many of `JavaView`'s `geometry` objects. The functions `give()` and its variant `give_dehomogenized()` are part of `polymake`. They are used for the communication between the rule's body and `polymake`'s kernel. For instance, `give_dehomogenized("VERTICES")` returns the full list of vertex coordinates in one text block. Items in the code related to error recovery and more advanced transaction control are omitted. The interface for the visualization of triangulations is similar and uses the same elements.

Currently, as a joint project with the `JavaView` group, we develop a more sophisticated XML based bi-directional interface which allows to feed results from the interactive manipulation of objects back into `polymake`. This will be used for the design of interactive Schlegel diagrams which are essential for the visualization of 4-polytopes.

Polytopes can be triangulated effectively, e. g., by the Beneath-And-Beyond algorithm whose primary goal is to compute the convex hull of the input points. However, on the way it automatically builds a triangulation which is used in the convex hull computation. If one gives vertices of a polytopes as input this way the algorithm produces a triangulation of the convex hull without additional points. The order in which the vertices are processed affects the triangulation produced.

Such triangulations can be quite intricate. Therefore the corresponding interface to `JavaView` is helpful in order to facilitate their examination. The polytope and each facet of the triangulation can be manipulated individually. For example, `polymake/JavaView` can automatically produce an animated explosion of the triangulation. The following construction gives the image shown in Figure 2.2: Let

$$\Pi_d = \operatorname{conv}\{(\sigma(1), \sigma(2), \ldots, \sigma(d+1)) \mid \sigma \in S_{d+1}\}$$

be the d-dimensional permutahedron, whose vertices correspond to the elements of the symmetric group of degree $d+1$. The lexicographic ordering on the vertices induces a regular triangulation with 42 tetrahedra, see Sturmfels[20], Example 8.5d.

2.3 TOPAZ module

As already pointed out above the `polymake` system is structured in a modular way, and the polytope specific knowledge is separate from the engine which handles the rule basis. That is to say, the same engine can also deal with objects other than polytopes. This is exactly what the new **TOPAZ** module does. **TOPAZ** stands for Topology Application Zoo; it is a collection of `polymake`

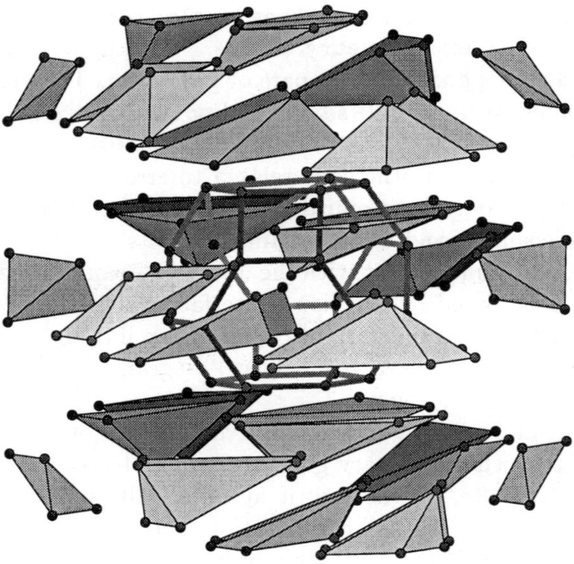

Figure 1. Lexicographic triangulation of Π_3. Explosion visualized with `JavaView`.

rules, clients, and interfaces to external programs treating finite simplicial complexes.

Main features of **TOPAZ** include the following:

- Simplicial homology and cohomology modules with integer coefficients. Moreover, bases for the (co-)homology can be determined.
- Intersection forms of closed oriented $4k$-manifolds.

We want to show how **polymake/TOPAZ** can be used. As an application we choose to examine the 16 vertex triangulation of the $K3$ surface due to Casella and Kühnel[8]. It is given as an abstract simplicial complex with 288 facets. The goal is to describe how **TOPAZ** can prove that this is indeed the triangulation of the $K3$ surface. We essentially follow the same steps in the proof as the original paper[8].

In the sequel we show how to use **polymake** commands. Each command is usually followed by one or more lines of output. The principal structure of the commands is always the same: **polymake** *file [list of properties]*.

The first step is to prove that the given complex is a combinatorial manifold. In order to do so it is necessary to prove that all vertex links are

3-spheres. Although there are algorithms due to Rubinstein[21], Thompson[22], and King[23] this seems to be hopeless from the practical point of view. However, — in this example — the simulated annealing heuristic of Björner and Lutz[24] can be used to show that each vertex link can be transformed into the boundary of the 4-simplex by a sequence of bistellar flips. For this **TOPAZ** relies on the program `bistellar`[25], which is integrated via an interface similar to the `JavaView` interface from Section 2.2.

```
> polymake K3-16.top CLOSED_MANIFOLD
CLOSED_MANIFOLD
```

The output starts with the name of the property, which is `CLOSED_MANIFOLD` in the example above, followed by more data (in most cases). Since a simplicial complex is either a closed manifold or not there is nothing more to say. If it turns out that a complex is not a (closed) manifold, the output would read `!CLOSED_MANIFOLD`.

In the next step we prove that our manifold is oriented. This can be verified by looking at the (reduced integer) homology modules, which can be read off the Smith normal forms of the simplicial boundary matrices. We have the following command line input and subsequent output.

```
> polymake K3-16.top HOMOLOGY
HOMOLOGY
({} 0)
({} 0)
({} 22)
({} 0)
({} 1)
```

The output lists the homology modules $\widetilde{H}_k(K)$, for $k \in \{0, 1, 2, 3, 4\}$. Each line starts with the list of torsion coefficients, which are all empty in this case, followed by the free rank. So we obtain $\widetilde{H}_2(K) \cong \mathbb{Z}^{22}$ and, in particular, $\widetilde{H}_4(K) \cong \mathbb{Z}$, that is, K is oriented.

Next we want to know whether K is simply connected. In this particular case, this is can be deduced from the f-vector, that is, the number of faces in each dimension.

```
> polymake K3-16.top F_VECTOR
F_VECTOR
16 120 560 720 288
```

From the output we see that K has $560 = \binom{16}{3}$ triangles, that is, the triangulation is neighborly and thus simply connected. Now a celebrated result of Freedman[26] says that the (smooth) simply connected closed oriented 4-manifolds are characterized by their intersection form. **TOPAZ** computes the intersection form Q_K as the restriction of the cup product

$$\cup : H^2(K) \otimes H^2(K) \to H^4(K) \cong \mathbb{Z}.$$

By looking at the diagonal elements one can tell whether the intersection form is even or odd. Finally, a diagonalization over the rationals gives the signature. For an introduction to intersection forms of 4-manifolds see Gompf and Stipsicz[27].

```
> polymake K3-16.top INTERSECTION_FORM
INTERSECTION_FORM
0
(3 19)
```

The first row of the output says that the Q_K is even. The second row contains the signature. Therefore, over the reals, Q_K is equivalent to the diagonal form $3\langle 1\rangle \oplus 19\langle -1\rangle$. From Theorem 1.2.21 in Gompf and Stipsicz[27] we conclude that, over the integers,

$$Q_K \cong -2E_8 \oplus 3H,$$

which is known to be the intersection form of the $K3$ surface. Hence this is the end of the proof.

3 On Computer Based Proofs

The mathematical software community is still confronted with doubt, when it comes to accepting proofs generated or supported by a computer. The most important reason is that it is often very hard to verify whether a computer proof is correct or not. Usually, there is a huge gap between what is published and what is necessary to reproduce the proof. One problem arises from the fact that there are no established forms for a peer-reviewed publication of mathematical software itself. And it seems hard to provide any: On the one hand, according to Rice's Theorem[28], formal verification methods are out of question.[a] On the other hand, a careful analysis of the source code of a large software project "by hand" can hardly be expected of any referee.

So it seems futile to ask for provably correct software. As the next best thing it is hence desirable to have several mutually independent implementations of an algorithm or, even better, independent algorithms to deal with the same problem. In order to facilitate that two software packages work on the same data a common framework is useful.

[a]The Mizar Project, see `http://mizar.org`, is devoted to (computer supported) formalized mathematical proofs.

We indicate how **polymake** can be used for the verification of results. Suppose someone is in doubt whether the crucial homology computation in the example above is actually correct: In spite of being simple in a principal way a reasonable implementation requires non-trivial data structures and is thus not as trivial as one might think.[b] Now there is a software package due to Heckenbach[30] [31] which also computes simplicial homology and which is entirely independent from **polymake**. Yet we designed an interface to make it available through our framework.

```
> polymake -verify K3-16.top HOMOLOGY heckenbach
```

In accordance with the common behavior of UNIX programs, **polymake** with the additional option **-verify** remains quiet if the verification process is successful. This shows that the result of the two computations, from **polymake**'s internal code, and from Heckenbach's code, respectively, coincide. **polymake** uses *pseudo-targets* to allow for calling specific programs. If there is a choice, the system chooses a *weight* minimal sequence of rules which yields the desired result; a weight can be explicitly assigned to each rule. Here the additional pseudo-target **heckenbach** enforces to call Heckenbach's implementation rather than the built-in.

The situation with computer based proofs is even worse. Even if someone is ready to trust a certain program, be it freely available on the Internet or a commercial product, the program itself cannot be expected to last. Hardware, operating systems, and also programming languages eventually die out to be replaced by others. That is to say, even if one is willing to accept a computer program as a valid proof this proof has a finite life span. This is contrary to what one expects from written proofs generated by "human" mathematicians. Moreover, essentially because hand written proofs are less formal, it is easier to incorporate them into more complex proofs by means of citation. Incorporating foreign code into ones own software on the other hand usually is non-trivial.

One way of rendering computer proofs more acceptable consists in making available the results of a computation in a well documented form. Here it is important to consider the fact that many interesting computer proofs are based on intermediate results far too large to be read by any human. So the only way out is to make them computer-readable. That is to say, "well documented" could be translated to "unambiguously formally specified." This brings up the idea of using the Extensible Markup Language (XML)[32] quite naturally. While MathML[33] is an XML dialect which primarily takes care of the representation of mathematical symbols, it is OpenMath[4] which pro-

[b]See Bredon[29], page 259.

poses a semantically enriched XML standard for mathematics. For a more detailed description of OpenMath and, in particular, its impact on web based mathematics see Cohen et al.[34].

Suppose some software writes an XML document[c] which represents a mathematical object. Later some other software reads the same XML document in order to compare the content to its own computational design of the same object. This could be seen as a delayed form of communication between software components, and it seems we are back to the discussion on distributed mathematical software as above. However, there is much more to it.

We need an interface to the "human" mathematician somewhere on the way. And, furthermore, it would be desirable to have the content of the XML document available for ever, or, slightly less demanding, beyond the life span of XML itself. We indicate a possible solution in the next section.

4 Electronic Geometry Models

Electronic Geometry Models, or EG-Models for short, is a hybrid. It is an electronic journal as well as an Internet based archive of geometric objects. It is freely accessible at `http://www.eg-models.de`. The mathematical content interprets the notion of geometry in a broad sense. It includes algebraic, algorithmic, convex, discrete, differentiable, and incidence geometry. Additionally, there are objects on the border to numerical mathematics and computer graphics. For an introduction to the technical background of EG-Models see[35]

Each article in the electronic journal corresponds to one or a few geometric objects, called *models*. The articles, together with these models, are peer reviewed like for a traditional mathematical journal. An editorial board decides whether to accept such a submission for publication on the EG-Models server.

Models are required to be submitted in one of currently four (ASCII based) file formats, including one XML format. The restriction of the file formats makes it easy to check whether a model file is syntactically correct. On a different account this is also the key to make the models last: If there are only a few file formats to take care of it is easy to write a few scripts to migrate to a new format. This way one can see these models as entities which are independent of their current physical form. This can be compared to math

---

[c]The author is convinced that XML will form *the* basis for all kinds of computer generated documents in the near feature. It seems likely that several dialects will be used in a mathematical context. For the rest of the paper we leave it open which dialect to choose for a particular purpose.

books which during their lifetime might rest on several different shelfs in the library. The content is unaffected.

As already pointed out above it does not suffice to please the machines alone. Therefore each model is accompanied by a description, which essentially is the article in the electronic journal. This description itself is a structured XML document. This is an advantage over most other current electronic math journals which have their articles in TeX, postscript, or PDF. For none of these formats it is easy to extract the actual information in order to convert it into a new format. As for now these XML files are automatically translated to HTML by means of transformation style sheets, and this is what is visible on the Internet. Future web browsers may be able to directly display XML documents.

Most of the models published in EG-Models come with interactive Java applets for visualization. However, it is essential to understand that visualization is not the primary goal. It is the data for the model itself that counts. All the rest is thrown in for better understanding the model.

As an example consider the Model 2000.09.028[36] by Ziegler. The problem which is addressed by the model is the following: Is there a 3-dimensional polytope which is combinatorially equivalent to the 3-dimensional cube, but with the extra condition that each pair of opposite facets is perpendicular? This question was raised in an algorithmic context, and it is answered affirmatively by the existence of this model. From an explicit coordinate description it is easy to verify the distinctive properties of this polytope. It may be even more convincing to see an interactive visualization, which is available on the EG-Models server. However, for this particular example it is possible to go beyond the virtual reality: In Figure 2 we show a planar layout from which, by cutting out and gluing together, one can obtain a real 3-dimensional cardboard model.

5 Perspectives

The author is convinced that mathematical software has the potential to contribute substantially to the progress of mathematics as a whole. However, in order to do so, it is necessary also to reflect on the weak points. This paper is intended to contribute to this reflection.

On a different account, this paper could be misunderstood as to imply that there is a wide gap between an inter-operational and an integrational approach towards distributed mathematical software systems. The contrary is true. The acceptance of a communication standard for mathematical software, such as OpenMath, would greatly simplify the development of an integrational

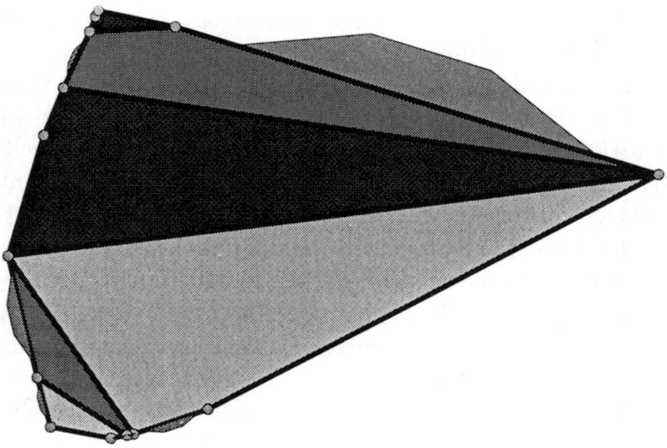

Figure 2. Planar layout (with extra flaps for gluing) of a 3-dimensional combinatorial cube such that each pair of opposite facets is perpendicular. Produced with `JavaView` from Ziegler's EG-Model 2000.09.028.

framework like `polymake`.

Acknowledgments

The author is indebted to his **polymake** and **TOPAZ** co-author Ewgenij Gawrilow. Further thanks to Thilo Schroeder and Niko Witte for their contributions to the **polymake/TOPAZ** code. The partial support by Deutsche Forschungsgemeinschaft (DFG), Sonderforschungsbereich 288 "Differential Geometry and Quantum Physics" is gratefully acknowledged.

References

1. Martin Schönert et al. *GAP - Groups, Algorithms, and Programming.* Lehrstuhl D für Mathematik, Rheinisch Westfälische Technische Hochschule, Aachen, Germany, fifth edition, 1995.
2. Ed Anderson et al. LAPACK 3.0. `http://www.netlib.org/lapack/`,

1999.

3. ILOG, Inc. CPLEX 7.0. `http://www.ilog.com/products/cplex/`, 2000.
4. OpenMath. `http://www.openmath.org/`.
5. Wolfram Research. MathLink. `http://www.wolfram.com/solutions/mathlink/`, 2002.
6. Waterloo Maple, Inc. Maple Powertools. `http://www.mapleapps.com/powertools/powertools.shtml`, 2002.
7. Andrew Solomon. Distributed computing for conglomerate mathematical systems. In Michael Joswig and Nobuki Takayama, editors, *Algebra, Geometry, and Software Systems*. Springer, to appear. 17 pages.
8. Mario Casella and Wolfgang Kühnel. A triangulated $K3$ surface with the minimum number of vertices. *Topology*, 40(4):753–772, 2001.
9. Günter M. Ziegler. *Lectures on Polytopes*. Springer, 1998. 2nd ed.
10. Ewgenij Gawrilow and Michael Joswig. `polymake` 1.5. `http://www.math.tu-berlin.de/diskregeom/polymake`, 2002.
11. Ewgenij Gawrilow and Michael Joswig. polymake: a framework for analyzing convex polytopes. In Gil Kalai and Günter M. Ziegler, editors, *Polytopes — Combinatorics and Computation*, pages 43–74. Birkhäuser, 2000.
12. Ewgenij Gawrilow and Michael Joswig. polymake: an approach to modular software design in computational geometry. In *Proceedings of the 17th Annual Symposium on Computational Geometry*, pages 222–231. ACM, 2001. June 3-5, 2001, Medford, MA.
13. Michael Joswig. Reconstructing a non-simple polytope from its graph. In Gil Kalai and Günter M. Ziegler, editors, *Polytopes — Combinatorics and Computation*, pages 167–176. Birkhäuser, 2000.
14. GNU Multiprecision Library 4.0.1, 2002. `http://www.swox.com/gmp/`.
15. David Avis. lrslib 4.1. `http://cgm.cs.mcgill.ca/~avis/C/lrs.html`, 2001.
16. Komei Fukuda. cddlib 0.91. `http://www.ifor.math.ethz.ch/staff/fukuda/cdd_home/cdd.html`, 2000.
17. Geomview 1.8.1, 2001. `http://www.geomview.org/`.
18. Persistence of Vision: POV-Ray 3.5b, 2002. `http://www.povray.org/`.
19. Konrad Polthier, Samy Khadem, Eike Preuß, and Ulrich Reitebuch. Javaview 2.21. `http://www.javaview.de/`, 2002.
20. Bernd Sturmfels. *Gröbner bases and convex polytopes*. American Mathematical Society, Providence, RI, 1996.
21. J. Hyam Rubinstein. Polyhedral minimal surfaces, Heegaard splittings and decision problems for 3-dimensional manifolds. In *Geometric topology*

(Athens, GA, 1993), pages 1–20. Amer. Math. Soc., Providence, RI, 1997.
22. Abigail Thompson. Thin position and the recognition problem for S^3. *Math. Res. Lett.*, 1(5):613–630, 1994.
23. Simon A. King. The size of triangulations supporting a given link. *Geom. Topol.*, 5:369–398 (electronic), 2001.
24. Anders Björner and Frank H. Lutz. Simplicial manifolds, bistellar flips and a 16-vertex triangulation of the Poincaré homology 3-sphere. *Experiment. Math.*, 9(2):275–289, 2000.
25. Anders Björner and Frank H. Lutz. BISTELLAR. `http://www.math.tu-berlin.de/~lutz/`, 1999–2002.
26. Michael H. Freedman. The topology of four-dimensional manifolds. *J. Differ. Geom.*, 17:357–453, 1982.
27. Robert E. Gompf and András I. Stipsicz. *4-manifolds and Kirby calculus.* American Mathematical Society, Providence, RI, 1999.
28. H. G. Rice. Classes of recursively enumerable sets and their decision problems. *Trans. Amer. Math. Soc.*, 74:358–366, 1953.
29. Glen E. Bredon. *Topology and geometry.* Springer-Verlag, New York, 1997. Corrected third printing of the 1993 original.
30. Frank Heckenbach. homology 3.0. `http://www.mi.uni-erlangen.de/~heckenb/`, 1999.
31. Frank Heckenbach, Jean-Guillaume Dumas, David Saunders, and Volkmar Welker. Computing simplicial homology based on efficient Smith normal form algorithms. In Michael Joswig and Nobuki Takayama, editors, *Algebra, Geometry, and Software Systems.* Springer, to appear. 28 pages.
32. XML. `http://www.w3.org/XML/`.
33. MathML 2.0, 2001. `http://www.w3.org/Math/`.
34. Arjeh M. Cohen, Hans Cuypers, Ernesto Reinaldo Barreiro, and Hans Sterk. Interactive mathematical documents on the web. In Michael Joswig and Nobuki Takayama, editors, *Algebra, Geometry, and Software Systems.* Springer, to appear. 19 pages.
35. Michael Joswig and Konrad Polthier. EG-Models — a New Journal for Digital Geometry Models. In Jonathan Borwein, Maria H. Morales, Konrad Polthier, and José F. Rodrigues, editors, *Multimedia Tools for Communicating Mathematics*, pages 165–190. Springer, 2002.
36. Günter M. Ziegler. Sharir's cube. *Electronic Geometry Models*, 2000. No. 2000.09.028, `http://www.eg-models.de/2000.09.028`.

SOFTWARE CARRY-SAVE FOR FAST MULTIPLE-PRECISION ALGORITHMS

DAVID DEFOUR AND FLORENT DE DINECHIN
ENS lyon, 46 allée d'Italie, 69364 Lyon, France

This paper presents the Software Carry Save (SCS) machine representations of multiple-precision (MP) numbers, geared toward simple and fast implementations. We observe, in the usual high-radix representations, that carry management accounts for a lot of the complexity of the core multiple-precision algorithms (addition and multiplication). A SCS representation therefore trades space for simplicity: the digits of the multiple-precision numbers are coded on less bits than the machine numbers can offer. The reserved bits are later filled with intermediate carries during MP addition and multiplication. This has two advantages: Firstly, it guarantees that no overflow or rounding can occur in any internal computation. This leads to simple and therefore fast algorithms. Secondly, the software carry-save representation exposes parallelism just like its hardware counterpart, and this parallelism can be exploited by modern superscalar processors. An initial implementation of a SCS library is shown to compare favorably to other multiple-precision libraries.

1 Introduction: An overview of multiple-precision formats

Most modern computers obey the IEEE-754 standard for floating-point (FP) arithmetic, which defines the well-known *single* and *double precision* FP formats, along with the behavior of the usual arithmetic operators in several rounding modes. For some applications, however, the precision provided by this standard (53 bits in double precision) is not enough. These applications include large-scale numerical simulations, but also very small computations where precision is paramount. For instance, previous research has shown that an internal precision of around 200 bits is needed to provide correctly rounded elementary functions in double precision[1].

In addition to scientific computing packages such as Maple or Mathematica, many general-purpose multiple-precision (MP) libraries exist. These libraries are widely used for numerical analysis, cryptography or computational geometry. Some offer arbitrary precision with static or dynamic precision control, other simply offer a fixed precision which is higher than that of the IEEE-754 standard.

All of these libraries internally represent MP numbers as arrays of *digits*, which are themselves *machine numbers*, i.e numbers in one of the native format on which the microprocessor can directly compute: integer, or IEEE-754 floating-point numbers.

1.1 Unevaluated sums of FP numbers

Briggs' *double double*[2], Bailey's *quad doubles*[3] and Daumas' *floating-point expansions*[4] consist of the unevaluated sum of several double-precision FP numbers. Arithmetic operations on these formats use only FP operations. The motivation here is that the floating-point units (FPUs) in modern processors are highly optimized: for example, in recent 64-bit architectures like UltraSparc and Itanium, floating-point multiplications are faster than their integer counterpart. Besides, super-scalar processors have several such units, and the trend is towards more FP computing power than integer.

One advantage of such format is that conversion from IEEE-754 floating point to such a format is trivial, and conversion back is relatively easy. Addition and multiplication, however, are more complicated due to the rounding errors inherent to FP formats: They requires sophisticated algorithms which have to be carefully proven, sometimes even requiring the assistance of automatic theorem checking[5]. One of our motivations will be to avoid such complexity if we are to use FP arithmetic.

1.2 Dense high-radix formats

These formats use a standard radix representation of MP number, where the digits are machine numbers. They can be FP numbers as in Bailey's MPFUN[6], or integers as in the GNU Multi Precision (GMP) library[7], on top of which several other libraries are built (see MPFR[8] or Arithmos[9]). In GMP the digits are integers of the native machine word size (currently 32 or 64 bits).

Arithmetic operations use algorithms very similar to those learnt in elementary schools, with sequences of digit-wise operations, which are operations on machine numbers. The main difficulty, in implementing these algorithms, is that the sum or product of two machine numbers is usually not a machine number: there may occur *rounding* in the case of FP digits, and *overflow* in the case of integer digits. The algorithms have to be designed to handle these rounding or overflow cases. In general, this is translated at the algorithm level as *carry propagations*, which account for a lot of the execution time of MP algorithms. These notions will be detailed in the sequel.

In addition, conversions between such formats and the IEEE-754 floating-point format are sometimes fairly expensive.

1.3 Sparse high-radix formats

Such a format codes a MP number as an array of integers or FP numbers, which we will also call *digits*. However, to avoid carry propagation, it ensures that no precision is lost in the addition or multiplication of two digits. To this purpose, the digits don't use all the precision available in the machine format: Some of the bits are reserved, to be used by the MP algorithms to store intermediate carries. In the sequel we will describe these formats as a software *carry-save*, a term borrowed from a similar idea in hardware.

The history of this idea is interesting. It is first found in Brent's MP library[10]. His motivation seems to have been *portability*: Although processors have long offered machine instructions allowing efficient multiple-precision software (for instance *add-with-carry*), many of these instructions cannot be accessed from high-level programming languages. As we will see in the sequel, carry-save MP only uses basic, and therefore portable, arithmetic.

Brent's MP was superseded by GMP, which has a different answer to portability: In GMP, the core loops (and only them) of both addition and multiplication are written in assembly language for most existing architectures, and thus make the best use of those machine instructions not available in high-level languages.

However, Brent's idea has been resurfacing recently. It seems to be used by Abraham Ziv in his IBM mathematical library with correct rounding[11], the difference being that they use FP digits. Unfortunately we could not find any article on the subject. Independently, the authors developed the library presented here, which can use either FP or integer digits. Our motivation was again portability, but also *efficiency*: we observe that carry-save MP allows carry-free algorithms which, in addition of being simpler, may exploit the increasingly parallel nature of modern processors.

1.4 Multiple precision on superscalar processor architectures

Current processors offer more and more integer and FP units which can operate in parallel, provided the compiler and/or the hardware can find independent instructions, *i.e.* instructions which can execute concurrently.

One of the main motivations of the proposed format was to express MP addition and multiplication with more parallelism in the atomic operations than is possible with GMP-like representation. We show in this paper that this format allows us to express MP algorithms with much less *data dependencies* between atomic operations. In other terms, a carry-save representation allows us to express more digit-grained parallelism, which, as our tests show, modern processors can exploit.

1.5 Outline of this paper

Section 2 presents in details the proposed representation, along with its main motivation in a MP multiplication. Section 3 details the algorithms used to perform MP operations and conversions. Section 4 presents some results of our initial implementation, comments and compares them to other available multiple-precision libraries. Section 5 concludes the paper.

2 A format for fast multiple-precision multiplication

2.1 Motivation: Overview of the standard MP multiplication algorithm

There are several multiple-precision multiplication algorithms. The simplest one is similar to the multiplication algorithm learnt in elementary school, and is depicted on Fig. 1. It should be pointed out that algorithms exist with a better asymptotic complexity, however they are usually considered for precisions much larger than those considered here.

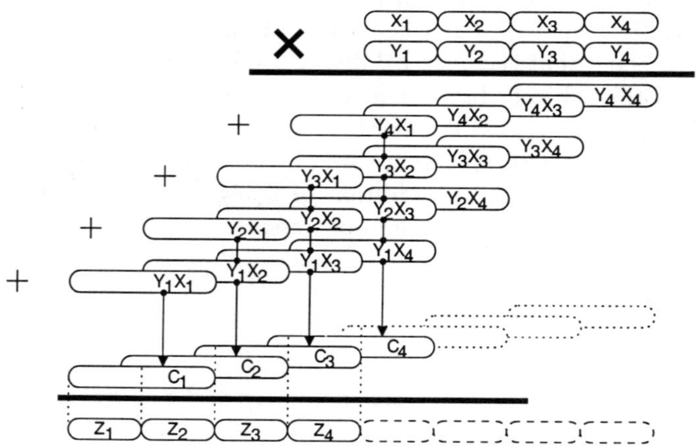

Figure 1. Multiple-Precision multiplication

The figure 1 represents the two input numbers X and Y, decomposed into their n digits x_i and y_i (with $n = 4$ on the figure). Each digit is itself coded in m bits of precision: These digits will be machine integers in GMP ($m = 32$ or 64), or FP numbers in Bailey's libraries ($m = 24$ or $m = 53$).

The multiplication algorithm involves three steps (which can be interleaved). First, an array of partial products $x_i y_j$, each a $2m$-bit number, is computed. Then these partial products have to be summed. There are at most n of them in a column, which means that the sum c_i in each column is a number of up to $2m + \lfloor \log_2 n \rfloor$ bits. Finally the column sums c_i are decomposed into numbers of only m bits. This is done in an iterative manner by summing from right to left the numbers of equivalent weight, storing the lower m bits as a digit of the result, and keeping the higher bits as a *carry* for the next iteration.

It is clear that any implementation using, as digits, full machine numbers (be it integer or floating-point) can nor represent in one machine number the partial products, neither the column sums. Algorithmic tricks are therefore needed to get these values coded on two or more digits. It implies in turn strong dependencies on the operations. In GMP, for instance, computing each line of partial products involves a *carry propagation* where each line is converted to a standard MP number. For a small n, this may lead to poorer performance than the processor can offer. In Bailey's library, the dependencies are not even regular.

2.2 The proposed format

The central idea behind our format is to code the digits of X and Y in such a way to ensure that each column sum c_i can fit on a machine number. If we denote m_I the precision offered by the machine format, m_r the precision used to code a digit, and n_r the number of digits, then a sufficient condition is:

$$m_I \geq \lfloor log_2(n_r) \rfloor + 2m_r \qquad (1)$$

This ensures that any intermediate computation in the multiplication algorithm is exact, in the sense that its result fits in a machine number without any overflow/rounding. The same then also holds for additions and subtractions.

Inequality (1) is a relation between the number m_r of bits stored on each machine number and the number of digits manipulated n_r. The overall precision is the product $n_r \times m_r$. Table 1 gives some values of these parameters for various machine word formats, using sensible values of m_r.

Our format is interesting only for moderate precisions (a few hundred bits), because the waste of space and time increases as m_r decreases for a given m_I. A detailed study of this tradeoff remains to be done. It should be noted that Brent[10] imposed a weaker condition, such that 8 partial products can be summed without rounding error. He thus still has to perform one carry propagation out of eight, but his format allows for arbitrary precision. Conversely, our format is not suited for arbitrary precision, however it allows for precisions of up to several thousand bits.

Machine word	m_I	m_r	n_r	max. precision
IEEE double precision	52+1 bits	26 Bits	15 terms	390 Bits
IEEE double precision	52+1 bits	25 Bits	63 terms	1575 Bits
Double-extended (x86)	64 bits	30 Bits	31 terms	930 Bits
Double-extended (x86)	64 bits	29 Bits	127 terms	3683 Bits
64-bit integer	64 bits	30 Bits	15 terms	450 Bits
64-bit integer	64 bits	29 Bits	63 terms	1827 Bits

Table 1. Some examples of precisions that can be reached with the proposed format.

2.3 Machine implementation

A MP number is represented in the proposed format as a *Software Carry Save* (SCS) structure R, depicted on Figure 2 and composed of the following fields:

R.digits[n_r] A table of n_r digits with m_r bits of precision, using either integer or FP machine numbers (depending of the architecture);

R.index An integer storing the index of the first digit in the range of representable numbers, as depicted on Figure 2;

R.sign A sign information.

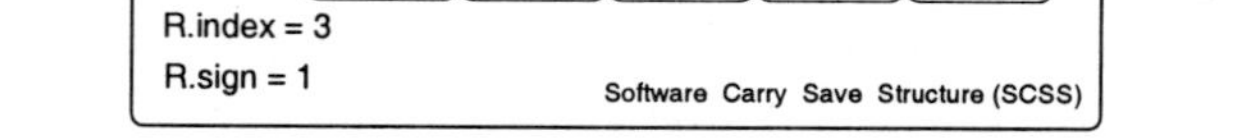

Figure 2. The proposed format

In other words, the value x of a representation R is:

$$x = R.sign \times \sum_{j=1}^{n_r} R.digits[j] \times 2^{m_r*(R.index-j)} \tag{2}$$

In such a *normal* SCS number R, the bits from m_I to m_r of the $R.digits$ fields are thus set to zero. They will be exploited by the algorithms to store temporary *carry* information, and are therefore called *carry-save* bits. An SCS number where these bits are non-zero is said to be non-normal.

In the case when FP machine numbers are used, their exponent field is unused: Using it would make alignment of mantissas, in the addition algorithm, much more complex. However, a trick allows to exploit their sign bit, along with the well-defined rounding properties of the IEEE-754 standard, to increase the precision of each word by one bit, which explains some values of Table 1. Due to lack of space we do not detail this trick here.

3 Arithmetic operations

In this section we present algorithms used to perform basic arithmetic operations. For easier comprehension, we adopt a syntax very similar to the C language syntax.

3.1 Conversion from double to SCS

A first method for converting a double precision floating point number d into an SCS representation is to extract the exponent d_{exp} from d, and then determine the corresponding $R.index$ as the integer part of $\frac{d_{exp}}{2^{m_r}}$.

Another method, described below, uses only a variable number of multiplications by 2^{m_r} or 2^{-m_r}. This method is faster than the previous one when the exponent of d is close to 0.

Algorithm 1 (double2SCS)

Input: *A double precision floating point number d*
Output: *A multi-precision number R*
Method: $R.index = 0;$
if $d \in \{Nan, +\infty, -\infty\}$ **then** $\{R = d;$ **return**; $\}$
if $d > 0$ **then** $\{ R.sign = 1; \}$
else $\{ R.sign = -1; \quad d = -d; \}$
while $d > 2^{m_r}$ **do** $\{ R.index++; \quad d = d \times 2^{-m_r}; \}$
while $d < 1$ **do** $\{ R.index--; \quad d = d \times 2^{m_r}; \}$
for$(i = 1 ; i \leq n_r ; i++)$
$\{ R.digits[i] = \lfloor d \rfloor; \quad d = (d - R.digits[i]) \times 2^{m_r}; \}$

3.2 Addition and subtraction

The addition of two SCS numbers of the same sign consists in aligning, then adding digits of the same order. Thanks to the carry-save bits, all these additions will be *exact* and *independent*. However the result will usually not be a normal SCS number: the sums will have overflown in the carry-save bits.

A *renormalization* procedure is presented in section 3.4 to propagate these carry bits and get again a normal SCS number. However, the advantage of SCS representation is that many SCS numbers can be summed before needing to perform this expensive step (the exact number depends on the value chosen for m_r, m_I and n_r). For example, with $m_I = 64$ and $m_r = 30$ (see Table 1), we can add approximately 8×10^9 normal SCS numbers between two renormalizations.

The subtraction (addition of two numbers of opposite signs) is very similar to the addition algorithm, although more depending on the choice of integer or FP format. It may also classically lead to a cancellation, which may need an update of the index of the result. Due to lack of space we do not detail this here.

Although all the digit operations are exact, the addition or subtraction of two numbers also classically involves a rounding error, due to aligning the digits of same magnitude.

3.3 Multiplication

The multiplication of two normal SCS numbers involves the operations depicted on the Figure 1: The partial products are computed (in parallel) and summed in columns. We already showed in section 2.2 that none of these operation overflows as long as equation 1 is satisfied. Again, the result is not a normal SCS number, and a renormalization procedure (described below) has to be applied to empty the carry bits. However, a few additions may follow a multiplication before this renormalization, which allows for further optimization of algorithms using SCS arithmetic. For instance, a polynomial evaluation can be implemented with a renormalization after one multiplication and one addition.

Here also, a rounding error is involved when two n_r-digit numbers are multiplied if the result is to fit on n_r digits. The actual implementation tests if the most significant digit (z_1 on Figure 1) is null, in which case the index of the result is that of z_2.

A further optimization is to avoid computing the columns of lower magnitude, at the expense of a known increase in the rounding error. The significance of such a tradeoff is application-dependent.

3.4 Renormalization (carry propagation)

Renormalization is a carry propagation from the low order to high order digits: Starting with an initially null carry, at each step, the previous carry is added to the current digit, and this sum is then split into two parts. The low m_r bits are a digit of the normalized result, and the upper part is the next carry.

The actual algorithm is a little bit more complex. The initial non-normal number may not be representable exactly as a normal SCS number, therefore the index of the normalized result has to be increased by one or two. Normalization thus again involves a rounding error.

To get the higher-order part of a floating-point digit, the classical method is to add, then subtract the constant $2^{m_I+m_r}+2^{m_I+m_r-1}$. For integer digits, a simple mask is used.

3.5 Conversion from SCS to floating-point

A few (4 in the worst case) multiplications and additions suffice to get the FP number closest to a SCS number. For instance, for $m_I=53$ and $m_r=26$, we need to compute $d=A.sign\times 2^{A.index\times m_r}\times(A.digits[0]+2^{-m_r}\times A.digits[1]+2^{-2.m_r}\times A.digits[2]+2^{-3.m_r}\times A.digits[3])$. The number $2^{A.index\times m_r}$ is build using integer masks. The actual implementation of this formula is slightly less simple, but this conversion is still very fast.

4 Results and timing

It is quite difficult to compare the performances of MP libraries. The cost of a machine multiplication or addition is very different from one processor to another one, and also depends on the context of this operation in today's sophisticated processors with deep pipelines, out-of-order execution, and cache hierarchies. Even the definition of an operation is machine-dependent (32- or 64-bit, availability or not of a fused multiply-and-add).

We have written (in C) and refined several implementations of this library, with various machine formats and instruction ordering. In our tests, we use the same code, compiled with the same compiler (gcc.2.95.4), the only change being the machine format which can be floating-point or integer (we chose the best). In both cases we ensure 200 bits of precision.

We compare our code with the reference multiple-precision library GMP[7] (more precisely it floating representation MPF), with Bailey's quad-double library[3], and also with Ziv's library[11]. Each result is obtained by measuring the execution times on 10^6 random values (the same values are used for all the libraries). For clarity we normalized results to SCS library.

In the SCS timings, each addition and multiplication is followed by a renormalization. Therefore the following results don't show all the potentiality of our methods in a more realistic context.

	Description	SCS lib.	Ziv lib.	Bailey lib.	GMP lib.
Conversion double ⇒ MP	Itanium	1	13.05	1.73	1.84
	Celeron	1	5.85	0.33	0.74
	UltraSparc	1	1.22	0.32	0.94
MP +	Itanium	1	8.74	25.1	1.31
	Celeron	1	2.35	9.0	1.41
	UltraSparc	1	2.97	26.33	1.39
MP ×	Itanium	1	7.84	37.71	1.97
	Celeron	1	0.81	8.45	0.32
	UltraSparc	1	3.28	38.41	2.37
Conversion MP ⇒ double	Itanium	1	7.15	0.66	7.28
	Celeron	1	6.0	0.19	2.37
	UltraSparc	1	8.64	0.46	4.36

Figure 3. Timing comparison of different MP libraries. The machines used for these tests are an Itanium 733 Mhz, an Intel Celeron 566 Mhz, and a Sun Ultra-5 400Mhz.

The multiplication times on Celeron and Itanium seem contradictory. An interpretation could be that we get the best results on the Itanium architecture, whose instruction set was designed with parallelism in mind, and therefore makes better use of our more parallel algorithms. However this hypothesis has to be taken with caution. A deeper investigation is needed to see what is the influence of the compiler and coding style on these results. It should also be mentionned that the quad-double library is written in C++, which may affect its performances.

5 Conclusion and future work

We have presented the software carry-save representation for multiple-precision numbers that allows simpler, more parallel, and therefore faster MP additions and multiplications. This representation can provide very high precision, however is probably most suited for limited precisions of up to a few hundred bits.

The main idea developed in this method is to avoid overflows/rounding/carries in the machine operations internal to a MP operation. Being parallel, the code is suited to modern highly super-scalar processors. Being simple, it is also fairly compact. Using plain integer or IEEE-754 normalized arithmetic, it is also fully portable.

After our first tests, the most obvious limitation of this method is the high dependence of the performance to the architecture. We will try to investigate the interaction between code, compiler, and processor architecture to write code which performs well on most architectures.

Future work also includes real-world tests of this library, in particular to compute transcendental functions such as cosine, sine, exponential.

References

1. V. Lefèvre, J.M. Muller, and A. Tisserand. Towards correctly rounded transcendentals. *IEEE Transactions on Computers*, 47(11):1235–1243, November 1998.
2. `www-epidem.plantsci.cam.ac.uk/~kbriggs/doubledouble.html`.
3. Yozo Hida, Xiaoye S. Li, and David H. Bailey. Algorithms for quad-double precision floating-point arithmetic. In Neil Burgess and Luigi Ciminiera, editors, *15th IEEE Symposium on Computer Arithmetic*, pages 155–162, Vail, Colorado, June 2001.
4. Marc Daumas. Expansions: lightweight multiple precison arithmetic. In *Architecture and Arithmetic Support for Multimedia*, page 14, Dagstuhl, Germany, 1998.
5. Sylvie Boldo and Marc Daumas. A mechanically validated technique for extending the available precision. In *35th Asilomar Conference on Signals, Systems, and Computers*, Pacific Grove, California, 2001.
6. David H. Bailey. A Fortran-90 based multiprecision system. *ACM Transactions on Mathematical Software*, 21(4):379–387, 1995.
7. GMP, the GNU multi-precision library. `http://swox.com/gmp/`.
8. MPFR. `http://www.loria.fr/equipes/polka/software.html`.
9. A. Cuyt, P. Kuterna, B. Verdonk, and J. Vervloet. *Arithmos: a reliable integrated computational environment.* 2001.
10. Richard P. Brent. A fortran multiple-precision arithmetic package. *ACM Transactions on Mathematical Software*, 4(1):57–70, 1978.
11. IBM accurate portable mathematical library. `http://oss.software.ibm.com/mathlib/`.

Acknowledgements

The support of Intel and HP through the donation of an Itanium based machine is gratefully acknowledged.

LINBOX: A GENERIC LIBRARY FOR EXACT LINEAR ALGEBRA

J.-G. DUMAS[a], T. GAUTIER[b], M. GIESBRECHT[c], P. GIORGI[d], B. HOVINEN[e], E. KALTOFEN[f], B.D. SAUNDERS[e], W.J. TURNER[f] AND G. VILLARD[d]

[a] www-lmc.imag.fr/lmc-mosaic/Jean-Guillaume.Dumas
[b] www-id.imag.fr/~gautier [c] www.uwaterloo.ca/~mwg
[d] www.ens-lyon.fr/{~pgiorgi, ~gvillard}
[e] www.cis.udel.edu/{~hovinen, ~saunders}
[f] www.math.ncsu.edu/{~kaltofen, ~wjturner}

1 Introduction

Black box techniques[12] are enabling exact linear algebra computations of a scale well beyond anything previously possible. The development of new and interesting algorithms has proceeded apace for the past two decades. It is time for the dissemination of these algorithms in an easily used software library so that the mathematical community may readily take advantage of their power. LinBox is that library. In this paper, we describe the design of this generic library, sketch its current range of capabilities, and give several examples of its use. The examples include a solution of Trefethen's "Hundred Digit Challenge" problem #7[14] and the computation of all the homology groups of simplicial complexes using the Smith normal form[8].

Exact black box methods are currently successful on sparse matrices with hundreds of thousands of rows and columns and having several million nonzero entries. The main reason large problems can be solved by black box methods is that they require much less memory in general than traditional elimination-based methods do. This fact is widely used in the numerical computation area. We refer for instance to the templates for linear system solution and eigenvalue problems[2,1]. This has also led the computer algebra community to a considerable interest in black box methods. Since Wiedemann's seminal paper[16], many developments have been proposed especially to adapt Krylov or Lanczos methods to fast exact algorithms. We refer to our paper[5] and references therein for a review of problems and solutions.

LinBox supplies efficient black box solutions for a variety of problems including linear equations and matrix normal forms with the guiding design principle of re-usability. The most essential and driving design criterion for LinBox is that it is generic with respect to the domain of computation. This is because there are many and various representations of finite fields each of which is advantageous to use for some algorithm under some circumstance. The integral and rational number capabilities depend heavily on modular techniques and hence on the capabilities over finite fields. In this regard, generic software methodology is a powerful tool.

Partly modeled on the STL, LinBox uses the C++ template mechanism as the primary tool to achieve the genericity. The library is inspired by the FoxBox black box and plug-and-play design objectives[6]. Projects with some similar goals include MTL [`http://www.osl.iu.edu/research/mtl`] in numerical linear algebra and SYNAPS [`http://www-sop.inria.fr/galaad/logiciels/synaps`] in symbolic computation. WiLiSys[10] is a forerunner of LinBox.

The following section presents design decisions and their motivation using the field and black box representations as examples. In Section 3 we discuss the current capabilities provided in LinBox and guiding principles for the implementation of their underlying algorithms. In Section 4 we illustrate the power of LinBox with some example solutions.

2 LinBox design

The overarching goal of our design is to create a software library that supports reuse and reconfiguration at a number of levels without sacrificing performance. At the top level, we provide algorithms for many standard problems in linear algebra. As input, these algorithms accept black box matrices, a notion that uniformly captures sparse and structured (and dense) matrices alike. Any object conformant with the specification for a black box matrix can be plugged into these algorithms. At a lower level, we want our code to operate over a multitude of coefficient domains and, for a given domain, a variety of implementations. For instance, into our algorithms one might plug any of several implementations of the integers modulo a prime number, e.g., when the field operations may be performed via a Zech logarithm table or via integer remaindering. We thus hope to capture future improvements on field arithmetic without rewriting our programs. One might also plug in a field of rational functions, or the floating point numbers, although the resulting methods may not be numerically stable. At every stage we have applied the principle commonly called *generic programming*. We realize this through C++ templates and virtual member functions.

LinBox provides what we call *archetype* classes for fields and black box matrices. An archetype serves three purposes: to define the common object interface, to supply one instance of the library as distributable compiled code, and to control code bloat. An archetype is an abstract class whose use is similar to a Java interface. It specifies exactly what methods and members an explicitly designed class must have to be a pluggable template parameter type. Through the use of pointers and virtual member functions, the field archetype, for instance, can hook into different LinBox fields. Thus the precompiled library code can be executed on a user supplied field.

Field design. The algorithms in LinBox are designed to operate with a variety of domains, particularly finite prime fields. To perform the required arithmetic, additional parameters, such as the modulus, must be available to the algorithm. One can store a pointer to the required parameters in each field element, but that would require too much memory. One can also use a global variable to store these parameters—as is done in NTL, for instance—but it is then impossible to operate over more than one field concurrently. Our approach is to have a separate field object whose methods include field arithmetic operations. For example, the call `F.add(x,y,z)` adds the elements `y` and `z` in the field `F` and stores the result in the element `x`. The field object stores the required parameters, and it is passed to each of the generic algorithms. Because of this design, we do not support traditional infix notation.

Given a field class `Field`, elements of this field are of the type `Field::element`. This may be a C++ `long`, for integers modulo a word size prime, or a more complicated data structure declared elsewhere in the library. The field interface requires only that the data type support C++ assignment and a copy constructor. Because elements by themselves do not have access to the field parameters, they are initialized by the field, as in `F.init(x,5)`. The field type contains methods to initialize and perform arithmetic on field elements and check their equality. In addition to standard arithmetic, in which the result is stored in a separate field element, we support "in-place" arithmetic, similar to C++ `+=`, `-=`, `*=`, and `/=`. Field types are also required to support assignment and equality checking of whole fields. For each field type, there exists a class that uniformly generates random elements of that field or an unspecified subset of a given cardinality. Many of the algorithms in LinBox depend on the availability of such random elements (see Section 3).

Whether or not a field requires parameters, such as a modulus, to perform arithmetic, its interface is the same. We provide a template *wrapper* class for the creation of an unparameterized field meeting the LinBox interface from a C++ data type that supports standard arithmetic and equality operations. For example, `unparam_field<ZZ_p> F` is a field of NTL modular integers. If a user-defined field implements a required method in a different manner, one can resort to partial template specialization in order to define the corresponding operation. The following example adjusts Victor Shoup's `inv` function of his `ZZ_p` class to the signature of LinBox's `inv` method.

```
template <> NTL::ZZ_p& unparam_field<NTL::ZZ_p>::inv (
```

```
                    NTL::ZZ_p& x, const NTL::ZZ_p& y) const{
                                  return x = NTL::inv(y);}
```

Thus we can easily adapt fields from other libraries to LinBox.

The field archetype defines the interface that all field implementations must satisfy. Any class that meets that interface can be hooked into a generic algorithm. A virtual copy method ("clone") is introduced via an abstract base class pointed to by the archetype thus yielding an STL-compliant copy constructor in the archetype. Generic algorithms can be instantiated with the archetype and compiled separately. Code making use of these algorithms can supply a field inheriting the abstract field type and link against this code, with a modest performance loss resulting from the inability to inline field operations and from additional memory indirection. Finally, LinBox provides a template class called `Field_envelope` that hooks any archetype-compliant field type onto this abstract class, so any field type may be used in this manner, then without any performance loss (see below).

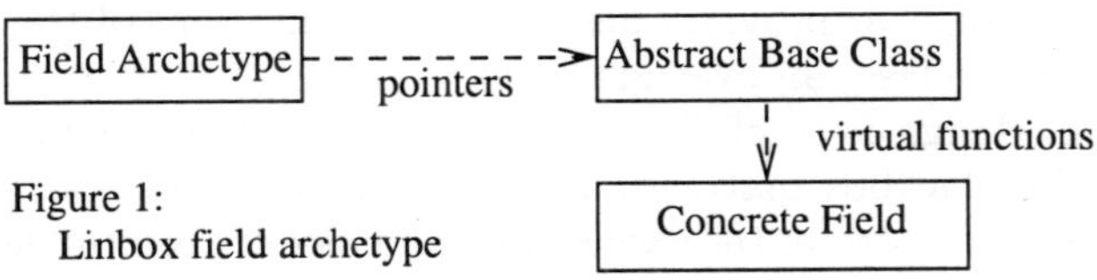

Figure 1:
Linbox field archetype

Black box design. The LinBox black box matrix archetype is simpler than the field archetype because the design constraints are less stringent. As with the field type, we need a common object interface to describe how algorithms are to access black box matrices, but it only requires functions to access the matrix's dimensions and to apply the matrix or its transpose to a vector. Thus our black box matrix archetype is simply an abstract class, and all actual black box matrices are subclasses of the archetype class. We note that the overhead involved with this inheritance mechanism is negligible in comparison with the execution time of the methods, unlike for our field element types.

The black box matrix archetype is template-parameterized by the vector type upon which it acts, but not by the field in which the arithmetic is done as we saw no necessity for this. The field of entries is bound directly to the black box matrix class and is available as an argument to our black box algorithms, which may perform additional coefficient field operations. Optionally, black box matrix classes can have the field type as a template parameter. In addition, variants of the `apply` method are provided through which one could pass additional information, including the field over which to operate.

LinBox currently has three types of vectors, *dense*, *sparse sequence*, and *sparse map*. The dense vectors store every entry of the vector and are generally implemented as `std::vector<element>`. Sparse vectors only store nonzero entries. Sparse sequence vectors have the archetype `std::list<std::pair<integer,element>>`, and sparse map vectors have the archetype `std::map<integer,element>`. The C++ `operator[]` is disallowed in the latter to avoid fill-in with zero values. By its data structure, a map entry access is logarithmic time.

We do not parameterize our algorithms with a black box type. We use the black box archetype directly in the algorithms. The caller provides a specific black box subclass. For convenience, some methods have default implementations in the archetype. For example, `apply` and `applyTranspose` each have three variants, which handle allocation of the input and output vectors differently. Only one variant is necessary:

```
Vector& apply (Vector& y, const Vector& x) const;
Vector& applyTranspose (Vector& y, const Vector& x) const;
```

The archetype base class provides default implementations of the other variants, but a derived class can override them.

Overall our design is carefully constructed to be generic with respect to the field and the blackbox matrices, and yet to provide high performance. A simple experiment illustrates the successful avoidance of performance degradation. In this test we compare using an accumulation by a sequence of additions and multiplications done using each of three different setups for the arithmetic. The first is use of `NTL::zz_p` arithmetic directly, the second is our wrapper of the `NTL::zz_p` class so that the arithmetic is done using LinBox's generic field interface, and the third is with use of the LinBox field archetype. This third interface involves an extra pointer dereference for each field operation, but allows separate compilation of field and algorithm. The results are 8.52 seconds for `NTL::zz_p` directly, 8.32 seconds for Linbox wrapped `NTL::zz_p`, and 11.57 seconds for Linbox archetype over the LinBox wrapper of `NTL::zz_p`. Thus there is *no performance cost* for the generic field interface, because the function signature changes have been resolved at compilation, but there is a cost for support of separate compilation (and code bloat avoidance). However, note that this overhead is relatively smaller for more expensive arithmetic, such as using the archetype over `NTL::ZZ_p` on integers larger than word size.

3 Black box algorithms

Wiedemann's paper[16] has led to many developments in the application of Krylov and Lanczos methods to fast exact problem solving. Here we present the main directions followed in LinBox regarding these methods. Linear algebra done over finite fields is the core of the library. Computations over the integers or rationals build upon this core.

Randomized algorithms, heuristics and checking. In black box linear algebra, the fastest known algorithms generally use random bits. Our library includes *Monte Carlo* algorithms (which produce correct results with controllably high probability), *Las Vegas* algorithms (which always produce correct results and are fast with high probability), and deterministic algorithms. Properties of these algorithms are proven under conditions on the source of random bits, e.g., that we are able to choose random elements uniformly from a sufficiently large subset of the ground field. This condition may be prohibitively costly in practice, e.g., if the field is small the use of an algebraic extension may be required. This has led us to a new implementation strategy. We relax theoretical conditions on our algorithms but introduce subsequent checks for correctness. We exploit randomized algorithms as heuristics even when the provable conditions for success are not satisfied. To complement this, specific checks are developed to certify the output. These checks may themselves be randomized (see below), in which case we certify only the probability of success. This strategy has been powerful in obtaining solutions for huge determinant and Smith form problems in Section 4.

Minimal polynomial and linear system solution over finite fields. For a matrix $A \in \mathbb{F}^{n \times n}$ over a field $\mathbb{F}$, Lanczos and Krylov subspace methods essentially compute the minimal polynomial of a vector with respect to A. These methods access A only via matrix-vector products $v = A \cdot u$ and $w = A^{\mathrm{T}} \cdot u$, i.e., they treat A as a black box function. Thus the library will only employ a black box for A, which can exploit the structure or sparsity of A.

The minimal polynomial $f^{A,u}$ of a vector u is computed as a linear dependency between the iterates $u, Au, A^2u, \ldots$. In the Krylov/Wiedemann approach, the dependency is found by applying the Berlekamp-Massey algorithm to the sequence $v^{\mathrm{T}}u, v^{\mathrm{T}}Au, v^{\mathrm{T}}A^2u, \ldots \in \mathbb{F}$, for a random vector v. This identifies the *minimum generating polynomial* $f_v^{A,u}(x) = x^d - f_{d-1}x^{d-1} - \cdots - f_1 x - f_0$, i.e., $v^{\mathrm{T}}A^{d+i}u = f_{d-1} \cdot v^{\mathrm{T}}A^{d+i-1}u + \cdots + f_1 \cdot v^{\mathrm{T}}A^{i+1}u + f_0 \cdot v^{\mathrm{T}}A^i u$, for all $i \geq 0$. Then $f_v^{A,u}(x)$ always divides $f^{A,u}$, and they are equal with high probability[16]. Berlekamp-Massey computes $f_v^{A,u}(x)$ after processing the first $2d$ elements of the sequence.

`LinBox::minpoly`: With high probability, the minimal polynomial of the sequence $\{v^{\mathrm{T}}A^i u\}_{0 \leq i \leq 2n-1}$ is the minimum polynomial f^A of A[16,11], for randomly chosen vectors u, v. This algorithm is randomized of the Monte Carlo type. As first observed by Lobo, the cost can be reduced by early termination:

as soon as the linear generator computed by the Berlekamp-Massey process remains the same for a few steps, it is likely the minimal polynomial. This argument is heuristic in general but provable for the Lanczos algorithm on preconditioned matrices over suitably large fields[9] [Eberly, private communication (2000)]. A Monte Carlo check of the early termination is implemented by applying the computed polynomial to a random vector.

Dominant costs in these algorithms are given in Table 1[13,7]. Terms between brackets give memory requirements (in field elements). Early termination and randomized Monte Carlo algorithms correspond to bi-orthogonal Lanczos algorithms with or without lookahead. In both approaches, the number of matrix-vector products may be cut in half if the matrix is symmetric. Since the update of the linear generator is computed by dot products instead of elementary polynomial operations, a Lanczos strategy has a slightly higher cost for computing the minimal polynomial.

Table 1. **Costs of Wiedemann and Lanczos algorithms** for f^A of degree d and for $Az = b$. A or A^{T} can be applied to a vector using at most ω operations.

	Early termin. f^A	Monte Carlo f^A	Sys. soln. $Az = b$
Wiedem. $[6n]$	$2d\omega + 4d(n+d)$	$2n\omega + 4n^2 + 2d(n+d)$	$+d\omega + 2dn$
Wiedem. $[O(dn)]$	$2d\omega + 4d(n+d)$	$2n\omega + 4n^2 + 2d(n+d)$	$+2dn$
Lanczos $[3n]$	$2d\omega + 8dn$	$2n\omega + 4n^2 + 4dn$	$+2dn$

`LinBox::linsolve`: For nonsingular A, a linear system $Az = b$ is solved by computing the minimal polynomial $f^{A,b}(x) = x^d + f_{d-1}x^{d-1} + \cdots + f_1 x + f_0$ of b; its coefficients directly give $z = -(1/f_0)(A^{d-1}b + f_{d-1}A^{d-2}b + \cdots + f_1 b)$. Checking $Az = b$ makes the system solution Las Vegas. The Lanczos approach allows one to compute z within the iterations for the minimal polynomial, thus the arithmetic and memory costs are only slightly greater than for basic Lanczos. The main drawback of the Wiedemann approach is that it needs to either store or recompute the sequence $\{A^i b\}_{0 \le i \le d-1}$.

For both minimal polynomial and system solution, we are developing block versions of the Wiedemann and Lanczos algorithms[15]. The choice between the strategies will rely on the same criteria as for the non-blocked versions[7]. We can also cover the case of singular systems.

Preconditioning for the rank and the determinant over finite fields. The computation of the rank and determinant of a black box matrix A reduce to the computation of the minimal polynomial of a matrix $\tilde{A}$, called a *preconditioning* of A, obtained by composing A with one or more other black boxes called *preconditioners*[16,5]. A first type of preconditioning involves multiplication of A by Toeplitz, sparse, or butterfly matrices, and typically incurs a cost of $O(n^2 \log n)$ or $O(n^2 \log^2 n)$ operations. For large n, this cost can be prohibitive. We have thus focused on diagonal preconditioners, i.e. *scalings*, which involve only n field elements and incur $O(n)$ cost. They are proven effective for large fields[9,5].

`LinBox::rank`: The minimal polynomial $f_v^{A,u}(x)$ of $\{v^{\mathrm{T}} A^i u\}_{i \geq 0}$ always reveals a lower bound for rank A. Whether this bound coincides with the rank depends on the spectral structure of A[16]. In a given application, if that structure is favorable then `LinBox::minpoly` is sufficient (for instance, full rank is certified by checking if $\deg f_v^{A,u}(x)$ is maximal). For large fields the preconditioning $\tilde{A} = D_1 A^{\mathrm{T}} D_2 A D_1$, with D_1 and D_2 two random diagonal matrices, probabilistically ensures a good structure[9]. For smaller fields, one might use a field extension, but we can also use this preconditioning in the ground field as a heuristic. The Lanczos approach allows an easy check. It can produce an orthogonal basis for the range of $\tilde{A}$, and with $O(n)$ dot products one may orthogonalize a random vector with respect to this basis. The rank is quickly certified by checking that the result is in the null space.

`LinBox::determinant`: The determinant is easily obtained from the constant term of the minimal polynomial of the preconditioning $\tilde{A} = DA$, where D is a random scaling[5]. This has been used for some of the experiments in Sections 4.

Integer computations. Most LinBox algorithms for integer and rational number computations are based on the finite field functionality. Minimal polynomial, determinant and system solution may be computed using Chinese remaindering, including early termination strategies[4,12]. Integer linear systems, especially sparse ones, may also be solved using the p-adic lifting approach combined with the use of a black box for the inverse matrix modulo p[16,11]. An efficient Monte Carlo rank determination is based on rank computations modulo random primes (see above). Specialized algorithms for Smith normal form computations[8] (see also Section 4) and diophantine problems served during the early development of LinBox to validate the design of the library. We plan to provide additional algorithms for integral problems, like a sparse diophantine linear system solver.

4 Computational experiences

What do we know in practice about the prospects for high performance exact linear algebra computations? In this section we cite some examples to illustrate the range of possibilities. The problems described here have been difficult to solve by other means. The alternative to LinBox's exact linear algebra is numeric approximation in some cases and combinatorial methods in others. These alternative approaches suffer fatal numerical instability and/or exponential memory demand, which are avoided here. The first example is not a real application, but rather a posed Challenge problem. Its solution is illustrative of LinBox's capability. A noteworthy fact is that a very long computation succeeded without exhausting memory. Computations with high time to memory ratio are rare in computer algebra.

Trefethen's Hundred Digits. Nick Trefethen has just posed a "Hundred Dollar, Hundred Digit Challenge[14]." Aimed at numerical analysts, the Challenge consists of 10 problems which have real number solutions. Ten digits of accuracy are asked in the answer to each. All of them are numerically difficult. Problem #7 is the computation of the $(1,1)$ entry of A^{-1}, where A is the 20000×20000 matrix whose entries are zero everywhere except for the primes $2, 3, 5, 7, \ldots, 224737$ along the main diagonal and the number 1 in all the positions a_{ij} with $|i-j| = 1, 2, 4, 8, \ldots, 16384$. We took this on as a challenge for LinBox to produce the *exact* solution. It is a rational number whose numerator and denominator each has approximately 100,000 digits. To compute this number is well beyond the capability of present day general purpose systems such as Maple and Mathematica on current processors and memories. Indeed, it is beyond the capabilities of any software we know of, save LinBox. We have computed this exact answer using approximately 2 years of CPU time (about 180 CPUs running for 4 days). Trefethen has asked that we not reveal the solution or the details of our method before the deadline of his Challenge, May 20, 2002. However, we intend to include a more complete description of this computation in the final version of this paper. Also, we are experimenting with several approaches to this problem and will expand on their relative performances.

Rank, competition with Gaussian elimination techniques. In table 2 we report some comparisons between Wiedemann's algorithm and elimination with reordering for computing the rank. For the first method, we compute the minimal polynomial of $D_1 A^T D_2 A D_1$, as seen in Subsection Preconditioning of Section 3. In order to show the usefulness of this approach, we compare it to elimination-based methods. As the matrices are sparse, some pre-elimination and on-the-fly reordering may be used to reduce fill-in and augment efficiency. We have chosen not to focus on this (Dumas's thesis[7] and references therein review some reordering heuristics and their application to the matrices). The timings in column "Gauß" of Table 2 have been obtained using the algorithm of the thesis[7], §5.1. The timings in column "SuperLU" are those of a generic version [`www-sop.inria.fr/galaad/logiciels/synaps`] of the SuperLU 2.0 numerical code [`www.nersc.gov/~xiaoye/SuperLU`]. For several cases, fill-in causes a failure of elimination. This is due to memory thrashing (MT).

Our experiments show that *as long as enough memory is available*, elimination is very often more efficient when the matrix is already nearly triangular (cyclic8_m11, from Gröbner basis computation), when the matrix has a very small number of elements per row (*nick*; ch*i*-*j*, from algebraic topology), or when the matrix is very unbalanced (bibd – Balanced Incomplete Block Design, from combinatorics). Nonetheless, when the mean number of elements per row is significant (greater than 5, say), Wiedemann's algorithm is superior, even on small matrices, and is sometimes the only practical solution.

Table 2. **Rank modulo 65521, Elimination vs. Wiedemann** on an Intel PIII, 1GHz, 1Gb. $n \times m$ is the shape, Ω the number of non-zero elements, r the integer rank of the matrix, timings are in seconds.

Matrix	Ω, $n \times m$, r	Gauß	SuperLU	Wiedem.
cyclic8_m11	2462970, 4562x5761, 3903	257.33	448.38	2215.36
bibd_22_8	8953560, 231x319770, 231	100.24	938.81	594.29
n4c6.b12	1721226, 25605x69235, 20165	188.34	1312.27	2158.86
mk13.b5	810810, 135135x270270, 134211	MT		44907.1
ch7-7.b5	211680, 35280x52920, 29448	2179.62	MT	2404.5
ch7-8.b5	846720, 141120x141120, 92959	5375.76	MT	29109.8
ch7-9.b5	2540160, 423360x317520, 227870	MT		210698
ch8-8.b5	3386880, 564480x376320, 279237	MT		363754
TF14	29862, 2644x3160, 2644	50.58	50.34	27.21
TF15	80057, 6334x7742, 6334	734.39	776.68	165.67
TF16	216173, 15437x19321, 15437	18559.40	15625.79	1040.36
TF17	586218, 38132x48630, 38132	MT	MT	7094.97

Smith Form, Simplicial Homology. Integer Smith Normal form, especially as applied to Simplicial Homology computation, was the earliest application of LinBox methods. A GAP share package implements this and the web site http://linalg.org offers both downloads and access to these Smith form computation using a web math server. Combinatorial conjectures have been disproven and others affirmed or inspired by computations using these tools. Matrices with hundreds of thousands of rows and columns and several million nonzero entries have proven tractable for computing Smith forms. See our paper[8], and Björner/Welker's[3] for some of the homology applications.

Acknowledgements. We are grateful to B. Mourrain and P. Trébuchet for their help in benchmarking their SuperLU code, to A. Lobo and J.-L. Roch for their comments, and to W. Eberly for his algorithmic improvements published elsewhere. This material is based on work supported in part by NSF (USA) under grants Nrs. DMS-9977392, CCR-9988177, and ITR/ASC-0113121 (Kaltofen) Nrs. CCR-0098284 and

ITR/ASC-0112807 (Saunders) and by CNRS (France) Actions Incitatives No 5929 et STIC LINBOX 2001 (Villard) and by NSERC (Canada, Giesbrecht).

References

1. Z. Bai, J. Demmel, J. Dongarra, A. Ruhe, and H. van der Vorst, editors. *Templates for the solution of Algebraic Eigenvalue Problems: A Practical Guide.* SIAM, Philadelphia, PA, 2000.
2. R. Barrett, M. Berry, T.F. Chan, J. Demmel, J. Donato, J. Dongarra, V. Eijkhout, R. Pozo, C. Romine, and H. Van der Vorst. *Templates for the Solution of Linear Systems: Building Blocks for Iterative Methods, 2nd Ed.* SIAM, 1994.
3. Anders Björner and Volkmar Welker. Complexes of directed graphs. *SIAM Journal on Discrete Mathematics*, 12(4):413–424, November 1999.
4. H. Brönnimann, I.Z. Emiris, V.Y. Pan, and S. Pion. Sign determination in residue number systems. *Theoret. Comput. Sci.*, 210(1):173–197, 1999. Special Issue on Real Numbers and Computers.
5. L. Chen, W. Eberly, E. Kaltofen, B.D. Saunders, W.J. Turner, and G. Villard. Efficient matrix preconditioners for black box linear algebra. *Linear Algebra and its Applications*, 343-344:119–146, 2002.
6. A. Díaz and E. Kaltofen. FoxBox a system for manipulating symbolic objects in black box representation. In O. Gloor, editor, *Proc. ISSAC '98*, pages 30–37, New York, N. Y., 1998. ACM Press.
7. J.-G. Dumas. Algorithmes parallèles efficaces pour le calcul formel: algèbre linéaire creuse et extensions algébriques. Thèse de Doctorat, Institut National Polytechnique de Grenoble, France, décembre 2000.
8. J-G. Dumas, B. D. Saunders, and G. Villard. On efficient sparse integer matrix Smith normal form computations. *Journal of Symbolic Computations*, 32(1/2):71–99, July–August 2001.
9. W. Eberly and E. Kaltofen. On randomized Lanczos algorithms. In *Proc. ISSAC 1997*, pages 176–183. ACM Press, 1997.
10. E. Kaltofen and A. Lobo. Distributed matrix-free solution of large sparse linear systems over finite fields. *Algorithmica*, 24(3–4):331–348, July–Aug. 1999. Special Issue on "Coarse Grained Parallel Algorithms".
11. E. Kaltofen and B.D. Saunders. On Wiedemann's method of solving sparse linear systems. In *Proc. AAECC-9*, LNCS 539, Springer, pages 29–38, 1991.
12. E. Kaltofen and B. Trager. Computing with polynomials given by black boxes for their evaluations: Greatest common divisors, factorization, separation of numerators and denominators. *J. Symb. Comp.*, 9(3):301–320, 1990.
13. R. Lambert. *Computational aspects of discrete logarithms.* PhD thesis, University of Waterloo, Ontario, Canada, 1996.
14. N. Trefethen. A Hundred-dollar, Hundred-digit Challenge. *SIAM News*, 35(1), 2002.
15. G. Villard. A study of Coppersmith's block Wiedemann algorithm using matrix polynomials. Rapport de Recherche 975 IM, `www.imag.fr` (Grenoble), April 1997. Extended abstract in Proc. ISSAC 1997, pages 32–39, ACM Press.
16. D. Wiedemann. Solving sparse linear equations over finite fields. *IEEE Transf. Inform. Theory*, IT-32:54–62, 1986.

BLOCK-ITERATIVE ALGORITHMS FOR IMAGE RECONSTRUCTION

MING JIANG

*School of Mathematical Science, Peking University, Beijing 100871, China**

E-mail: jiangm@math.pku.edu.cn, jiangm@ct.radiology.uiowa.edu

GE WANG

Department of Radiology, University of Iowa, Iowa City, IA 52242, USA

E-mail: WangG@mail.medicine.uiowa.edu, ge-wang@ieee.org

In this paper, we first introduce a general iterative scheme for image reconstruction and then apply it to CT image reconstruction with practical data. This scheme can be readily formulated in the simultaneous and block-iterative formats. This provides a facility to obtain new iterative algorithms from known methods and also permits more flexibility in task scheduling for parallel implementation. It can be shown that the simultaneous algebraic reconstruction technique (SART), is a special example of the general scheme. Based on the convergence properties of the general scheme under mild conditions on the relaxation parameters, in the consistent or inconsistent cases, the convergence of the SART and its block-iterative version can be easily derived. The block-iterative SART is applied to reconstruct CT image from practical data to demonstrate its performance. Some issues related to the parallel implementation are also addressed.

1 Introduction

Most of the imaging systems can be modeled by the following equation

$$Ax = b \tag{1}$$

where the observed data $b = (b^1 \cdots b^M)^{\mathrm{T}} \in \mathbf{R}^M$, the original image $x = (x_1 \cdots x_N)^{\mathrm{T}} \in \mathbf{R}^N$, $A = (A_{i,j})$ is a non-zero $M \times N$ matrix. The image reconstruction problem is to compute the image x from the observed data b. A direct solution is usually not feasible because of the ill-posedness, the noise and huge data dimension in practice. The large number of unknowns and equations in application also makes it impractical to solve (1) with straightforward implementation of general purpose optimization algorithms[1].

The iterative approach has been important in this field because of their superior performance[1]. With rapid development of the computer technology, iterative algorithms receive increasingly more attention. In this paper, we

*The first author performed this work while he was a visiting professor with CT/Micro-CT Laboratory, Department of Radiology, University of Iowa, Iowa City, IA 52242, USA.

consider the following scheme to solve (1):

$$x^{(n+1)} = x^{(n)} + \lambda_n V^{-1} A^* W(b - Ax^{(n)}) \tag{2}$$

where $\lambda_n > 0$ is the relaxation coefficient and V and W are two positive definite *diagonal* matrices of order N and M, respectively.

In recent years, great interests are in the *block-iterative* (BI) or *ordered-subset* version of (2), which invokes only part of the data b to update the iteration[2,1,3]. The BI version of (2) can be formulated as follows. Assume that the index set $B = \{1, \cdots, M\}$ is partitioned into T nonempty subsets B_t such that

$$B = \bigcup_{1 \le t \le T} B_t, \text{ where } B_t = \{i_1^t, \cdots, i_{M(t)}^t\}. \tag{3}$$

We assume that the subsets B_t are disjoint in this paper. Detailed discussions can be found in[4]. For $n \ge 0$, let $[n] \triangleq n \mod (T) + 1$. Let

$$A_t = \begin{pmatrix} A^{i_1^t} \\ \vdots \\ A^{i_{M(t)}^t} \end{pmatrix}, \ b_t = \begin{pmatrix} b^{i_1^t} \\ \vdots \\ b^{i_{M(t)}^t} \end{pmatrix} \text{ and } W_t = \begin{pmatrix} W^{i_1^t} & & \\ & \ddots & \\ & & W^{i_{M(t)}^t} \end{pmatrix}, \tag{4}$$

where A^i is the i-th row of A, W^i is the i-th diagonal element of W and b^i is the i-th component of b, respectively. Then the BI version of (2) can be written as

$$x^{(n+1)} = x^{(n)} + \lambda_n V^{-1} A_{[n]}^* W_{[n]}(b_{[n]} - A_{[n]} x^{(n)}). \tag{5}$$

When $T = 1$, (5) reduces to the simultaneous version (2). It has been shown that some of the widely used algorithms are special cases of (2) or (5)[4].

The iteration process from $n = kT$ to $n = (k+1)T - 1$ in (5) is called one BI cycle. We assume that λ_n is constant during one BI cycle, i.e., λ_n for $n = kT + t$ with $t = 0, \cdots, T-1$ are the same; for notation simplicity and without confusion, we will write this constant as λ_k. This general iterative scheme and its BI version were studied in [?]. The main feature of the results is the convergence of the *whole* sequence for the BI versions in the *inconsistent* case, rather than the convergence of the cyclic sub-sequences [2,3,5,6].

In this paper, we apply the general convergence results from [4] to establish the convergence of the SART, one well-known image reconstruction algorithm, in the consistent or inconsistent cases, for the simultaneous or BI versions. The BI versions of the SART is new and is applied to reconstruct CT image from practical data to demonstrate its performance. The convergence of SART in the inconsistent cases was studied in [3,5,6,7], but the convergence of it BI version is established first time in this work. Some issues related to the parallel implementation are discussed in the last section.

The authors thank Y. Censor and T. Elfving for sharing their paper [3].

2 Preliminaries

2.1 Some Linear Algebra

The canonical inner product and norm on $\mathbf{R}^p$ are denoted by $< \cdot, \cdot >$ and $||\cdot||$, respectively. The zero element of $\mathbf{R}^p$ is denoted by θ. For a $q \times p$ matrix U, U^* denotes the transpose of U. $N(U)$ and $(R(U)$ denote the kernel and the image of U, respectively. The orthogonal projection from $\mathbf{R}^p$ to $N(U)$ is denoted by $P(U)$.

Given a positive definite matrix D of order p, it induces the following inner product on $\mathbf{R}^p$: $< x, x >_D = < Dx, x >$, for $x \in \mathbf{R}^p$. The corresponding induced norm, called D-norm, is denoted as $||\cdot||_D$. We use $\mathbf{R}^p$ to denote the space $\mathbf{R}^p$ with the canonical inner product $< \cdot, \cdot >$ and $\mathbf{R}^p_D$ the space $\mathbf{R}^p$ with the inner product $< \cdot, \cdot >_D$, respectively. The orthogonal projection from $\mathbf{R}^p_D$ to $N(U)$ is denoted by $P_D(U)$.

Let V and W be two positive definite matrices of order p and q, respectively. The operator norm of a $q \times p$ matrix U as a linear operator from $\mathbf{R}^p_V$ to $\mathbf{R}^q_W$, is defined as

$$||U||_{V,W} = \sup_{x \in \mathbf{R}^p_V \backslash \{\theta\}, ||x||_V = 1} ||Ux||_W \tag{6}$$

If V and W are identity matrices, $||U||_{V,W}$ reduces to the conventional spectral norm[8] and we write it as $||U||$. In this paper, if not explicitly stated, the inner product and norm are the canonical ones.

In the following, let V and W be two positive definite *diagonal* matrices of order N and M as in § 1, respectively. Let L_W be the following weighted least square functional

$$L_W(x) = \frac{1}{2} < W(b - Ax), b - Ax > = \frac{1}{2}||b - Ax||^2_W \tag{7}$$

for $x \in \mathbf{R}^N$. The gradient of L_W is $\nabla L_W(x) = -A^*W(b - Ax)$. All the minimizers of L_W satisfy the following normal equation $\nabla L_W(x) = \theta$:

$$A^*WAx = A^*Wb. \tag{8}$$

It can be shown: (a) there is always a solution to (8); (b) each solution of (8) is a minimizer of L_W, and vice versa; (c) among all the solutions of (8), there is a unique solution of the minimal V-norm, which is denoted by $x_{V,W}(A, b)$.

If there is a solution of (1), it is called *consistent*. Otherwise, it is called *inconsistent*. If (1) is consistent, it is easy to prove that (1) and (8) have exactly the same solutions. Hence, the minimal V-norm solutions of (1) and (8) are the same in the consistent case.

2.2 SART

The simultaneous algebraic reconstruction technique (SART) was proposed as a major refinement of the ART [9]:

$$x_j^{(n+1)} = x_j^{(n)} + \frac{\lambda_n}{\sum_{i=1}^{M} |A_{i,j}|} \sum_{i=1}^{M} \frac{A_{i,j}}{\sum_{j=1}^{N} |A_{i,j}|} \left(b^i - A^i x^{(n)} \right). \tag{9}$$

Let the diagonal matrices V and W be defined as

$$V^j = A_{+,j} = \sum_{i=1}^{M} |A_{i,j}|, \quad \text{for } j = 1, \cdots, N, \tag{10}$$

$$\frac{1}{W^i} = A_{i,+} = \sum_{j=1}^{N} |A_{i,j}|, \quad \text{for } i = 1, \cdots, M, \tag{11}$$

Then the SART (9) can be written in the form of (2).

The BI version of the SART, with the partition set (3), is

$$x_j^{(n+1)} = x_j^{(n)} + \frac{\lambda_n}{V^j} \sum_{i \in B_{[n]}} W^i A_{i,j} \left(b^i - A^i x^{(n)} \right). \tag{12}$$

3 Convergence Results

The following results were proved in [4].

Theorem 3.1. *Assume that there exists $\rho > 0$ such that $||A_t||_{V,W_t} \leq \rho$ for $t = 1, \cdots, T$ and $0 \leq \rho^2 \lambda_k \leq 2$.*

If (1) is consistent and

$$\sum_k \min(\rho^2 \lambda_k, 2 - \rho^2 \lambda_k) = \infty, \tag{13}$$

then the sequence $x^{(n)}$ generated by (5) converges to $x_{V,W}(A, b) + P_V(A) x^{(0)}$;

If the partition (3) is disjoint and

$$\lim_k \lambda_k = 0 \quad \text{and} \quad \sum_k \lambda_k = \infty, \tag{14}$$

then the sequence $x^{(n)}$ *generated by (5) converges to* $x_{V,W}(A,b) + P_V(A)x^{(0)}$, *even if (1) is inconsistent.*

In either case, the sequence $x^{(n)}$ *converges to a solution of the normal equation (8).*

Theorem 3.2. *Assume that* $\rho = ||A||_{V,W} > 0$ *and* $0 \leq \rho^2\lambda_k \leq 2$. *Then the conclusions of Theorem 3.1 hold.*

The convergence for the simultaneous scheme (2) can be established with the same conditions for the consistent case.

Theorem 3.3. *Let* $\rho = ||A||_{V,W}$. *Assume* $0 \leq \rho^2\lambda_k \leq 2$ *and (13) holds, then the sequence* $x^{(n)}$ *generated by (2) converges to* $x_{V,W}(A,b) + P_V(A)x^{(0)}$.

4 Applications

4.1 Convergence Property of the SART

In the following, we will show the convergence of the SART, by estimating $||A||_{V,W}$, with the corresponding choices of the matrices V and W as in (10) and (11).

By the convexity of the function $t \mapsto t^2$,

$$\begin{aligned}
&||Ax||_W^2 \\
&= \sum_{i=1}^{M} \frac{1}{A_{i,+}} \left| \sum_{j=1}^{M} A_{i,j} x_j \right|^2 \leq \sum_{i=1}^{M} A_{i,+} \left(\sum_{j=1}^{N} \frac{|A_{i,j}|}{A_{i,+}} |x_j| \right)^2 \\
&\leq \sum_{i=1}^{M} A_{i,+} \sum_{j=1}^{N} \frac{|A_{i,j}|}{A_{i,+}} |x_j|^2 = \sum_{j=1}^{N} A_{+,j} |x_j|^2 = ||x||_V^2.
\end{aligned}$$

Therefore, $||A||_{V,W} \leq 1 = \rho$. Hence, we have the following theorem:

Theorem 4.1. *Assume that* $0 \leq \lambda_k \leq 2$.

A *If (1) is consistent and*

$$\sum_k \min(\lambda_k, 2 - \lambda_k) = \infty, \tag{15}$$

then the sequence $x^{(n)}$ *generated by (12) converges to* $x_{V,W}(A,b) + P_V(A)x^{(0)}$;

B *If the partition (3) is disjoint and*

$$\lim_k \lambda_k = 0 \quad \textit{and} \quad \sum_k \lambda_k = \infty, \tag{16}$$

then the sequence $x^{(n)}$ *generated by (12) converges to* $x_{V,W}(A,b) + P_V(A)x^{(0)}$*, even if (1) is inconsistent.*

C *If*

$$\sum_k \min(\lambda_k, 2-\lambda_k) = \infty, \tag{17}$$

then the sequence $x^{(n)}$ *generated by (9) converges to* $x_{V,W}(A,b) + P_V(A)x^{(0)}$*, even if (1) is inconsistent.*

In all the cases, the sequence $x^{(n)}$ *converges to a solution of the normal equation (8).*

Theorem 4.1 A and B are about the convergence of the BI version of the SART; C is about the convergence of the SART, the condition is general than those in [3,5,6,7].

4.2 Experiment

To demonstrate the correctness of the BI SART, the filtered back-projection, SART and BI SART algorithms were coded and integrated as a computed tomography (CT) simulator in the programming language C on an SGI workstation (Model O2, Silicon Graphics, Inc.; Mountain View, CA, USA). In the simulator, we assumed the fan-beam imaging geometry and projection data format obtained from the Toshiba America Medical Systems (Tustin, CA, USA). For the details about CT, please refer to [10].

Figure 1. A CT performance phantom.

A CT performance phantom (Toshiba America Medical Systems, Tustin, CA, USA) was used as the test object, as shown in Figure 1. The CT phantom consists of a cylindrical Plexiglas container and five embedded solid plastic cylinders of various CT values, which are clinically relevant. Before being scanned, the phantom was filled with water to mimic the soft tissue background.

A phantom scan was performed using a Toshiba Xpress/SX CT scanner (Toshiba America Medical Systems, Tustin, CA, USA) with 120 kVp, 200 mAs, 2 millimeter detector collimation, and 1.5 second scan time. Reconstruction was done using 300 projections, 256 detectors per projection, from a 360o scanning.

The BI SART is applied with the following setting: the size of all the subsets was set to 10. A random ordering was performed of the subsets; the relaxation sequence can be readily determined to meet the convergence requirements, for example, $\lambda_k = \frac{1}{1+\delta(k-1)}$, $k \geq 1$ and δ is a fraction. A heuristics for such a selection is that the decreasing speed of the relaxation coefficients should not be faster than $O(\frac{1}{k})$ so that $\sum_k \lambda_k = \infty$, while the speed should not be slower than $O(\frac{1}{k})$ so that the change with each iteration will quickly become sufficiently small, which may suggest practical convergence. Additionally, the constant δ allows a significant degree of control over the relaxation process. An intuitive guideline is that the higher the noise level of data, the smaller the δ value; for noise-free data, $\delta = 1$.

To quantify the performance of the BI SART relative to the SART for image reconstruction, the Euclidean distance in the image domain were computed from the images in the CT phantom experiment, with the filtered back-projection image being the gold standard.

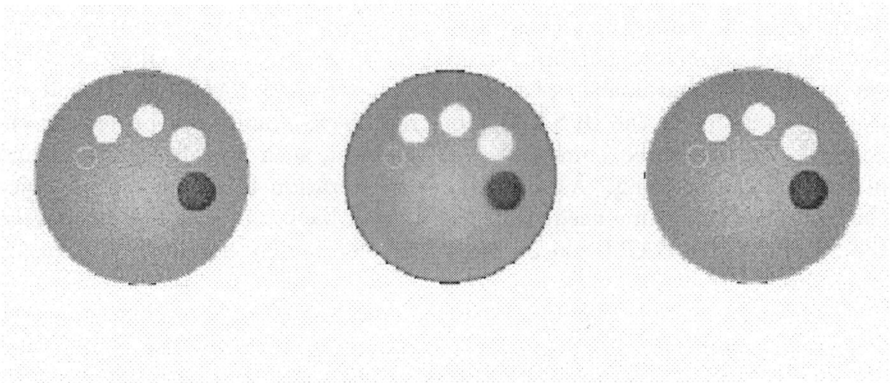

Figure 2. Reconstructed Phantom Images.

Figure 2 presents CT phantom images reconstructed using the filtered back-projection method, SART and BI SART ($\delta = 0.5$). It can be observed in Figure 2 that the image reconstructed using filtered back-projection (Figure 2, the left image) is in an excellent agreement with that reconstructed using BI SART (Figure 2, the right image). Furthermore, the BI SART produced a better image quality (Figure 2, the right image) at a speed one order of magnitude higher than that (Figure 2, the center image) of the SART. Actually, with the BI algorithm, the image quality depends on selection of the size and ordering of subsets. Generally, it was found that the smaller the size of subsets, the faster the convergence, but at the same time, the more residual artifacts. Quantitatively, the BI SART ran 10 iterations, taking about 50 minutes and giving a mean square error of about 1.25 of the full dynamic range of 255 gray levels, while the SART spent 100 iterations with a total time of 8 hours 20 minutes and a mean error of 1.45.

5 Discussions and Conclusion

As discussed in depth[1], the iterative algorithms of the form (2) or (5), either simultaneous or BI, lead to parallel computing algorithms. The iteration scheme (2) can be executed concurrently using up to M parallel processors. "However, the introduction of blocks permits more flexibility in task scheduling. The blocks can be chosen so as to ensure that all processors receive tasks of the same computational difficulty. Hence, all processors complete their work in roughly the same amount of time. Delays while processors wait for each other are minimized. The specification of block size will typically depends on the computer architecture." [1]

We recognize that our implementation of the current simulator has not yet utilized the parallelism of the SART and the BI SART. In each iteration of the SART or the BI SART, both reprojection and back-projection are needed with associated computational overhead, such as subset control and intermediate processing. As a result, every iteration takes more than twice the computational time needed for the filtered back-projection. Fortunately, the SART and BI SART can be greatly accelerated by a multi-processor architecture.

Currently, we are planning to extend the BI SART from fan-beam to cone-beam geometry. Recently, with introduction of multi-slice systems the CT scanner began a transition to spiral/helical cone- beam geometry. Cone-beam spiral CT allows larger scanning range in shorter time with higher image resolution, and has important medical and other applications. We hypothesize that our BI algorithm should perform well in cone-beam geometry, since the SART algorithm has been shown to be effective in cone-beam geometry[11,12,13,14,15]. Further work along this line will be reported later.

Acknowledgment

The authors thank Yusuke Toki with Toshiba Corporation, Kyung Han with BioImaging Research, Inc., and James Pitcher from the University of Iowa for important technical assistance on scanner geometry, raw data format and conversion.

References

1. Y. Censor and S. A. Zenios. *Parallel Optimization: Theory, Algorithms and Applications.* Oxford University Press, 1997.
2. Y. Censor, P. P. B. Eggermont, and D. Gordon. Strong underrelaxation in Kaczmarz's method for inconsistent systems. *Numerische Mathematik*, 41:83 – 92, 1983.
3. Y. Censor and T. Elfving. Block-iterative algorithms with diagonally scaled oblique projections for the linear feasibility problem. *SIAM Journal on Matrix Analysis and Applications.* to appear.
4. M. Jiang and G. Wang. Convergence studies on iterative algorithms for image reconstruction. Technical report, Micro-CT Laboratory, Department of Radiology, University of Iowa, Iowa City, IA, 2002.
5. M. Jiang and G. Wang. Convergence of the simultaneous algebraic reconstruction technique (SART). In *Proceedings of the 35th Asilomar Conference on Signals, Systems, and Computers*, Pacific Grove, CA, 2001. invited talk.
6. M. Jiang and G. Wang. Convergence of the simultaneous algebraic reconstruction technique (SART). *IEEE Trans. Image Processing*, 2001. accepted for publication.
7. C. L. Byrne and J. Graham-Eagle. A short proof of the convergence of the SART algorithm. Technical report, Department of Mathematical Sciences, University of Massachusetts, Lowell, 2001.
8. P. Lancaster and M. Tismenetsky. *The Theory of Matrices.* Academic Press, second edition, 1985.
9. A. H. Andersen and A. C. Kak. Simultaneous algebraic reconstruction technique (SART): a superior implementation of the ART algorithm. *Ultrasonic Imaging*, 6:81 – 94, 1984.
10. F. Natterer. *The Mathematics of Computerized Tomography.* John Wiley & Sons, New York, 1986.
11. G. Wang, D. L. Snyder, J. A. O'Sullivan, and M. W. Vannier. Iterative deblurring for CT metal artifact reduction. *IEEE Trans. Medical Imaging*, 15:657–664, 1996.

12. G. Wang, M. W. Vannier, M. W. Skinner, M. G. P. Cavalcanti, and G. Harding. Spiral CT image deblurring for cochlear implantation. *IEEE Trans. Medical Imaging*, 17:251 – 262, 1998.
13. G. Wang, G. D. Schweiger, and M. W. Vannier. An iterative algorithm for X-ray CT fluoroscopy. *IEEE Trans. Medical Imaging*, 17:853–856, 1998.
14. K. Mueller and R. Yagel et al. Anti-aliased three-dimensional cone-beam reconstruction of low-contrast objects with algebraic methods. *IEEE Trans. Medical Imaging*, 18(6):519 – 537, 1999.
15. K. Mueller and R. Yagel et al. Fast implementations of algebraic methods for three-dimensional reconstruction from cone-beam data. *IEEE Trans. Medical Imaging*, 18(6):538 – 548, 1999.

THE SNAP PACKAGE FOR ARITHMETIC WITH NUMERIC POLYNOMIALS

C.-P. JEANNEROD
LIP, Ecole Normale Supérieure de Lyon,
46 Allée d'Italie, 69364 Lyon Cedex 07, France
E-mail: Claude-Pierre.Jeannerod@ens-lyon.fr

G. LABAHN
Department of Computer Science, University of Waterloo,
Ontario N2L 3G1 Canada
E-mail: glabahn@scg.math.uwaterloo.ca

In this paper we describe the SNAP package, a new MAPLE package for algebraic manipulation of univariate numeric polynomials. The package includes commands for quotient, remainder, greatest common divisors and related operations. We discuss the methods used in the package along with the key issues that were encountered during the implementation.

1 Introduction

There are a number of application areas which can be formally described in terms of algebraic operations on polynomials and matrix polynomials such as division, remainders and greatest common divisors. Areas which make use of such an approach include control theory, linear systems theory and signal processing[14]. While an algebraic polynomial formalism is extremely useful for theoretical studies of properties, it has drawbacks when one tries to take advantage for computations. The primary problem is that many applications use inexact data and finite precision computations while the formalism assumes exact information and error-free computations[16,20].

In this paper we describe the SNAP (Symbolic-Numeric Algorithms for Polynomials) package for computing with polynomials having inexact coefficients. This package is a first attempt to provide the standard functionalities for inexact polynomials that exist for exact polynomials, including the taking of quotients and remainders, determining if two polynomials are relatively prime and finding greatest common divisors (GCDs). The package is included in the coming release of the MAPLE computer algebra system.

Several algorithms exist for performing such operations, in particular for computing the various notions of the so-called approximate GCDs. These algorithms include modifications of the Euclidean algorithm[10,16,19], optimization techniques[5,12,13], subresultant-based algorithms[6,7], and polynomial root approximation[17]. For the most part, the previously mentioned algorithms resort to either infinite or adaptive precision. In our case, the software we present works under the customary model of numerical computation using fixed precision floating-point arithmetic and the underlying algorithms benefit from round-off error analysis which ensures numerical stability[9].

This approach, based on estimating the distance to the closest pair of polynomials having a common factor[3], also allows for computations that typically run in quadratic time. Also, the two approximate GCDs that we focus on in the package (quasi-GCD[19] and ϵ-GCD[3]) are certified in the sense of Emiris *et al* [7].

The remainder of the paper is organized as follows. The next section gives a description of the commands in the SNAP package. Since the numerical computation of GCDs plays such a significant role in the package, we summarize in Section 3 the Beckermann-Labahn algorithm, which is the main tool used for this computation and for a number of commands inside the package. Section 4 then gives some details of the implementation used while Section 5 gives some of the practical features of the package. The paper ends with a conclusion along with topics for future research.

2 Overview of the SNAP package

The SNAP package has been designed for univariate polynomials with real floating-point coefficients. The commands it offers are described below, where (a, b) denotes a pair of such numeric polynomials:

- `AreCoprime`: coprimeness test for (a, b) when known up to a given ϵ;
- `DistanceToCommonDivisors`: estimate the distance between (a, b) and the set of polynomial pairs with at least one nontrivial common divisor;
- `DistanceToSingularPolynomials`: estimate the distance between a and the set of polynomials with at least one multiple root;
- `EpsilonGCD`: compute a polynomial g and a quantity ϵ such that g is an ϵ-GCD for (a, b);
- `EuclideanReduction`: return a degree reduced basis that is numerically equivalent to the basis (a, b);
- `IsSingular`: decide whether a has at least one mutiple root when known up to a given ϵ;

- `QuasiGCD`: compute a polynomial g and a quantity ϵ such that g is a quasi-GCD with precision ϵ for (a, b);
- `Quotient`: compute the quotient of a divided by b;
- `Remainder`: compute the remainder of a divided by b.

A typical use of a SNAP command is illustrated through the simple example below (with 10 digits) where one investigates the relative primeness of two polynomials up to a tolerance of 0.5 and 0.1:

```
> with(SNAP):
> a := 0.1*z^2+1.5*z-0.2:
> b := 0.2*z^3+0.15:
> AreCoprime(a,b,z,0.5);
```

false

```
> AreCoprime(a,b,z,0.1);
```

true

Since for the latter tolerance the above polynomials are coprime, their Bézout coefficients for the associated linear diophantine equation can also be obtained using an optional parameter:

```
> AreCoprime(a,b,z,0.1,output='BC');
```

true,

$$[-0.8710041008177655581z^2 - 0.1122329072639631 63z - 0.05851459268,\\ 0.4355020504088827 34z + 6.588647210]$$

3 Solving polynomial Diophantine equations numerically

Let $a(z)$, $b(z)$ be two univariate polynomials over $\mathbb{C}$ with degrees m, n. Except for `Quotient` and `Remainder`, the majority of the current functions in the SNAP package are based on the following approach, developed by Beckermann and Labahn[2,3]: compute polynomial solutions $u(z)$, $v(z)$, $\underline{u}(z)$, $\underline{v}(z)$ to the Diophantine equations

$$a(z)v(z) + b(z)u(z) = 1, \quad \deg u < m, \quad \deg v < n, \tag{1}$$

$$a(z)\underline{v}(z) + b(z)\underline{u}(z) = z^{m+n-1}, \quad \deg \underline{u} < m, \quad \deg \underline{v} < n, \tag{2}$$

in a fast and numerically stable way. Here, "fast" means arithmetic complexity in $O((m+n)^2)$ and "stable" means weakly stable in the sense of Bunch[4].

This approach provides a numerical coprimeness test together with Bézout coefficients. Additionally, a solution $(u, v, \underline{u}, \underline{v})$ to Equations (1), (2) yields a sharp estimation of the distance $\epsilon(a, b)$ to the set of polynomial pairs with a common root. Indeed, denoting by $|| \; ||$ the 1-norm for the space of matrix polynomials over $\mathbb{C}$, one has

$$\epsilon(a,b) = \min\{||(a-a^*, b-b^*)|| : \deg\gcd(a^*, b^*) > 0, \deg a^* \leq m, \deg b^* \leq n\}.$$

If we let $S(a, b)$ be the Sylvester matrix associated with $a(z)$, $b(z)$ then it can be shown[2] that

$$\frac{1}{||S(a,b)^{-1}||} \leq \frac{1}{\kappa} \leq \epsilon(a,b) \quad \text{where} \quad \kappa = \left|\left|\begin{bmatrix} v & \underline{v} \\ u & \underline{u} \end{bmatrix}\right|\right|. \tag{3}$$

This improves upon the well known lower bound $1/||S(a,b)^{-1}||$ for $\epsilon(a, b)$.

Of course, a solution to Equations (1), (2) may not exist. However, the COPRIME[3] algorithm of Beckermann and Labahn always returns a reduced pair of polynomials from which one often can deduce a solution of (under additional assumptions) some approximate GCDs.

3.1 The algorithm COPRIME

We may assume[3,11] w.l.o.g. that $m > n$ and that the input numerical polynomials have been scaled to satisfy $1/2 \leq ||(a, b)|| \leq 1$. Starting with $(a^{(0)}, b^{(0)}, c^{(0)}) = (a(z), b(z), -z^{n-1})$, the COPRIME algorithm computes for $k \leq m$ a numerical polynomial remainder sequence $(a^{(k)}, b^{(k)}, c^{(k)})$ by means of so-called **unimodular reductions** $U^{(k)}(z) \in \mathbb{C}[z]^{2\times 2}$ **of order** k together with **associated vectors** $\underline{U}^{(k)}(z) \in \mathbb{C}[z]^{2\times 1}$. More precisely,

$$(a^{(k)}, b^{(k)}) = (a, b) \cdot U^{(k)}, \quad c^{(k)} = (a, b) \cdot \underline{U}^{(k)} - z^{n+k-1} \tag{4}$$

with $U^{(k)}$ unimodular and where, for $k \geq 1$,

$$n \geq \deg a^{(k)} = m - k > \deg b^{(k)}, \quad \deg c^{(k)} < m - k - 1,$$

$$\deg U^{(k)} \leq \begin{bmatrix} n-m+k-1 & n-m+k \\ k-1 & k \end{bmatrix}, \quad \deg \underline{U}^{(k)} \leq \begin{bmatrix} n-m+k-1 \\ k-1 \end{bmatrix}.$$

The two main features of this Euclidean-like reduction process are summarized in the theorem below.

Theorem 3.1 *The pairs (a, b) and $(a^{(k)}, b^{(k)})$ define two different bases of the same ideal with $a^{(k)}$ of smaller degree than a. When $k = m$, a solution $(u, v, \underline{u}, \underline{v})$ to Equations (1), (2) is given by*

$$\begin{bmatrix} v \\ u \end{bmatrix} = \frac{1}{a^{(m)}(0)} U^{(m)} \cdot \begin{bmatrix} 1 \\ 0 \end{bmatrix}, \quad \begin{bmatrix} \underline{v} \\ \underline{u} \end{bmatrix} = \underline{U}^{(m)}. \tag{5}$$

In finite precision arithmetic, some remainders defined by (4) may correspond to ill-conditioned subproblems[3]. In order to ensure numerical stability, such remainders are discarded by a "look-ahead" strategy: one jumps from $(a^{(k)}, b^{(k)}, c^{(k)})$ to the first $(a^{(k+s)}, b^{(k+s)}, c^{(k+s)})$ with $s \geq 1$ such that

$$|\det U^{(k+s)}(0)| > \tau \quad \text{and} \quad ||\underline{U}^{(k+s)}|| < 1/\tau. \tag{6}$$

Here, τ is a threshold parameter of order the cubic root of the machine precision and $(a^{(k+s)}, b^{(k+s)})$ is the first basis after $(a^{(k)}, b^{(k)})$ that is **numerically well-behaved** in the sense of (6). The set $\mathcal{A}$ of the indices of all such accepted bases clearly satisfies $\mathcal{A} \subset \{0, 1, 2, \ldots, m\}$.

On the other hand, $U^{(k+s)}$ and $\underline{U}^{(k+s)}$ are determined from $a^{(k)}$, $b^{(k)}$, $c^{(k)}$, $U^{(k)}$, $\underline{U}^{(k)}$ in a numerically stable manner as follows. Consider the $2s \times 2s$ complex matrix

$$M_s^{(k)} = \left[\begin{array}{cccc|cccc} a_{m-k}^{(k)} & & & & 0 & & & \\ a_{m-k-1}^{(k)} & a_{m-k}^{(k)} & & & b_{m-k-1}^{(k)} & \ddots & & \\ \vdots & & \ddots & & \vdots & & \ddots & 0 \\ \vdots & & & a_{m-k}^{(k)} & \vdots & & & b_{m-k-1}^{(k)} \\ \vdots & & & \vdots & \vdots & & & \vdots \\ a_{m-k-2s+1}^{(k)} & \cdots & \cdots & a_{m-k-s}^{(k)} & b_{m-k-2s+1}^{(k)} & \cdots & & b_{m-k-s}^{(k)} \end{array}\right] \tag{7}$$

and solve the three linear systems $M_s^{(k)} x_i = y_i$ with

$$y_1 = (0, \ldots, 0, 1)^T, \quad y_2 = -\left(b_{m-k-j}^{(k)}\right)_{j=1,..,2s}, \quad y_3 = \left(c_{m-k-j}^{(k)}\right)_{j=1,..,2s}. \tag{8}$$

Setting up the 2×2 and 2×1 polynomial matrices

$$U^{(k,k+s)}(z) = \begin{bmatrix} z^{s-1} & \cdots & z^1 & z^0 & 0 & \cdots\cdots & 0 \\ 0 & \cdots\cdots & 0 & z^{s-1} & \cdots & z^1 & z^0 \end{bmatrix} \cdot \left(x_1, x_2\right) + \begin{bmatrix} 0 & 0 \\ 0 & z^s \end{bmatrix}$$

and

$$\underline{U}^{(k,k+s)}(z) = \begin{bmatrix} z^{s-1} & \cdots & z^1 & z^0 & 0 & \cdots\cdots & 0 \\ 0 & \cdots\cdots & 0 & z^{s-1} & \cdots & z^1 & z^0 \end{bmatrix} \cdot x_3,$$

one finally obtains[3]

$$U^{(k+s)} = U^{(k)} \cdot U^{(k,k+s)} \quad \text{and} \quad \underline{U}^{(k+s)} = z^s \underline{U}^{(k)} - U^{(k)} \cdot \underline{U}^{(k,k+s)}.$$

When $k + s \in \mathcal{A}$, additional scaling ensures that $1/2 \leq ||U^{(k+s)}|| \leq 1$. The resulting algorithm, called COPRIME, is given in Table 1.

Method:	Construct (scaled) unimodular reductions $U^{(k)}$ of (a,b) of order $k \in \mathcal{A}$ together with associated vectors $\underline{U}^{(k)}$ for $1 \leq k \leq m$.
Input:	Two polynomials a, b with $\deg a = m > \deg b$. A stability parameter τ.
Output:	If $m \in \mathcal{A}$: RETURN $1/\kappa$ given by (3) and (5). If $m \notin \mathcal{A}$: message, for $1/\kappa$ does not exist or is "too small".
Initialization:	$k = 0$, $\mathcal{A} = \{\}$, $U^{(0)} = \begin{bmatrix} 1 & 0 \\ 0 & 1 \end{bmatrix}$ and $\underline{U}^{(0)} = \begin{bmatrix} 0 \\ 0 \end{bmatrix}$.
Single Step:	For $s = 1, 2, \ldots$:
	Compute $U^{(k,k+s)}$, $\underline{U}^{(k,k+s)}$ by solving linear systems (7-8) (if $\det M_s^{(k)} = 0$ then increase s and restart). Rescale $U^{(k)} \cdot U^{(k,k+s)}$ to obtain $U^{(k+s)}$.
Exit s–loop:	if (6) holds. In this case $k \leftarrow k + s$ and $\mathcal{A} \leftarrow \mathcal{A} \cup \{k\}$.
Exit ALGO:	If $k + s = m$.

Table 1. The algorithm COPRIME

Theorem 3.2 *The algorithm COPRIME is weakly stable and requires in most cases $O((m+n)^2)$ flops.*

Quadratic complexity can be achieved mainly because for most dense polynomial pairs (a, b) only small jumps $(s \leq 3)$ and thus only small linear systems are typically encountered. Note further that the products of (4) can be replaced with the "shorter" products

$$(a^{(k+s)}, b^{(k+s)}) = (a^{(k)}, b^{(k)}) \cdot U^{(k,k+s)}, \quad c^{(k+s)} = z^s c^{(k)} - (a^{(k)}, b^{(k)}) \cdot \underline{U}^{(k,k+s)}.$$

When $m \in \mathcal{A}$ we conclude that (a, b) are numerically coprime up to perturbations of order $\epsilon < 1/\kappa$. When $m \notin \mathcal{A}$ then the last accepted basis $(a^{(k)}, b^{(k)})$ may still provide various approximate GCDs.

3.2 Approximate GCDs from the last accepted basis

Recall that a polynomial g is a **quasi-GCD**[19] **with precision** ϵ for a, b if there exist polynomials u_1, v_1, u_2, v_2 such that

$$||(a,b) - g(u_2, v_2)|| < \epsilon, \quad ||av_1 + bu_1 - g|| < \epsilon ||g||, \quad \deg u_1 < m, \quad \deg v_1 < n.$$

Also, g is an **ϵ-GCD**[3] for a, b if there exist polynomials $\tilde{a}$, $\tilde{b}$ with degrees at most m, n such that $||(a,b) - (\tilde{a},\tilde{b})|| \leq \epsilon$ and $g = \gcd(\tilde{a},\tilde{b})$ has maximal degree.

All the computations in the COPRIME algorithm are done using finite precision arithmetic. As such there are residual error polynomials $\alpha^{(k)}$ and $\beta^{(k)}$ so that

$$(a,b) = U^{(k)} \cdot (a^{(k)}, b^{(k)}) + (\alpha^{(k)}, \beta^{(k)})$$

with $\deg a^{(k)} = m - k > \deg b^{(k)}$. Additionally, let $\rho_l(a,b)$ be the minimum of the set of all products $||(a,b)|| \cdot ||(g_a, g_b)^T||$ where the polynomial pair (g_a, g_b) is such that $\deg g_a < l$, $\deg g_b < l$ and

$$z^n a(z) g_a(z) + z^m b(z) g_b(z) = z^{m+n+l-1} + O(z^{m+n-1})_{z\to\infty}. \tag{9}$$

Beckermann and Labahn point out that when the last accepted basis has "small" errors then it can often still be used for computing either a quasi-GCD or an ϵ-GCD.

Theorem 3.3 *Let $U^{(k)}$ be the last well-behaved unimodular reduction computed by the algorithm COPRIME. Then*

(a) *If $||b^{(k)}|| + ||(\alpha^{(k)}, \beta^{(k)})|| < \epsilon\, |\det U^{(k)}(0)|/12$ with $0 < \epsilon < 1/6$ then $a^{(k)}$ is a quasi-GCD with precision ϵ.*

(b) *If $2||b^{(k)}|| + (2 + 4\rho_{n-m+2k}(a,b)) \cdot ||(\alpha^{(k)}, \beta^{(k)})|| \leq \epsilon\, |\det U^{(k)}(0)|$ and $|\det U^{(k)}(0)| > 4\rho_{n-m+2k}(a,b) \cdot \epsilon$ then $a^{(k)}$ is an ϵ-GCD.*

4 Implementation details

4.1 Fast and stable update of QR factorizations

It may happen that the look-ahead procedure has several jumps of order $s = O(m)$. In this case, successively computing QR decompositions of $M_1^{(k)}, \ldots, M_s^{(k)}$ by the classical method costs $\sum_{i=1}^{s} i^3$ flops, resulting in an $O((m+n)^4)$ algorithm. However, by taking advantage of the recursive structure of the matrix $M_s^{(k)}$ of (7) it is possible to reduce the above complexity to $O((m+n)^3)$. Indeed, we can determine the ith QR decomposition from the $(i-1)$th one in quadratic rather than cubic time[8]. This is precisely why such intermediate linear systems are solved via QR decomposition rather than say Gaussian elimination[1]. The details of such QR updates can be found in the SNAP user's guide[11].

4.2 Using MAPLE hardware floats

MAPLE has two floating point systems, hardware float and software float[15]. Software floats are used for extended precision arithmetic while hardware float makes use of double precision arithmetic available on all computers. Hardware floating point is much faster than software float, but is limited in its usage. Our implementation of COPRIME was done is such a way that it could run in either setting, making use of hardware floats where valid. We remark that the structures allowed inside procedures that can work with hardware floating point are limited (for example sets, lists and sequences are not allowed).

4.3 Determining $\rho_l(a,b)$ via least squares

In order to check the conditions of Theorem 3.3, we need to estimate $\rho_l(a,b)$ (see Equation (9)). Computing a 2-norm equivalent of $\rho_l(a,b)$ allows to reduce to the following least squares problem.

When $l \geq 0$, consider the $l \times 2l$ complex matrix

$$M = \left[\begin{array}{ccc|ccc} a_m & & & b_n & & \\ \vdots & \ddots & & \vdots & \ddots & \\ a_{m-l+1} & \cdots & a_m & b_{n-l+1} & \cdots & b_n \end{array}\right].$$

In the case where $l < 0$, it suffices to replace l with $-l$ and $a(z)$, $b(z)$ with $z^m a(z^{-1})$, $z^n b(z^{-1})$ respectively. We then compute the Moore-Penrose inverse M^+ of M by singular value decomposition[8]. We then take for $\rho_l(a,b)$ the product $||M^+(.,1)||_2 \cdot ||(a,b)||_2$ where $M^+(.,1)$ denotes the first column of M^+.

The SNAP commands `EpsilonGCD` and `QuasiGCD` are based on this approach.

5 Practical features of SNAP

5.1 Efficient coprimeness test

The SNAP package allows for numerical coprimeness to be detected efficiently for most inputs. Dense polynomials of degree of order 1000 can typically be handled by our implementation of COPRIME within a few minutes (on a Pentium III 800 MHz with 512MB of RAM, under Linux). This is illustrated in the table below. For comparison, we also give the time (in seconds) required to estimate $\epsilon(a,b)$ via setting up the Sylvester matrix $S(a,b)$ and solving the associated linear systems by LU decomposition (with the MAPLE command `LinearAlgebra[LinearSolve]`):

$m = n+1$	Sylvester	COPRIME
50	0.33	0.34
100	1.30	1.34
250	8.6	8.7
500	43	33
10^3	307	126
$2 \cdot 10^3$	3360	515

5.2 Certified approximate GCDs

Another important feature of the package is the ability to compute approximate GCDs at the same cost as a coprimeness test. Such GCDs are certified by Theorem 3.3. As an illustration, we consider the following example due to Rupprecht[18]. Let

$$a(z) = (z^4 - 1.000001)(z^3 - 3.000001z + 0.99999999)$$

and

$$b(z) = (z^4 - 0.9999999)(z^4 - 3.0000003z - 2.9999999).$$

Calling `SNAP[EpsilonGCD](a,b,z)` with 10 digits (the default in MAPLE) yields the 0.0006481143605-GCD

$$z^4 + 5.198691325\,10^{-8}z^3 + 5.804634175\,10^{-7}z^2 \\ +9.873862873\,10^{-7}z - 1.000000584$$

after normalization of the leading coefficient.

5.3 Satisfactory answers from Euclidean reduction

When the conditions of Theorem 3.3 are not satisfied, the SNAP command `Euclideanreduction` proves in general to be very useful for two reasons: in most cases, the returned polynomial pair has much smaller degrees that the input; also, the first component can be seen as a satisfactory candidate for an approximate GCD. For example, let

$$a(z) = z^6 - 12.4z^5 + 62.53z^4 - 163.542z^3 + 232.9776z^2 - 170.69184z + 50.18112$$

and

$$b(z) = z^5 - 17.6z^4 + 118.26z^3 - 372.992z^2 + 538.3333z - 274.09272$$

Calling `SNAP[EuclideanReduction](a,b,z,tau=1e-8)` with 10 digits then yields the reduced pair

$$0.2500000000z^2 - 0.8750003765z + 0.6600005057,$$
$$-0.90726410^{-7}z + 0.133529610^{-6}.$$

Degrees have decreased from $(6,5)$ to $(2,1)$ but the second component has 1-norm of order $O(10^{-7})$ and should be interpreted as a zero remainder.

6 Concluding remarks

In this paper we have given a description of the SNAP package, a new MAPLE package for arithmetic with univariate numeric polynomials. This package is a first version and it is by no means complete with its primary focus being on the operations of quotient, remainder and GCDs. As such there are a number of additional commands that need to be investigated and implemented for future versions of the package. These include commands for numerical division of univariate polynomials, reduction of numeric rational expressions and others. Furthermore, commands that allow for numerically correct arithmetic with multivariate polynomials will also be needed.

In this paper, we have focused only on two types of approximate GCDs. The package would thus gain by the inclusion of the algorithms which follow an adaptive precision model, for example as found in Corless *et al* [5] and Karmarkar and Lakshman[13]. It would also be interesting to compare our available software with implementations of the methods of Noda and Sasaki[16], Pan[17] and Zarowski *et al* [21]. Because of these various definitions and algorithms for approximate GCDs, it may be of interest to provide a user with an automatic selection facility, depending on the problem to be solved.

Finally, the package would also gain significant use if it included algebraic operations for matrix polynomials, particularly in such application areas as control theory and linear systems theory. The delay in this aspect of the package has to do with the lack of numerically stable algorithms for computing such operations as numeric matrix greatest common divisor or even numeric matrix normal forms.

Acknowledgements

We would like to thank Bernhard Beckermann for providing some related MATLAB code, and Robert Corless and Lihong Zhi for their early helpful comments on the SNAP package.

References

1. B. Beckermann, The stable computation of formal orthogonal polynomials, *Numerical Algorithms* **11** (1996) 1-23.
2. B. Beckermann and G. Labahn, When are two numerical polynomials relatively prime? *Journal of Symbolic Computation* **26** (1998) 677-689.
3. B. Beckermann and G. Labahn, A fast and numerically stable Euclidean-like algorithm for detecting relatively prime numerical polynomials, *Journal of Symbolic Computation* **26** (1998) 691-714.
4. J.R. Bunch, The weak and strong stability of algorithms in numerical linear algebra, *Lin. Alg. Appl.* **88-89** (1987) 49-66.
5. R.M. Corless, P.M. Gianni, B.M. Trager and S.M. Watt, The singular value decomposition for polynomial systems, Proceedings ISSAC'95, ACM Press (1995) 195-207.
6. I. Z. Emiris, Symbolic-numeric algebra for polynomials. Survey available at `http://www-sop.inria.fr/galaad/emiris` (1997).
7. I. Z. Emiris, A. Galligo and H. Lombardi, Certified approximate univariate GCDs, *J. Pure and Applied Algebra* **117** (1997) 229-251.
8. G.H. Golub and C. Van Loan, Matrix computations, Johns Hopkins University Press (1989).
9. N.J. Higham, *Accuracy and Stability of Numerical Algorithms* (SIAM, Philadelphia, 1996).
10. V. Hribernig and H.J. Stetter, Detection and validation of clusters of polynomial zeros, J. Symbol. Comput. 24 (1997) 667-681.
11. C.P. Jeannerod and G. Labahn, SNAP User's Guide (2002).
12. N. Karmarkar and Y.N. Laksmann, Approximate polynomial greatest common divisors and nearest singular polynomials, Proceedings ISSAC'96, ACM Press (1996) 35-43.
13. N. Karmarkar and Y.N. Laksmann, On approximate GCDs of univariate polynomials, *Journal of Symbolic Computation* **26** (1998) 653-666.
14. T. Kailath, Linear systems, Prentice–Hall (1980).
15. M.B. Monagan, K.O. Geddes, K.M. Heal, G. Labahn, S.M. Vorkoetter, J. McCarron and P. DeMarco, *Maple 7 Programming Guide,* Toronto: Waterloo Maple Inc., 2001.
16. M.-T. Noda & T. Sasaki, Approximate GCD and its applications to ill–conditioned algebraic equations, *J. Comput. Appl. Math.* **38** (1991) 335-351.
17. V.Y. Pan, Numerical computations of a polynomial GCD and extensions. Technical Report 2969, INRIA, Sophia-Antipolis (1996).
18. D. Rupprecht, Eléments de géométrie algébrique approchée: étude du PGCD et de la factorisation, Thèse de l'Université de Nice-Sophia Antipolis (2000).
19. A. Schönhage, Quasi–GCD computations, *J. Complexity* **1** (1985) 118-137.
20. H. Stetter, Numerical Polynomial Algebra: Concepts and Algorithms, ATCM 2000, Proceed. 5th Asian Technology Conf. in Math. (Eds.: W.-Ch. Yang, S.-Ch. Chu, J.-Ch. Chuan), 22-36, ATCM Inc. USA (2000).
21. C.J. Zarowski, X. Ma and F.W. Fairman, QR-factorization method for computing the greatest common divisor of polynomials with inexact coefficients, *IEEE Trans. Signal Processing* **48(11)** (2000) 3042-3051.

ALGORITHMS OF COMPOSITE RATIONAL INTERPOLATION BASED ON CONTINUED FRACTIONS*

JIEQING TAN AND SHUO TANG
INSTITUTE OF APPLIED MATHEMATICS, HEFEI UNIVERSITY OF TECHNOLOGY
193 TUNXI ROAD, 230009 HEFEI, P. R. CHINA
E-MAIL: JQTAN@MAIL.HF.AH.CN

Abstract

For some data given on a rectangular grid, sometimes one finds it hard to build up the rational interpolants on the whole grid. Therefore " divide and conquer" technique has to be used to tackle the problem. In this paper, we present the recursive algorithms for composite rational interpolation formed by means of the hybrid of the Thiele's continued fractions and Newton's polynomials which work in the cases where our previous blending rational interpolants fail.

Key words: Blending interpolation, algorithm, composite scheme.
AMS Subject Classification: 41A20, 65D05.

1 Introduction

Suppose

$$\Pi_{m,n} = \{(x_i, y_j)|i = 0, 1, \ldots, m; j = 0, 1, \ldots, n\}$$

is a two dimensional set of points and $f(x, y)$ is a bivariate function defined on the domain D enclosing $\Pi_{m,n}$, then there are many ways to construct a bivariate rational function such that it interpolates $f(x, y)$ on $\Pi_{m,n}$. We refer to Cuyt and Verdonk ([1] and Siemaszko ([2]). In [3] and [4], the author introduced the blending rational interpolants, Newton-Thiele's rational interpolants and Thiele-Newton's rational interpolants respectively, which are the hybrids of the Thiele's continued fractions and the Newton's polynomials. A Newton-Thiele's rational interpolant (NTRI) is a blending rational function of the

*THIS SUBJECT IS SUPPORTED BY THE NATIONAL NATURAL SCIENCE FOUNDATION OF CHINA UNDER GRANT NO. 10171026 AND IN PART BY THE FOUNDATION FOR EXCELLENT YOUNG TEACHERS OF THE MINISTRY OF EDUCATION OF CHINA

following form

$$NT_{m,n}(x,y) = T_0(y) + (x-x_0)T_1(y) + \cdots + (x-x_0)\cdots(x-x_{m-1})T_m(y), \quad (1)$$

and for $i = 0, 1, \ldots, m$

$$T_i(y) = \psi[x_0,\ldots,x_i;y_0] + \frac{y-y_0 \,|}{|\, \psi[x_0,\ldots,x_i;y_0,y_1]} + \cdots + \frac{y-y_{n-1} \,|}{|\, \psi[x_0,\ldots,x_i;y_0,\ldots,y_n]}, \quad (2)$$

where $\psi[x_0,\ldots,x_i;y_0,\ldots,y_j]$ is so-called blending difference of Newton-Thiele type of $f(x,y)$ at the set of points $\{x_0,\ldots,x_i\} \times \{y_0,\ldots,y_j\}$ which is to be computed by the following recursive formulae.

$$\psi[x;y] = f(x,y), \ \forall (x,y) \in D, \quad (3)$$

$$\psi[x_i;y_p,y_q] = \frac{y_q - y_p}{\psi[x_i;y_q] - \psi[x_i;y_p]}, \ 0 \le p < q \le n, \quad (4)$$

$$\psi[x_i;y_0,\ldots,y_j] = \frac{y_j - y_{j-1}}{\psi[x_i;y_0,\ldots,y_{j-2},y_j] - \psi[x_i;y_0,\ldots,y_{j-1}]}, \quad (5)$$

$$\psi[x_0,\ldots,x_i;y_j] = \frac{\psi[x_0,\ldots,x_{i-2},x_i;y_j] - \psi[x_0,\ldots,x_{i-1};y_j]}{x_i - x_{i-1}}, \quad (6)$$

$$\psi[x_0,\ldots,x_i;y_0,\ldots,y_j] = \frac{y_j - y_{j-1}}{\psi[x_0,\ldots,x_i;y_0,\ldots,y_{j-2},y_j] - \psi[x_0,\ldots,x_i;y_0,\ldots,y_{j-1}]}. \quad (7)$$

A Thiele-Newton's rational interpolant (TNRI) is a blending rational function of the form

$$TN_{m,n}(x,y) = N_0(y) + \frac{x-x_0 \,|}{|\, N_1(y)} + \frac{x-x_1 \,|}{|\, N_2(y)} + \cdots + \frac{x-x_{m-1} \,|}{|\, N_m(y)}, \quad (8)$$

and for $i = 0, 1, \ldots, m$

$$N_i(y) = \varphi[x_0,\ldots,x_i;y_0] + (y-y_0)\varphi[x_0,\ldots,x_i;y_0,y_1] + \cdots + (y-y_0)\cdots(y-y_{n-1})\varphi[x_0,\ldots,x_i;y_0,\ldots,y_n], \quad (9)$$

where $\varphi[x_0,\ldots,x_i;y_0,\ldots,y_j]$ is so-called blending difference of Thiele-Newton type of $f(x,y)$ at the set of points $\{x_0,\ldots,x_i\} \times \{y_0,\ldots,y_j\}$, which is to be computed with the aid of the following recursive formulae

$$\varphi[x;y] = f(x,y), \ \forall (x,y) \in D, \quad (10)$$

$$\varphi[x_i;y_p,y_q] = \frac{\varphi[x_i;y_q] - \varphi[x_i;y_p]}{y_q - y_p}, \ 0 \le p < q \le n, \quad (11)$$

$$\varphi[x_i;y_0,\ldots,y_j] = \frac{\varphi[x_i;y_0,\ldots,y_{j-2},y_j] - \varphi[x_i;y_0,\ldots,y_{j-1}]}{y_j - y_{j-1}}, \quad (12)$$

$$\varphi[x_0,\ldots,x_i;y_j] = \frac{x_i - x_{i-1}}{\varphi[x_0,\ldots,x_{i-2},x_i;y_j] - \varphi[x_0,\ldots,x_{i-1};y_j]}, \tag{13}$$

$$\varphi[x_0,\ldots,x_i;y_0,\ldots,y_j] = \frac{\varphi[x_0,\ldots,x_i;y_0,\ldots,y_{j-2},y_j] - \varphi[x_0,\ldots,x_i;y_0,\ldots,y_{j-1}]}{y_j - y_{j-1}}. \tag{14}$$

It is shown(see [3] and [4])

$$TN_{m,n}(x_i,y_j) = NT_{m,n}(x_i,y_j) = f(x_i,y_j), \forall (x_i,y_j) \in \Pi_{m,n}. \tag{15}$$

Obviously the blending rational interpolants strongly depend on the existence of the corresponding blending differences. Let us take a look at the following example

Example 1. Let (x_i, y_j) and $f(x_i, y_j)$ be given in Table 1

	$x_0 = 0$	$x_1 = 1$	$x_2 = 2$
$y_0 = 0$	$f(x_0, y_0) = 6$	$f(x_1, y_0) = 6$	$f(x_2, y_0) = 0$
$y_1 = 1$	$f(x_0, y_1) = 6$	$f(x_1, y_1) = 6$	$f(x_2, y_1) = 6$
$y_2 = 2$	$f(x_0, y_2) = 6$	$f(x_1, y_2) = 6$	$f(x_2, y_2) = 4$

Table 1. Interpolation data

Then neither TNRIs nor NTRIs exist in this case. We present a solution scheme to such problems as in the above example.

2 Composite interpolants and their algorithms

Let us divide the square $\Pi_{n,n}$ into two subgrids, the lower triangular grid S_L and the upper triangular grid S_U, namely,

$$S_L = \{(x_i,y_j)|i = 0,1,\ldots,n;\ j = i, i+1,\ldots,n\},$$
$$S_U = \{(x_i,y_j)|i = 1,2,\ldots,n;\ j = 0,1,\ldots,i-1\}.$$

Then with S_L one may construct the blending rational interpolant Thiele-Newton type

$$TN_n^L(x,y) = N_0^L(y) + \frac{x-x_0|}{|N_1^L(y)} + \frac{x-x_1|}{|N_2^L(y)} + \cdots + \frac{x-x_{n-1}|}{|N_n^L(y)} \tag{1}$$

and the blending rational interpolant of Newton-Thiele type

$$NT_n^L(x,y) = T_0^L(y) + T_1^L(y)(x-x_0) + \cdots + T_n^L(y)(x-x_0)\cdots(x-x_{n-1}), \tag{2}$$

where for $i = 0, 1, \ldots, n$

$$\begin{aligned} N_i^L(y) &= a_{i,i} + a_{i,i+1}(y - y_i) + a_{i,i+2}(y - y_i)(y - y_{i+1}) \\ &\quad + \cdots + a_{i,n}(y - y_i)\cdots(y - y_{n-1}) \end{aligned} \tag{3}$$

and

$$T_i^L(y) = b_{i,i} + \frac{y - y_i}{\lfloor b_{i,i+1}} + \frac{y - y_{i+1}}{\lfloor b_{i,i+2}} + \cdots + \frac{y - y_{n-1}}{\lfloor b_{i,n}}. \tag{4}$$

Where the definitions of a_{ij} and b_{ij} are clearly seen from Step 5 of Algorithm 1 and Algorithm 3 respectively. With S_U one may construct the blending rational interpolant of Thiele-Newton type

$$TN_n^U(x, y) = N_0^U(y) + \frac{x - x_n}{\lfloor N_1^U(y)} + \frac{x - x_{n-1}}{\lfloor N_2^U(y)} + \cdots + \frac{x - x_2}{\lfloor N_{n-1}^U(y)} \tag{5}$$

and the blending rational interpolant of Newton-Thiele type

$$NT_n^U(x, y) = T_0^U(y) + T_1^U(y)(x - x_n) + \cdots + T_{n-1}^U(y)(x - x_n)(x - x_{n-1})\cdots(x - x_2), \tag{6}$$

where for $i = 0, 1, \ldots, n-1$

$$\begin{aligned} N_i^U(y) &= c_{i,0} + c_{i,1}(y - y_0) + c_{i,2}(y - y_0)(y - y_1) \\ &\quad + \cdots + c_{i,n-i-1}(y - y_0)\cdots(y - y_{n-i-2}) \end{aligned} \tag{7}$$

and

$$T_i^U(y) = d_{i,0} + \frac{y - y_0}{\lfloor d_{i,1}} + \frac{y - y_1}{\lfloor d_{i,2}} + \cdots + \frac{y - y_{n-i-2}}{\lfloor d_{i,n-i-1}}. \tag{8}$$

Where c_{ij} and d_{ij} are to be determined by specific algorithms, we refer to Step 4 of Algorithm 2 and Algorithm 4 respectively for the details. Let

$$v_{i,j} = \begin{cases} f(x_i, y_j)/U(x_i, y_j), \text{ for } (x_i, y_j) \in S_L, \\ f(x_i, y_j)/L(x_i, y_j), \text{ for } (x_i, y_j) \in S_U, \end{cases} \tag{9}$$

where $L(x, y)$ and $U(x, y)$ are polynomials such that

$$\begin{aligned} L(x_i, y_j) &= 0, \ \forall (x_i, y_j) \in S_L, \\ L(x_i, y_j) &\neq 0, \ \forall (x_i, y_j) \in S_U, \\ U(x_i, y_j) &= 0, \ \forall (x_i, y_j) \in S_U, \\ U(x_i, y_j) &\neq 0, \ \forall (x_i, y_j) \in S_L. \end{aligned}$$

Now one may make use of the four blending rational interpolants $TN_n^L(x,y)$, $NT_n^L(x,y)$, $TN_n^U(x,y)$ and $NT_n^U(x,y)$ to yield four composite schemes as follows

$$TNTN_n(x,y) = U(x,y)TN_n^L(x,y) + L(x,y)TN_n^U(x,y), \tag{10}$$
$$TNNT_n(x,y) = U(x,y)TN_n^L(x,y) + L(x,y)NT_n^U(x,y), \tag{11}$$
$$NTTN_n(x,y) = U(x,y)NT_n^L(x,y) + L(x,y)TN_n^U(x,y), \tag{12}$$
$$NTNT_n(x,y) = U(x,y)NT_n^L(x,y) + L(x,y)NT_n^U(x,y). \tag{13}$$

As we know, the computation of the four composite schemes (10) to (13) reduces to that of the four blending rational interpolants $TN_n^L(x,y), TN_n^U(x,y), NT_n^L(x,y)$ and $NT_n^U(x,y)$, which corresponds to the following four algorithms.

Algorithm 1. This algorithm is aimed to compute the blending rational interpolant $TN_n^L(x,y)$ of Thiele-Newton type over the lower triangular subgrid S_L and is composed of the following steps in terms of matrix transformation

Step 1. Let

$$A = \begin{bmatrix} \alpha_{0,0}^{(0,0)} & & & \\ \alpha_{0,1}^{(0,0)} & \alpha_{1,1}^{(0,0)} & & \\ \vdots & \vdots & \ddots & \\ \alpha_{0,n}^{(0,0)} & \alpha_{1,n}^{(0,0)} & \cdots & \alpha_{n,n}^{(0,0)} \end{bmatrix}$$

where for $i = 0, 1, \ldots, n; j = i, i+1, \ldots, n$

$$\alpha_{i,j}^{(0,0)} = v_{i,j}.$$

Step 2. Let

$$A_1 = \begin{bmatrix} \alpha_{0,0}^{(0,0)} & & & \\ \alpha_{0,1}^{(0,0)} & \alpha_{1,1}^{(1,0)} & & \\ \vdots & \vdots & \ddots & \\ \alpha_{0,n}^{(0,0)} & \alpha_{1,n}^{(1,0)} & \cdots & \alpha_{n,n}^{(n,0)} \end{bmatrix}$$

where for $j = 1, 2, \ldots, n; i = 1, 2, \ldots, j; p = 1, 2, \ldots, i$

$$\alpha_{i,j}^{(p,0)} = \frac{x_i - x_{p-1}}{\alpha_{i,j}^{(p-1,0)} - \alpha_{p-1,j}^{(p-1,0)}}$$

Step 3. Let

$$A_2 = \begin{bmatrix} \alpha_{0,0}^{(0,0)} & & & \\ \alpha_{0,1}^{(0,0)} & \alpha_{1,1}^{(1,1)} & & \\ \vdots & \vdots & \ddots & \\ \alpha_{0,n}^{(0,0)} & \alpha_{1,n}^{(1,1)} & \cdots & \alpha_{n,n}^{(n,n)} \end{bmatrix}$$

where for $i = 1, 2, \ldots, n; j = i, i+1, \ldots, n$

$$\alpha_{i,j}^{(i,i)} = \alpha_{i,j}^{(i,0)}$$

Step 4. Let

$$A_3 = \begin{bmatrix} \alpha_{0,0}^{(0,0)} & & & \\ \alpha_{0,1}^{(0,1)} & \alpha_{1,1}^{(1,1)} & & \\ \vdots & \vdots & \ddots & \\ \alpha_{0,n}^{(0,n)} & \alpha_{1,n}^{(1,n)} & \cdots & \alpha_{n,n}^{(n,n)} \end{bmatrix}$$

where for $i = 0, 1, \ldots, n-1; j = i+1, \ldots, n; q = i+1, \ldots, j$

$$\alpha_{i,j}^{(i,q)} = \frac{\alpha_{i,j}^{(i,q-1)} - \alpha_{i,q-1}^{(i,q-1)}}{y_j - y_{q-1}}.$$

Step 5. For $i = 0, 1, \ldots, n$, let

$$N_i^L(y) = \alpha_{i,i}^{(i,i)} + \alpha_{i,i+1}^{(i,i+1)}(y - y_i) + \alpha_{i,i+2}^{(i,i+2)}(y - y_i)(y - y_{i+1}) + \cdots + \alpha_{i,n}^{(i,n)}(y - y_i)\cdots(y - y_{n-1}).$$

Step 6. Let

$$TN_n^L(x,y) = N_0^L(y) + \frac{x - x_0|}{|N_1^L(y)} + \cdots + \frac{x - x_{n-1}|}{|N_n^L(y)},$$

then $TN_n^L(x,y)$ satisfies

$$TN_n^L(x_i, y_j) = v_{i,j}, \ \forall (x_i, y_j) \in S_L.$$

Algorithm 2. This algorithm is aimed to compute the blending rational interpolant $TN_n^U(x,y)$ of Thiele-Newton type over the upper triangular subgrid S_U and is composed of the following steps in terms of matrix transformation

Step 1. Let

$$B = \begin{bmatrix} \beta_{n-1,0}^{(0,0)} & \beta_{n-2,0}^{(0,0)} & \cdots & \beta_{0,0}^{(0,0)} \\ & \beta_{n-2,1}^{(0,0)} & \cdots & \beta_{0,1}^{(0,0)} \\ & & \ddots & \vdots \\ & & & \beta_{0,n-1}^{(0,0)} \end{bmatrix}$$

where for $i = 0, 1, \ldots, n-1; j = 0, 1, \ldots, n-i-1$

$$\beta_{i,j}^{(0,0)} = v_{n-i,j}.$$

Step 2. Let

$$B_1 = \begin{bmatrix} \beta_{n-1,0}^{(n-1,0)} & \beta_{n-2,0}^{(n-2,0)} & \cdots & \beta_{0,0}^{(0,0)} \\ & \beta_{n-2,1}^{(n-2,0)} & \cdots & \beta_{0,1}^{(0,0)} \\ & & \ddots & \vdots \\ & & & \beta_{0,n-1}^{(0,0)} \end{bmatrix}$$

where for $j = 0, 1, \ldots, n-2; i = 1, 2, \ldots, n-j-1; p = 1, 2, \ldots, i$

$$\beta_{i,j}^{(p,0)} = \frac{x_{n-i} - x_{n-p+1}}{\beta_{i,j}^{(p-1,0)} - \beta_{p-1,j}^{(p-1,0)}}$$

Step 3. Let

$$B_2 = \begin{bmatrix} \beta_{n-1,0}^{(n-1,0)} & \beta_{n-2,0}^{(n-2,0)} & \cdots & \beta_{0,0}^{(0,0)} \\ & \beta_{n-2,1}^{(n-2,1)} & \cdots & \beta_{0,1}^{(0,1)} \\ & & \ddots & \vdots \\ & & & \beta_{0,n-1}^{(0,n-1)} \end{bmatrix}$$

where for $i = 0, 1, \ldots, n-2; j = 1, 2, \ldots, n-i-1; q = 1, 2, \ldots, j$

$$\beta_{i,j}^{(i,q)} = \frac{\beta_{i,j}^{(i,q-1)} - \beta_{i,q-1}^{(i,q-1)}}{y_j - y_{q-1}}.$$

Step 4. For $i = 0, 1, \ldots, n-1$, let

$$N_i^U(y) = \beta_{i,0}^{(i,0)} + \beta_{i,1}^{(i,1)}(y - y_0) + \cdots + \beta_{i,n-i-1}^{(i,n-i-1)}(y - y_0) \cdots (y - y_{n-i-2}).$$

Step 5. Let

$$TN_n^U(x, y) = N_0^U(y) + \frac{x - x_n}{\left| N_1^U(y) \right.} + \frac{x - x_{n-1}}{\left| N_2^U(y) \right.} + \cdots + \frac{x - x_2}{\left| N_{n-1}^U(y) \right.},$$

then $TN_n^U(x,y)$ satisfies

$$TN_n^U(x_i,y_j)=v_{i,j},\ \forall(x_i,y_j)\in S_U.$$

Algorithm 3. This algorithm is aimed to compute the blending rational interpolant $NT_n^L(x,y)$ of Newton-Thiele type over the lower triangular subgrid S_L and is composed of the following steps

Step 1. For $i=0,1,\ldots,n; j=i,i+1,\ldots,n$, let

$$\xi_{i,j}^{(0,0)}=v_{i,j}.$$

Step 2. For $j=1,2,\ldots,n; i=1,2,\ldots,j; p=1,2,\ldots,i$, let

$$\xi_{i,j}^{(p,0)}=\frac{\xi_{i,j}^{(p-1,0)}-\xi_{p-1,j}^{(p-1,0)}}{x_i-x_{p-1}}$$

Step 3. For $i=1,2,\ldots,n; j=i,i+1,\ldots,n$, let

$$\xi_{i,j}^{(i,i)}=\xi_{i,j}^{(i,0)}$$

Step 4. For $i=0,1,\ldots,n-1; j=i+1,\ldots,n; q=i+1,\ldots,j$, let

$$\xi_{i,j}^{(i,q)}=\frac{y_j-y_{q-1}}{\xi_{i,j}^{(i,q-1)}-\xi_{i,q-1}^{(i,q-1)}}.$$

Step 5. For $i=0,1,\ldots,n$, let

$$T_i^L(y)=\xi_{i,i}^{(i,i)}+\frac{y-y_i|}{|\xi_{i,i+1}^{(i,i+1)}}+\frac{(y-y_{i+1})|}{|\xi_{i,i+2}^{(i,i+2)}}+\cdots+\frac{y-y_{n-1}|}{|\xi_{i,n}^{(i,n)}}.$$

Step 6. Let

$$NT_n^L(x,y)=T_0^L(y)+T_1^L(y)(x-x_0)+\cdots+T_n^L(y)(x-x_0)\cdots(x-x_{n-1}),$$

then $NT_n^L(x,y)$ satisfies

$$NT_n^L(x_i,y_j)=v_{i,j},\ \forall(x_i,y_j)\in S_L.$$

Algorithm 4. This algorithm is designed to compute the blending rational interpolant $NT_n^U(x,y)$ of Newton-Thiele type over the upper triangular subgrid S_U and is composed of the following steps

Step 1. For $i=0,1,\ldots,n-1; j=0,1,\ldots,n-i-1$, let

$$\eta_{i,j}^{(0,0)}=v_{n-i,j}.$$

Step 2. For $j=0,1,\ldots,n-2; i=1,2,\ldots,n-j-1; p=1,2,\ldots,i$, let

$$\eta_{i,j}^{(p,0)}=\frac{\eta_{i,j}^{(p-1,0)}-\eta_{p-1,j}^{(p-1,0)}}{x_{n-i}-x_{n-p+1}}$$

Step 3. For $i = 0, 1, \ldots, n-2; j = 1, 2, \ldots, n-i-1; q = 1, 2, \ldots, j$, let

$$\eta_{i,j}^{(i,q)} = \frac{y_j - y_{q-1}}{\eta_{i,j}^{(i,q-1)} - \eta_{i,q-1}^{(i,q-1)}}.$$

Step 4. For $i = 0, 1, \ldots, n-1$, let

$$T_i^U(y) = \eta_{i,0}^{(i,0)} + \frac{y - y_0|}{|\eta_{i,1}^{(i,1)}} + + \cdots + \frac{y - y_{n-i-2}|}{|\eta_{i,n-i-1}^{(i,n-i-1)}}.$$

Step 5. Let

$$NT_n^U(x,y) = T_0^U(y) + T_1^U(y)(x - x_n) + T_2^U(y)(x - x_n)(x - x_{n-1}) + \cdots + T_{n-1}^U(y)(x - x_n) \cdots (x - x_2),$$

then $NT_n^U(x,y)$ satisfies

$$NT_n^U(x_i, y_j) = v_{i,j}, \ \forall (x_i, y_j) \in S_U.$$

3 Numerical example

Let $\Pi_{2,2}$ and $f(x_i, y_j)$ be given as in Example 1. Since $\Pi_{2,2}$ is uniform, we may set

$$L(x,y) = (x - y)(x - y + 1)(x - y + 2),$$
$$U(x,y) = (x - y - 1)(x - y - 2),$$

then according to (9), the values $v_{i,j}$ to be interpolated over S_L and S_U can be displayed respectively in Table 2 and Table 3

	$x_0 = 0$	$x_1 = 1$	$x_2 = 2$
$y_0 = 0$	3		
$y_1 = 1$	1	3	
$y_2 = 2$	1/2	1	2

Table 2. Interpolation data over S_L

With the interpolation data offered in Table 2, $TN_2^L(x,y)$ can be obtained via Algorithm 1 as follows

$$TN_2^L(x,y) = \frac{3y^2 - 11y + 12}{4} + \frac{x|}{|\frac{3y-2}{2}} + \frac{x-1|}{|-\frac{3}{2}}$$

	$x_1 = 1$	$x_2 = 2$
$y_0 = 0$	1	0
$y_1 = 1$		1

Table 3. Interpolation data over S_U

$$= \frac{3y^2 - 11y + 12}{4} + \frac{6x}{9y - 4x - 2}$$

With the interpolation data offered in Table 3, $TN_2^U(x, y)$ can be obtained via Algorithm 2 as follows

$$TN_2^U(x, y) = y + \frac{x - 2\rfloor}{\lceil -1} = y - x + 2.$$

Consequently we obtain

$$\begin{aligned} TNTN_2(x, y) &= U(x, y)TN_2^L(x, y) + L(x, y)TN_2^U(x, y) \\ &= (x - y - 1)(x - y - 2)\left(\frac{3y^2 - 11y + 12}{4} + \frac{6x}{9y - 4x - 2}\right) \\ &+ (x - y)(x - y + 1)(x - y + 2)(y - x + 2) \end{aligned}$$

Acknowledgements. The authors would like to thank the referees for their good suggestions.

References

1. Cuyt,A. and Verdonk,B., Multivariate Rational Interpolation. *Computing*, 34(1985), 41-61.
2. Siemaszko,W., Thiele-type branched continued fractions for two variable functions. *J. Comput. Appl. Math.*,9(1983), 137–153.
3. Tan J. Q. and Fang Y., Newton-Thiele's rational interpolants, *Numerical Algorithms*, 24(2000)141–157.
4. Tan J. Q., Bivariate blending rational interpolants. *Approx. Theory & Its Appl.*, 15(2)1999, 74–83.

FAST AND SPACE-EFFICIENT ALGORITHMS FOR DECIDING SHELLABILITY OF SIMPLICIAL COMPLEXES OF LARGE SIZE USING h-ASSIGNMENTS

SONOKO MORIYAMA AYUMU NAGAI HIROSHI IMAI

Department of Computer Science, University of Tokyo
7-3-1 Hongo, Bunkyo-ku, Tokyo 113-0033, Japan
E-mail: {moriso,nagai,imai}@is.s.u-tokyo.ac.jp

Although it is important to decide whether a simplicial complex is shellable or not, no efficient algorithm is known for the problem. We introduce an innovative software *SHIDE* (SHellability decIDEr) which decides shellability of large-size simplicial complexes. SHIDE traverses the facet-ridge incidence graph of a simplicial complex to find a shelling in a depth-first search fashion which is space-efficient. SHIDE also uses a hashing technique that efficiently checks whether a subproblem has already been solved or not to avoid redundant computation during the traversal. Experimental results show that shellability of almost all simplicial complexes with about 40 facets are decided regardless the dimension within 3 hours using a conventional workstation.

1 Introduction

Shellability has been extensively studied in old days in the field of topology, especially in relation with the characterization of spheres, but after Bruggesser and Mani's work relating to the boundary complexes of polytopes, its importance has been realized also in the field of combinatorics. However, it is hard to show whether a simplicial complex is shellable or not. Deciding shellability of a simplicial complex is one of difficult problems which are unknown to be solved in polynomial time or not, thus, we try in this paper to study whether a d-dimensional simplicial complex is decided practically to be shellable or not.

By introducing an *h-assignment* as a new concept, we can develop an innovative algorithm to decide shellability of larger-size simplicial complexes, which cannot be solved by the previous algorithms. [6] This algorithm is the *hash-DFS method.* If an h-assignment satisfies some condition, it is called an *appropriate* h-assignment. An appropriate h-assignment is compatible with a subset of shellings and shellability of a simplicial complex $\mathcal{C}$ is equivalent to the existence of appropriate h-assignments. Hence, the hash-DFS method decides whether there exists an appropriate h-assignment in a simplicial complex $\mathcal{C}$.

We implement the hash-DFS method as *SHIDE* (SHellability decIDEr) by C++. SHIDE is the first system to decide shellability of larger-size simplicial complexes, which cannot be solved by previous algorithms [6]. Experimental results show that SHIDE decides shellability of almost all simplicial complexes in dimension four to seven with about 40 facets within 3 hours using a conventional workstation.

2 Definitions and Basic Properties

An (*abstract*) *simplicial complex* is a set $\mathcal{C}$ consisting of subsets of a finite set $V=\{1,\ldots,n\}$ such that if $\sigma \in \mathcal{C}$, $\tau \subset \sigma$ then $\tau \in \mathcal{C}$. An element of $\mathcal{C}$ is a *face*. A *facet* is a face maximal with respect to set inclusion. The *dimension* $\dim \sigma$ of a face σ is $|\sigma|-1$. The *dimension* of a simplicial complex $\mathcal{C}$ is $\max_{\sigma \in \mathcal{C}} \dim \sigma$. A simplicial complex is *pure* if all facets have the same dimension. A *ridge* of a pure simplicial complex is a face having dimension $\dim \mathcal{C} - 1$. From now on, all simplicial complexes are pure. A *boundary ridge* is a ridge contained in only one facet, and a facet containing a boundary ridge is a *boundary facet.* A d-dimensional simplicial complex will be denoted as a d-simplicial complex.

The f-vector of $\mathcal{C}$ is $f(\mathcal{C}) = (f_{-1}(\mathcal{C}), f_0(\mathcal{C}), f_1(\mathcal{C}), \ldots, f_d(\mathcal{C}))$, where $f_i(\mathcal{C})$ denotes the number of i-dimensional faces of $\mathcal{C}$. The h-vector $h(\mathcal{C}) = (h_0(\mathcal{C}), h_1(\mathcal{C}), h_2(\mathcal{C}), \ldots, h_{d+1}(\mathcal{C}))$ is defined by $h_k := \sum_{i=1}^{k} (-1)^{k-i} \binom{d+1-i}{d+1-k} f_{i-1}$ $(0 \leq k \leq d+1)$. The f-vector and the h-vector are *uniquely* determined for a simplicial complex.

A *shelling* of a d-simplicial complex $\mathcal{C}$ is a sequence $F_1, \ldots, F_t$ of all facets satisfying

$$\overline{F_i} \cap \left(\bigcup_{j=1}^{i-1} \overline{F_j} \right) \text{ is a pure } (d-1)\text{-simplicial complex} \quad (2 \leq i \leq t),$$

where $\overline{\sigma} = \{\tau : \tau \subset \sigma\}$. A simplicial complex is *shellable* if it has a shelling.

If a simplicial complex $\mathcal{C}$ is shellable, it is generally known that there exists a relation between the h-vector of $\mathcal{C}$ and a shelling. In a shelling $F_1, F_2, \ldots F_t$, the *restriction* $Res(F_j)$ of the face F_j is defined as the set of all vertices $v \in V$ such that $F_j - \{v\}$ is contained in one of the earlier facets: $Res(F_j) := \{v \in F_j : F_j - \{v\} \subseteq F_i \text{ for some } 1 \leq i < j\}$. The main observation is that when we build up $\mathcal{C}$ according to this shelling, the new faces at the jth step are exactly the faces G with $Res(F_j) \subseteq G \subseteq F_j$. Here, let h_i denote the number of faces whose restriction set has size i: $|\{j : |Res(F_j)| = i, \ 1 \leq j \leq t\}| = h_i$. The h-vector of $\mathcal{C}$ equals to this sequence $(h_0, h_1, \ldots, h_d, h_{d+1})$.

3 An h-assignment and Its Properties

Details of this section are described in Moriyama [7].

3.1 A New Concept: An h-assignment

Taking notice of the property of h-vectors of shellable simplicial complexes, we introduce a new concept an h-assignment, and we define a shellable h-assignment as follows:

Definition 1 (h-assignments). *An h-assignment is an assignment of label $0 \le i \le d+1$ to each facet such that the number of facets with the label i equals to $h_i(\mathcal{C})$.*

Definition 2 (A shellable h-assignment). *An h-assignment of $\mathcal{C}$ is compatible with a shelling if each facet is labeled by the size of its restriction set in the shelling. A shellable h-assignment is an h-assignment which is compatible with some shelling of $\mathcal{C}$.*

That is, deciding shellability of an h-assignment A is to decide whether A is compatible with a shelling.

Theorem 3 (Deciding shellability of an h-assignment). *Let $\mathcal{C}$ be a d-simplicial complex with t facets, and A be an h-assignment attached to $\mathcal{C}$. It takes $O(t)$ to decide whether A is compatible with some shelling. That is, shellability of an h-assignment can be decided in $O(t)$ time.*

3.2 An appropriate h-assignment and Its Characterization

Definition 4 (A facet-ridge incidence graph). *Let $\mathcal{C}$ be a simplicial complex with facets $\{F_i\}$ and ridges $\{R_j\}$. The facet-ridge incidence graph of $\mathcal{C}$ is a graph with the following nodes: v_{F_i} corresponding to the facets F_i, v_{R_j} corresponding to the ridges R_j, and v_G corresponding to the exterior of $\mathcal{C}$, and the following edges: $v_{F_i}v_{R_j}$ if $F_i \supset R_j$, and $v_G v_{F_i}$ if F_i is a boundary facet.*

If a facet-ridge incidence graph can be oriented according to the following conditions, these orientations are an *appropriate orientation*, and an h-assignment corresponding to such an orientation is an *appropriate h-assignment.*

Definition 5 (Appropriate Orientations). *For an h-assignment A, an orientation of $G(\mathcal{C})$ is* appropriate *if it satisfies the following degree conditions:*

- $\deg_{in}(v_G) = \#\{boundary\ facet\}$ *and* $\deg_{out}(v_G) = 0$,

- $\deg_{in}(v_{F_i}) = A(F_i)$ *and* $\deg_{out}(v_{F_i}) = d+1-A(F_i)$,
- $\deg_{in}(v_{R_j}) = 1$ *and* $\deg_{out}(v_{R_j}) = \#\{F_i \mid F_i \supset R_j\} - 1$,
- *The orientation is* acyclic.

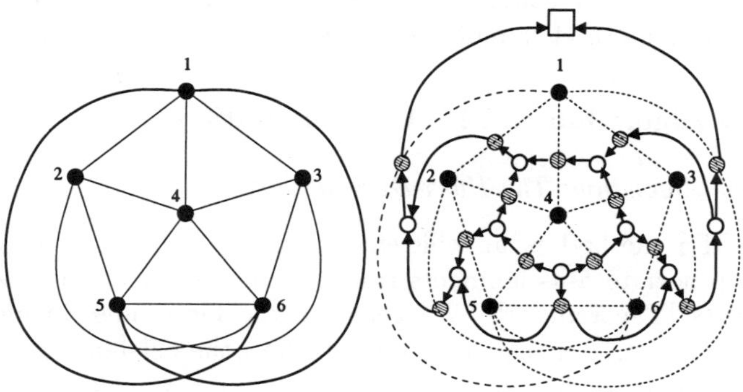

Figure 1. Appropriate orientations [right] of $\mathcal{C} = \{124, 126, 134, 135, 245, 256, 346, 356, 456\}$ [left] for an h-assignment $A(\mathcal{C}) = \{2, 2, 2, 1, 1, 2, 1, 2, 0\}$

Definition 6 (Appropriate h-assignments). *An appropriate h-assignment is an h-assignment which has an appropriate orientation.*

We can characterize appropriate h-assignments as follows, hence the existence of an appropriate h-assignment directly corresponds to shellability of $\mathcal{C}$:

Theorem 7 (Characterization of Appropriate h-assignments). *Let $\mathcal{C}$ be a simplicial complex, and A be an h-assignment attached to $\mathcal{C}$. An h-assignment A is appropriate if and only if A is shellable.*

4 Previous Algorithms for Deciding Shellability

The trivial algorithm to decide shellability of a simplicial complex with t facets is to generate $t!$ permutations and check whether each permutation is a shelling by the definition of shellability, that is, this algorithm takes $O(t! \cdot t)$.

There exist the following two algorithms which are a little more efficient than the trivial algorithm: *the backtrack method* to find a shelling by backtracking, and *the reverse backtrack method* [6] to find a reverse shelling by backtracking. The backtrack method is a *simple* backtrack method according to the definition of shellability, but the reverse backtrack method is more efficient using some specific property. If there exists a facet with k boundary ridges when $h_{d+1-k} \geq 1$, we can generate a reverse shelling explicitly by *putting* the facet to the previous-facets *stack* in each step. Thus, in this case, we can construct a shelling by determining the last facet in each step. The problem is that a facet selected in each step is not always correct, because this property depends on a necessary condition. Thus, we need to use backtracking, and it is why even this method cannot decide shellability of $\mathcal{C}$ with over 40 facets.

5 A New Algorithm and Its Implementation

5.1 A New Algorithm: The Hash-DFS Method

The hash-DFS method is an efficient method to check directly the existence of appropriate h-assignments from the characterization of appropriate h-assignments [Theorem 7]. We generate the facet-ridge incidence graph $G(\mathcal{C})$ [Definition 4], and search on $G(\mathcal{C})$ to find the existence of appropriate orientations corresponding to an appropriate h-assignment. In each step of this search, we attach a label to a facet F_i with no labels according to the degree condition [Definition 5] to generate a new *partial h-assignment*:

Definition 8 (Partial h-assignments). [7] *A partial h-assignment is a partial assignment of labels $0 \leq i \leq d+1$ to facets s.t. the number of facets with the label i equals to $h_i(\mathcal{C})$ or less. A facet with no labels in a partial h-assignment is an* open facet.

That is, an h-assignment is a partial h-assignment with all facets labeled. Hence, in order to get an appropriate h-assignment, we define the rule to expand a partial h-assignment together with all orientations in $G(\mathcal{C})$ as the *expansion rule*, and we also define the search tree for the hash-DFS method using the expansion rule:

Definition 9 (The expansion rule).

- *Step 1: Orientations for F_i*
  We call facets F_i s.t. $\mathrm{deg}_{\mathrm{in}}(F_i) = k$ *and $h_{d+1-k} \leq 1$ for some k the* candidate facets. *Select one of the candidate facets F_i and orient the rest edges incident to v_{F_i} out-going way.*

- *Step 2: Complementary orientations for R_j*
  If $\mathrm{deg}_{out}(v_{R_j}) = \#\{F_i | F_i \supset R_j\} - 1$ *in $G(\mathcal{C})$, the rest edge connecting with v_{F_i} is oriented into v_{R_j}.*

As an initial condition of this search, each of edges connecting with v_G in $G(\mathcal{C})$ is oriented into v_G. And, for $\{R_j\}$ connecting with v_G in $G(\mathcal{C})$, if

$\deg_{out}(v_{R_j}) = \#\{F_i | F_i \supset R_j\}$, the rest edge connecting with v_{F_i} is oriented into v_{R_j}. These orientations in $G(\mathcal{C})$ are the *initial orientations.*

Definition 10 (The search tree of the hash-DFS method). *The search tree* $Tree(\mathcal{C})$ *of the hash-DFS method is composed of the following nodes:*

- *The root node with a partial* h*-assignment with no labels and the initial orientations in* $G(\mathcal{C})$*,*
- *Each of the other nodes with a partial* h*-assignment* A *and partial orientation of* $G(\mathcal{C})$*.*

A child node in $Tree(\mathcal{C})$ *is a node expanded from a parent node by the expansion rule.*

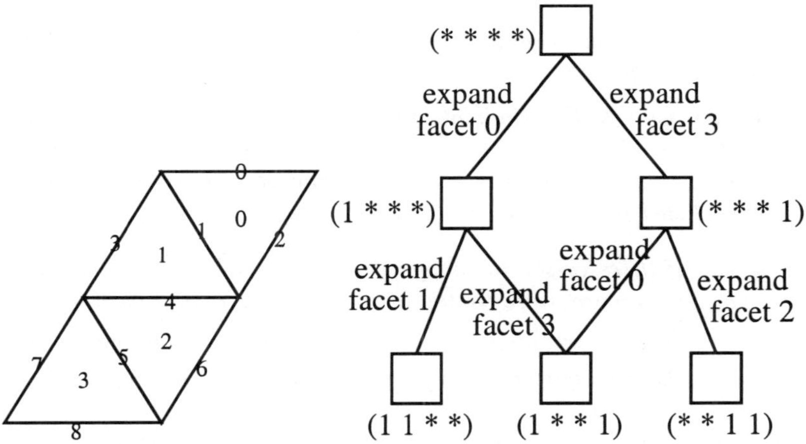

Figure 2. An example of a simplicial complex with $h = (1, 3, 0, 0)$ [left] and the search tree of the hash-DFS method [right]

The hash-DFS method searches on $Tree(\mathcal{C})$ by the depth first search (DFS). If there exist no candidate facets in a partial h-assignment attached to a node in $Tree(\mathcal{C})$, $Tree(\mathcal{C})$ cannot be expanded no more. Such a partial h-assignment is called an *inappropriate* partial h-assignment. An inappropriate partial h-assignment has the following property:

Theorem 11 (Inappropriate Partial h-assignments). *A partial* h*-assignment including an inappropriate partial* h*-assignment* A_E *is also inappropriate.*

If a partial h-assignment including an inappropriate partial h-assignment A_E appears, or A_E itself appears when $Tree(\mathcal{C})$ is expanded, we can cut the branch of $Tree(\mathcal{C})$ using Theorem 11 to avoid useless searching. And, when a partial h-assignment A_E is determined to be in appropriate during the search, if another inappropriate partial h-assignment $A_{E'}$ including A_E is already registered in a hash, then we exchange the registered one $A_{E'}$ by A_E because A_E can cover the role of the old one $A_{E'}$. This is the *exchange rule.*

Algorithm 12 (Algorithm : The Hash-DFS Method). [6] *Let $\mathcal{C}$ be a d-simplicial complex with facets $\{F_i\}$ and ridges $\{R_j\}$, $G(\mathcal{C})$ be the facet-ridge incidence graph [Definition 4], and $Tree(\mathcal{C})$ be the search tree of $\mathcal{C}$ [Definition 10].*

1. *The initial partial h-assignment of $Tree(\mathcal{C})$ is set as a partial h-assignment with no labels,*
2. *If there exist some candidate facets which have not been selected in a partial h-assignment,*
   - *then we select a candidate facet F_i, which has not been selected in this search, to generate a new partial h-assignment A by the expansion rule. And, if there exists an inappropriate partial h-assignment included in A in a hash, then go to step 3, else go to step 4,*
   - *else if there exists an inappropriate partial h-assignment including a new inappropriate partial h-assignment A_E in a hash,*
     - *then we exchange the registered one by A_E, and go to step 3,*
     - *else we register A_E into a hash, and go to step 3,*
3. *We return to the parent node for backtracking. If we search throughout $Tree(\mathcal{C})$, then we decide to be nonshellable, and end this algorithm, else return to step 2,*
4. *If there exist no open facets, then we return this partial h-assignment as an appropriate h-assignment, decide to be shellable, and end this algorithm, else return to step 2,*

The hash-DFS method can be seen as an improvement of the reverse backtrack method.

5.2 Implementation Issue of a Hash

In the hash-DFS algorithm, we register an inappropriate partial h-assignment into a hash to avoid the useless search. If $Tree(\mathcal{C})$ cannot be expanded to get an inappropriate partial h-assignment A_E, we have to preserve the information of A_E. However, we cannot register all inappropriate partial h-assignments into a hash, because the size of a hash is restricted. Hence, we need to rearrange a hash to leave useful ones and throw useless ones away, which enables SHIDE to work under a restricted memory, while a hash with a larger memory is more functional. We adopt open addressing hash, and the algorithm to remove the old entries is as follows:

- If the load factor of the hash table is over P%, at least Q% of all entries are removed by SmallTreeGC. SmallTreeGC is a method to remove entries whose depth is relatively large.
- If the load factor is under P%, one of the entry is replaced with the new entry by SmallTreeReplacement. SmallTreeReplacement is a method to look at all R candidates to record, select the entry with the largest depth, and replace it with the new entry.

This hashing technique is a universal method to check efficiently whether a subproblem has already been solved or not to avoid redundant computation [8].

6 The result of the experiment

Our experimental environment is Sun Ultra 60 UPA/PCI (UltraSPARC-II 450MHz) and 2048 Megabytes memory.

In our program, each of parameters of the hash is set as $P = 85\%, Q = 30\%$, and $R = 6$.

6.1 Data Specifications

To see the behavior of the hash-DFS method, we apply this method to many simplicial complexes. We used `rand_sphere` utility which is provided with the POLYMAKE package [2] to get the data of random convex polytopes, and convert them into lists of facets using POLYMAKE. We know that these data are all shellable a priori because convex polytopes are shellable. We generate the following examples for the experiment:

- the dimension is $4, 5, 6, 7$, and the number of facets is from 20 to 80,
- each number of d-simplicial complexes with n facets is about 30,

and checked whether SHIDE can determine each of the data is shellable within a practical time.

On the other hand, it is difficult to generate a nonshellable simplicial complex with the nonnegative h-vector which is not decided to be nonshellable a priori. Hence, we use the existing nonshellable simplicial complexes in *Simplicial Complex Library* [3] where famous simplicial complexes are listed, to apply the hash-DFS method to each of them.

6.2 The Application of The Hash-DFS Method

Figure 3 shows the result of the application of the hash-DFS method to shellable simplicial complexes generated by POLYMAKE. The x-axis (# facet) shows the number of facets, and the y-axis (ratio [%]) shows how many % of all data can be decided within 3 hours. Almost all examples with around 40 facets are decided regardless of the dimension by the hash-DFS method, while they cannot be decided by the reverse backtrack method.

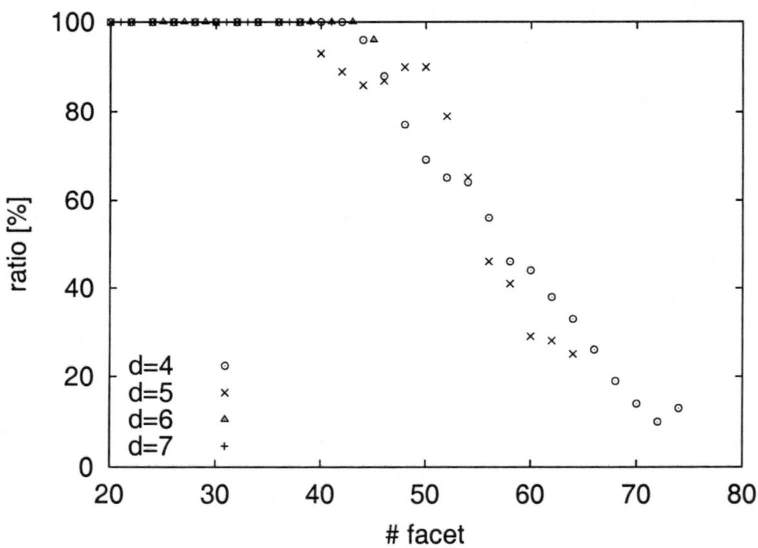

Figure 3. The ratio of decision

Table 1 shows the result of the application of the hash-DFS method and the reverse backtrack method to nonshellable simplicial complexes. The reverse backtrack method is implemented in the *Simplicial Complex Calculator* `scc` written by Hachimori [4] as a function named **`shellability_f`**. It is more difficult to decide shellability of nonshellable simplicial complexes, because we check the whole search space by DFS. Examples in Table 1 other than

Data	f-vector	h-vector	time1 [s]	time2 [s]
c-ns3.dat	$(1, 10, 31, 22)$	$(1, 7, 14, 0)$	0.03	1.34
c-ns1.dat	$(1, 12, 37, 26)$	$(1, 9, 16, 0)$	0.01	1.30
c-ns2.dat	$(1, 13, 39, 27)$	$(1, 10, 16, 0)$	—	18.32
Ziegler's Ball	$(1, 10, 38, 50, 21)$	$(1, 6, 14, 0, 0)$	0.04	1.31
Gruenbaum's Ball	$(1, 14, 54, 70, 29)$	$(1, 10, 18, 0, 0)$	0.05	1.31
Rudun's Ball	$(1, 14, 66, 94, 41)$	$(1, 10, 30, 0, 0)$	0.04	1.28
Poincare's Sphere	$(1, 16, 106, 180, 90)$	$(1, 12, 64, 12, 1)$	—	—

Table 1. Comparison of the computational time between the reverse backtrack method (time1) and the hash-DFS method (time2): "—" means over 24 hours.

c-ns2.dat and Poincare's sphere are decided to be noshellable by the reverse backtrack method rapidly. This is because the h-vector of each of such examples become negative when one facet is removed by this method and decided to be nonshellable at that time. However, in the case of other examples, c-ns2.dat and Poincare's sphere, the h-vector of each example does not become negative until many steps ahead in the reverse backtrack method, hence it takes very much time to compute these examples. But the hash-DFS method can decide one of these examples, c-ns2.dat, much more rapidly than the reverse backtrack method, by the sake of the branch cutting using the hash.

References

1. Gopal Danaraj and Victor Klee. *A Representation of 2-dimensional pseudomanifolds and its use in the design of a linear-time shelling algorithm.* Annals of Discrete Mathematics2(1978) 53–63.
2. Ewgenij Gawrilow and Michael Joswig. *POLYMAKE.* http://www.math.tu-berlin.de/diskregeom/polymake/doc/(1997).
3. Masahiro Hachimori. *Simplicial Complex Library.* http://www.sk. tsukuba.ac.jp/~hachi/math/library/index_eng.html(2001).
4. Masahiro Hachimori. *Simplicial Complex Calculator Ver.0.81.* http://www.sk. tsukuba.ac.jp/~hachi/math/scc/index_eng.html(2001).
5. Sonoko Moriyama. *Extendable shellability in dimension two.* Bachelor thesis, University of Tokyo, Tokyo(2000).
6. Sonoko Moriyama. *Deciding shellable simplicial complexes.* Master thesis, University of Tokyo, Tokyo(2002).
7. Sonoko Moriyama. *Antimatroidal Decomposition of Shellings by h-*

assignments.(in preparation).
8. Ayumu Nagai. *Df-pn Algorithm for Searching AND/OR Trees and Its Applications.* Doctor thesis, University of Tokyo, Tokyo(2002).
9. Richard P. Stanley. *Combinatorics and Commutative Algebra.* Progress in Mathematics 41 Birkhäuser(1996) 84.
10. Günter M. Ziegler. *Lectures on Polytopes.* Graduate Texts in Mathematics 152 Springer-Verlag (1994).

COMPUTING GCLF USING SYZYGY ALGORITHM

MINGSHENG WANG
State Key Laboratory of Information security
Institute of Software, Chinese Academy of Sciences
E-mail: mswang@acae.cuhk.edu.hk

C. P. KWONG
Department of Automation and Computer-Aided Engineering
The Chinese University of Hong Kong
E-mail: cpkwong@acae.cuhk.edu.hk

This paper discusses problem of computing a greatest common left factor (GCLF) for two n-D($n > 2$) polynomial matrices with the same number of rows. We give a partial answer by proposing a new algorithm based on syzygy computation. Some examples are computed using computer algebra softwares **Singular** and **CoCoA**. It demonstrates the fundamental importance of Gröbner basis and syzygy algorithm.

1 Introduction

Youla and Gnavi [14] studied basic structures of n-D system theory, and proposed three important and inequivalent notions of polynomial matrices primeness, namely, zero primeness, minor primeness, and factor primeness, for $n > 2$. Since then, there exists a number of research papers towards studying primeness of n-D rational matrices, [3,9,10,11,12,13,15] because of their wide applications in circuits, systems, controls, and signal processing.

There are some concrete algorithms for some special class of polynomial matrices, and some conjectures and open problems were proposed. [3,10] An important problem is to find a greatest common left (right) factor of two polynomial matrices. In this paper, we study problem of how to extracting a greatest common left factor of two n-D polynomial matrices with the same number of rows. For $n \leq 2$, this problem have been solved, [7,12] however, for $n > 2$, this problem seems to have some difficulties to solve. [1,2,3] In this paper, we attempt to give a partial answer using symbolic computation algebra.

The paper is organized as follows. Section 2 recalls some basic concepts concerning n-D polynomial matrices, and states main research problem. Section 3 gives main algorithm for finding a greatest common left factor (GCLF) of two n-D polynomial matrices with the same number of rows. Section 4 is devoted to giving some concrete examples to illustrate our algorithm using famous symbolic computation softwares, for example, Singular and CoCoA.

2 The Problem

Let $\mathbb{C}$ be the complex number field, and $\mathbb{C}[z_1, \cdots, z_n]$ be the polynomial ring in variables $z_1, \cdots, z_n$, which we briefly denote as $\mathbb{C}[z]$ throughout the paper. Also, we denote the ring of all the $l \times m$-matrices with entries in $\mathbb{C}[z]$ by $\mathbb{C}^{l\times m}[z]$.

First, we recall the following definition:

Definition 2.1 A greatest common left factor (GCLF) of two n-D polynomial matrices A, B with the same number of rows is any n-D matrix R with the following properties:

(1) R is a left factor of A and B; i.e., there exist polynomial matrices A_1, B_1 such that

$$A = RA_1, B = RB_1$$

(2) If R_1 is any other left factor of A and B, then R_1 is a left factor of R; i.e., there exists a polynomial matrix W such that $R = R_1 W$.

The following definition gives an important invariant for n-D polynomial matrix. [9,10]

Definition 2.2 Let $F \in \mathbb{C}^{l\times m}[z]$ be of full row rank with $m > l$, and let $a_1, \ldots, a_\beta$ denote the $l \times l$ minors of the matrix F in some fixed order, where $\beta = \binom{m}{l}$. Let d be the greatest common divisor (gcd) of $a_1, \ldots, a_\beta$, and b_i such that

$$a_1 = db_i, \quad i = 1, \cdots, \beta.$$

Then $b_1, \cdots, b_\beta$ are called the "reduced minors" of F.

Also, the following basic definition is a starting point for multidimensional system theory. [14,1,10]

Definition 2.3 Let $F \in \mathbb{C}^{l\times m}[z]$ with $l \leq m$. Then F is said to be:

(i) zero left prime (ZLP) if there exists no n- tuple z_0 which is a common zero of all the $l \times l$ minors of F;

(ii) minor left prime (MLP) if all the $l\times l$ minors of F are relatively prime; i.e., their greatest common polynomial divisor (gcd) is a nonzero constant;

(iii) factor left prime (FLP) if in any polynomial matrix decomposition $F = F_1 F_2$ with $F_1 \in \mathbb{C}^{l\times l}[z]$, F_1 is necessarily a unimodular matrix; i.e., det F_1 is a non-zero constant in $\mathbb{C}$.

Similarly, we can define ZRP, MRP, FRP for n-D polynomial matrices with number of rows larger than number of columns.

Now we can state the following problem:

Let $A \in \mathbb{C}^{l\times m_1}[z]$ and $B \in \mathbb{C}^{l\times m_2}[z]$ be two polynomial matrices with $l \leq m_1 + m_2 = m$, and $C = [A|B]$, the following open problem was proposed by N. K. Bose: [2,3]

Open problem [3] Under the assumption that a GCLF of A and B exists and the reduced minors of C have common zeros, find an algorithm for computing this GCLF.

3 A Partial Solution

In this section, we will propose an algorithm for finding the greatest common left factor of two polynomial matrices with the same number of rows. Thus, we give a partial answer for the open problem at least in some theoretical sense.

In order to give our algorithm, we must give some preliminaries for the convenience of reader.

Let $F \in \mathbb{C}^{l\times m}[z]$ with $l \leq m$, and d be a polynomial in $\mathbb{C}[z]$.

Let $M = \{[u,v] \in \mathbb{C}^{1\times(l+m)}[z] | uF = dv, u \in \mathbb{C}^{1\times l}[z], v \in \mathbb{C}^{1\times m}[z]\}$. It is obvious that M is a submodule of the free module $\mathbb{C}^{1\times(l+m)}[z]$, and $M = \{[u,v] \in \mathbb{C}^{1\times(l+m)}[z] | [u,v] \begin{bmatrix} F \\ -dI_m \end{bmatrix} = 0, u \in \mathbb{C}^{1\times l}[z], v \in \mathbb{C}^{1\times m}[z]\}$.

Thus, we recall the following important concept: [4]

Definition 3.1 Let $(f_1, \cdots, f_t)$ be an ordered t-tuple of elements of $\mathbb{C}^{1\times m}[z]$. The set of all $(a_1, \cdots, a_t) \in \mathbb{C}^{1\times t}[z]$ such that $a_1 f_1 + \cdots + a_t f_t = 0$ is an $\mathbb{C}[z]$-submodule of $\mathbb{C}^{1\times t}[z]$, called the syzygy module of $(f_1, \cdots, f_t)$, and denoted as $\mathrm{Syz}(f_1, \cdots, f_t)$.

Similar definition can be defined for column vectors.

Definition 3.2 Let N be a finitely generated module over the ring $\mathbb{C}[z]$, define the minimal number of generators of the module N to be the smallest number of elements in any generating set of N, denoted as $\mu(N)$.

For $R \in \mathbb{C}^{l\times m}[z]$, let $s_1, \cdots, s_l$ be all the rows of R. We also denote $\mathrm{Syz}(s_1, \cdots, s_l)$ by $\mathrm{Syz}(R)$. [11] In terms of definition 3.1, the above submodule $M = \mathrm{Syz}(\begin{bmatrix} F \\ -dI_m \end{bmatrix})$.

Let $\mu(M) = \nu$, and $f_1, \cdots, f_\nu \in \mathbb{C}^{1\times(l+m)}[z]$ be a system of generators of $\mathrm{Syz}(\begin{bmatrix} F \\ -dI_m \end{bmatrix})$. We set $S = \begin{bmatrix} f_1 \\ \vdots \\ f_\nu \end{bmatrix} \in \mathbb{C}^{\nu\times(l+m)}[z]$ be a matrix formed by f_i, $i = 1, \cdots, \nu$. S is said to be a generating matrix of submodule M. [11] The generating matrix has the following important properties:

Lemma 3.1 With the above notation, the following properties hold:

(i) Let $S \in \mathbb{C}^{\nu\times(l+m)}[z]$ be a generating matrix of M. Then for any polynomial matrices $D \in \mathbb{C}^{l\times l}[z]$ and $N \in \mathbb{C}^{l\times m}[z]$ with $DF = dN$, there exists a polynomial matrix $T \in \mathbb{C}^{l\times \nu}[z]$ such that $[D|N] = TS$.

(ii) Let S_1 and S_2 be two generating matrices of M. Then there exist two $\nu \times \nu$-polynomial matrices T_1 and T_2 such that $S_1 = T_1S_2$, and $S_2 = T_2S_1$.

(iii) If $\nu = \mu(M) = l$, and F is of full row rank, and T_1 and T_2 are as in (ii). Then we have

$$T_1T_2 = T_2T_1 = I_l,$$

where I_l denotes the identity matrix.

A direct corollary of lemma 3.1 is that when $\mu(M) = l$, if there is a generating matrix S of M is ZLP (MLP or FLP) then all the generating matrices are also ZLP(MLP or FLP).

We also need the following lemma, [11] which is important to problems concerning n-D rational polynomial matrices.

Lemma 3.2 [11] Let $P = \begin{bmatrix} -N \\ D \end{bmatrix} \in \mathbb{C}^{t_2\times t_1}[z]$ with $t_1 < t_2$ and $D \in \mathbb{C}^{t_1\times t_1}[z]$ being nonsingular, also let $r = t_2 - t_1$. Then $\mu(\mathrm{Syz}(P)) = r$ if and only if there exists a MLP matrix $H \in \mathbb{C}^{r\times t_2}[z]$ such that $HP = 0$, where $\mathrm{Syz}(P)$ is the syzygy module of row vectors of P.

Remark 3.1 Lemma 3.2 holds for the real number field $\mathbb{R}$. Thus all the results are true when the coefficient field $\mathbb{C}$ is replaced by the real number field $\mathbb{R}$.

With above preliminaries, we now begin with our descriptions:

Let $A \in \mathbb{C}^{l\times m_1}[z]$ and $B \in \mathbb{C}^{l\times m_2}[z]$ be two polynomial matrices with one of them being full of row rank, and $l \le m = m_1+m_2$. Let $F = [A|B]$, d be the greatest common divisor of all the $l \times l$-minors of F. If $\mu(\mathrm{Syz}(\begin{bmatrix} F \\ -dI_m \end{bmatrix})) = l$, then we can find a greatest common left factor of matrices A and B, where $\mathrm{Syz}(\begin{bmatrix} F \\ -dI_m \end{bmatrix})$ denotes the syzygy module of the row vectors of $\begin{bmatrix} F \\ -dI_m \end{bmatrix}$.

In fact, since $\mu(\mathrm{Syz}(\begin{bmatrix} F \\ -dI_m \end{bmatrix})) = l$, by lemma 3.2, there exists a MLP generating matrix $[G|H]$ with $G \in \mathbb{C}^{l\times l}[z]$ and $H \in \mathbb{C}^{l\times m}[z]$. Hence we have

$$[G|H]\begin{bmatrix} F \\ -dI_m \end{bmatrix} = 0.$$

That is, we have

$$GF = dH.$$

Since every row of $[dI_l|F]$ belongs to $\text{Syz}(\begin{bmatrix} F \\ -dI_m \end{bmatrix})$, by lemma 3.1, we have

$$[dI_l|F] = C[G|H],$$

where C is a $l \times l$ polynomial matrix. Thus $dI_l = CG$, and $F = CH$.

First, we prove the following claim:

Claim 3.1 With the above notation, det $C = dk$, and $C = dG^{-1}$, where k is a nonzero constant.

Proof Let c_i be all the $l \times l$-minors of $[G|H]$, $i = 1, \cdots, \gamma$, $\gamma = \binom{m+l}{l}$. Because $[G|H]$ is a MLP matrix, the greatest common divisor of c_i is 1.

Let c'_i be all the $l \times l$-minors of $[dI_l|F]$, $i = 1, \cdots, \gamma$, and $\gamma = \binom{m+l}{l}$, and r be the greatest common divisor of c'_i. It is easy to see that the $l \times l$-minors of $[dI_l|F]$ have the following three cases:

(i) det dI_l, namely d^l;

(ii) the $l \times l$-minors of F; i.e., db_j, $j = 1, \cdots, \beta$, $\beta = \binom{m}{l}$. So, in this case, $b_1, \cdots, b_\beta$ have no non-trial common divisor;

(iii) the $l \times l$-minors that are determinants of matrices composed of t rows of dI_l and $l - t$ rows of F, $0 < t < l$, which are of form: $d^t e_k$, where e_k are polynomials;

Hence we have $r = d$ up to a non-zero constant, thus, we may assume $r = d$. Let $c'_i = dc''_i$, $i = 1, \cdots, \gamma$, then all the c''_i are relatively prime.

By considering all the $l \times l$-minors of two sides of $[dI_l|F] = C[G|H]$, an application of Cauchy-Binet formula, we have the following formulae:

$$c'_i = dc''_i = c_i \det C, i = 1, \cdots, \gamma.$$

Since all the c_i are relatively prime, $i = 1, \cdots, \gamma$, we get $c_i = kc''_i$, $i = 1, \cdots, \gamma$, and $kd = \det C$, where k is a non-zero constant.

Thus, by claim 3.1, without loss of generality, we can assume $k = 1$. So det $C = d$, and $C = dG^{-1}$.

Now we can prove the following claim:

Claim 3.2 C is a greatest common left factor of A and B.

Proof Let $C' \in \mathbb{C}^{l \times l}[z]$ be a common left factor of A and B. Then there exist $A' \in \mathbb{C}^{l \times m_1}[z]$ and $B' \in \mathbb{C}^{l \times m_2}[z]$ such that

$$F = [A|B] = C'[A'|B'].$$

Let e be the greatest common divisor of all the $l \times l$-minors of $[A'|B']$. Then by computing all the $l \times l$-minors of the above formulae, we can prove $d = e \cdot \det C' \cdot k_1$, where k_1 is a non-zero constant of $\mathbb{C}$. Without loss of generality, we may assume $k_1 = 1$, hence $d = e \cdot \det C'$.

Since $(C')^{-1} = \frac{1}{\det C'}\text{adj}C'$, where adj$C'$ is the adjoint matrix of C'. We can get the following formula:

$$\text{adj}C' \cdot F = \det C'[A'|B'].$$

Hence we have

$$(e \cdot \text{adj}C') \cdot F = (e \cdot \det C')[A'|B] = d[A'|B'].$$

Let $F' = [A'|B']$. Since $[G|H]$ is a generating matrix of Syz$(\begin{bmatrix} F \\ -dI_m \end{bmatrix})$, there exists a $G_1 \in \mathbb{C}^{l\times l}[z]$ such that

$$[e \cdot \text{adj}C'|F'] = G_1[G|H].$$

Hence we get

$$e \cdot \text{adj}C' = G_1 \cdot G.$$

By adj$C' = \det C' \cdot (C')^{-1}$, and $G = dC^{-1}$, we get the following formula:

$$e \cdot \det C' \cdot (C')^{-1} = d \cdot G_1 \cdot C^{-1}.$$

Since $d = e \cdot \det C'$, we have $(C')^{-1} = G_1C^{-1}$; i.e., $C = C'G_1$. Thus, we have proved that C is a greatest common right factor of A and B.

Thus, by claim 3.2, we have proved the following main theorem:

Theorem 3.1 Let $A \in \mathbb{C}^{l\times m_1}[z]$ and $B \in \mathbb{C}^{l\times m_2}[z]$ be two matrices with one of them being full row rank, and $l \leq m_1 + m_2 = m$. Let $F = [A|B]$, d be the greatest common divisor of all the $l \times l$-minors of F. If $\mu(\text{Syz}(\begin{bmatrix} F \\ -dI_m \end{bmatrix})) = l$, then matrices A and B have a greatest common left factor, and there exists an algorithm to find it.

In the following, we give an algorithm to find the greatest common left factor of two polynomial matrices A and B provided that the conditions of theorem 3.1 are satisfied.

Algorithm 3.1.

(i) Compute the greatest common divisor d of all the $l \times l$-minors of $F = [A|B]$.

(ii) Compute $\mu(\text{Syz}(\begin{bmatrix} F \\ -dI_m \end{bmatrix}))$, if this number is l, then go to (iii), otherwise, stop!

(iii) From the minimal generating set of the syzygy module, we form its generating matrix $[G|H]$ such that $G \in \mathbb{C}^{l\times l}[z]$ and $H \in \mathbb{C}^{l\times m}[z]$. Hence $d \cdot G^{-1}$ is the greatest common left factor

The following corollary shows that algorithm 3.1 has some significance for it gives a new method for finding GCLF when $n \leq 2$.

Corollary 3.1 Let $n \leq 2$, then algorithm 3.1 give a complete algorithm for finding GCLF for two polynomial matrices with the same number of rows.

Proof Let $A \in \mathbb{C}^{l\times m_1}[z]$ and $B \in \mathbb{C}^{l\times m_2}[z]$ with $m_1 + m_2 = m \geq l$ and $[A|B]$ being full row rank. Let d be the greatest common divisor of all the $l \times l$-minors of $F = [A|B]$.

If C is a GCLF of A and B, then there are matrices A' and B' such that $A = CA'$ and $B = CB'$. Hence

$$F = [A|B] = C[A'|B']$$

Let $F' = [A'|B']$, we can prove that F' is a FLP by means of definition 2.1. Hence F' is a MLP [14,15] because of $n \leq 2$. Thus by comparing all the $l \times l$-minors of $F = CF'$, we can infer that det $C = dk$, where k is a nonzero constant.

Note that $C^{-1} = \frac{G}{kd}$, where G is the adjoint matrix of C, then we get $GF = kdF'$. Let $G_1 = \frac{G}{k}$, then we get

$$[G_1|F'] \begin{bmatrix} F \\ -dI_m \end{bmatrix} = 0$$

Since F' is a MLP, $[G_1|F']$ is also a MLP. Thus by lemma 3.2, we have $\mu(\text{Syz}(\begin{bmatrix} F \\ -dI_m \end{bmatrix})) = l$. Hence the condition of theorem 3.1 is satisfied, by theorem 3.1, we can make use of algorithm 3.1 to find a GCLF for A and B.

Remark 3.2 With the similar proofs as those in theorem 3.1, we can prove that $\mu(\text{Syz}(\begin{bmatrix} F \\ -dI_m \end{bmatrix})) = l$ if and only if rational matrix $\frac{1}{d}F$ has a MLP matrix fraction description $N_1 D_1^{-1}$. Thus the most interesting matrices in n-D system theory should have this property. [15]

Remark 3.3 When $n \leq 2$, algorithm 3.1 gives another algorithm for finding GCLF of two polynomial matrices with the same number of rows. When $n = 1$, elementary column (row) operations for polynomial matrices may be used to find the GCLF(GCRF). [7] For $n = 2$, a similar method as that of $n = 1$ can be used by using pseudo-division for finding GCLF, [1,12] however these methods can not be generalized to higher dimensional cases.

Remark 3.4 We do not need to know whether or not the reduced minors have common zeros in algorithm 3.1.

Remark 3.5 From algorithm 3.1, how to decide whether or not $\mu(\text{Syz}(\begin{bmatrix} F \\ -dI_m \end{bmatrix})) = l$ is an important problem. We may use command "Minimalized()" in **CoCoA** [5] to find a minimal set of generators, but how to find $\mu(M)$ is still not solved completely.

4 Computational examples

In this section we will compute some examples to explain algorithm 3.1. First, we consider the following example. [3]

Example 4.1 Let $A = \begin{bmatrix} z_1(z_1-1) \\ z_1 - 1 \end{bmatrix}$, $B = \begin{bmatrix} z_1 - z_2z_3 - z_3 & z_1z_3 - 2z_1 - z_3 \\ z_2^2 + 2z_2 + 2 & z_2 + z_3 - 1 \end{bmatrix}$.

Let $F = [A|B]$. All the 2×2-minors are computed as follows:

$$\begin{aligned} a_1 &= (z_1-1)(z_2+1)(z_1z_2+z_1+z_3) \\ a_2 &= (z_1-1)(z_1z_2+z_1+z_3) \\ a_3 &= (-z_2z_3 - z_3 + 2z_2 + 3)(z_1z_2+z_1+z_3) \end{aligned}$$

Thus, the greatest common divisor of all the 2×2-minors of F is $d = z_1z_2 + z_1 + z_3$, and the reduced minors of F have common zeros. [3] Hence, existing algorithm [3] can not be applied for this case!

Note that

$$\begin{bmatrix} F \\ -dI_3 \end{bmatrix} = \begin{bmatrix} z_1(z_1-1) & z_1 - z_2z_3 - z_3 & z_1z_3 - 2z_1 - z_3 \\ z_1 - 1 & z_2^2 + 2z_2 + 2 & z_2 + z_3 - 1 \\ -(z_1z_2+z_1+z_3) & 0 & 0 \\ 0 & -(z_1z_2+z_1+z_3) & 0 \\ 0 & 0 & -(z_1z_2+z_1+z_3) \end{bmatrix}.$$

Let s_i be the ith row of the above matrix, $i = 1, \cdots, 5$. By computing the syzygy module of $s_1, \cdots, s_5$ under module term ordering (dp, C) using **Singular**, [6] we find the syzygy module is generated by the following two elements:

$f_1 = z_1e_2 + z_2e_4 + e_5 + e_4 - e_1$, $f_2 = z_1e_3 + z_2e_1 + z_3e_5 + z_3e_2 - 2e_5 + e_4 - e_3 + e_1$.

Thus, $f_1 = [-1, z_1, 0, z_2+1, 1]$, $f_2 = [z_2+1, z_3, z_1-1, 1, z_3-2]$.

By algorithm 3.1, we know that $D = d \cdot G^{-1}$ is a greatest common left factor, where $G = \begin{bmatrix} -1 & z_1 \\ z_2+1 & z_3 \end{bmatrix}$. Thus $D = \begin{bmatrix} -z_3 & z_1 \\ z_2+1 & 1 \end{bmatrix}$.

Example 4.2 Let $A = \begin{bmatrix} (z_1+1)z_2z_4 & z_1z_4 \\ z_2(z_3+1) & z_3 - 1 \end{bmatrix}$, $B = \begin{bmatrix} -z_4^2 \\ -2z_4 \end{bmatrix}$.

Let $F = [A|B]$. All the 2×2-minors of F is the following:

$$\begin{aligned} a_1 &= z_2z_4(-2z_1 + z_3 - 1) \\ a_2 &= z_2z_4^2(-2z_1 + z_3 - 1) \\ a_3 &= z_4^2(-2z_1 + z_3 - 1) \end{aligned}$$

Hence the greatest common divisor of a_1, a_2, a_3 is $d = z_4(z_1 - \frac{1}{2}z_3 + \frac{1}{2})$ and the reduced minors of F have common zeros in $\mathbb{C}^3$, thus the existing algorithm [3] can not be used for this example. (We have chosen degree reverse lexicographic ordering on the polynomial ring $\mathbb{Q}[z_1, z_2, z_3, z_4]$).

Let s_i be the ith row of $\begin{bmatrix} F \\ -dI_3 \end{bmatrix}$, $i = 1, \cdots, 5$. Then we can write s_i as follows:

$$\begin{aligned} s_1 &= [(z_1+1)z_2z_4, z_1z_4, -z_4^2] \\ s_2 &= [z_2(z_3+1), z_3-1, -2z_4] \\ s_3 &= [-z_4(z_1 - \tfrac{1}{2}z_3 + \tfrac{1}{2}), 0, 0] \\ s_4 &= [0, -z_4(z_1 - \tfrac{1}{2}z_3 + \tfrac{1}{2}), 0] \\ s_5 &= [0, 0, -z_4(z_1 - \tfrac{1}{2}z_3 + \tfrac{1}{2})] \end{aligned}$$

We now compute the syzygy module of $s_1, s_2, \cdots, s_5$ using **CoCoA** 4.1 under default module term ordering (DegRevLex and ToPos), [5,8] and it has two generators h_1, h_2. Hence algorithm 3.1 can be used, and we list h_1, h_2 as follows:

$$\begin{aligned} h_1 &= [1, -\tfrac{1}{2}z_4, z_2, 1, 0] \\ h_2 &= [z_3+1, -z_1z_4 - z_4, 0, 2, 2z_4] \end{aligned}$$

Let $G = \begin{bmatrix} 1 & -\frac{1}{2}z_4 \\ z_3+1 & -z_1z_4 - z_4 \end{bmatrix}$, then $D = dG^{-1}$ is a GCLF of A and B, namely, $D = \begin{bmatrix} z_1z_4 + z_4 & -\frac{1}{2}z_4 \\ z_3+1 & -1 \end{bmatrix}$.

Example 4.3 Let $A = \begin{bmatrix} z_1+1 & -z_2 \\ z_2 & z_1-1 \end{bmatrix}$, $B = \begin{bmatrix} z_1 - z_2^2 + 1 \\ z_1z_2 \end{bmatrix}$.

Let $F = [A|B]$, its all the 2×2-minors a_1, a_2, a_3 are computed as follows:

$$\begin{aligned} a_1 &= z_1^2 + z_2^2 - 1 \\ a_2 &= z_2(z_1^2 + z_2^2 - 1) \\ a_3 &= -(z_1^2 + z_2^2 - 1) \end{aligned}$$

Thus $d = z_1^2 + z_2^2 - 1$ is the gcd of a_1, a_2, a_3, and the reduced minors of F have no common zeros in $\mathbb{C}^2$.

Let s_i be all the rows of F, $i = 1, \cdots, 5$, we list these s_i as follows:

$$\begin{aligned} s_1 &= [z_1+1, -z_2, z_1 - z_2^2 + 1] \\ s_2 &= [z_2, z_1-1, z_1z_2] \\ s_3 &= [-(z_1^2 + z_2^2 - 1), 0, 0] \\ s_4 &= [0, -(z_1^2 + z_2^2 - 1), 0] \\ s_5 &= [0, 0, -(z_1^2 + z_2^2 - 1)] \end{aligned}$$

Now, we compute the syzygy module of $s_1, s_2, \cdots, s_5$ using **CoCoA** 4.1 under default module term ordering (DegRevLex and ToPos), [5,8] and it has two generators g_1, g_2. Hence theorem 3.1 can be used, and we list g_1, g_2 as follows:

$$g_1 = [-z_1 + 1, -z_2, -1, 0, -1]$$
$$g_2 = [-z_2, z_1 + 1, 0, 1, z_2]$$

Let $G = \begin{bmatrix} -z_1 + 1 & -z_2 \\ -z_2 & z_1 + 1 \end{bmatrix}$, then $D = dG^{-1} = -\begin{bmatrix} z_1 + 1 & z - 2 \\ z_2 & -z_1 + 1 \end{bmatrix}$ is a greatest common left factor (GCLF) of A and B.

5 Conclusions

We have proposed an algorithm for finding the greatest common left factor of two polynomial matrices A and B with entries in the polynomial ring in n variables with coefficients in the complex number field provided that they satisfy certain conditions. Although our results are stated in the complex number field $\mathbb{C}$, all the results are true in the real number field $\mathbb{R}$ because lemma 3.2 is still true in this case. Examples are given to explain the algorithm. The algorithm is useful in testing examples from literatures. Although the problem for finding the greatest common left factor of two polynomial matrices is still not settled completely, the algorithm proposed in this paper seems to be useful for computing the most interesting examples by means of powerful symbolic computation softwares.

Acknowledgements

This work is partially supported by national key basic research projects G1998030600 and G1999035802. The authors are very grateful to anonymous referees for their valuable suggestions.

References

1. N. K. Bose, Applied Multidimensional Systems Theory. Van Nostrand Reinhold, New York, 1982.
2. N. K. Bose (Edited), Multidimensional Systems: Theory and Applications. IEEE Press, New York, 1979.

3. C. Charoenlarpnopparut and N. K. Bose, Multidimensional FIR filter Bank Design using Gröbner Basis. IEEE Trans. Circuits Syst. –II: Analog and Digital Signal Processing **46**, 1475-1486 (1999).
4. D. Cox, L. Little and D. O'Shea, Using Algebraic Geometry. GTM 185. Springer-Verlag, New York, 1998.
5. A. Capani, G. Niesi, L. Robbiano, CoCoA, a system for doing Computations in Commutative Algebra. Available via anonymous ftp from: cocoa.dima.unige.it.
6. G. M. Greuel etl, Singular Manual for Singular Version 2.0.0. Http://www.singular.uni-kl.de/Manual/2.0.0, 2001.
7. T. Kailath, Linear Systems. Englewood Cliffs, Prentice Hall, 1980.
8. M. Kreuzer and L. Robbiano, Computational Commutative Algebra I. Springer-Verlag, New York, 2000.
9. Z. Lin, On Matrix Fraction Descriptions of Multivariable Linear n-D Systems. IEEE Trans. Circuits Syst. **35 (10)**, 1317-1322 (1988).
10. Z. Lin, Notes on n-D Polynomial Matrix Factorization. Multidimensional System and Signal Processing **10**, 379-393 (1999).
11. Z. Lin, On Syzygy Modules for Polynomial Matrices. Linear Algebra and Its Applications **298**, 73-86 (1999).
12. M. Morf, B. C. Levy and S. Y. Kung, New Results in 2-D Systems Theory, Part I: 2-D Polynomial Matrices, Factorizations and Coprimeness. Proc. IEEE **65**, 861-872 (1977).
13. J. Wood, E. Rogers and H. Owens, A Formal Theory of Matrix Primeness. Mathematics of Control, Signals, and Systems **11**, 40-78 (1988).
14. D. C. Youla and G. Gnavi, Notes on n-Dimensional System Theory. IEEE Trans. Circuits Systems **26 (2)**, 105-111 (1979).
15. E. Zerz, Primeness of Multivariate Polynomial Matrices. System and Control Letter **29**, 139-145 (1996).

EFFICIENT COOPERATIVE SOLVERS FOR NONLINEAR CONTINUOUS CONSTRAINT PROBLEMS *

YI SHANG, MARKUS P.J. FROMHERZ, AND LARA S. CRAWFORD
Palo Alto Research Center
3333 Coyote Hill Road
Palo Alto, CA 94304
{yshang,fromherz,lcrawford}@parc.com

In the past decade, significant progress has been made in understanding problem complexity of discrete constraint problems. In contrast, little similar work has been done for constraint problems in the continuous domain. In this paper, we study the complexity of several popular methods for nonlinear constraint problems and present cooperative solvers with improved performance. The methods include a sequential quadratic programming (SQP) method, a penalty method with a fixed penalty function, a penalty method with a sequence of penalty functions, and an augmented Lagrangian method. In our experiments, we apply these methods to solve a series of continuous constraint problems with increasing constraint-to-variable ratios and obtain novel results on complexity phase transition phenomena. Specifically, for constraint satisfaction problems, the SQP method is the best on weakly constrained problems, whereas the augmented Lagrangian method is the best on highly constrained ones. Although the static penalty method performs poorly by itself, by combining it with the SQP method, we show a cooperative solver that is significantly better than any of the individual methods on problems with moderate to large constraint-to-variable ratios.

1 Introduction

Many important real-world applications in domains such as control, planning, and diagnosis of physical systems can be regarded as constraint satisfaction (CSPs) and constrained optimization problems (COPs) [1]. Effective constraint-based techniques should be aware of the complexity of real-world constraint problems by dynamically selecting and adapting solvers to the structure of the problem. Toward this end, this paper studies the complexity of solution methods on continuous constraint problems.

In the past decade, significant progress has been made in understanding problem complexity of discrete constraint satisfaction and combinatorial optimization problems, such as satisfiability (SAT), planning, graph coloring, and the traveling salesman problem. One particularly well-studied phenomenon is the *complexity phase transition* [2]. For example, in solving randomly gen-

*THIS WORK HAS BEEN SUPPORTED IN PART BY DARPA UNDER CONTRACT F33615-01-C-1904.

erated SAT problems, the average time to find a solution or determine that none exists is short when the ratio of clauses to variables is small (i.e., the problem is weakly constrained) as well as when this ratio is large (where the problem is highly or overconstrained), while solving time is largest in the intermediate case. This phenomenon has been observed for a variety of complete and incomplete search algorithms, including the Davis-Putnam method [3] and GSAT [4].

Towards understanding the complexity of continuous constraint problems, in this paper, we empirically study the complexity phase transition phenomena of several popular methods in solving these problems with varying constraint-to-variable ratios. The methods include a sequential quadratic programming (SQP) method, a penalty method with a fixed penalty function, a penalty method with a sequence of penalty functions, and an augmented Lagrangian method. They represent major approaches for nonlinear constraint satisfaction and constrained optimization problems: solving a sequence of quadratic programming sub-problems, or minimizing one or a sequence of unconstrained function. The test cases consist of randomly generated nonlinear constraints in the form of trigonometric functions, which are similar to problems appearing in robot control applications. In our experiments, we study the complexity of the methods in solving a series of continuous constraint problems with increasing constraint-to-variable ratios. In addition, we propose cooperative solvers that combine multiple methods and show their performance improvement.

The rest of this paper is structured as follows. In Section 2, we first introduce individual methods, and then discuss cooperative solvers. In Section 3, we present the test-case generators for generating continuous constraint problems used in our study. In Section 4, we show experimental results on problems with different constraint-to-variable ratios (or constraint ratios). Finally, in Section 5, we conclude the paper and discuss future work.

2 Constraint Solvers

Different methods have different complexity behaviors in solving continuous constraint problems. In this section, we introduce some popular methods and discuss alternatives of combining multiple methods in a cooperative solver.

2.1 Quadratic Penalty Method

One fundamental approach to constrained optimization is to replace the original problem by a penalty function or a sequence of penalty functions [5]. The approach is popular because of its simplicity and intuitive appeal. One of the simplest penalty function is the quadratic penalty function. For the inequality-constrained problem (the type of problem we tested in the experiments)

$$\min_x f(x) \quad \text{subject to} \quad c_i(x) \le 0, \quad i = 1, \cdots, m \tag{1}$$

where $x = (x_1, x_2, \ldots, x_n)$, the quadratic penalty function $Q(x, \mu)$ is

$$Q(x, \mu) = f(x) + \frac{1}{2\mu} \sum_{i=1}^{m} (\max(0, c_i(x)))^2 \tag{2}$$

where $\mu > 0$ is the penalty parameter. Based on the penalty function, the quadratic penalty method is (Section 17.1 of *Numerical Optimization* [5])

Given $\mu_1 > 0$, $r \in (0, 1)$, starting point x^1
for $k = 1, \cdots, k_{max}$
  Find a minimizer $x^{k'}$ of $Q(x, \mu_k)$, starting at x^k
  if a solution satisfying the requirement is found, stop with $x^{k'}$
  $\mu_{k+1} = r * \mu_k$, $x^{k+1} = x^{k'}$
end

In this paper, when $k_{max} = 1$ we call it *static penalty method* (SPM), i.e., solving a fixed penalty function. When $k_{max} > 1$, we call it *dynamic penalty method* (DPM), i.e., solving a sequence of penalty functions with different penalty parameters.

In our experiments, we use a quasi-Newton solver, fminunc from the Matlab Optimization Toolbox, to find a minimizer of $Q(x, \mu_k)$ in each iteration. For nonlinear constrained problems, the penalty method may be trapped by local optima and may not be able to find a feasible solution. When that happens, we restart it from another random point.

2.2 Augmented Lagrangian Method

The augmented Lagrangian method (ALM) improves the quadratic penalty function by using explicit Lagrange multiplier estimates to avoid ill-conditioning. For the inequality-constrained problem in Eq. (1), the augmented Lagrangian function $L_A(x, \lambda, \mu)$ is (Section 17.4 of *Numerical Optimization* [5])

$$L_A(x, \lambda, \mu) = f(x) + \sum_{i=1}^{m} \Psi(c_i(x), \lambda_i, \mu) \tag{3}$$

where $\lambda = (\lambda_1, \cdots, \lambda_m)$ are Lagrange multipliers, $\mu > 0$ is the penalty parameter, and

$$\Psi(c_i(x), \lambda_i, \mu) = \begin{cases} \lambda_i c_i(x) + \frac{1}{2\mu} c_i^2(x) & \text{if } c_i(x) + \mu\lambda_i \ge 0 \\ -\frac{\mu}{2}\lambda_i^2 & \text{otherwise} \end{cases} \tag{4}$$

Thus, the augmented Lagrangian method is

Given $\mu_1 > 0$, $r \in (0,1)$, starting point x^1 and λ^1

for $k = 1, \cdots, k_{max}$

Find a minimizer $x^{k'}$ of $L_A(x, \lambda^k, \mu_k)$, starting at x^k

if a solution satisfying the requirement is found, stop with $x^{k'}$

Update Lagrange multipliers: $\lambda_i^{k+1} = \max(\lambda_i^k + c_i(x^{k'})/\mu_k, 0)$

$\mu_{k+1} = r * \mu_k$, $x^{k+1} = x^{k'}$

end

In our experiments, we use fminunc to find a minimizer of $L_A(x, \lambda, \mu)$ in each iteration. For nonlinear constrained problems, the augmented Lagrangian method may not be able to find a feasible solution. When that happens, we restart it from another random point.

2.3 Sequential Quadratic Programming Method

Sequential quadratic programming (SQP) methods are among the most effective methods for nonlinearly constrained optimization. The essential idea of SQP is to model the constrained problem at each iteration as a quadratic programming subproblem and to use the minimizer of this subproblem to define a new iteration. One example is to apply Newton's method to the Karush-Kuhn-Kucker optimality conditions for the constrained problem. This iteration is repeated until a constrained locally optimal solution is found.

The SQP solver used in our experiments is fmincon from the Matlab Optimization Toolbox. An fmincon iteration consists of three main stages: updating of the Hessian matrix of the Lagrangian function, quadratic programming problem solution, and line search and merit function calculation. When fmincon fails to find a feasible solution, we restart it from another random point in the experiments.

2.4 Cooperative Solvers

To take advantage of the respective strengths of multiple methods, various forms of meta-heuristics have been proposed to combine them. One popular approach is to combine global and local search methods, in which local search may follow global search in sequence, or local search may be used as a refinement component in global search. The goal of the former strategy is to let the global solver move towards feasible regions or global optimal solutions, from which the local solver converges quickly to local minima [6]. The second strategy is for the global solver to use a local solver to improve its intermediate points to true local optima [7].

Another approach is to combine different types of methods, such as penalty methods and SQP methods. This is the approach we take in this paper. In our experiments on constraint satisfaction problems, we found that the static penalty method can find solutions with small penalty values very quickly, even for highly constrained problems. However, it has difficulty in finding a solution that satisfies all constraints simultaneously. On the other hand, the SQP method are very effective when starting from an initial points close to the feasible region. Therefore, we propose a cooperative solver (called *SPM+SQP*) that combines the *static penalty method* and the *SQP method* to solve CSPs as follows:

1. Run the static penalty method from a starting point x_s and obtain a solution x'. If x' is a feasible solution, then stop.
2. Run the SQP method from x' and obtain a solution x''. If x'' is a feasible solution, then stop.
3. Generate a new random starting point and go to step 1.

In a similar way, we propose cooperative solvers *DPM+SQP* and *ALM+SQP* that combine the *dynamic penalty method* or the *augmented Lagrangian method* with the *SQP method*, respectively.

The procedure of SPM+SQP is slightly different for COPs as follows:

1. Run the static penalty method from a starting point x^s to minimize constraint violation and obtain a solution x'.
2. Run the SQP method from x' and obtain a solution x''. If x'' is a feasible local minimum, then stop.
3. Generate a new random point and go to step 1.

The difference is that for COPs, SQP is always run to optimize the objective since the static penalty method only finds a feasible solution. Another variation of the procedure is to let the static penalty method minimize the penalty function in Eq. (2) that includes the original objective.

3 Test Problems

Since our performance analysis is experimental in nature, it is important to choose appropriate problem ensembles for testing algorithms, as is done in the study of discrete problems [2]. Regarding continuous constraint problems, there is an unfulfilled need for generic benchmark problem sets, in spite of the work in the mathematical programming and evolutionary computing communities. In the mathematical programming field, test problem sets such as CUTE (Constrained and Unconstrained Testing Environment) [8] and COPS

(Constrained Optimization ProblemS) [1] provide very limited ability to generate problems with different characteristics by changing problem parameters. In the evolutionary computing field, a flexible test-case generator proposed recently [9] has the major limitation that it defines a landscape that consists of a collection of functions that are only piece-wise continuous and are defined on different subspaces of equal size. Because derivatives do not exist on the boundary of two subspaces, methods requiring continuous derivatives do not work well on these problems.

To study constraint problems of different sizes with various numbers and types of constraints, we need a scalable test suite, in which many of the problem parameters can be changed easily. It should allow one to easily compare different optimization techniques developed in different fields. It should be useful in investigating the relationships between various characteristics (e.g., constraint ratio) and complexity of constraint problems, as well as the computational complexity of various solving techniques.

In this section, we present a flexible test-case generator for continuous constraint problems. It is based on trigonometric functions and generates constraint problems similar to those appearing in robot control applications. It also borrows ideas from randomly generated SAT problems and generates constraints using either a fixed-clause-length model or a constant-density model [3]. In the fixed-clause-length model, each constraint has a fixed number of variables that are randomly chosen. In contrast, in the constant-density model, each variable may appear in a constraint with some equal probability.

Our test-case generator $G(n, m, k_s, k_p, p_s, p_p, t)$ is mainly parameterized to generate different forms of constraints, with optional objectives. In this paper, the objective $f(x)$ of the test cases is either 0 or a quadratic function

$$\min_{x_i} f(x) = \sum_{i=1}^{n} (x_i - a_i)^2 \tag{5}$$

where a_i is either a constant or a random number. When $f(x)$ is 0, the problem becomes a constraint satisfaction problem.

The constraints of the test cases are generated according to a single base function with random coefficients. For the experiments in this paper, the base function of the test cases is a nonlinear function, in the form of sums of products of sine and cosine functions:

$$\frac{1}{\sum_{i=1}^{NS} \prod_{j=1}^{NP_i} c_{ij}} \sum_{i=1}^{NS} \prod_{j=1}^{NP_i} c_{ij} \sin\left(2\pi x_{r_{ij}} + 2\pi\delta_{ij}\right) \leq \theta \tag{6}$$

where $r_{ij} \in \{1, \ldots, n\}$ is the index of a randomly chosen variable; $-1 \leq c_{ij} \leq 1$ is a random number specifying the magnitude of the sine function; $0 \leq \delta_{ij} \leq 1$

is a random number specifying the phase; NS is the number of sum terms; NP_i is the number of product terms in each sum term; and $-1 \leq \theta \leq 1$ specifies the threshold for satisfying the constraint.

Based on constraints defined in Eq. (6), the parameters of the generator $G(n, m, k_s, k_p, p_s, p_p, t)$ are the number of variables (n), the number of constraints (m), the maximum number of sum terms in a constraint ($k_s, k_s \geq 1$), the maximum number of product terms in any sum term of a constraint ($k_p, k_p \geq 1$), the probability determining the number of sum terms (up to k_s) in a constraint (p_s), the probability determining the number of product terms (up to k_p) in a sum term (p_p), and the average tightness of the constraints ($t, 0 \leq t \leq 1$). The tightness t of a (continuous) constraint is defined as the ratio of the size of the space made infeasible by the constraint to the size of the total problem space. For example, $t = 0$ means that the constraint leaves the entire problem space feasible, and $t = 1/2$ means that it makes half of the problem space infeasible. The parameter θ in Eq. (6) is determined based on the tightness t. The θ value that achieves the average tightness specified by t is found experimentally through random sampling.

All the parameters of the generator can be changed independently. Constraints are generated with randomly chosen coefficients. By changing the tightness parameter t we can control the difficulty of the constraints, which in turn affects the complexity of the whole problem.

4 Experimental Results

In the experiments, the test cases are generated by $G(n, m, 3, 3, 1, 1, t)$. A simple bound $-1 \leq x \leq 1$ is imposed. To guarantee a randomly generated problem has feasible solutions, we let the origin be a feasible point of all its constraints. When we generate a constraint for which the origin is not a feasible point, the constraint is simply discarded.

Since the constraint ratio has been identified as a key indicator of complexity for discrete CSPs, in this paper we compare the performance of solvers on continuous problems with varying constraint ratios. For each constraint ratio, 100 random instances are generated and the performance of the solvers is compared based on the median numbers of function evaluations. We use medians instead of means because the means of the number of function evaluations are heavily influenced by a small number of extremely large values. Our interest is in what the majority of instances from the distribution are like, not in the unusual or extreme cases. As the median is more robust in the presence of such outliers, it appears to be a more informative statistic for our current purpose.

4.1 *Results for CSPs*

This set of experiments is based on CSPs with $n = 25$ and 50, $t = 1/4$ and $1/8$, and varying m. The initial and restart points are generated with the following pre-sampling scheme: n random points are sampled within the bounds of the constrained problems, i.e., $[-1, 1]^n$, and the best of them is chosen as the initial or restart point. The best sample point is the one with the smallest constraint violation, which may be close to or even inside a feasible region. The pre-sampling scheme improves the solvers' performance over just picking a single random point within the bounds [6].

Four individual methods (SQP, static penalty, dynamic penalty, and augmented Lagrangian method) and three cooperative solvers (SPM+SQP, DPM+SQP, and ALM+SQP) are used to solve the problems. In the static penalty method, $\mu = 1/2$. In both the dynamic penalty method and the augmented Lagrangian method, $k_{max} = 5$, $\mu_1 = 10$, and $r = 0.1$. $\lambda^1 = 0$ in the augmented Lagrangian method. All methods run using only constraint and/or objective values, without user supplied gradients.

Figure 1 shows the numbers of function evaluations and restarts used by the four individual methods in solving 25 and 50 variable instances of tightness $t = 1/4$ or $1/8$. Their complexity behaviors are quite different. In terms of number of restarts, they all have a phase transition from flat, or 1, to exponential. The SQP and the static penalty method have similar transition points, whereas the dynamic penalty and augmented Lagrangian method have similar transition points, which are much larger than the ones of the SQP and the static penalty method. In the exponential phase, the number of restarts used by the static penalty method increases at a much slower rate than that of SQP, but much larger than those of the dynamic penalty and the augmented Lagrangian method. In terms of number of function evaluations, they also have phase transitions from polynomial to exponential, corresponding to their phase transitions in number of restarts. For problems of smaller constraint ratios, SQP uses the fewest number of function evaluations, while the other three methods are similar. On the other hand, for problems of larger constraint ratios, SQP is the worst, the static penalty method is in the middle, and the augmented Lagrangian and the dynamic penalty method are much better. Similar results were obtained on 50 variable instances. For these methods, when the tightness parameter doubles, i.e., from 1/8 to 1/4, the problems become harder and the number of function evaluations needed is nearly doubled.

Figure 1 also shows the performance improvement of the cooperative solver SPM+SQP over the individual methods on highly constrained prob-

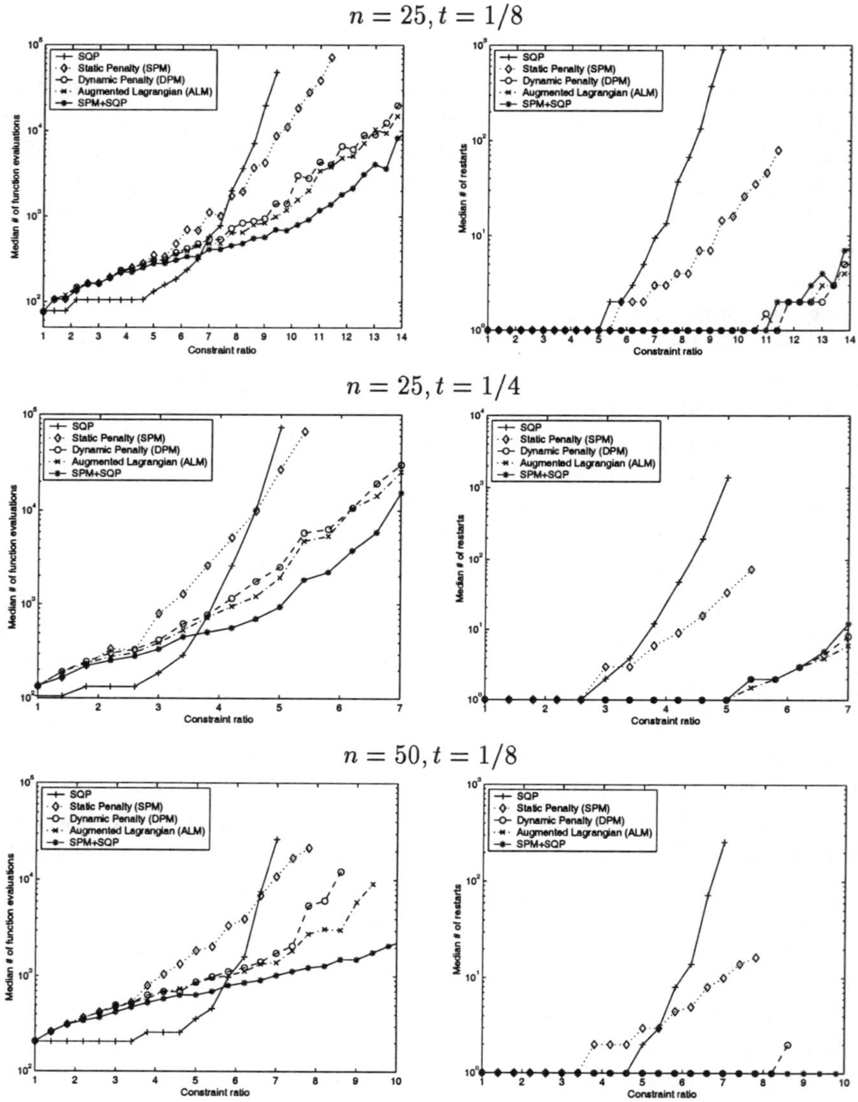

Figure 1. Complexity of four individual methods and one cooperative solver SPM+SQP in solving CSPs of 25 and 50 variables. The constraint tightness parameter t is 1/8 or 1/4. SPM+SQP is the best on highly constrained problems.

lems. Compared with the best individual method, the augmented Lagrangian method, SPM+SQP has about the same number of restarts, but uses much fewer function evaluations.

Similarly we have tried two other cooperative solvers, DPM+SQP and ALM+SQP. They improve over the dynamic penalty method and the augmented Lagrangian method, respectively, on highly constrained problems by reducing the number of restarts needed and the number of function evaluations.

4.2 Results for COPs

In this section, we show results on COPs with a simple quadratic objective function in Eq. (5), where each a_i is a random number in [-0.5, 0.5]. The constraints are generated in the same way as in the last section.

Figure 2 compares the complexity of SQP and SPM+SQP on COPs with increasing constraint ratios. In terms of function evaluations, their cost curves have a up-down-up shape. One explanation is that the initial up phase happens when feasible regions are large and easy to find. The solvers spend a long time on optimizing the objective. As the constraint ratio increases, the boundaries of feasible regions become more complicated, requiring more effort in finding a feasible local minimum. In the down phase, feasible regions get fewer and smaller, and their boundaries may become simpler. It becomes easier to find a feasible local minimum. No restart is necessary so far. In the second up phase, feasible regions become rare and hard to find. The solvers used increasingly more restarts.

The cost curves of the two solvers are almost identical on problems with small constraint ratios. The cooperative solver is slightly better on problems with large constraint ratios. They are similar in number of restarts.

5 Conclusions

In this paper, we study the complexity of several representative methods for continuous constraint problems and obtain various novel observations. By combining two methods, we propose some cooperative solvers that improve over existing methods on harder problems by reducing the number of restarts. We show that problem parameters such as constraint ratio and tightness are important indicators of the difficulty of the constraint problems. Based on these parameters, adaptive solvers that select appropriate methods dynamically and automatically may achieve the best-possible performance for the problem at hand.

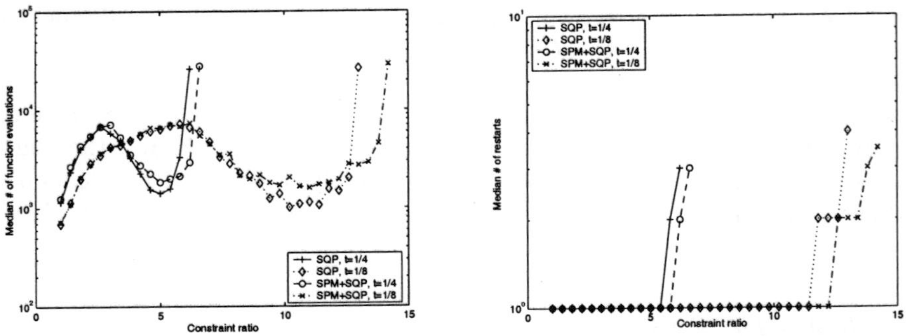

Figure 2. Comparison of SQP and SPM+SQP on 25-variable COPs, where the tightness parameter $t = 1/8$ or $1/4$.

References

1. A. S. Bondarenko, D. M. Bortz, and J. J. Moré. COPS: Large-scale nonlinearly constrained optimization problems. Technical Memorandum ANL/MCS-TM-237, Argonne National Laboratory, Argonne, Illinois, 1998.
2. Tad Hogg, Bernardo A. Huberman, and Colin Williams. Phase transitions and the search problem. *Artificial Intelligence*, 81:1–15, 1996.
3. Bart Selman, David Mitchell, and Hector J. Levesque. Generating hard satisfiability problems. *Artificial Intelligence*, 81:17–29, 1996.
4. M. Yokoo. Why Adding More Constraints Makes a Problem Easier for Hill-Climbing Algorithms: Analyzing Landscapes of CSPs. In *Proc. of CP'97, number 1330 in LNCS*, pages 357–370. Springer-Verlag, 1997.
5. J. Nocedal and S. J. Wright. *Numerical Optimization.* Springer, 1999.
6. Y. Shang, M. P. J. Fromherz, T. Hogg, and W. B. Jackson. Complexity of continuous, 3-SAT-like constraint satisfaction problems. In *Proc. IJCAI-01 Workshop on Stochastic Search Algorithms*, pages 49–54, Seattle, WA, 2001.
7. W. E. Hart. *Adaptive global optimization with local search.* PhD thesis, University of California, San Diego, 1994.
8. CUTE. Constrained and unconstrained testing environment, 2001. http://www.cse.clrc.ac.uk/Activity/CUTE.
9. Z. Michalewicz, K. Deb, M. Schmidt, and T. Stidsen. Test-case generator for constrained parameter optimization techniques. *IEEE Transactions on Evolutionary Computation*, 4(3):197–215, September 2000.

PSEUDO-DIVISION MACHINE (I): A MODEL OF SYMBOLIC COMPUTATION

SHE ZHIKUN, XIA BICAN AND ZHENG ZHIMING
School of Mathematical Sciences, Peking University, Beijing 100871, China
E-mail: zkshe@mail1.math.pku.edu.cn, xbc@math.pku.edu.cn, zzheng@pku.edu.cn

Through careful study of the theory of real Turing machine and the algorithm of pseudo-division, we define a new machine, called pseudo-division machine, which is a formal model of computation over ring. Especially, it applies to symbolic computation and it represents algorithms naturally. Some fundamental concepts are defined and some primary results presented.

1 Introduction

The study of computability and computational complexity originated in the work of logicians such as Gödel, Turing, Church, Kleene, and Post in the 1930s. The model of computation that developed in the following decades, the Turing machine, has been extraordinarily successful in giving the foundations and framework for theoretical computer science [3,6]. Based on the point of view that "the Turing model with its dependence on 0s and 1s is fundamentally inadequate for giving such a foundation to the theory of modern scientific computation, where most of the algorithms are real number algorithms", L. Blum, M. Shub and S. Smale [1,2] presented a formal model of real computation , which is now called "the real Turing machine" or "the BSS machine". Their theory preserves the Turing theory as a special case. Furthermore, by providing a bridge between complexity theory and the fields of analysis, geometry, and topology, the real Turing machine may give us possibilities for new attack on the classical problems of complexity theory, which heretofore have been attacked using only the tools of logic and combinatorics.

The computation on computers includes two aspects: numerical computation and symbolic computation. The study of symbolic computation originated in the 1950s and has gained wide interests since the 1970s. In 1978, Wen-tsün Wu [8,9] proposed a decision procedure for proving geometry theorems of "equality type", i.e. the hypotheses and conclusions of the statements consist of polynomial equations only. This is a very efficient method for mechanically proving elementary geometry theorems (of equality type) [5]. Now, Wu's method has been one of the most famous algorithms in the field of symbolic computation. On the other hand, the complexity of Wu's method has rarely been studied. To discuss the computation complexity of Wu's method, as well as many other famous algorithms such as Gröbner basis method [4], from the viewpoint of theoretical computer science is our motivation.

The point of view of this paper is that the symbolic computation, which is over the objects having some basic algebraic structures, is quite different from the numerical computation, which is over numbers. Therefore, although the BSS machine, by its definition, would apply to computation over any ring or field, it is not suitable for our study of the complexity of Wu's method. For example, the pseudo-division of two polynomials in several variables with respect to one main variable can not be represented in the frame of the BSS machine naturally. We need a new formal model that would describe the essential properties of symbolic computation better. The goal of our work is to develop a formal theory of symbolic computation so that we can study the complexity of algorithms defined under the framework. The work presented in this paper is our primary results.

2 Pseudo-Division

Let $\mathbb{D}$ be a *unique factorization domain* (UFD), A be a polynomial in $\mathbb{D}[x_1, ..., x_n]$ and x_k a fixed variable. While considered as a polynomial in x_k, A can be written as

$$\begin{aligned} A &= A_0 x_k^a + A_1 x_k^{a-1} + \cdots + A_a, \\ A_i &\in \mathbb{D}[x_1, ..., x_{k-1}, x_{k+1}, ..., x_n], \end{aligned} \tag{1}$$

where a is the degree of A in x_k and denoted by $\deg(A, x_k)$, A_0 is the *leading coefficient* of A in x_k, denoted by $\mathrm{lc}(A, x_k)$.

Let A and B be two polynomials in $\mathbb{D}[x_1, ..., x_n]$ and $B \neq 0$, $b = \deg(B, x_k)$, $a = \deg(A, x_k)$. For pseudo-dividing A by B, considered as polynomials in x_k, we have a division algorithm as follows [7]. Let $R = A$; repeat the following process until $r = \deg(R, x_k) < b$: $R \leftarrow B_0 R - R_0 x_k^{r-b} B$, where $R_0 = \mathrm{lc}(R, x_k)$. Finally, one obtains two polynomials Q and R in $\mathbb{D}[x_1, ..., x_n]$ satisfying the relation

$$I^s A = QB + R, \tag{2}$$

where $I = \mathrm{lc}(B, x_k)$, $s = \max(a - b + 1, 0)$, $\deg(R, x_k) < b$. In case $b = 0$, $R = 0$ and $Q = B^a A$.

Remark 1. *Two* standard *ways to represent polynomials are* dense *representation (where every monomial is specified) and* sparse *representations (which disregard terms whose coefficient is* 0*). For pseudo-division, polynomials are regarded as univariate and every term is specified.*

Let A and B be two polynomials in the form of Eq. (1) and denoted by $A = (A_0, ..., A_a)$ and $B = (B_0, ..., B_b)$, respectively. From the viewpoint of fraction-free Gauss elimination, we have a **prem** algorithm as follows. Let $(P, Q, t) \leftarrow (A, B, a - b)$; repeat the following two steps until $t < 0$:

1. $(P, Q, t) \leftarrow (g(P, Q), Q, t - 1)$, where $P \leftarrow g(P, Q)$ means
$$P_i \leftarrow Q_0 P_i - P_0 Q_i,\ 0 \leq i \leq b + t,\ Q_j = 0 \text{ if } j > b; \tag{3}$$
2. $(P, Q, t) \leftarrow \text{ShiftL}(P, Q, t)$, which means $P_i \leftarrow P_{i+1}\,(0 \leq i \leq b + t - 1)$, $Q_j \leftarrow Q_j$, $t \leftarrow t$.

Return P while $t < 0$. We may represent the **prem** algorithm schematically as in Fig. (1).

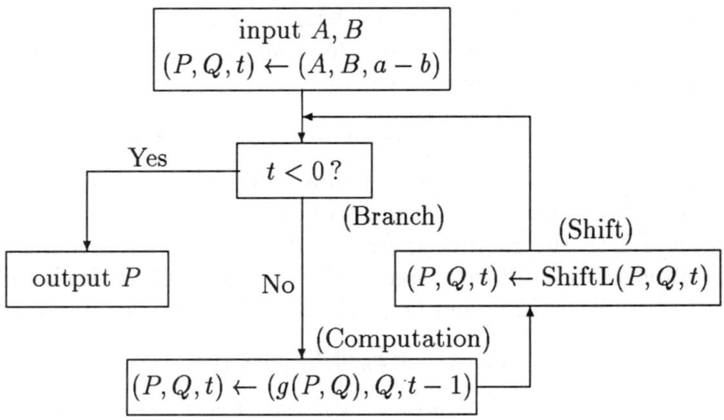

Figure 1. A Machine For Pseudo-Division

3 The Pseudo-Division Machine

Suppose R be a ring. We list some notations defined in [1]. Let

$$R^\infty = \bigsqcup_{n \geq 0} R^n,$$

where for $n > 0$, R^n is the standard n-dimensional space over R and R^0 is the 0-dimensional space with just one point $\mathbf{0}$. For $x \in R^n \subset R^\infty$, we call

n the *length* of x. We denote by R_∞ the *bi-infinite direct sum* space over R. Elements of R_∞ have the form

$$x = (..., x_{-2}, x_{-1}, x_0.x_1, x_2, ...),$$

where $x_i \in R$ for all integers i, $x_k = 0$ for $|k|$ sufficient large, and . is a distinguished marker between x_0 and x_1. In fact, R_∞ can be viewed as a work tape of a machine. Suppose $h : R^m \mapsto R$ is a polynomial (or rational) function of degree d over R. Then h defines a polynomial (or rational) function $\widehat{h} : R_\infty \mapsto R$ on R_∞ of dimension m and degree d by letting $\widehat{h}(x) = h(x_1, ..., x_m)$ for each $x \in R_\infty$. Suppose $g_i : R^m \mapsto R$, $i = 1, ..., m$, are polynomial (or rational) functions of maximum degree d over R. Then the g_i, $i = 1, ..., m$, defines a polynomial (or rational) map on R_∞

$$\widehat{g} : R_\infty \mapsto R_\infty$$

of dimension m and degree d by letting $(\widehat{g}(x))_i = \widehat{g_i}(x)$, $i = 1, ..., m$ and $(\widehat{g}(x))_i = x_i$ for $i < 1$ or $i > m$. The space R_∞ has natural *shift* operations, *shift left* σ_l and *shift right* σ_r, where

$$\sigma_l(x)_i = x_{i+1} \text{ and } \sigma_r(x)_i = x_{i-1}. \tag{4}$$

In order to define pseudo-division machine, we define the *state space* as $R_\infty \times \mathbb{Z}^{k_M}$, where k_M is a positive integer. We call R_∞ the *computation space* and $\mathbb{Z}^{k_M}$ the *counter space*. That is to say, our machine will have two work tapes. We relate the space R^∞ and the state space by defining maps $I_\infty : R^\infty \mapsto R_\infty \times \mathbb{Z}^{k_M}$ and $O_\infty : R_\infty \times \mathbb{Z}^{k_M} \mapsto R^\infty$ as follows.

$$I_\infty(x) = (..., 0, 0.x_1, ..., x_n, 0, 0, ...) \times (v_1, ..., v_{k_M}) \text{ for } x \in R^n, \tag{5}$$

where every v_i, $i = 1, ..., k_M$, is a linear function in n, or, if x belongs to a product space, is a linear function in the dimensions of all spaces in the product space, and

$$O_\infty((..., x_0.x_1, ...) \times (w_1, ..., w_{k_M})) = \begin{cases} 0 \in R^0 & \text{if } w_1 = 0 \\ (x_1, ..., x_{w_1}) \in R^{w_1} & \text{otherwise.} \end{cases} \tag{6}$$

Definition 1. *A pseudo-division machine M over R consists of a finite directed connected graph with five types of nodes: input, computation, shift, branch and output. The unique input node has no incoming edges and only one outgoing edge. All other nodes have possibly several incoming edges. Computation and shift nodes have only one outgoing edge, branch nodes exactly two, Yes and No, and output nodes none.*

In addition, the machine has three spaces: an input space $\mathcal{I}_M = R^\infty$, state space $\mathcal{S}_M = R_\infty \times \mathbb{Z}^{k_M}$, and output space $\mathcal{O}_M = R^\infty$, where k_M is a positive *integer. Associated with each node of the graph are maps of these spaces and next node assignments.*

I. *Associated with the input node is the map I_∞ defined by Eq. (5) and a unique next node β_1.*

II. *Each computation node η has an associated computation map, $g_\eta : \mathcal{S}_M \longmapsto \mathcal{S}_M$, and a unique next node β_η. The computation map g_η is defined as follows.*

$$g_\eta = (p_\eta \circ \pi_1, f_\eta) : R_\infty \times \mathbb{Z}^{k_M} \longmapsto R_\infty \times \mathbb{Z}^{k_M}$$
$$y = (x, v) \quad \mapsto \quad (p_\eta \circ \pi_1(y), f_\eta(y)) = (p_\eta(x), f_\eta(y))$$

where $p_\eta(x) : R_\infty \mapsto R_\infty$ is a polynomial (or rational) function on R_∞, f_η is a map from the state space to the counter space and $\pi_1 : \mathcal{S}_M \to R_\infty$ is the projection onto R_∞.

III. *Associated with each shift node η is a map $g_\eta \in \{\widehat{\sigma}_l, \widehat{\sigma}_r\}$ of the state space to itself and a unique next node β_η. The shift maps $\widehat{\sigma}_l$ and $\widehat{\sigma}_r$ are defined as follows. For $(x, v) \in \mathcal{S}_M$,*

$$\widehat{\sigma}_l(x, v) = (\sigma_l(x), v) \text{ and } \widehat{\sigma}_r(x, v) = (\sigma_r(x), v),$$

where σ_l and σ_r are defined by Eq. (4).

IV. *Each branch node η has an associated branching function $h_\eta : \mathcal{S}_M \longmapsto \mathbb{Z}$. The next node along the Yes outgoing edge, β_η^+, is associated with the condition $h_\eta(y) \geq 0$ and the next node along the No outgoing edge, β_η^-, with the condition $h_\eta(y) < 0$.*

V. *Finally, each output node η has an associated map O_∞ defined by Eq. (6).*

Remark 2. *The* dimension K_M *and* degree D_M *of M are, respectively, the maximum of the dimensions and the degrees of all maps associated with its computation and branch nodes.*
Remark 3. *We can test if $h(y) = 0$ by a subroutine that uses the comparisons $h(y) \geq 0$ and $-h(y) \geq 0$ (see Fig. (2))*
Remark 4. *For convenience, we assume all branch nodes are* standard, *that is, that $h_\eta(y) = v_{i_\eta}$ for some $1 \leq i_\eta \leq k_M$ for each branch node η and $y = (x, v) \in \mathcal{S}_M$. It is easy to see that this is without loss of generality of our theory.*
Remark 5. *It is sometimes convenient to define computations that involve non-positive coordinates. And this can be easily achieved by combining some shift operations with computations.*

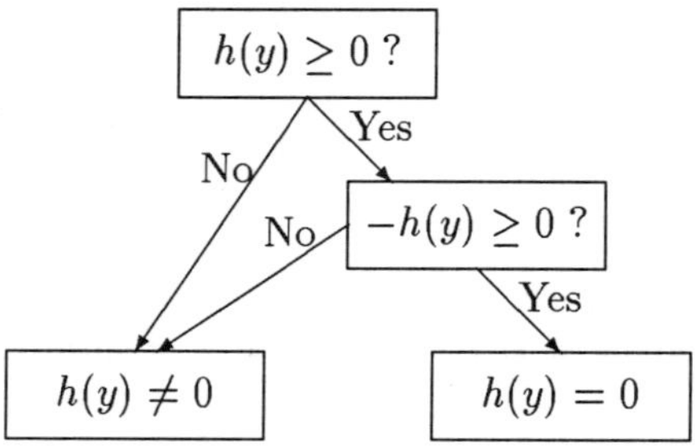

Figure 2. Testing for Equality

Remark 6. *It is often convenient to stipulate that a machine M over R has a unique output node. We then can identify the nodes of the machine with the labels $1, \ldots, N$ where 1 denotes the input node and N the output node.*
Remark 7. *It's not difficult to see that a classical Turing machine is a pseudo-division machine over $\mathbb{Z}_2$.*

4 Relations to the BSS Machines

In this section, we will demonstrate by two examples that the BSS machine can be simulated by the pseudo-division machine defined in last section. For all the definitions and notations related to the BSS machine, please refer to [1].
Example 1. *Let R be $\mathbb{R}$ or $\mathbb{Q}$, that is, the field of the real numbers or rational numbers. Let $M^{(1)}$ be a BSS machine in its* normal form *over R. We construct a pseudo-division machine $M^{(2)}$ over R as follows.*

I. *The input spaces are both R^∞. Let $k_{M^{(2)}} = 1$. For any $x \in R^n \subseteq R^\infty$, the input map of $M^{(1)}$ is $I_\infty^{(1)}(x) = (\ldots, 0, 0, \overbrace{1, \ldots, 1}^{n}.x_1, \ldots, x_n, 0, 0, \ldots)$. Let the input map of $M^{(2)}$ be $I_\infty^{(2)}(x) = ((\ldots, 0.x_1, \ldots, x_n, 0, 0, \ldots), n)$. Furthermore, let the input node of $M^{(2)}$ be followed immediately by a computation node η with $g_\eta((\ldots, 0.x_1, \ldots, x_n, 0, 0, \ldots), n) = ((\ldots, 0, 0, \overbrace{1, \ldots, 1}^{n}.x_1, \ldots, x_n, 0, 0, \ldots), n)$.*

II. *For each computation node $\eta^{(1)}$ in $M^{(1)}$, suppose $g_{\eta^{(1)}}$ be its associated polynomial (or rational) map, construct a computation node $\eta^{(2)}$ in $M^{(2)}$ with associated map $g_{\eta^{(2)}}$ defined as $g_{\eta^{(2)}}(x, v) = (g_{\eta^{(1)}}(x), v)$.*

III. *For each shift node in $M^{(1)}$, construct a shift node in $M^{(2)}$ with the same* *shift operation.*

IV. *For each branch node* $\eta^{(1)}$ *in* $M^{(1)}$*, we construct a computation node* $\eta_1^{(2)}$ *and a branch node* $\eta_2^{(2)}$ *in* $M^{(2)}$ *as follows. Suppose* $h_{\eta^{(1)}}$*, a polynomial map, be the branching function associated with* $\eta^{(1)}$*, we define the associated map with* $\eta_1^{(2)}$ *to be* $g_{\eta_1^{(2)}}(x,v) = (x, \text{sign}(h_{\eta^{(1)}}(x)))$*, where, for any* $a \in R$*,* $\text{sign}(a)$ *is* 1 *if* $a > 0$*;* 0 *if* $a = 0$*;* -1 *if* $a < 0$*. And we define the next node to be* $\eta_2^{(2)}$*. As to the branch node* $\eta_2^{(2)}$*, let the associated branching function* $h_{\eta_2^{(2)}}(y) = v_1$*.*

V. *For the output node* $N^{(1)}$ *in* $M^{(1)}$*, we construct a computation node* $\eta_{N^{(2)}}$ *and the output node* $N^{(2)}$ *in* $M^{(2)}$ *as follows.* $g_{\eta_{N^{(2)}}} : (x,v) \mapsto (x, f(x))$*, where* $f(x) = \min_{i\geq 0}\{x_{-i} = 0\}$*. And the next node is* $N^{(2)}$*. The output map of* $N^{(2)}$ *is defined by Eq. (6).*

It's easy to see that $M^{(2)}$ *simulates* $M^{(1)}$*.*

Remark 8. *If* R *is an ordered ring and* $M^{(1)}$ *a BSS machine over* R *with branching condition* $h(x) < 0$*, we can construct a pseudo-division machine over* R *which simulates* $M^{(1)}$ *by doing the same procedure as above and letting* $\text{sign}(a) : R \mapsto \mathbb{Z}$ *to be defined as above.*

Example 2. *Let* R *be* $\mathbb{C}$ *or* $\mathbb{Z}_p$ *where* p *is a positive integer , that is, the field of the complex numbers or the ring of integers modulo* p*. Let* $M^{(1)}$ *be a BSS machine over* R *with the branching condition* $h(x) = 0$ *, We can construct a pseudo-division machine* $M^{(2)}$ *over* R *to simulate* $M^{(1)}$ *by almost the same procedure as in last example except replacing* $\text{sign}(a)$ *by* $h^B : R \mapsto \mathbb{Z}$*, where* $h^B(a)$ *is 0 if* $a = 0$*; 1 otherwise.*

For $m > 0$ let $\mathcal{S}_m = \{((x_{-m+1}, ..., x_0, x_1, ..., x_m), v) \mid x_i \in R \text{ and } v \in \mathbb{Z}^{k_M}\} = R^{2m} \times \mathbb{Z}^{k_M}$. Naturally, we have the injections $\widetilde{i} : \mathcal{S}_m \longmapsto \mathcal{S}$ and $\widetilde{i_1} : R^{2m} \longmapsto R_\infty$, and the projections $\widetilde{\pi} : \mathcal{S} \longmapsto \mathcal{S}_m$, $\widetilde{\pi_1} : \mathcal{S}_m \longmapsto R^{2m}$ and $\widetilde{\pi_1'} : R_\infty \longmapsto R^{2m}$, defined, respectively, by

$$\widetilde{i}((x_{-m+1}, ..., x_0, x_1, ..., x_m), v) = ((..., 0, x_{-m+1}, ..., x_0.x_1, ..., x_m, 0, ...), v),$$

$$\widetilde{i_1}((x_{-m+1}, ..., x_0, x_1, ..., x_m)) = (..., 0, x_{-m+1}, ..., x_0.x_1, ..., x_m, 0, ...),$$

$$\widetilde{\pi}(x, v) = ((x_{-m+1}, ..., x_0, x_1, ..., x_m), v),$$

$$\widetilde{\pi_1}((x_{-m+1}, ..., x_0, x_1, ..., x_m), v) = (x_{-m+1}, ..., x_0, x_1, ..., x_m),$$

and $\widetilde{\pi_1'}(x) = (x_{-m+1}, ..., x_0, x_1, ..., x_m)$. Then, we associate with the input map I_∞ the *modified input map* $\widetilde{I_\infty} = \widetilde{\pi} \circ I_\infty : \mathcal{I}_M \longmapsto \mathcal{S}_m$, and with each

$g_\eta = (p_\eta \circ \pi_1, f_\eta)$ the *modified computation map* $\widetilde{g_\eta} = (\widetilde{p_\eta} \circ \widetilde{\pi_1}, \widetilde{f_\eta}) : \mathcal{S}_m \longmapsto \mathcal{S}_m$, where $\widetilde{p_\eta} = \widetilde{\pi_1' \circ p_\eta \circ i_1} : R^{2m} \longmapsto R^{2m}$, $\widetilde{f_\eta} = f_\eta \circ \widetilde{i} : \mathcal{S}_m \longmapsto \mathbb{Z}^{k_M}$.

Now suppose γ is a computation path $\eta^0, \eta^1, ..., \eta^k, ...$ At each step k in the path γ, M evaluates a map $\widetilde{P}_{\gamma(k)} : \mathcal{I}_M \longmapsto R^{2m}$ defined by $\widetilde{P}_{\gamma(k)} = \widetilde{p_{\eta^k}} \circ \cdots \circ \widetilde{p_{\eta^0}} \circ \widetilde{\pi_1} \circ \widetilde{I_\infty}$, and a map $\widetilde{F}_{\gamma(k)} : \mathcal{I}_M \longmapsto \mathbb{Z}^{k_M}$ defined by $\widetilde{F}_{\gamma(k)} = \widetilde{f_{\eta^k}} \circ \widetilde{g_{\eta^{k-1}}} \cdots \circ \widetilde{g_{\eta^0}} \circ \widetilde{I_\infty}$. Restricted to R^n, where $m \geq \max(K_M, n) + k$, $\widetilde{P}_{\gamma(k)}$ is an ordinary polynomial function.

We also associate with the output map O_∞ the *modified output map* $\widetilde{O_\infty} = O_\infty \circ \widetilde{i} : \mathcal{S}_m \longmapsto R^\infty$. Let $\widetilde{G}_{\gamma(k)} = (\widetilde{P}_{\gamma(k)}, \widetilde{F}_{\gamma(k)})$, then, for $\gamma \in \Gamma_T$, $\Phi_M|_{\mathcal{V}_\gamma^n} = \widetilde{O_\infty} \circ \widetilde{G}_{\gamma(T-1)}$. It's easy to see that the input-output map Φ_M, restricted to the n-dimensional component of the path set $\mathcal{V}_\gamma$, is a polynomial map composed with the output map $\widetilde{O_\infty}$.

Likewise, suppose η_k is a standard branch node (see Remark 4), M evaluates the step-k branching function $\widetilde{F}^B_{\gamma(k)} : \mathcal{I}_M \longmapsto \mathbb{Z}$ defined by $\widetilde{F}^B_{\gamma(k)} = \Pi_{j_\eta} \circ \widetilde{F}_{\gamma(k)}$, where $\Pi_{j_\eta} : \mathbb{Z}^{k_M} \longmapsto \mathbb{Z}$ is the projection onto the j_ηth $(1 \leq j_\eta \leq k_M)$ coordinate.

We note that $\mathcal{V}_{\gamma(k)}$ is determined by the branching conditions along the path $\gamma(k)$. Suppose m is large enough, for example $m = \max(K_M, n) + k$, let

$$\mathtt{L}_{\gamma(k)} = \{\widetilde{F}^B_{\gamma(k')} | k' < k,\ k' \text{ a branch step in } \gamma, \text{ and } \eta^{k'+1} = \beta^-(\eta^{k'})\}$$
$$\mathtt{R}_{\gamma(k)} = \{\widetilde{F}^B_{\gamma(k')} | k' < k,\ k' \text{ a branch step in } \gamma, \text{ and } \eta^{k'+1} = \beta^+(\eta^{k'})\}$$

Theorem 1. *For any pseudo-division machine M, the following properties hold.*

(1) *For a computation path $\gamma(k)$,*

$$\mathcal{V}^n_{\gamma(k)} = \{x \in R^n | f(x) < 0, g(x) \geq 0, f \in \mathtt{L}_{\gamma(k)}, g \in \mathtt{R}_{\gamma(k)}\}.$$

(2) *For $\gamma \in \Gamma_T$, $\Phi_M|_{\mathcal{V}_\gamma^n} = \widetilde{O_\infty} \circ \widetilde{G}_{\gamma(T-1)} = \widetilde{O_\infty} \circ (\widetilde{P}_{\gamma(T-1)}, \widetilde{F}_{\gamma(T-1)})$. Because $\widetilde{P}_{\gamma(T-1)}$, restricted to the n-dimensional component of the path set $\mathcal{V}_\gamma$, is a polynomial map, the input-output map Φ_M, restricted to the n-dimensional component of the path set $\mathcal{V}_\gamma$, is a polynomial map composed with the output map $\widetilde{O_\infty}$.*

Remark 9. *For any map $f \in \mathtt{L}_{\gamma(k)} \bigcup \mathtt{R}_{\gamma(k)}$, if it can be expressed as $f = f_1 \circ f_2$, where $f_2 : \mathcal{I}_M \mapsto R$ is a polynomial function and $f_1 : R \mapsto \mathbb{Z}$ satisfies that $f_2(x) <_R 0$ if and only if $f(x) < 0$, then $\mathcal{V}^n_{\gamma(k)}$ is a basic semi-algebraic set. Here, $<_R$ means the ordering in R. So, from Examples 1 and 2, for a pseudo-division machine simulating a BSS machine, we have the same* path decomposition theorem *as in* [1].

5 An Example

In this section, we will give a formal, detailed construction of a pseudo-division machine that performs pseudo-division of two polynomials like the `prem` algorithm in Section 2. And then, we point out some essential differences between our machine and the BSS machine.

Let $\mathbb{D}$ be a UFD, A and B be two polynomials in $\mathbb{D}[u_1, ..., u_s]$ and u_k a fixed variable. We pseudo-divide A by B with respect to u_k. Let $R = \mathbb{D}[u_1, ..., u_{k-1}, u_{k+1}, ..., u_s]$, $k_M = 5$ and $A = (a_1, ..., a_n)$, $B = (b_1, ..., b_m)$, where $a_i, b_j \in R$.

Input $(A, B) = (a_1, ..., a_n, b_1, ..., b_m) \in R^\infty$, let $t = n - m$ and

$$I_\infty(A, B) = (x, v) = ((..., 0.a_1, ..., a_n, b_1, ..., b_m, 0, ...), (n, m, t, n, m - 1)).$$

Then, by a sequence of machine operations, we get a sequence of states:

$$\vdots$$
$$((...0.a_1, 0, a_1, ..., a_n, b_1, ..., b_m, 0, ...), (n, m, t, n, m - 1))$$
$$\vdots$$
$$((...0, a_1, ..., a_n.a_1, b_1, b_1, ..., b_m, 0, ...), (n, m, t, 0, m - 1))$$
$$((...0, a_2, ..., a_n.a_1, b_1, b_1, ..., b_m, 0, ...), (n, m, t, n - 1, m - 1))$$
$$\vdots$$
$$((...0.a_1, b_1, a_2b_1, ..., a_nb_1, b_1, ..., b_m, 0, ...), (n, m, t, 0, m - 1))$$
$$((...0.a_1, b_1, a_2b_1, ..., a_nb_1, b_1, ..., b_m, 0, ...), (n, m, t, n, m - 1))$$
$$\vdots$$
$$((...0, a_2b_1, ..., a_nb_1, b_1.a_1, b_2, b_2..., b_m, 0, ...), (n, m, t, 0, m - 2))$$
$$((...0, a_2b_1, ..., a_nb_1, b_1.a_1, b_2, b_2..., b_m, 0, ...), (n, m, t, n - 1, m - 2))$$
$$\vdots$$
$$((...0, a_2b_1 - a_1b_2.a_1, b_2, a_3b_1, ..., a_nb_1, b_1, b_2..., b_m, 0, ...), (n, m, t, 0, m - 2))$$
$$\vdots$$
$$((..., a_2b_1 - a_1b_2, ..., a_mb_1 - a_1b_m.a_1, b_m, a_{m+1}b_1, ..., a_nb_1, b_1, ..., b_m, ...), (n, m, t, 0, 0))$$
$$\vdots$$
$$((...0.a_1, b_m, a_2b_1 - a_1b_2, ..., a_mb_1 - a_1b_m, a_{m+1}b_1, ..., a_nb_1, b_1, ..., b_m, ...), (n, m, t, 0, 0))$$
$$((...0.a_2^{(1)}, 0, a_2^{(1)}, ..., a_n^{(1)}, b_1, ..., b_m, ...), (n - 1, m, t - 1, n - 1, m - 1)),$$

where

$$a_i^{(1)} = \begin{cases} a_ib_1 - a_1b_i & \text{if } 2 \le i \le m \\ a_ib_1 & \text{if } m < i \le n. \end{cases}$$

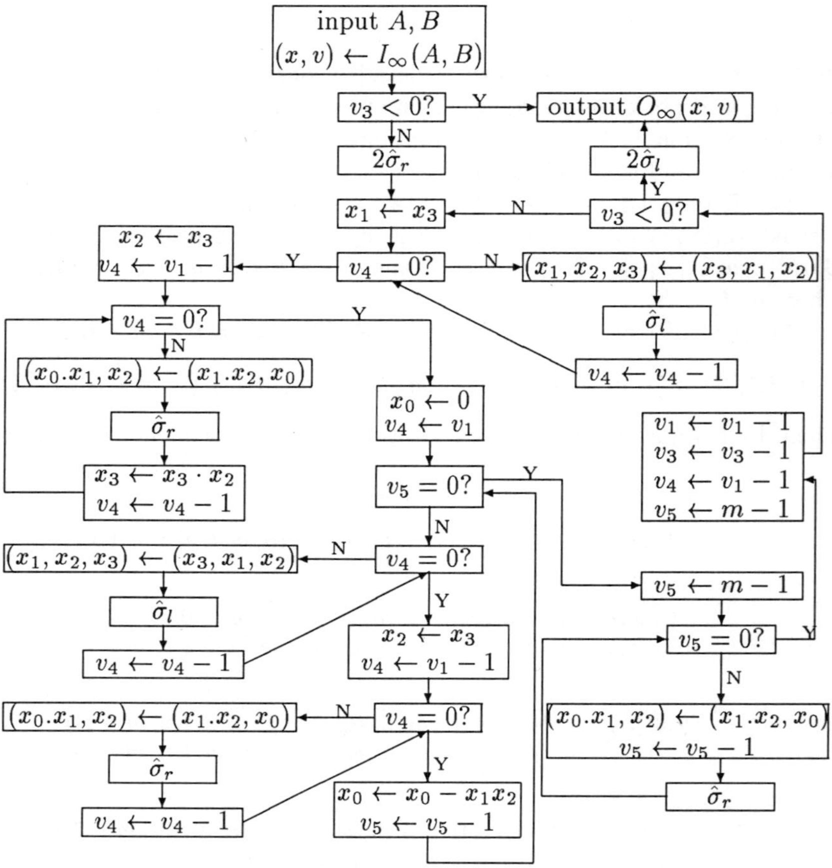

Figure 3. The machine doing pseudo-division

Now proceeding as before, the machine produces in turn the states:

$$((\ldots 0.a_3^{(2)}, 0, a_3^{(2)}, \ldots, a_n^{(2)}, b_1, b_2 \ldots, b_m, 0, \ldots), (n-2, m, t-2, n-2, m-1))$$

$$\vdots$$

$$((\ldots 0.a_{t+2}^{(t+1)}, 0, a_{t+2}^{(t+1)}, \ldots, a_n^{(t+1)}, b_1, \ldots, b_m, 0, \ldots), (m-1, m, -1, m-1, m-1)),$$

where

$$a_i^{(j)} = \begin{cases} a_i^{(j-1)} b_1 - a_1^{(j-1)} b_i & \text{if } i \le m \\ a_i^{(j-1)} b_1 & \text{if } m < i \le n. \end{cases}$$

And finally, after two shift left operations, output the first $m-1$ elements of the computation space, that is $(a_{n-m+2}^{(n-m+1)}, \ldots, a_n^{(n-m+1)})$. See Fig.3.

There are some essential differences between our machine and the BSS machine. First of all, we separate the computations on the ring R from those on $\mathbb{Z}$. When the characteristic of R is $p > 0$, it seems difficult and unnatural for a BSS machine to manage computations over R while the computations in our machine are natural and easy to understand. Secondly, the branching conditions are now tested in $\mathbb{Z}$ other than R. This is necessary in most computations over such rings as rings of polynomials in several variables. For pseudo-division, in any BSS machine one has to simulate the test of branching condition $n - m < 0$ since it cannot be checked under the ordering of $R = \mathbb{D}[u_1, \ldots, u_{k-1}, u_{k+1}, \ldots, u_s]$.

Acknowledgments

The authors are indebted to Professor H. Ganzinger for helpful discussion and to NKBRSF-(G1998030600) for support. Also, one of the authors, Xia, thanks the Max-Planck-Institut für Informatik for hospitality.

References

1. L. Blum, F. Cucker, M. Shub and S. Smale, *Complexity and Real Computation*, Springer-Verlag, New York, 1997.
2. L. Blum, M. Shub and S. Smale, *On a theory of computation and complexity over the real numbers: NP-completeness, recursive functions and universal machines*, Bulletin (New Series) of the American Mathematical Society, 1989, 21(1): 1-46.
3. D. S. Bridges, *Computability*, New York: Springer-Verlag, 1997.
4. B. Buchberger, Gröbner bases: An algorithmic method in polynomial ideal theory, In *Multidimensional Systems Theory*, (Edited by N.K. Bose), pp. 184-232, Reidel, Dordrecht, 1985.
5. S. C. Chou, *Mechanical geometry theorem proving*, Reidel, Dordrecht, 1988.
6. M. Garey and D. Johnson, *Computers and Intractabilitiy: A Guide to the Theory of NP-Completeness*, Freedman, 1979.
7. D.M. Wang, *Elimination Methods*, Springer-Verlag, New York, 2001.
8. W. T. Wu, On the decision problem and the mechanization of theorem-proving in elementary geometry, *Sci. Sinica* **21**: 159-172 1978.
9. W.T. Wu, *Mechanical theorem proving in geometries: Basic principles* (translated by X. Jin and D. Wang), Springer, New York, 1994.

XSSR: AN EXPERIMENTAL SYSTEM FOR STRING REWRITING – DECISION PROBLEMS, ALGORITHMS, AND IMPLEMENTATION

DIETER HOFBAUER
FRIEDRICH OTTO
Fachbereich Mathematik/Informatik, Universität Kassel, D-34109 Kassel
E-mail: {hofbauer,otto}@theory.informatik.uni-kassel.de

CHRISTOPH KÖGL
KLAUS E. MADLENER
Fachbereich Informatik, Universität Kaiserslautern, D-67663 Kaiserslautern
E-mail: christoph@familie-koegl.de, madlener@informatik.uni-kl.de

Here we present the *e***X***perimental* **S***ystem for* **S***tring* **R***ewriting*, **XSSR**, which is a special purpose computer algebra system for combinatorial group and semigroup theory. It is intended for effectively performing computations in monoids and groups that are given through finite presentations in terms of generators and defining relations. This system, which has been developed in a joint research project at the University of Kaiserslautern and the University of Kassel, is based on rewriting techniques, adopted to the special situation of string rewriting systems at hand.

1 Presentations and Decision Problems

String rewriting systems have played a prominent role in Automata Theory, Formal Language Theory, Recursion Theory, and Combinatorial Group and Semigroup Theory under the names of *semi-Thue systems*, *Thue systems*, *sets of productions*, or *defining relations*. Also they are a special case of the so-called *term rewriting systems*, which occur in presentations of equational theories.

Here we are concerned with the capability of string rewriting systems to present semigroups, monoids, and groups. Let Σ be a finite alphabet, that is, a set of *generators*, and let $R \subseteq \Sigma^* \times \Sigma^*$ be a finite string rewriting system, that is, a set of *defining relations*. Then R induces a *congruence* $\leftrightarrow_R^*$ on the free monoid Σ^*, which is simply the reflexive, symmetric, and transitive closure of the *single-step reduction relation* $\to_R := \{ (xuy, xvy) \mid (u,v) \in R, x, y \in \Sigma^* \}$. By M_R we denote the factor monoid $M_R := \Sigma^*/\leftrightarrow_R^*$. As the monoid M_R is uniquely determined by Σ and R up to isomorphisms, $(\Sigma; R)$ is called a *presentation* of any monoid that is isomorphic to M_R.

The most fundamental decision problem associated with a presentation $(\Sigma; R)$ is the *word problem*: Given two strings $u, v \in \Sigma^*$, decide whether u

and v represent the same element of the monoid M_R, that is, whether $u \leftrightarrow_R^* v$ holds. Other problems of a similar form are the *(left-) conjugacy problem* [1], the problems of deciding *Green's relations* [2], and the problems of *recognizing certain elements* of M_R like those that are *(left-, right-) invertible.*

Other decision problems deal with submonoids of M_R. An example is the *submonoid problem*, which is the monoid version of the *generalized word problem* considered in combinatorial group theory: Given a finite subset $U \subseteq \Sigma^*$ and a string $w \in \Sigma^*$, decide whether w belongs to the submonoid of M_R that is generated by U. Finally, there are the *uniform versions* of all these problems, where the finite presentation considered is part of the problem instance. Actually, these uniform problems can be stated for each class of presentations, for example, the *uniform word problem* for a class $\mathcal{C}$ of finite presentations is the following decision problem: Given a presentation $(\Sigma; R)$ from $\mathcal{C}$ and two strings $u, v \in \Sigma^*$, decide whether u and v represent the same element of the monoid M_R presented by $(\Sigma; R)$. Also one can ask for various algebraic properties of a monoid M that is given through the presentation from $\mathcal{C}$, e.g., is M a free monoid, is it finite, or is it a group?

Unfortunately, in general one cannot extract much information about a monoid from a given presentation, as there are finite presentations with undecidable word problem, and similar for the other decision problems. On the other hand, many problems have been solved effectively for certain classes of presentations. For example, the word problem and also many other problems are easily solved for monoids that are generated by a single element. For those groups that can be defined by a single defining relation, the word problem is decidable [3], and the same is true for those monoids that can be defined by a single relation of the form $w = \lambda$, where λ denotes the empty string [4].

By restricting the set of defining relations to satisfy certain *small cancellation conditions* the so-called *small cancellation groups* are obtained, for which the word problem is solved by *Dehn's algorithm* [5]. This algorithm is of particular interest as it is a purely syntactical algorithm that transforms a given string stepwise by repeatedly replacing substrings by strings that are strictly shorter until no further such replacement is possible. The small cancellation conditions then guarentee that a string is transformed into the empty string in this way if and only if it represents the identity element of the group considered.

2 Rewriting Techniques

The *e***X***perimental* **S***ystem for* **S***tring* **R***ewriting*, **XSSR**, is a computer algebra system for combinatorial group and semigroup theory that can be used to effectively perform computations in and with monoids and groups that are given through finite presentations. It contains implementations of algorithms for solving various decision problems including those listed above. As these problems are all undecidable in general, these algorithms only apply to certain restricted types of presentations.

Most of these algorithms are based on a rewrite approach that can be seen as an extension of the strategy underlying Dehn's algorithm. Let $(\Sigma; R)$ be a finite presentation. The process of replacing the lefthand side u of a rule (u, v) of R within a string $w = xuy$ by the righthand side v, which results in the string xvy, is to be interpreted as a *simplification*. Obviously the process of repeatedly simplifying a string should eventually terminate with a 'simplest' string, which can then be interpreted as a *normal form* of the original string. Therefore, we require that the *reduction relation* $\rightarrow_R$ that is induced by the string rewriting system R is *terminating.*

In general, a string may have several different normal forms with respect to the above simplification process. If the congruence class of a string w contains a single normal form only, then the reduction relation $\rightarrow_R$ is *confluent* on this congruence class. We will require that the relation $\rightarrow_R$ is confluent on certain congruence classes, which leads to presentations that are λ-*confluent* (that is, confluent on the class of the empty string) or *weakly confluent* (that is, confluent on the class of a for all generators a and the empty string)[6,7], or we will require that it is confluent everywhere, which leads to the *confluent* presentations.

For the class of all finite, terminating and λ-confluent presentations the problems of deciding whether the monoid defined by the given presentation is a free monoid or is a group are decidable [8]. For example, in order to decide that the presentation $(\Sigma; R)$ defines a group, a string $w_a \in \Sigma^*$ is determined for each generator $a \in \Sigma$ such that w_a is a candidate for the inverse of a in the monoid M_R. This can be done in such a way that M_R is a group if and only if aw_a represents the identity element for each $a \in \Sigma$, that is, if and only if these candidates are in fact representatives for the inverses of the given generators. If R is terminating and λ-confluent, then we only have to check whether all the strings aw_a reduce to the empty string.

For the finite, terminating and confluent presentations also the uniform word problem and the finiteness problem are decidable. Many additional decidability results are obtained when the syntactical form of the rewriting rules is further restricted. For example, for the finite systems that are length-reducing (or just weight-reducing) and confluent, the various forms of the uniform conjugacy problem become decidable [9], and the existence of nontrivial idempotents and of nontrivial elements of finite order are even decidable in polynomial time [10]. Also the submonoid problem has been solved for certain restricted classes of confluent and λ-confluent presentations, and it has been shown that the validity of the so-called *linear sentences* of Book [11], which are a class of severely restricted logical formulae that allow to express certain properties of congruences and monoids, can be solved in polynomial space for various classes of finite monadic presentations [6,11,12]. Here a presentation $(\Sigma; R)$ is called *monadic* if R is length-reducing and $v \in \Sigma \cup \{\lambda\}$ for each rule (u, v) of R. Properties that can be expressed by linear sentences are, e.g., the *independent set problem* (that is, is a given finite set $\{u_1, \ldots, u_m\} \subset \Sigma^*$ independent in the monoid $(\Sigma; R)$), Green's relations $\mathcal{R}$, $\mathcal{L}$, and $\mathcal{I}$, and certain generalized word problems.

For an exposition concerning the various decidability results we refer to the survey paper [13] and to the monograph [14].

In general a given presentation will not be confluent, but there are completion procedures that attempt to transform a given presentation into an equivalent one that is. For general confluence this is the Knuth-Bendix completion procedure [15] and its variants, but also for λ-confluence and weak confluence corresponding procedures have been obtained [16,7].

3 Functionality

XSSR incorporates algorithms for deciding various properties of monoids and groups and for effectively performing calculations in and with presentations.

Internally a monoid or a group is described by a collection of finite presentations. These presentations are either entered by the user, or they are generated from given presentations through the use of the operations provided by the system. These operations include the Tietze transformations and various Knuth-Bendix style procedures for completion, λ-completion and weak completion. Also some algebraic operations like free or direct products are available to construct new monoids or groups from given ones. However, in order to ensure that the database of presentations is free from contradictions, only presentations generated by the system can be added to the collection of presentations describing a certain monoid. If the user wants to add another presentation of the same monoid or group, then this must be done in such a way that the system can verify that the new presentation does indeed present the same object. In particular, the system determines and stores the morphisms that realize the isomorphisms between the various presentations that belong to the same collection of presentations of a monoid or group.

To analyze a specific monoid or group the user can ask questions about properties of elements, subsets, or the monoid or group as such. For example, he can ask whether the monoid considered is actually a group. If the information is already available, or if an algorithm for answering this question is known for one of the presentations contained in the collection of presentations for the monoid or group considered, the system will provide an answer immediately; otherwise, it will state that the answer is not yet known, and it will suggest ways of trying to find the answer. For example, if asked whether the monoid M given by a finite special presentation $(\Sigma; R)$ is a group, where a presentation is called *special* if each rule is of the form (u, λ) for some nonempty string u, the system will use the algorithm for deciding the group property if the string rewriting system R is λ-confluent; otherwise, it will state that the answer is unknown, but that the user may try to obtain a special λ-confluent presentation for M by running the λ-completion procedure, and that an answer will be obtained in case the completion procedure succeeds. In addition, if the monoid is indeed a group, the system will also determine a function that realizes the operation of forming inverses in that group for future use (see the above discussion of this problem).

4 System Architecture

For modelling the application domain 'Computing in monoids and groups presented by finite string rewriting systems' an object-oriented approach was used. First the central notions like 'letter,' 'alphabet,' 'string,' 'string equation,' 'rewrite rule,' 'string rewriting system,' 'monoid presentation,' 'reduction ordering' etc. were analyzed with respect to their relationship and with respect to the operations involving them. In this way those aspects of these notions that are of concern for performing efficient calculations in finitely presented monoids and groups were isolated and documented. Through additional refinements class descriptions were obtained, which were produced with the help of the Aonix software packages StP/OMP and StP/UML. Further subdomains like 'completion procedures for finite presentations,' 'regular languages,' 'decision procedures for finitely presented monoids and groups' and infra-structural components were established during this process. The subdomains were then modelled in the same way. Of course, this was not a clear top-down process, as often additional requirements became necessary or conflicts were discovered, not just during the design phase, but even during the implementation phase. By adding missing aspects and by resolving conflicts the modelling of the domain went on concurrently with the implementation. In particular through the use of design patterns [17] we have obtained a modelling that is both natural and easily extendible.

XSSR is characterized by a *multi-tier architecture*:

- The data-structures and functions concerning the algebraic operations and decision algorithms are integrated in a kernel modul.
- Application facades enable a simple coupling between kernel functionalities and user interfaces. By these facades the development process of user interfaces is decoupled from the details of the dozens of classes of the kernel and the details of the protocols governing their use. In this way the facades give a more abstract view of the funtionalities of the kernel.
- Several file-based and text-based user interfaces are incorporated in **XSSR**, in particular for the use in the script-controlled processing of extensive, homogeneous tasks or for test purposes. Also **XSSR** has an intuitive graphical user interface, that supports a natural way of working with several monoids and their hierarchical collection of presentations. Finally, there is graphical user interface particularly for the variant of the Knuth-Bendix completion procedure implemented in **XSSR**, which can be configured by the expert user in multiple ways. These include the choice of various types of orderings for ensuring termination, the choice of different reduction strategies, or the possibility to choose between different strategies to handle critical pairs.
- The storage of data and the exchange of data with other software tools is realized by a collection of services for serializing the data created at run time. Here a data format is used that is specified in XML.

5 Implementation

XSSR has been implemented in C++. The reasons for this choice were the existence of sophisticated development tools, the multi-paradigmatic approach of C++ supporting high-level concepts of object-oriented programming as well as low-level system-oriented programming as well as generic programming, and the availability of numerous libraries. Of special importance has been the C++-standard library, and in particular that part known as STL (*standard template library*). **XSSR** is heavily based on this collection of generic data structures and efficient algorithms, but quite a few basic data structures had to be newly developed nevertheless.

The functionalities of **XSSR** are organised in the form of a class library, which can also be used for other software projects involving string rewriting systems. The components of this library have been designed and implemented with a special emphasis on reusability, at least as long as this did not contradict runtime and space efficiency. By using the software tool Doxygen[18] a detailed documentation of the class interfaces was created concurrently with the implementation. This documentation is extracted from structured comments contained in the source code and by Doxygen converted into HTML, LaTeX, RTF, or other chosen formats. During this process class diagrams, that describe the relationships between the various classes, are generated automatically together with other useful diagrams. This was motivated by a desire to obtain a documentation that enables system developers and application programmers to easily extract information on the services and interfaces of single classes and on the interactions between these classes.

An additional aspect of the implementation of **XSSR** has been the desire to attain as much platform independence as possible. By exclusively using ISO-C++ [19] it is guaranteed that **XSSR** can be compiled by every C++-compiler that adheres to the C++ standard. Also Qt, the class library of Trolltech AS [20] which is used for the implementation of the graphical user interfaces, is available on all prevalent platforms.

6 Other Systems

At the University of Kaiserslautern the system MRC has been implemented, which is a prototyp of a computer algebra system for performing computations in monoid and group rings [21,22]. This system determines Gröbner bases with respect to prefix-reduction in monoid and group rings over the field $\mathbb{Q}$ of rational numbers [23]. Eventually this system will be integrated into **XSSR**, thereby extending the range and variety of applications for **XSSR** considerably.

The latest version of the system GAP [24] (version 4.2) also includes an implementation of the Knuth-Bendix completion procedure, which, however, is much less efficient than that of **XSSR**, as it is implemented in the GAP-language, which is interpreted. Also the decision procedures based on rewriting techniques that are the main focus of **XSSR** are not available in GAP.

Derek Holt's system KBMAG2 [25] contains a highly efficient implementation of the Knuth-Bendix completion procedure and algorithms for *automatic groups*, but no algorithms for dealing with finitely presented monoids and groups. The completion procedure of KBMAG2 can be configured in various ways. The run time efficiency is comparable to that of **XSSR**, but for each of the two systems there are examples that the one system can do faster than the other.

The system Magma [26] of the Magma group of John Cannon at the University of Sydney, which emerged from the system Cayley, contains some algorithms for finitely presented semigroups, monoids, and groups. It seems that no completion procedure is implemented, but Magma contains a confluence test. There are some simple algorithms (a finiteness test, an algorithm for computing the order of a finite monoid, etc.), but the rewrite based algorithms of **XSSR** are not implemented.

Finally the system Magnus [27] of the Group Theory Cooperative at the City College of the City University of New York has a graphical user interface for performing computations in groups. Magnus assigns graphical symbols to all artefacts that are generated by the user or by the system. These artefacts are then used for communication between the user and the system. Depending on the type of group considered (abelian, free, free nilpotent, nilpotent, one-relator, etc.) various algorithms for deciding algebraic properties are provided. Partially it incorporates other special purpose systems, for example KBMAG for the completion procedure, ace for the Todd-Coxeter procedure, and TietzeTrek for the heuristic search for Tietze transformations. These systems are triggered from the user interface, and Magnus then processes the output provided by them. However, the Magnus group is creating more and more implementations of their own for these functionalities. The ability of Magnus to work with finitely presented groups is rather limited, the rewrite based group algorithms of **XSSR** are not included, and Magnus contains no algorithms for finitely presented monoids.

7 Availability

More detailed information on **XSSR** can be found at the **XSSR** home page (http://www-madlener.informatik.uni-kl.de/ag-madlener/research/xssr/). A first release of **XSSR** is currently being tested at the University of Kaiserslautern and at the University of Kassel. This release runs under Solaris 2.7. Currently it consists of about 250 C++ classes and 200.000 lines of C++ source code (including about 50.000 lines of comments and embedded documentation).

It is intended to make this release available to interested researchers throughout the world. However, as there are only very limited resources available for maintenance, it has not yet been decided whether a web service will be established that can be used to test **XSSR** by simply running it remotely on our machines, or whether the binaries will be distributed.

Acknowledgment

The development of **XSSR** was supported by a grant from the *Deutsche Forschungsgemeinschaft* (DFG). A preliminary announcement of the system was made at a workshop on *Gröbner Bases and Rewriting Techniques* affiliated with the 1999 Federated Logic Conference, FLoC'99, at Trento, Italy.

References

1. F. Otto. Conjugacy in monoids with a special Church-Rosser presentation is decidable. *Semigroup Forum*, 29:223–240, 1984.
2. G. Lallement. *Semigroups and Combinatorial Applications.* Wiley-Interscience, New York, 1979.
3. W. Magnus, A. Karrass, and D. Solitar. *Combinatorial Group Theory.* Dover Publications, New York, 2nd revised edition edition, 1976.
4. S.I. Adjan. Defining relations and algorithmic problems for groups and semigroups. In *Proc. Steklov Institute of Mathematics 85.* American Mathematical Society, Providence, RI, 1966.
5. R.C. Lyndon and P.E. Schupp. *Combinatorial Group Theory.* Springer-Verlag, Berlin, 1977.
6. F. Otto and L. Zhang. Decision problems for finite special string-rewriting systems that are confluent on some congruence class. *Acta Informatica*, 28:477–510, 1991.
7. K. Madlener, P. Narendran, F. Otto, and L. Zhang. On weakly confluent monadic string-rewriting systems. *Theoretical Computer Science*, 113:119–165, 1993.
8. F. Otto. On deciding whether a monoid is a free monoid or is a group. *Acta Informatica*, 23:99–110, 1986.
9. P. Narendran and F. Otto. Complexity results on the conjugacy problem for monoids. *Theoretical Computer Science*, 35:227–243, 1985.
10. P. Narendran and F. Otto. Elements of finite order for finite weight-reducing and confluent Thue systems. *Acta Informatica*, 25:573–591, 1988.
11. R.V. Book. Decidable sentences of Church-Rosser congruences. *Theoretical Computer Science*, 24:301–312, 1983.
12. K. Madlener and F. Otto. Decidable sentences for context-free groups. In C. Choffrut and M. Jantzen, editors, *Proceedings STACS'91*, Lecture Notes in Computer Science 480, pages 160–171. Springer-Verlag, Berlin, 1991.
13. F. Otto. A survey on the computational power of some classes of finite monoid presentations. In J.C. Birget, S. Margolis, J. Meakin, and M. Sapir, editors, *Algorithmic Problems in Groups and Semigroups*, Trends in Mathematics, pages 171–194. Birkhäuser, Boston, 2000.
14. R.V. Book and F. Otto. *String-Rewriting Systems.* Springer-Verlag, New

York, 1993.
15. D. Knuth and P. Bendix. Simple word problems in universal algebras. In J. Leech, editor, *Computational Problems in Abstract Algebra*, pages 263–297. Pergamon Press, New York, 1970.
16. F. Otto. Completing a finite special string-rewriting system on the congruence class of the empty word. *Applicable Algebra in Engineering, Communication and Computing*, 2:257–274, 1992.
17. E. Gamma, R. Helm, R. Johnson, and J. Vlissides. *Design Patterns - Elements of Reusable Object-Oriented Software*. Addison-Wesley, Reading, MA, 1995.
18. D. van Heesch. Doxygen. URL: http://www.stack.nl/~dimitri/doxygen/.
19. ISO/IEC 14882. International Standard. Programming Language C++, 1998.
20. Trolltech AS. URL: http://www.trolltech.com/products/qt/qt.html.
21. B. Reinert and D. Zeckzer. User reference for MRC: A system for computing Gröbner bases in monoid and group rings. Fachbereich Informatik, Universität Kaiserslautern, 1998.
22. B. Reinert and D. Zeckzer. MRC - data structures and algorithms for computing in monoid and group rings. *Applicable Algebra in Engineering, Communication and Computing*, 10:41–78, 1999.
23. K. Madlener and B. Reinert. Computing Gröbner bases in monoid and group rings. In M. Bronstein, editor, *Proc. ISSAC'93*, pages 254–263. ACM, 1993.
24. The GAP Group. GAP - Groups, Algorithms, and Programming, Version 4.2. Aachen, St Andrews, 2000. URL: http://www.gap-system.org/.
25. D. Holt. KBMAG2. URL: http://www.maths.warwick.ac.uk/~dfh/.
26. The Magma Computational Algebra System.
URL: http://magma.maths.usyd.edu.au/magma/.
27. Magnus. URL: http://zebra.sci.ccny.cuny.edu/web/magnushome.html.

THE DIFFERENTIAL STRUCTURE OF THE BRIESKORN LATTICE

MATHIAS SCHULZE

University of Kaiserslautern, Germany
E-mail: mschulze@mathematik.uni-kl.de

The Brieskorn lattice H'' of an isolated hypersurface singularity with Milnor number μ is a free $\mathrm{C}\{\{s\}\}$-module of rank μ with a differential operator $t = s^2\partial_s$. Based on the mixed Hodge structure on the cohomology of the Milnor fibre, M. Saito constructed $\mathrm{C}\{\{s\}\}$-bases of H'' for which the matrix of t has the form $A = A_0 + A_1 s$. We describe an algorithm to compute the matrices A_0 and A_1. They determine the differential structure of the Brieskorn lattice, the spectral pairs and Hodge numbers, and the complex monodromy of the singularity.

1 The Milnor Fibration

Let $f : (\mathrm{C}^{n+1}, \underline{0}) \longrightarrow (\mathrm{C}, 0)$ be a holomorphic function germ with an isolated critical point and Milnor number $\mu = \dim_{\mathrm{C}} \mathrm{C}\{\underline{x}\}/\langle \underline{\partial}(f)\rangle$ where $\underline{x} = x_0, \ldots, x_n$ is a complex coordinate system of $(\mathrm{C}^{n+1}, 0)$ and $\underline{\partial} = \partial_{x_0}, \ldots, \partial_{x_n}$. By the finite determinacy theorem, we may assume that $f \in \mathrm{C}[\underline{x}]$. By E.J.N Looijenga[7(2.B)], for a good representative $f : X \longrightarrow T$ where $T \subset \mathrm{C}$ is an open disk at the origin, the restriction $f : X' \longrightarrow T'$ to $T' = T\backslash\{0\}$ and $X' = X\backslash f^{-1}(0)$ is a C^∞ fibre bundle unique up to diffeomorphism, the Milnor fibration. By J. Milnor[9(6.5)], the general fibre $X_t = f^{-1}(t)$, $t \in T'$, is homotopy equivalent to a bouquet of μ n-spheres and, in particular, its reduced cohomology is $\widetilde{\mathrm{H}}^k(X_t) \cong \delta_{k,n}\mathrm{Z}^\mu$ where δ is the Kronecker symbol. Since T' is locally contractible, the n-th cohomologies $\mathrm{H}(U) = \mathrm{H}^n(X_U)$ of $X_U = f^{-1}(U)$ form a locally free Z-sheaf of rank μ and $\mathrm{H}_{\mathrm{C}} = \mathrm{H} \otimes_{\mathrm{Z}} \mathrm{C}$ is a complex local system of dimension μ. Hence, the sheaf of holomorphic sections $\mathcal{H} = \mathrm{H} \otimes_{\mathrm{Z}} \mathcal{O}_{T'}$ of H_{C} is a locally free $\mathcal{O}_{T'}$-sheaf of rank μ, the cohomology bundle. By P. Deligne[4(2.23)], there is a natural flat connection $\nabla : \mathcal{H} \longrightarrow \mathcal{H} \otimes_{\mathcal{O}_{T'}} \Omega^1_{T'}$ on $\mathcal{H}$ with sheaf of flat sections $\mathrm{H} = \ker \nabla$, the Gauss-Manin connection.

2 The Monodromy Representation

Let t be a complex coordinate of $T \subset \mathrm{C}$, $i : T' \longrightarrow T$ the canonical inclusion, and $u : T^\infty \longrightarrow T'$ the universal covering of T' defined by $u(\tau) = \exp(2\pi i\tau)$ for a complex coordinate τ of $T^\infty \subset \mathrm{C}$. Then the covariant derivative ∇_{∂_t} of ∇ along ∂_t induces a differential operator ∂_t on $i_*\mathcal{H}$ and the pullback $f^\infty : X^\infty = X' \times_{T'} T^\infty \longrightarrow T^\infty$ is a C^∞ fibre bundle with $X^\infty_\tau = X_{u(\tau)}$, the (canonical) Milnor fibre. Since T^∞ is contractible, the n-th cohomologies $H(U) = \mathrm{H}^n(X_U^\infty)$ of $X_U^\infty = (f^\infty)^{-1}(U)$ form a free Z-sheaf of rank μ and u_*H is the sheaf of multivalued sections of H. Lifting closed paths in T' along sections of H defines the monodromy representation $\pi_1(T', t) \longrightarrow \mathrm{Aut}(\mathrm{H}_t)$ on H_t inducing the monodromy representation $\pi_1(T') \longrightarrow \mathrm{Aut}(H)$ on the cohomology H of the Milnor fibre. The image M of the counterclockwise generator of $\pi_1(T')$ is called the monodromy operator and fulfills $\mathrm{M}(s)(\tau) = s(\tau+1)$ for $s \in H$. The sheaf H is determined by the monodromy representation up to isomorphism. The following well known theorem is due to E. Brieskorn[2(0.6)] and others.

Theorem 1 (Monodromy Theorem). *The eigenvalues of the monodromy are roots of unity and its Jordan blocks have size at most* $(n+1) \times (n+1)$ *and size at most* $n \times n$ *for eigenvalue* 1.

3 The Gauss-Manin Connection

Let $\mathrm{M} = \mathrm{M}_s\mathrm{M}_u$ be the decomposition of M into semisimple part M_s and unipotent part M_u and let $\mathrm{N} = -\frac{\log \mathrm{M}_u}{2\pi\mathrm{i}}$ be the nilpotent part of M. Note that $-2\pi\mathrm{i}\mathrm{N} \in \mathrm{End}_{\mathrm{Q}}(H_{\mathrm{Q}})$ where $H_{\mathrm{Q}} = H \otimes_{\mathrm{Z}} \mathrm{Q}$. Let $H_{\mathrm{C}} = \bigoplus_\lambda H_{\mathrm{C}}^\lambda$ be the decomposition of $H_{\mathrm{C}} = H \otimes_{\mathrm{Z}} \mathrm{C}$ into generalized λ-eigenspaces H_{C}^λ of M and $\mathrm{M}^\lambda = \mathrm{M}|_{H_{\mathrm{C}}^\lambda}$. Note that $H_{\mathrm{Q}} = H_{\mathrm{Q}}^1 \oplus H_{\mathrm{Q}}^{\neq 1}$ where $H_{\mathrm{Q}}^1 \otimes_{\mathrm{Q}} \mathrm{C} = H_{\mathrm{C}}^1$ and $H_{\mathrm{Q}}^{\neq 1} \otimes_{\mathrm{Q}} \mathrm{C} = \bigoplus_{\lambda \neq 1} H_{\mathrm{C}}^\lambda$. Then there is an inclusion

$$H_{\mathrm{C}}^{e^{-2\pi i\alpha}} \xrightarrow{\psi_\alpha} (i_*\mathrm{H})_0$$

defined by $\psi_\alpha(A) = t^{\alpha+\mathrm{N}}A = t^\alpha \exp(N\log(t))$ with image $C^\alpha = \mathrm{im}(\psi_\alpha)$. In particular, the operators M and N act on C^α. The following lemma is an immediate consequence of the definition of ψ_α.

Lemma 2.

1. $t \circ \psi_\alpha = \psi_{\alpha+1}$ *and* $\partial_t \circ \psi_\alpha = \psi_{\alpha-1} \circ (\alpha + \mathrm{N})$.

2. $t : C^\alpha \longrightarrow C^{\alpha+1}$ *is bijective and* $\partial_t : C^\alpha \longrightarrow C^{\alpha-1}$ *is bijective if* $\alpha \neq 0$.

3. *On* C^α, $t\partial_t - \alpha = \mathrm{N}$ *and* $\exp(-2\pi\mathrm{i}t\partial_t) = \mathrm{M}^{e^{-2\pi i\alpha}}$.

4. $C^\alpha = \ker(t\partial_t - \alpha)^{n+1}$.

Definition 3. *We call* $G = \bigoplus_{-1<\alpha\leq 0} \mathrm{C}\{t\}[t^{-1}]C^\alpha \subset (i_*\mathrm{H})_0$ *the local Gauss-Manin connection.*

The local Gauss-Manin connection is a μ-dimensional $\mathrm{C}\{t\}[t^{-1}]$-vectorspace and a regular holonomic $\mathrm{C}\{t\}[\partial_t]$-module. The generalized α-eigenspaces C^α of the operator $t\partial_t$ define the decreasing filtration on G by free $\mathrm{C}\{t\}$-modules

$$V^\alpha = \bigoplus_{\alpha\leq\beta<\alpha+1} \mathrm{C}\{t\}C^\beta, \quad V^{>\alpha} = \bigoplus_{\alpha<\beta\leq\alpha+1} \mathrm{C}\{t\}C^\beta$$

of rank μ, the V-filtration. In contrast to the ψ_α and C^α, the V^α are independent of the coordinate t. The C^α define a splitting

$$C^\alpha \cong V^\alpha/V^{>\alpha} = \mathrm{gr}_V^\alpha G$$

of the V-filtration and we denote by lead_V the leading term with respect to this splitting. The ring $\mathrm{C}\{t\}$ is a free module of rank 1 over the ring

$$\mathrm{C}\{\{s\}\} = \left\{\sum_{k=0}^{\infty} a_k s^k \in \mathrm{C}[\![s]\!] \,\middle|\, \sum_{k=0}^{\infty} \frac{a_k}{k!} t^k \in \mathrm{C}\{t\}\right\}$$

where $s = \int_0^1 \mathrm{d}t$ acts by integration. This fact is generalized by the following lemma.[13(1.3.11)]

Lemma 4. *The action of* $s = \partial_t^{-1}$ *on* $V^{>-1}$ *extends to a* $\mathrm{C}\{\{s\}\}$*-module structure and* $V^{>-1}$ *is a free* $\mathrm{C}\{\{s\}\}$*-module of rank* μ.

Since $[\partial_t, t] = 1$, $[t, s] = s^2$ and hence

$$t = s^2\partial_s, \quad \partial_t t = s\partial_s.$$

We call a free $\mathrm{C}\{\{s\}\}$-submodule of $V^{>-1}$ of rank μ a $\mathrm{C}\{\{s\}\}$-lattice and call a $t\partial_t$-invariant $\mathrm{C}\{\{s\}\}$-lattice saturated. A basis $\underline{e}$ of a $\mathrm{C}\{\{s\}\}$-lattice defines a matrix $A = \sum_{k\geq 0} A_k s^k$ of t by $t\underline{e} = \underline{e}A$ such that

$$t \cong A + s^2\partial_s$$

is the basis representation of t.

4 The Brieskorn Lattice

The description of cohomology in terms of holomorphic differential forms by the de Rham isomorphism leads to the definition of the Brieskorn lattice

$$H'' = \Omega^{n+1}_{X,0}/\mathrm{d}f \wedge \mathrm{d}\Omega^{n-1}_{X,0}.$$

By E. Brieskorn[2(1.5)] and M. Sebastiani[15], the Brieskorn lattice is the stalk at 0 of a locally free O_T-sheaf H'' of rank μ with $\mathrm{H}''|_{T'} \cong \mathrm{H}$ and hence $H'' \subset (i_*\mathrm{H})_0$. The regularity of the Gauss-Manin connection proved by

E. Brieskorn[2(2.2)] implies that $H'' \subset G$. B. Malgrange[8(4.5)] improved this result by the following theorem.

Theorem 5. $H'' \subset V^{-1}$.

By E. Brieskorn[2(1.5)], the Leray residue formula can be used to express the action of ∂_t in terms of differential forms by $\partial_t[\mathrm{d}f \wedge \omega] = [\mathrm{d}\omega]$. In particular, $sH'' \subset H''$ and

$$H''/sH'' \cong \Omega^{n+1}_{X,0}/\mathrm{d}f \wedge \Omega^n_{X,0} \cong \mathbb{C}\{\underline{x}\}/\langle\underline{\partial}(f)\rangle.$$

Since the $V^{>-1}$ is a $\mathbb{C}\{\{s\}\}$-module, theorem 5 implies that H'' is a free $\mathbb{C}\{\{s\}\}$-module of rank μ and the action of s can be expressed in terms of differential forms by

$$s[\mathrm{d}\omega] = [\mathrm{d}f \wedge \omega].$$

For computational purposes, we may restrict our attention to the completion of the Brieskorn lattice. E. Brieskorn[2(3.4)] proved the following theorem.

Theorem 6. *The $\mathfrak{m}_{X,0}$- and $\mathfrak{m}_{T,0}$-adic topologies on H'' coincide.*

While the proof of theorem 6 is highly non-trivial, the analogous statement for the $\mathbb{C}\{\{s\}\}$-structure of the Brieskorn lattice is quite elementary.[13(1.5.4)]

Proposition 7. *The $\mathfrak{m}_{X,0}$- and $\mathfrak{m}_{\mathbb{C}\{\{s\}\}}$-adic topologies on H'' coincide.*

We call the completion $\widehat{H}''$ of H'' the formal Brieskorn lattice. Since completion is faithfully flat, $\widehat{H}''$ is a free $\mathbb{C}[\![s]\!]$-module of rank μ with a differential operator $t = s^2\partial_s$. The equality $[\underline{\partial}(f)\overline{g}\mathrm{d}\underline{x}] = s[\underline{\partial}(\overline{g})\mathrm{d}\underline{x}]$ motivates to consider the differential relation $\underline{\partial}(f) - s\underline{\partial}$. It is not difficult to prove that it defines the formal Brieskorn lattice as a quotient of $\mathbb{C}[\![s,\underline{x}]\!]$.[13(1.5.6)]

Proposition 8.

$$\mathbb{C}[\![s,\underline{x}]\!] \xrightarrow{\pi_H} \mathbb{C}[\![s,\underline{x}]\!]/\langle\underline{\partial}(f) - s\underline{\partial}\rangle\mathbb{C}[\![s,\underline{x}]\!] \cong_{\mathbb{C}[\![s]\!]} \widehat{H}''.$$

Proposition 8 is the starting point for an algorithmic approach to the local Gauss-Manin connection. Let $<_{\underline{x}}$ be a local degree ordering on $\mathbb{C}[\![\underline{x}]\!]$ such that $\deg(\underline{x}) < \underline{0}$ and $\deg(\underline{\partial}) = -\deg(\underline{x}) > \underline{0}$. One can compute a polynomial standard basis $\underline{g}$ of the Jacobian ideal $\langle\underline{\partial}(f)\rangle$ and a polynomial transformation matrix $B = (\overline{b}^j)^j$ such that $\underline{g} = \underline{\partial}(f)B$. By Nakayama's lemma, $\underline{m} = (\underline{x}^{\underline{\beta}})_{\underline{x}^{\underline{\beta}} \notin \langle \mathrm{lead}(\underline{g})\rangle}$ represents a $\mathbb{C}[\![s]\!]$-basis $[\underline{m}]$ of $\widehat{H}''$. Let $<_s$ be the local degree ordering on $\mathbb{C}[\![s]\!]$ and let $< = (<_s, <_{\underline{x}})$ be the block ordering of $<_s$ and $<_{\underline{x}}$ on $\mathbb{C}[\![s,\underline{x}]\!]$.

Definition 9.

1. $\underline{h} = \left((g_j - s\underline{\partial}\overline{b}^j)\underline{x}^{\underline{\beta}}\right)_{j,\underline{\beta}}$.

2. $\deg(s) = \min\deg(\underline{m}) + 2\min\deg(\underline{x}) < 0$.
3. $N = (N_K)_{K\geq 0}$ *with* $N_K = K\deg(s) - 2\min\deg(\underline{x})$.
4. $V = (V_K)_{K\geq 0}$ *with* $V_K = \{p \in \mathrm{C}[\![s,\underline{x}]\!] \,|\, \deg(p) < N_K\} + \langle s\rangle^K \subset \mathrm{C}[\![s,\underline{x}]\!]$.

Since $\widehat{H}''$ is a free $\mathrm{C}[\![s]\!]$-module, $\underline{h}$ is a standard basis of the $\mathrm{C}[\![s]\!]$-module $\langle\underline{\partial}(f) - s\underline{\partial}\rangle\mathrm{C}[\![s,\underline{x}]\!]$. The following lemma is technical but not very deep and can be generalized to formal differential deformations.[13(2.2.10)]

Lemma 10. *$V = (V_K)_{K\geq 0}$ is a basis of the $\langle s,\underline{x}\rangle$-adic topology of $\mathrm{C}[\![s,\underline{x}]\!]$ with $\pi_H(V_K) = \langle s\rangle^K \widehat{H}''$. If $s^\alpha \mathrm{lead}(h_{j,\beta}) \in V_K$ then $s^\alpha h_{j,\beta} \in V_K$.*

Lemma 10 leads to a normal form algorithm for the Brieskorn lattice.[13(2.2.12)] It computes a normal form with respect to $\underline{h}$ and hence the $[\underline{m}]$-basis representation in H''. The normal form computation up to a given degree can be continued up to any higher degree without additional computational effort. The normal form algorithm for the Brieskorn lattice is a special case of a modification of Buchberger's normal form algorithm[3] for power series rings where termination is replaced by adic convergence.[13(2.1.19)]

5 Mixed Hodge Structure

By lemma 2, there is a C-isomorphism

$$H_{\mathrm{C}} = \bigoplus_{-1<\alpha\leq 0} H_{\mathrm{C}}^{e^{-2\pi i\alpha}} \xrightarrow{\psi} \bigoplus_{-1<\alpha\leq 0} C^\alpha \cong V^{>-1}/sV^{>-1}$$

defined by $\psi = \bigoplus_{-1<\alpha\leq 0}\psi_\alpha$ and the monodromy M on H_{C} corresponds to $\exp(-2\pi i t\partial_t)$ on $\bigoplus_{-1<\alpha\leq 0} C^\alpha$.

The Hodge filtration $F = (F_k)_{k\in\mathrm{Z}}$ on $V^{>-1}$ defined by J. Scherk and J.H.M. Steenbrink[14] is the increasing filtration by the free $\mathrm{C}\{\{s\}\}$-modules

$$F_k = F^{n-k} = (s^{-k}H'') \cap V^{>-1}$$

of rank μ. Via the splitting $C^\alpha \cong \mathrm{gr}_V^\alpha V^{>-1}$, the Hodge filtration induces an increasing Hodge filtration FC^α by C-vectorspaces on C^α and, via ψ, on H_{C}. The nilpotent operator $-2\pi i\mathrm{N} \in \mathrm{End}_{\mathrm{Q}}(H_{\mathrm{Q}})$ defines an increasing weight filtration $W = (W_k)_{k\in\mathrm{Z}}$ centered at n resp. $n+1$ on $H_{\mathrm{Q}}^{\neq 1}$ resp. H_{Q}^{1}.

Theorem 11. *The weight filtration W on H_{Q} and the Hodge filtration F on H_{C} define a mixed Hodge structure on the cohomology H of the Milnor fibre and the operator* N *is a morphism of mixed Hodge structures of type* $(-1,-1)$.

The mixed Hodge structure on the cohomology of the Milnor fibre was discovered by J.H.M. Steenbrink[16] and described in terms of the Brieskorn lattice by A.N. Varchenko[17].

The nilpotent operator N on C^α defines an increasing weight filtration $W = (W_k)_{k\in \mathrm{Z}}$ centered at n on C^α. By definition N commutes with ψ_α and hence

$$\psi_\alpha\left(WH_{\mathrm{C}}^{\mathrm{e}^{-2\pi i\alpha}}\right) = \begin{cases} WC^\alpha, & \alpha \notin \mathrm{Z}, \\ W[-1]C^\alpha, & \alpha \in \mathrm{Z}. \end{cases}$$

The weight filtration $W = \bigoplus_{-1<\alpha\leq 0} \mathrm{C}\{\{s\}\}WC^\alpha$ on $V^{>-1}$ by free $\mathrm{C}\{\{s\}\}$-modules induces WC^α via the splitting $C^\alpha \cong \mathrm{gr}_V^\alpha V^{>-1}$.

The spectral pairs are those pairs $(\alpha, l) \in \mathrm{Q} \times \mathrm{Z}$ with positive multiplicity

$$d_l^\alpha = \dim_{\mathrm{C}} \mathrm{gr}_l^W \mathrm{gr}_V^\alpha \mathrm{gr}_0^F V^{>-1}.$$

Via the isomorphism ψ, they correspond to the Hodge numbers

$$h_\lambda^{p,l-p} = \dim_{\mathrm{C}} \mathrm{gr}_F^p \mathrm{gr}_l^W H_{\mathrm{C}}^\lambda$$

by $d_l^{\alpha+p} = h_{\mathrm{e}^{-2\pi i\alpha}}^{n-p,l-n+p}$ for $-1 < \alpha < 0$ and $d_l^p = h_1^{n-p,l+1-n+p}$ and inherit the symmetry properties

$$d_l^\alpha = d_l^{2n-l-1-\alpha}, \quad d_l^\alpha = d_{2n-l}^{\alpha-n+l}, \quad d_l^\alpha = d_{2n-l}^{n-1-\alpha}$$

from the mixed Hodge structure. The spectral numbers are those numbers $\alpha \in \mathrm{Q}$ with positive multiplicity

$$d^\alpha = \dim_{\mathrm{C}} \mathrm{gr}_V^\alpha \mathrm{gr}_0^F V^{>-1} = \sum_{l\in\mathrm{Z}} d_l^\alpha$$

and have the symmetry property $d^\alpha = d^{n-1-\alpha}$.

6 M. Saito's Basis

By P. Deligne[5(1.2.8)], a morphism of mixed Hodge structures is strict for the Hodge filtration. In particular, by theorem 11, N is strict for the Hodge filtration on H_{C} and on $\mathrm{gr}_V V^{>-1}$. Hence, there is a direct sum decomposition $F_k C^\alpha = \bigoplus_{j\leq k} C^{\alpha,j}$ such that $\mathrm{N}(C^{\alpha,k}) \subset C^{\alpha,k+1}$, and $sC^{\alpha,k} \subset C^{\alpha+1,k-1}$. By definition of the Hodge filtration,

$$\mathrm{lead}_V(H'') = \sum_{\alpha\in\mathrm{Q}} \sum_{k\leq 0} \mathrm{C}\{\{s\}\}C^{\alpha,k} = \bigoplus_{\alpha\in\mathrm{Q}} \mathrm{C}\{\{s\}\}G^\alpha$$

where $G^\alpha = C^{\alpha,0}$. Let $<_{\mathrm{Q}\times\mathrm{Z}} = (>_{\mathrm{Q}}, >_{\mathrm{Z}})$ be the block ordering of $>_{\mathrm{Q}}$ and $>_{\mathrm{Z}}$ on the index set $\mathrm{Q} \times \mathrm{Z}$. Then the Hodge filtration defines a refinement of the V-filtration on $V^{>-1}$ by free $\mathrm{C}\{\{s\}\}$-modules $V^{\alpha,k} = F_k C^\alpha \oplus V^{>\alpha}$ of rank μ and the $C^{\alpha,k}$ define a splitting of this refined filtration compatible with s. We call the refinement the Hodge refinement and the splitting a Hodge splitting. The following lemma follows essentially from the fact that $\mathbb{C}\{\{s\}\}$ is a discrete valuation ring.[13(1.10.5,1.10.10)]

Lemma 12. *Let H be a $\mathbb{C}\{\{s\}\}$-lattice and $C^{\alpha,k}$ a splitting of a refinement of the V-filtration compatible with s. Then a minimal standard basis of H is a $\mathbb{C}\{\{s\}\}$-basis and there is a reduced minimal standard basis of H.*

In particular, there is a reduced minimal standard basis of H'' for a Hodge splitting. The following proposition follows essentially from lemma 2.3.[13(1.10.12)]

Proposition 13. *Let $\underline{h}$ be a reduced minimal standard basis of H'' for a Hodge splitting. Then the $\underline{h}$-matrix A of t has degree 1. In particular,*

$$(H'', t) \xleftarrow{\underline{h}} (\mathbb{C}\{\{s\}\}^{\mu}, A_0 + A_1 s + s^2\partial_s)$$

is an isomorphism. Moreover, A_1 is semisimple with eigenvalues the spectral numbers of f added by 1 and $\mathrm{gr}_V(A_0)$ can be identified with N.

Note that the matrices A_0 and A_1 in proposition 13 determine the differential structure of the Brieskorn lattice. M. Saito[10] first constructed a $\mathbb{C}\{\{s\}\}$-basis of H'' as in proposition 13 without calling it a reduced minimal standard basis.

7 The Algorithm

We describe an algorithm to compute A_0 and A_1 as in proposition 13.[13] This algorithm can be simplified to compute the complex monodromy, the spectral numbers, or the spectral pairs only.[13]

The normal form algorithm for the Brieskorn lattice in section 4 computes the $[\underline{m}]$-matrix $A = \sum_{k\geq 0} A_k s^k$ of t defined by $t[\underline{m}] = [f\underline{m}] = [\underline{m}]A$ up to any degree. We identify the columns of a matrix H with the generators of a submodule $\langle H\rangle \subset \mathbb{C}[\![s]\!]^{\mu}$ and denote by E the unit matrix. Then $\langle E\rangle$ is the $[\underline{m}]$-basis representation of $\widehat{H}''$. Hence, the following two statements hold for $\underline{h} = [\underline{m}]$ with $\kappa = 0$ and $H = E$.

($H_{\underline{h}}$) One can compute $\kappa \geq 0$ and a $\mu \times \mu$-matrix H with coefficients in $\mathbb{C}[s]$ of degree at most κ such that $\langle H\rangle$ is the $\underline{h}$-basis representation of $\widehat{H}''$ and $s^{\kappa}\langle E\rangle \subset \langle H\rangle$.

($A_{\underline{h}}$) One can compute the $\underline{h}$-matrix A of t up to any degree.

Step by step, we improve the $\mathbb{C}[\![s]\!]$-basis $\underline{h}$ and show that ($H_{\underline{h}}$) and ($A_{\underline{h}}$) hold. After the last step, A_0 and A_1 as in proposition 13 can be computed by a basis transformation of A to a reduced minimal standard basis of $\langle H\rangle$ up to a certain degree bound.

We call the canonical projection $\mathrm{jet}_k : \mathbb{C}[\![s]\!] \longrightarrow \bigoplus_{j=0}^{k} \mathbb{C}s^j$ the k-jet. Let the monomial ordering on $\mathbb{C}[\![s]\!]^\mu = \mathbb{C}[\![s]\!] \otimes_{\mathbb{C}} \mathbb{C}^\mu$ be the block ordering $<= (<_s, >_\mu)$ of the local degree ordering $<_s$ on $\mathbb{C}[\![s]\!]$ and the inverse ordering $>_\mu$ on the indices of the basis elements of $\mathbb{C}^\mu$.

7.1 The Saturation of H''

In this step, we show that $(H_{\underline{h}})$ and $(A_{\underline{h}})$ hold for a $\mathbb{C}[\![s]\!]$-basis $\underline{h}$ of a saturated $\mathbb{C}[\![s]\!]$-lattice.

The increasing sequence of $\mathbb{C}[\![s]\!]$-lattices defined by

$$\widehat{H}''_0 = \widehat{H}'', \quad \widehat{H}''_{k+1} = s\widehat{H}''_k + t\widehat{H}''_k \subset \widehat{H}''$$

is stationary since $\widehat{H}''$ is noetherian. Hence, the saturation $\widehat{H}''_\infty = \bigcup_{k\geq 0} \widehat{H}''_k$ of $\widehat{H}''$ is a saturated $\mathbb{C}[\![s]\!]$-lattice. The $[\underline{m}]$-basis representation $\langle H_k\rangle$ of $\widehat{H}''_k$ can be computed by

$$H_0 = Q_{-1} = E, \quad Q_k = \big(\mathrm{jet}_k(A) + s^2\partial_s\big)Q_{k-1}, \quad H_{k+1} = (sH_k|Q_k).$$

We successively compute the H_k and check in each step if $\langle Q_k\rangle \subset \langle H_k\rangle$ by a standard basis and normal form computation. If $\langle Q_k\rangle \subset \langle H_k\rangle$ then we stop the computation and set $\kappa = k$ and $H_\infty = H_\kappa$. Then $\langle H_\infty\rangle$ is the $[\underline{m}]$-basis representation of $\widehat{H}''_\infty$. We replace H_∞ by a minimal standard basis of $\langle H_\infty\rangle$. Then $\underline{h} = s^{-\kappa}\underline{h}H_\infty$ is a $\mathbb{C}[\![s]\!]$-basis of a saturated $\mathbb{C}[\![s]\!]$-lattice. By a normal form computation with respect to H_∞ up to degree κ, we compute the $\underline{h}$-basis representation $\langle H_\infty^{-1}s^\kappa E\rangle = \langle \mathrm{jet}_\kappa(H_\infty^{-1}s^\kappa E)\rangle$ of $\widehat{H}''$. Since $\langle H_\infty\rangle \subset \langle E\rangle$, $s^\kappa\langle E\rangle \subset \langle H_\infty^{-1}s^\kappa E\rangle$. By a normal form computation with respect to H_∞ up to degree $\kappa + k$, one can compute the k-jet

$$\mathrm{jet}_k\big(H_\infty^{-1}(A - \kappa sE + s^2\partial_s)H_\infty\big) = \mathrm{jet}_k\big(H_\infty^{-1}\big(\mathrm{jet}_{\kappa+k}(A - \kappa sE) + s^2\partial_s\big)H_\infty\big)$$

of the $\underline{h}$-matrix of t for any $k \geq 0$.

7.2 The V-Filtration

In this step, we show that $(H_{\underline{v}})$ and $(A_{\underline{v}})$ hold for a $<_{\mathbb{Q}}$-increasingly ordered $\mathbb{C}[\![s]\!]$-basis $\underline{v}$ of a $\widehat{V}^\alpha$ compatible with the direct sum decomposition $\widehat{V}^\alpha/s\widehat{V}^\alpha \cong \bigoplus_{\alpha\leq\beta<\alpha+1} C^\beta$.

Since $\underline{h}$ is a $\mathbb{C}[\![s]\!]$-basis of a saturated $\mathbb{C}\{\{s\}\}$-lattice, $A_0 = 0$ and, by theorem 1, the eigenvalues of A_1 are rational. In order to compute the eigenvalues of A_1, we transform A_1 to Hessenberg form and factorize the characteristic polynomials of its blocks. Then we compute a constant $\mathbb{C}[\![s]\!]$-basis transformation such that $A_1 = \mathrm{diag}(\alpha_1, \ldots, \alpha_\mu) + N$ with $\alpha_1 \leq \cdots \leq \alpha_\mu$ where $\mathrm{diag}(\alpha_1, \ldots, \alpha_\mu)$ denotes the diagonal matrix with entries $\alpha_1, \ldots, \alpha_\mu$. If $\alpha_\mu - \alpha_1 < 1$ then $\underline{v} = \underline{h}$ is a $<_{\mathbb{Q}}$-increasingly ordered $\mathbb{C}[\![s]\!]$-basis $\underline{v}$ of a $\widehat{V}^\alpha$ compatible with the direct sum decomposition $\widehat{V}^\alpha / s\widehat{V}^\alpha \cong \bigoplus_{\alpha \leq \beta < \alpha+1} C^\beta$. If $\alpha_\mu - \alpha_1 \geq 1$ then we proceed as follows. Let

$$A = \begin{pmatrix} A^{1,1} & A^{1,2} \\ A^{2,1} & A^{2,2} \end{pmatrix}$$

such that $A_0 = 0$, $A_1^{1,2} = 0$, $A_1^{2,1} = 0$, and the eigenvalues of $A_1^{1,1}$ are the eigenvalues α of A_1 with $\alpha < \alpha_1 + 1$. Then the $\mathbb{C}[\![s]\!][s^{-1}]$-basis transformation

$$H \mapsto \begin{pmatrix} s & 0 \\ 0 & 1 \end{pmatrix}^{-1} H, \quad A \mapsto \begin{pmatrix} s & 0 \\ 0 & 1 \end{pmatrix}^{-1} (A + s^2 \partial_s) \begin{pmatrix} s & 0 \\ 0 & 1 \end{pmatrix} = \begin{pmatrix} A^{1,1} + s & s^{-1} A^{1,2} \\ s A^{2,1} & A^{2,2} \end{pmatrix}$$

decreases $\alpha_\mu - \alpha_1$ and the degree up to which A is computed by 1 and increases κ by 1. After at most n such transformations, $\alpha_\mu - \alpha_1 < 1$.

7.3 The Canonical V-Splitting

In this step, we show that $(H_{\underline{c}})$ and $(A_{\underline{c}})$ hold for a $<_{\mathbb{Q}}$-increasingly ordered $\mathbb{C}$-basis $\underline{c}$ of a direct sum $\bigoplus_{\alpha \leq \beta < \alpha+1} C^\beta$ compatible with the direct sum.

Let $\underline{c}$ be the image of $[\underline{v}]$ under the splitting $\widehat{V}^\alpha / s\widehat{V}^\alpha \cong \bigoplus_{\alpha \leq \beta < \alpha+1} C^\beta$. By Nakayama's lemma, $\underline{c}$ is a $\mathbb{C}$-basis of $\bigoplus_{\alpha \leq \beta < \alpha+1} C^\beta$ compatible with the direct sum. The eigenvalues of the commutator $[\cdot, A_1] \in \mathrm{End}_{\mathbb{C}}(\mathbb{C}^{\mu^2})$ are the differences of the eigenvalues of A_1. Since $\alpha_\mu - \alpha_1 < 1$, $[\cdot, A_1] - k \in \mathrm{GL}_{\mu^2}(\mathbb{C})$ for $k \geq 1$. Let $U = \sum_{j=0}^{\infty} U_j s^j$ be the $\mathbb{C}[\![s]\!]$-basis transformation defined by $\underline{c} = \underline{v} U$. Then $U_0 = E$ and $U A_1 s = (A + s^2 \partial_s) U$ or equivalently

$$U_k = \big([\cdot, A_1] - k\big)^{-1} \sum_{j=0}^{k-1} A_{k-j+1} U_j$$

for $k \geq 1$ and hence one can compute U up to any degree. Since $U_0 = E$ and $\kappa \geq 0$, $\mathrm{jet}_\kappa(U)$ is a minimal standard basis of $\langle E \rangle$. By a normal form computation with respect to U up to degree κ, we compute the $\underline{c}$-basis representation $\langle U^{-1} H \rangle = \langle \mathrm{jet}_\kappa(\mathrm{jet}_\kappa(U)^{-1} H) \rangle$ of $\widehat{H}''$ and $A_1 s$ is the $\underline{c}$-matrix of t.

7.4 A Hodge Splitting

In this step, we show that $(H_{\underline{f}})$ and $(A_{\underline{f}})$ hold for a $<_{\mathbb{Q}\times\mathbb{Z}}$-decreasingly ordered $\mathbb{C}$-basis $\underline{f}$ of a direct sum $\bigoplus_{\alpha \leq \beta < \alpha+1} \bigoplus_{k \in \mathbb{Z}} C^{\beta,k}$ compatible with the direct sum and that one can compute A_0 and A_1 as in proposition 13.

We compute a standard basis of H up to degree κ in order to compute the $\underline{c}$-basis representation of the Hodge filtration F. The nilpotent part of A_1 is the $\underline{c}$-basis representation of N. By computing images and quotients of C-vectorspaces, we compute the $\underline{c}$-basis representation of a Hodge splitting $F_k C^\beta = \bigoplus_{j\leq k} C^{\beta,j}$. Then we compute a constant $\mathbb{C}[\![s]\!]$-basis transformation $\underline{f} = \underline{c}U$ such that $\underline{f}$ is a $<_{\mathbb{Q}\times\mathbb{Z}}$-decreasingly ordered C-basis of the direct sum $\bigoplus_{\alpha\leq\beta<\alpha+1}\bigoplus_{k\in\mathbb{Z}} C^{\beta,k}$ compatible with the direct sum.

We replace H by a reduced minimal standard basis of $\langle H\rangle$ up to degree $\kappa+1$. By a normal form computation with respect to H up to degree $\kappa+1$, we compute the 1-jet

$$\mathrm{jet}_1\big(H^{-1}(A+s^2\partial_s)H\big) = \mathrm{jet}_1\big(\mathrm{jet}_{\kappa+1}(H)^{-1}\big(\mathrm{jet}_{\kappa+1}(A)+s^2\partial_s\big)\mathrm{jet}_{\kappa+1}(H)\big)$$

of the $\underline{c}H$-matrix A of t in order to compute A_0 and A_1 as in proposition 10.

8 An Example

The algorithm in section 7 is implemented in the computer algebra system SINGULAR[6] in the procedure **tmatrix** in the library **gaussman.lib**[12]. In an example SINGULAR session, we compute the differential structure of the Brieskorn lattice of the singularity of type $T_{2,5,5}$ defined by the polynomial $f = x^2y^2 + x^5 + y^5$.

First, we load the SINGULAR library **gaussman.lib**:

```
> LIB "gaussman.lib";
```

Then, we define the local ring $R = \mathbb{Q}[x,y]_{\langle x,y\rangle}$ with the local degree ordering **ds** as monomial ordering and the polynomial $f = x^2y^2+x^5+y^5 \in R$:

```
> ring R=0,(x,y),ds;
> poly f=x2y2+x5+y5;
```

Finally, we compute A_0 and A_1 as in proposition 10:

```
> list A=tmatrix(f);
```

The result is the list `A=A[1],A[2]` such that $\mathtt{A[i+1]} = A_i$ and

$$A_0 = \begin{pmatrix} 0 & 0 & \cdots & 0 \\ \vdots & \vdots & & \vdots \\ 0 & 0 & \cdots & 0 \\ 1 & 0 & \cdots & 0 \end{pmatrix}, \quad A_1 = \mathrm{diag}\Big(\frac{1}{2}, \frac{7}{10}, \frac{7}{10}, \frac{9}{10}, \frac{9}{10}, 1, \frac{11}{10}, \frac{11}{10}, \frac{13}{10}, \frac{13}{10}, \frac{3}{2}\Big).$$

By proposition 10, $(H'',t) \cong (\mathbb{C}\{\{s\}\}^\mu, A_0 + sA_1 + s^2\partial_s)$ and the spectral pairs are $(-\frac{1}{2},2)$, $(-\frac{3}{10},1)^2$, $(-\frac{1}{10},1)^2$, $(0,1)$, $(\frac{1}{10},1)^2$, $(\frac{3}{10},1)^2$, $(\frac{1}{2},0)$.

References

1. V.I. Arnold, S.M. Gusein-Zade, and A.N. Varchenko. *Singularities of Differentiable Maps*, volume II. Birkhäuser, 1988.
2. E. Brieskorn. Die Monodromie der isolierten Singularitäten von Hyperflächen. *Manuscr. Math.*, 2:103–161, 1970.
3. B. Buchberger. *Gröbner bases: an algorithmic method in polynomial ideal theory.* In N.K. Bose, editor, *Recent trends in multidimensional system theory.* Reidel, 1985.
4. P. Deligne. *Equations différentielles à points singuliers réguliers*, volume 163 of *Lect. Notes Math.* Springer, 1970.
5. P. Deligne. Théorie de Hodge, II. *Publ. Math. I.H.E.S*, 40:5–57, 1972.
6. G.-M. Greuel, G. Pfister, and H. Schönemann. SINGULAR 2.0.3. A Computer Algebra System for Polynomial Computations, Centre for Computer Algebra, University of Kaiserslautern, 2002. `http://www.singular.uni-kl.de`.
7. E.J.N Looijenga. *Isolated Singular Points on Complete Intersections*, volume 77 of *LMS Lect. Note Series.* Cambr. Univ. Press, 1984.
8. B. Malgrange. Intégrales asymptotiques et monodromie. *Ann. scient. Ec. Norm. Sup.*, 7:405–430, 1974.
9. J. Milnor. *Singular Points on Complex Hypersurfaces*, volume 61 of *Ann. Math. Stud.* Princ. Univ. Press, 1968.
10. M. Saito. On the structure of Brieskorn lattices. *Ann. Inst. Fourier Grenoble*, 39:27–72, 1989.
11. M. Schulze. Algorithms for the Gauss-Manin connection. *Journ. Symb. Comp.*, 32(5):549–564, 2001.
12. M. Schulze. `gaussman.lib`. SINGULAR 2.0.3 library, Centre for Computer Algebra, University of Kaiserslautern, 2001. `http://www.singular.uni-kl.de`.
13. M. Schulze. *Algorithmic Gauss-Manin Connection.* PhD thesis, University of Kaiserslautern, 2002.
14. J. Scherk and J.H.M. Steenbrink. On the mixed Hodge structure on the cohomology of the Milnor fibre. *Math. Ann.*, 271:641–655, 1985.
15. M. Sebastiani. Preuve d'une conjecture de Brieskorn. *Manuscr. Math.*, 2:301–308, 1970.
16. J. Steenbrink. Mixed Hodge structure on the vanishing cohomology. In *Real and complex singularities*, pages 525–562. Nordic summer school, Oslo, 1976.
17. A.N. Varchenko. Asymptotic Hodge structure in the vanishing cohomology. *Math. USSR Izvestija*, 18(3):496–512, 1982.

AN EFFICIENT MODULAR ALGORITHM FOR COMPUTING THE GLOBAL B-FUNCTION

MASAYUKI NORO
Kobe University, Rokko, Kobe 657-8501, Japan
email : noro@math.kobe-u.ac.jp

1 Introduction

The *n-dimensional Weyl algebra over a field* $\mathbf{K}$

$$D = \mathbf{K}\langle x_1, \ldots, x_n, \partial_1, \ldots, \partial_n \rangle$$

is an associative algebra generated by $x_1, \ldots, x_n$, $\partial_1, \ldots, \partial_n$ over $\mathbf{K}$ with relations $x_i x_j = x_j x_i$, $\partial_i \partial_j = \partial_j \partial_i$, $\partial_j x_i = x_i \partial_j (i \neq j)$ and $\partial_i x_i = x_i \partial_i + 1$. D is generated by $\{x^\alpha \partial^\beta | \alpha, \beta \in \mathbf{Z}^n_{\geq 0}\}$ as a $\mathbf{K}$ vector space. We call a left ideal in a Weyl algebra a *D-ideal*. For a D-ideal I, its *characteristic variety* $\mathrm{ch}(I)$ is a closed subvariety in $\mathbf{K}^{2n}$ and $\dim \mathrm{ch}(I) \geq n$ always holds. A D-ideal I is called *holonomic* if $\dim \mathrm{ch}(I) = n$. A univariate polynomial called a *b-function* is associated with a holonomic ideal. The (local) b-function was originally introduced in the context of analytic continuation of $f_+(x)^s$ for a real-valued real analytic function $f(x)$. Malgrange[1] pointed out a relation between the local monodromy of a hypersurface $f^{-1}(0)$ and the b-function of f. Recently Oaku[3] presented an algorithm for computing the b-function, and the roots of b-functions are used as bounds for D-module-related algorithms presented by Oaku and Takayama.[4,5] Therefore the efficient computation of b-function is now of practical importance. In this paper we present an efficient modular algorithm for computing global b-functions. First we review several necessary notions due to Saito et al.[7]

Definition 1 *Let D be an n-dimensional Weyl algebra. A real vector $(u, v) = (u_1, \ldots, u_n, v_1, \ldots, v_n) \in \mathbf{R}^{2n}$ is called a weight vector for D if $u_i + v_i \geq 0$ for $i = 1, \ldots, n$.*

We consider two types of weight vectors: $(-w, w)$ and $(0, e)$, where $e = (1, \ldots, 1)$. We can express $p \in D$ as

$$p = \sum_{(\alpha,\beta) \in E} c_{\alpha\beta} x^\alpha \partial^\beta,$$

where E is a finite subset of $\mathbf{Z}^{2n}_{\geq 0}$ and $c_{\alpha\beta} \in \mathbf{K} \setminus \{0\}$. For p expressed in this manner we define two types of initial form.

Definition 2 *Put* $m = \max_{(\alpha,\beta)\in E}(-w\alpha + w\beta)$. *We define the initial form* $\mathrm{in}_{(-w,w)}(p) \in D$ *of* p *with respect to* $(-w, w)$ *as*

$$\mathrm{in}_{(-w,w)}(p) = \sum_{(\alpha,\beta)\in E, -w\alpha+w\beta=m} c_{\alpha\beta}x^{\alpha}\partial^{\beta}.$$

For a D-ideal I, $\mathrm{in}_{(-w,w)}(I) = \mathbf{K}\cdot\{\mathrm{in}_{(-w,w)}(p) | p \in I\}$ *is a D-ideal.*

Definition 3 *Put* $m = \max_{(\alpha,\beta)\in E}(e\beta)$. *We define the initial form* $\mathrm{in}_{(0,e)}(p) \in \mathbf{K}[x,\xi] = \mathbf{K}[x_1,\ldots,x_n,\xi_1,\ldots,\xi_n]$ *of* p *with respect to* $(0,e)$ *as*

$$\mathrm{in}_{(0,e)}(p) = \sum_{(\alpha,\beta)\in E, e\beta=m} c_{\alpha\beta}x^{\alpha}\xi^{\beta}.$$

For a D-ideal I, $\mathrm{in}_{(0,e)}(I) = \mathbf{K}\cdot\{\mathrm{in}_{(0,e)}(p) | p \in I\}$ *is an ideal in* $\mathbf{K}[x,\xi]$.

Definition 4 *For D-ideal* I, *the characteristic variety* $\mathrm{ch}(I)$ *is defined as the zero set of* $\mathrm{in}_{(0,e)}(I)$ *in* $\mathbf{K}^{2n}$ *and is denoted by* $\mathrm{ch}(I)$. *A D-ideal* I *is called holonomic if* $\dim(\mathrm{ch}(I)) = n$.

Definition 5 *Let* I *be a holonomic D-ideal. For* $w \in \mathbf{R}^n \setminus \{0\}$ *we set* $s = w_1\theta_1 + \cdots + w_n\theta_n$, *where* $\theta_i = x_i\partial_i$. *Then* $\mathrm{in}_{(-w,w)}(I) \cap \mathbf{K}[s]$ *is a nonzero principal ideal in* $\mathbf{K}[s]$. *The monic generator* $b(s)$ *of the principal ideal is called the global b-function of* I *with respect to a weight vector* w.

Definition 6 *Consider a D-ideal* I_f *in an* $(n+1)$*-dimensional Weyl algebra* $D\langle t, \partial_t\rangle$:

$$I_f = \mathrm{Id}\left(t - f, \partial_1 + \frac{\partial f}{\partial x_1}\partial_t, \ldots, \partial_n + \frac{\partial f}{\partial x_n}\partial_t\right).$$

I_f *is known to be holonomic (cf. Saito* et al.[7]*). Let* $w = (1,0,\ldots,0) \in \mathbf{R}^{n+1}$ *be a weight vector such that the weight of* ∂_t *is 1, and* $B(s)$ *the b-function of* I_f *with respect to* w. *We call* $b(s) = B(-s-1)$ *the global b-function of* f.

2 Algorithms for computing b-functions

According to the definition, the b-function of a holonomic D-ideal I with respect to the weight $(-w, w)$ can be computed as follows:

1. Compute $J = \mathrm{in}_{(-w,w)}(I)$.
2. Compute the minimal polynomial of s in D/J.

For Step 1, we propose a weighted homogenization, which improves the efficiency for $I = I_f$. For Step 2, we present an algorithm for computing the minimal polynomial by the method of indeterminate coefficients.

2.1 Computation of $\mathrm{in}_{(-w,w)}(I)$

According to Saito et al.[7], $\mathrm{in}_{(-w,w)}(I)$ can be computed as follows:

1. Define a (non-term) monomial order $<_{(-w,w)}$ via a term order $<$:

$$x^\alpha \partial^\beta <_{(-w,w)} x^\gamma \partial^\delta \Leftrightarrow -w\alpha + w\beta < -w\gamma + w\delta$$
$$\text{or } (-w\alpha + w\beta = -w\gamma + w\delta \text{ and } x^\alpha\partial^\beta < x^\gamma \partial^\delta)$$

   Compute a Groebner basis G of I with respect to $<_{(-w,w)}$.

2. $G_{(-w,w)} = \{\mathrm{in}_{(-w,w)}(g) | g \in G\}$ is a Groebner basis of $\mathrm{in}_{(-w,w)}(I)$ with respect to the term order $<$.

For computing a Groebner basis with respect to a non-term order $<_{(-w,w)}$, Saito et al.[7] proposed a homogenization with $\partial_i x_i = x_i \partial_i + h^2$. We generalize this homogenization in the following manner. Let $u, v \in \mathbf{R}^n_{>0}$. $D^{(h)}_{(u,v)}$ is an associative algebra generated by $x_1, \ldots, x_n$, $\partial_1, \ldots, \partial_n, h$ over $\mathbf{K}$ with fundamental relations $x_i x_j = x_j x_i$, $\partial_i \partial_j = \partial_j \partial_i$, $h\partial_i = \partial_i h$, $hx_i = x_i h$, $\partial_j x_i = x_i \partial_j (i \neq j)$, and $\partial_i x_i = x_i \partial_i + h^{u_i + v_i}$. For $p = \sum c_{\lambda\alpha\beta} h^\lambda x^\alpha \partial^\beta$, the weighted total degree is defined as $\deg_{(u,v)}(p) = \max\{\lambda + u\alpha + v\beta | c_{\lambda\alpha\beta} \neq 0\}$, and for $p = \sum c_{\alpha\beta} x^\alpha \partial^\beta$ the weighted homogenization is defined as

$$H_{(u,v)}(p) = \sum c_{\alpha\beta} h^{\deg_{(u,v)}(p) - (u\alpha + v\beta)} x^\alpha \partial^\beta.$$

The product of two weighted homogeneous elements is again weighted homogeneous. For a monomial order $<$ in D, we define a term order $<^h$ in $D^{(h)}_{(u,v)}$ as follows:

$$t <^h s \Leftrightarrow \deg_{(u,v)}(t) < \deg_{(u,v)}(s)$$
$$\text{or } (\deg_{(u,v)}(t) = \deg_{(u,v)}(s) \text{ and } t|_{h=1} < s|_{h=1})$$

Then we have the following theorem (cf. Saito et al.[7], Theorem 1.2.4):
Theorem 7 *Let F be a finite subset of D. If G^h is a Groebner basis of* $\mathrm{Id}(H_{(u,v)}(F))$ *with respect to $<^h$, then $G^h|_{h=1}$ is a Groebner basis of* $\mathrm{Id}(F)$ *with respect to $<$.*
Since $<^h$ is a term order, we can compute a Groebner basis of $\mathrm{Id}(H_{(u,v)}(F))$ via the Buchberger algorithm. However, the execution of the Buchberger algorithm over $\mathbf{Q}$ often suffers from large intermediate coefficients during normal form computations. Although Traverso's trace lifting[8] may reduce such costs, nevertheless the final check that the obtained Groebner basis candidate is in fact a Groebner basis may be difficult. But, if the input set is a Groebner

basis with respect to some term order, we can make use of the modular change of ordering algorithm.[2] Let $\mathbf{Z}_{(p)} = \{a/b | a \in \mathbf{Z}, b \notin p\mathbf{Z}\}$ be the localization of $\mathbf{Z}$ at $p\mathbf{Z}$ and ϕ_p the canonical projection from $\mathbf{Z}_{(p)}$ onto $GF(p)$.

Algorithm 8
ModularChangeOfOrdering$(G_0, <_0, <)$
Input: a Groebner basis $G_0 \subset D$ *with respect to* $<_0$
with each element having a monic head term
a term order $<$
Output: the reduced Groebner basis G *of* $\mathrm{Id}(G_0)$ *with respect to* $<$
do
$p \leftarrow$ *a new prime such that* $G_0 \subset \mathbf{Z}_{(p)}[x, \partial]$
$G_p \leftarrow$ *the reduced Groebner basis with respect to* $<$
If there exists $G \subset \mathrm{Id}(G_0)$ *such that* $\phi_p(G) = G_p$ *then return* G
end do

If G_0 is not zero dimensional, we cannot apply FGLM-type algorithms for computing the inverse image G. Instead we can apply the trace lifting algorithm without the final check, which will also improve the performance in some cases. We note that we can apply Algorithm 8 repeatedly, that is we can traverse multiple term orderings without the final check.

2.2 A weight vector for efficient computation of the global b-function of a polynomial

Here we give a weight vector that is convenient for efficient computation of the global b-function of a polynomial f. For a polynomial f and $u \in \mathbf{R}^n_{>0}$, $\deg_u(f)$ denotes the weighted total degree of f with respect to u. Let $(\hat{u}, \hat{v})$ be a weight vector in $D\langle t, \partial_t\rangle = \mathbf{K}\langle t, x_1, \ldots, x_n, \partial_t, \partial_1, \ldots, \partial_n\rangle$, where $\hat{u} = (\deg_u(f), u_1, \ldots, u_n)$ and $\hat{v} = (1, \deg_u(f) - u_1 + 1, \deg_u(f) - u_2 + 1, \ldots, \deg_u(f) - u_n + 1)$. Then $\hat{u} + \hat{v} = d \cdot (1, \ldots, 1)$ and $\partial_t t = t\partial_t + h^d$, $\partial_i x_i = x_i \partial_i + h^d$ hold, where $d = \deg_u(f) + 1$. Let $F = \{t - f, \partial_1 + \frac{\partial f}{\partial x_1}\partial_t, \ldots, \partial_n + \frac{\partial f}{\partial x_n}\partial_t\}$ be the generator set of I_f. If we set $\hat{f} = H_{(\hat{u},\hat{v})}(f)$, then $H_{(\hat{u},\hat{v})}(F) = \{t - \hat{f}, \partial_1 + \frac{\partial \hat{f}}{\partial x_1}\partial_t, \ldots, \partial_n + \frac{\partial \hat{f}}{\partial x_n}\partial_t\}$.

Lemma 9 *For any term order* $<_0$ *such that* $\mathrm{in}_{<_0}(H_{(\hat{u},\hat{v})}(F)) = \{t, \partial_1, \ldots, \partial_n\}$, $H_{(\hat{u},\hat{v})}(F)$ *is a Groebner basis of* $\mathrm{Id}(H_{(\hat{u},\hat{v})}(F))$ *with respect to* $<_0$.

Proof We set $g_t = t - \hat{f}$, $g_i = \partial_i + \frac{\partial \hat{f}}{\partial x_i}\partial_t$ $(i = 1, \ldots, n)$. By direct calculation, we have $g_t g_i = g_i g_t$, $g_i g_j = g_j g_i$, and

$$\mathrm{Spoly}(g_t, g_i) = \partial_i g_t - t g_i = -(g_i - \partial_i) g_t + (g_t - t) g_i,$$

$$\mathrm{in}_{<_0}((g_i - \partial_i) g_t), \mathrm{in}_{<_0}((g_t - t) g_i) <_0 \mathrm{lcm}(\mathrm{in}_{<_0}(g_t), \mathrm{in}_{<_0}(g_i)),$$

$$\mathrm{Spoly}(g_i, g_j) = \partial_j g_i - \partial_i g_j = -(g_j - \partial_j) g_i + (g_i - \partial_i) g_j,$$
$$\mathrm{in}_{<_0}((g_j - \partial_j) g_i), \mathrm{in}_{<_0}((g_i - \partial_i) g_j) <_0 \mathrm{lcm}(\mathrm{in}_{<_0}(g_i), \mathrm{in}_{<_0}(g_j)),$$

which imply that $H_{(\hat{u},\hat{v})}(F)$ is a Groebner basis with respect to $<_0$. ∎

By Lemma 9, the Groebner basis computation of $\mathrm{Id}(H_{(u,v)}(F))$ with respect to $<^h$ is a change of ordering and we can apply Algorithm 8.

Example 10 *For $d(f_5)$ in Section 3, we can compute the Groebner basis candidate of $H_{(\hat{u},\hat{v})}(F)$ in 50 seconds on a PC with the trace lifting algorithm, whereas it takes 240 seconds to verify that the candidate is a Groebner basis. By applying Algorithm 8 we can reduce such time cost.*

If f is weighted homogeneous with respect to u, so is F with respect to $(\hat{u}, \hat{v})$ and we have $H_{(\hat{u},\hat{v})}(F) = F$. Later we will show examples that can be computed efficiently by the weighted homogenization in this case.

2.3 Computation of the minimal polynomial in D/J

For computing the minimal polynomial in D/J the following method is proposed by Oaku[3] and Saito et al.[7]:

1. Compute $J' = J \cap \mathbf{K}[\theta_1, \ldots, \theta_n]$.

2. Compute $J' \cap \mathbf{K}[s]$.

In general the computation of an elimination ideal J' is hard, but in this case we already have a Groebner basis with respect to a term order, which enables us to apply the method of indeterminate coefficients to compute the minimal polynomial because the existence of the global b-function is guaranteed. In the following, $\mathrm{NF}(f, G)$ denotes the normal form of f with respect to a Groebner basis G.

Algorithm 11

MinimalPolynomial$(G, <, P)$
Input: *a Groebner basis G of a D-ideal J with respect to a term order $<$*
    $P \in D$ such that $J \cap \mathbf{K}[P] \neq \{0\}$
Output: *$b(s) \in \mathbf{K}[s]$ such that $J \cap \mathbf{K}[P] = \mathrm{Id}(b(P))$*
$i \leftarrow 1$
do {
  If there exist $a_{i-1}, \ldots, a_0 \in \mathbf{K}$ such that
    $\mathrm{NF}(P^i, G) + \sum_{j=0}^{i-1} a_j \mathrm{NF}(P^j, G) = 0$
  then return $s^i + \sum_{j=0}^{i-1} a_j s^j$
  else $i \leftarrow i + 1$
}

This algorithm searches a minimal degree polynomial $b(s)$ such that $b(P) \in J$. The condition $b(P) \in J$ is equivalent to $\mathrm{NF}(b(P), G) = 0$, and we have a system of linear equations for the a_js. When the ground field is $\mathbf{Q}$, it is practical to incorporate a modular method, again to avoid unnecessarily large coefficients during the Gaussian elimination. Let D be a Weyl algebra over $\mathbf{Q}$, J a D-ideal, and $P \in D$ such that $J \cap \mathbf{Q}[P] = \mathrm{Id}(b(P))$ for $b(s) \in \mathbf{Z}[s]$. We may suppose that P has integral coefficients and b is primitive over $\mathbf{Z}$. Let G be a Groebner basis with respect to a term order $<$ such that each element in G is monic with respect to $<$.

Lemma 12 *For a prime p such that $G \subset \mathbf{Z}_{(p)}\langle x, \partial\rangle$, $\phi_p(G)$ is a Groebner basis of $\mathrm{Id}(\phi_p(G))$ with respect to $<$ and $\phi_p(b(P)) \in \mathrm{Id}(\phi_p(G))$.*

Proof For a polynomial with $\mathbf{Z}_{(p)}$ coefficients, its normal form computation modulo G can be done over $\mathbf{Z}_{(p)}$. Then we can apply ϕ_p to the reduction process and thus obtain the reduction over $GF(p)$ by $\phi_p(G)$. ∎

Since $\phi_p(b(s))$ is nonzero and $\phi_p(b(P)) \in \mathrm{Id}(\phi_p(G))$, there exists the minimal polynomial of $\phi_p(P)$ in $\phi_p(D)/\mathrm{Id}(\phi_p(G))$. Thus $b_p(s)|\phi_p(b(s))$ holds, which implies $\deg(b_p(s)) \leq \deg(b(s))$.

Theorem 13 *Let $b_p(s)$ be the minimal polynomial of $\phi_p(P)$ in $\phi_p(D)/\mathrm{Id}(\phi_p(G))$. If there exists $f \in \mathbf{Z}[s]$ such that $\deg(f(s)) = \deg(b_p(s))$ and $f(P) \in \mathrm{Id}(G)$, then $f(s) = b(s)$.*

Proof $f(\phi_p(P)) \in \mathrm{Id}(\phi_p(G))$ implies $b(s)|f(s)$. By the assumption, $\deg(f(s)) = \deg(b_p(s)) \leq \deg(b(s))$ holds. Hence $b(s) = f(s)$. ∎

Now we have the following algorithm:

Algorithm 14

ModularMinimalPolynomial$(G, <, P)$

Input: *A Groebner basis G of a D-ideal J with respect to a term order $<$,*
  each element in G being monic with respect to $<$
  $P \in D$ such that P is integral and $J \cap \mathbf{Q}[P] \neq \{0\}$

Output: *$b(s) \in \mathbf{Q}[s]$ such that $J \cap \mathbf{Q}[P] = \mathrm{Id}(b(P))$*

start:

$p \leftarrow$ *a new prime such that* $G \in \mathbf{Z}_{(p)}\langle x, \partial\rangle$

$i \leftarrow 1$

do {

  If there exist $a_{i-1}, \ldots, a_0 \in GF(p)$ *such that*

(L_p) $\phi_p(\mathrm{NF}(P^i, G)) + \sum_{j=0}^{i-1} a_j \phi_p(\mathrm{NF}(P^j, G)) = 0$

  then {

    If there exist $a_{i-1}, \ldots, a_0 \in \mathbf{Q}$ *such that*

(L) $\mathrm{NF}(P^i, G) + \sum_{j=0}^{i-1} a_j \mathrm{NF}(P^j, G) = 0$

      return $s^i + \sum_{j=0}^{i-1} a_j s^j$

else goto **start**

} *else* $i \leftarrow i + 1$

}

In this algorithm, the system of linear equations is solved over $\mathbf{Q}$ only when it has a solution over $GF(p)$. If p is sufficiently large, the condition $\deg(b(s)) = \deg(b_p(s))$ is established with high probability and solving over $\mathbf{Q}$ succeeds in most cases. The computation of the minimal polynomial over $GF(p)$ can be done efficiently by triangularizing $\{\mathrm{NF}(\phi_p(P^j), \mathrm{Id}(\phi_p(G)))|j = 1, \ldots i\}$ as linear forms of monomials. If p does not divide the leading coefficient of $b(s) \in \mathbf{Z}[s]$, f in Theorem 13 exists, whence Algorithm 14 terminates. If (L_p) has a solution for the first time at $i = i_0$, it is unique and is equal to $b_p(s)$. If (L) also has a solution at $i = i_0$, the solution is unique because of the uniqueness of $b_p(s)$. For solving (L) we can apply the Hensel lifting method. (See Noro and Yokoyama[2] for the details.)

2.4 An algorithm for computing the global b-function of a polynomial

Combining the methods described in this section, we have an algorithm for computing the global b-function of an integral polynomial. We use the following settings in the algorithm:

1. A weight vector $(\hat{u}, \hat{v})$ for the weighted homogenization.

   We set $\hat{u} = (d, u_1, \ldots, u_n)$, $\hat{v} = (1, d - u_1 + 1, \ldots, d - u_n + 1)$, where $u = (u_1, \ldots, u_n)$ and $d = \deg_u(f)$. By default we set $u = (1, 1, \ldots, 1)$, or we may set u heuristically according to the form of f.

2. A term order $<$ to be used as a tie breaker for the weight $(-w, w)$.

   We use a weighted graded reverse lexicographic (wgrlex) order with a weight vector $(\hat{u}, \hat{v})$ such that $t > x_1 > x_2 > \cdots > x_n > \partial_t > \partial_1 > \partial_2 > \cdots > \partial_n$.

Algorithm 15

Input: an integral polynomial $f(x_1, \ldots, x_n)$

an optional weight vector u

Output: the global b-function of f

If u *is not given then* $u \leftarrow (1, 1, \ldots, 1)$

$d \leftarrow \deg_u(f)$

$w \leftarrow (1, 0, \ldots, 0)$

$\hat{u} \leftarrow (d, u_1, \ldots, u_n)$

$\hat{v} \leftarrow (1, d + 1 - u_1, \ldots, d + 1 - u_n)$

$\hat{f} \leftarrow H_{(\hat{u}, \hat{v})}(f)$

$\hat{G}_f \leftarrow \{t - \hat{f}, \partial_1 + \frac{\partial \hat{f}}{\partial x_1}\partial_t, \ldots, \partial_n + \frac{\partial \hat{f}}{\partial x_n}\partial_t\}$
$<_0 \leftarrow$ *a term order such that* $\hat{G}_f$ *is a Groebner basis with respect to* $<_0$
$< \leftarrow$ *the wgrlex order with respect to* $(\hat{u}, \hat{v})$
$<^h_{(-w,w)} \leftarrow$ *the homogenized term order with tie breaker* $<$
$G^h \leftarrow$ *ModularChangeOfOrdering*$(\hat{G}_f, <_0, <^h_{(-w,w)})$
$G \leftarrow \mathrm{in}_{(-w,w)}(G^h|_{h=1})$
$B(s) \leftarrow$ *ModularMinimalPolynomial*$(G, <, t\partial_t)$
return $B(-s-1)$

3 Experiments

Algorithm 15 was implemented in Risa/Asir[11] using Weyl algebra arithmetics and the Buchberger algorithm. In the following tables we display the results of the computation of the global b-functions for various examples taken mainly from Yano.[9,10] Measurements were made on a PC with PentiumIII 1.26GHz and 2GB of memory. Times including garbage collection time, are given in seconds. Table 1 shows the results for $x^n + xy^{m-1} + y^m$. The wgrlex order is chosen for the weight vector $u = (1, 1)$ and $x > y$.

n, m	$\deg(b(s))$	time
4,5	13	2.5
4,6	13	1.5
4,7	19	2.1
4,8	11	0.8
4,9	25	3.1
5,5	8	0.1
5,6	21	21
5,7	25	11
5,8	29	10
5,9	33	12
6,5	20	34
6,6	10	0.2
6,7	31	91

n, m	$\deg(b(s))$	time
6,8	25	26
6,9	21	16
7,5	16	31
7,6	30	217
7,7	12	0.5
7,8	43	440
7,9	49	181
8,5	26	32
8,6	34	103
8,7	42	835
8,8	14	0.9
8,9	57	1220

Table 1. $x^n + xy^{m-1} + y^m$

Table 2 shows the results for $x^{n_1} + y^{n_2} + z^{n_3} + x^{m_1}y^{m_2}z^{m_3}$. The variable order is fixed for $x > y > z$ and several weight vectors are used.

n, m	$\deg(b(s))$	u:time
(6,6,6),(3,3,3)	12	$(1,1,1)$:12
(6,6,6),(4,4,4)	13	$(1,1,1)$:166
(7,7,7),(2,2,2)	28	$(1,1,1)$:185
(7,7,7),(2,2,3)	17	$(1,1,1)$:102
(8,8,8),(2,2,2)	23	$(1,1,1)$:38
(9,9,9),(3,3,3)	23	$(1,1,1)$:199
(6,6,7),(2,2,2)	60	$(1,1,1)$:$> 2h$, $(2,1,1)$:3428, $(2,2,1)$:5100
(6,6,7),(3,3,3)	49	$(1,1,1)$:39, $(2,2,1)$:88
(6,6,7),(4,4,4)	53	$(1,1,1)$:21, $(2,2,1)$:73
(6,7,7),(2,2,2)	62	$(2,1,1)$:2604

Table 2. $x^{n_1} + y^{n_2} + z^{n_3} + x^{m_1} y^{m_2} z^{m_3}$

Table 3 shows the results for polynomials with non-isolated singularities. From Table 3, we can guess that the global b-function of $(x_1x_2)^k + (x_3x_4)^k + (x_5x_6)^k$ is equal to $(s+1)(s+\frac{3k-2}{k})^3(s+\frac{3k-1}{k})^2(s+3)\cdot\prod_{j=3}^{3k-3}(s+\frac{j}{k})^4$.

polynomial	$b(s)$	time
$(x_1x_2)^2 + (x_3x_4)^2 + (x_5x_6)^2$	$(s+1)(s+\frac{3}{2})^4(s+2)^3(s+\frac{5}{2})^2(s+3)$	0.4
$(x_1x_2)^3 + (x_3x_4)^3 + (x_5x_6)^3$	$(s+1)^5(s+\frac{4}{3})^4(s+\frac{5}{3})^4(s+2)^4$ $(s+\frac{7}{3})^3(s+\frac{8}{3})^2(s+3)$	15
$(x_1x_2)^4 + (x_3x_4)^4 + (x_5x_6)^4$	$(s+\frac{3}{4})^4(s+1)^5(s+\frac{5}{4})^4(s+\frac{3}{2})^4(s+\frac{7}{4})^4$ $(s+2)^4(s+\frac{9}{4})^4(s+\frac{5}{2})^3(s+\frac{11}{4})^2(s+3)$	447
$(x_1x_2)^5 + (x_3x_4)^5 + (x_5x_6)^5$	$(s+\frac{3}{5})^4(s+\frac{4}{5})^4(s+1)^5(s+\frac{6}{5})^4(s+\frac{7}{5})^4$ $(s+\frac{8}{5})^4(s+\frac{9}{5})^4(s+2)^4(s+\frac{11}{5})^4$ $(s+\frac{12}{5})^4(s+\frac{13}{5})^3(s+\frac{14}{5})^2(s+3)$	40770

Table 3. $(x_1x_2)^n + (x_3x_4)^n + (x_5x_6)^n$

Table 4 shows the results for polynomials constructed as follows. Let $f_n(t) = t^n + x_1t^{n-1} + \cdots + x_{n-1}t + x_n$ be a monic polynomial of t with n parameters $x_1, \ldots, x_n$. We compute the global b-function of the discriminant $d(f_n) = \mathrm{resultant}_t(f_n, \frac{df_n}{dt})$. We use $u = (1, 2, \ldots, n)$ because f_n is weighted homogeneous with respect to this weight vector. When we applied Algorithm 15 to this case, the cost of computing $\mathrm{NF}((t\partial_t)^i, G)$ was very high in Algorithm 14. One reason for this is that t and ∂_t have higher orders than other variables, so we heuristically employed the following variant:

After obtaining $G = \mathrm{in}_{(-w,w)}(G^h|_{h=1})$, we compute a Groebner basis G_1

with respect to another wgrlex order such that $\partial_t <_1 \partial_n <_1 \cdots <_1 \partial_1 <_1 t <_1 x_n <_1 \cdots <_1 x_1$ *by Algorithm 8. Then we apply Algorithm 14.*

For $n = 4$ we also show the time required for the simple nonweighted homogenization, clearly demonstrating that the weighted homogenization is more suitable for inputs of this type. Yano[9] computes by micro-local calculus the local b-function at 0, which is equal to

$$\prod_{2 \le k \le n} \prod_{1 \le m \le k-1} (s + \frac{1}{2} + \frac{m}{k})$$

for $n = 5$. Our experiment shows that the global b-functions of $d(f_4)$, $d(f_5)$, $d(f_6)$ are represented by the above formula. For $n = 6$, almost all time is spent on the computation of $\mathrm{in}_{(-w,w)}(I_f)$ and it is a kind of intermediate expression swell. It seems that further improvements are necessary to handle such cases efficiently.

n	$b(s)$	u:time
4	$(s+\frac{3}{4})(s+\frac{5}{6})(s+1)^2(s+\frac{7}{6})(s+\frac{5}{4})$	$(1,2,3,4)$:0.4, simple homo:3110
5	$(s+\frac{7}{10})(s+\frac{3}{4})(s+\frac{5}{6})(s+\frac{9}{10})$ $(s+1)^2(s+\frac{11}{10})(s+\frac{7}{6})(s+\frac{5}{4})(s+\frac{13}{10})$	$(1,2,3,4,5)$:83
6	$(s+\frac{2}{3})(s+\frac{7}{10})(s+\frac{3}{4})(s+\frac{5}{6})^2$ $(s+\frac{9}{10})(s+1)^3(s+\frac{11}{10})(s+\frac{7}{6})^2$ $(s+\frac{5}{4})(s+\frac{13}{10})(s+\frac{4}{3})$	$(1,2,3,4,5,6)$:5days

Table 4. Discriminant of $u^n + x_i u^{n-1} + \cdots + x_{n-1} u + x_n$

4 Conclusion and future work

In this paper we have given a modular algorithm for computing the global b-function of a holonomic D-ideal. In the algorithm the global b-function is computed directly as the minimal polynomial of an element. For computing the global b-function of a polynomial, we introduced a weighted homogenization associated with the ideal I_f, which makes the subsequent Groebner basis computation a change of ordering and thus permits the Groebner basis check to be omitted in the trace lifting algorithm. We also applied a modular method for the minimal polynomial computation to avoid unnecessarily large coefficients. OpenXM package[11] provides many functions for computing D-module-related invariants, which require the roots of b-functions as bounds. In the isolated singularity case, the roots of the local b-function can be computed as eigenvalues of the local monodromy matrix and the implementation by Shulze[6] in SINGULAR computes them very efficiently. However, at present the algorithm by Oaku[3] is the only one that can compute the b-function for a general input and thus it is important that its efficiency is improved. As seen in Section 3, our new method enables us to compute the global b-functions for many nontrivial inputs, but we often have to set weights or variable orders heuristically. As future work we will seek an algorithm to set these parameters automatically. In addition, there are cases where the normal form computations are very involved when the number of variables or the total degree of the input polynomial is large. Some such cases are handled efficiently by the heuristic change of ordering described in Section 3, but this is too simplistic to apply more generally so another improvement in the method will be required to treat these difficult cases.

References

1. Malgrange, B. (1974): Intégrales asymptotiques et monodromie. *Ann. Sci. ENS* **4**, 405–430.
2. Noro, M., Yokoyama, K. (1999): A Modular Method to Compute the Rational Univariate Representation of Zero-Dimensional Ideals. *J. Symb. Comp.*, **28**,1, 243–263.
3. Oaku, T. (1997): Algorithm for the b-function and D-modules associated to a polynomial. *J. Pure Appl. Algebra*, **117** & **118**, 495–518.
4. Oaku, T., Takayama, N. (1999): An algorithm for de Rham cohomology groups of the complement of an affine variety via D-module computation. *J. Pure Appl. Algebra*, **139**, 201–233.
5. Oaku, T., Takayama, N. (2001): Algorithms for D-modules — restriction, tensor product, localization, and local cohomology groups. *J. Pure Appl. Algebra*, **164**, 199-220.
6. Schulze, M. (2001): Algorithms for the Gauss-Manin connection. *J. Symb. Comp.*, **32**,5, 549–564.
7. Saito, M., Sturmfels, B., Takayama, N. (2000): *Gröbner deformations of hypergeometric differential equations.* Algorithms and Computation in Mathematics **6**. Springer-Verlag.
8. Traverso, C. (1988): Groebner trace algorithms. LNCS **358** (Proceedings of ISSAC'88), 125–138. Springer-Verlag.
9. Yano, T. (1977): On the Holonomic System of f^s and b-functions. *Publ. RIMS Kyoto Univ.* **12** Suppl., 469–480.
10. Yano, T. (1978): On the theory of b-functions. *Publ. RIMS Kyoto Univ.* **14**, 111–202.
11. `http://www.openxm.org/`.

A METHOD FOR CONSTRUCTING HOLONOMIC SYSTEMS FOR ALGEBRAIC LOCAL COHOMOLOGY CLASSES WITH SUPPORT ON A ZERO DIMENSIONAL VARIETY

YAYOI NAKAMURA

Graduate School of Humanities and Sciences, Ochanomizu University, 2-1-1 Ohtsuka Bunkyo-ku, Tokyo, 112-8610 Japan.
E-mail: nakamura@math.ocha.ac.jp

SHINICHI TAJIMA

Department of Information Engineering, Niigata University, 2-8050, Ikarashi, Niigata, 950-2181 Japan.
E-mail: tajima@ie.niigata-u.ac.jp

In this paper, we consider zero-dimensional algebraic local cohomology classes in the context of $\mathcal{D}_X$-modules and give a method for computing annihilating differential operators of algebraic local cohomology classes. There is an algorithm due to T.Oaku [3] that computes the $\mathcal{D}_X$-module structures of algebraic local cohomology classes. This algorithm works on any codimensional case and gives the Gröbner basis of the annihilating ideal for algebraic local cohomology classes. For the zero dimensional case, it is known that the associated holonomic system is simple as $\mathcal{D}_X$-module, while the explicit computation of annihilators in the algebra is of great importance in many applications. We thus consider the zero dimensional case from a computational point of view. By using the zero dimensionality, we derive an effective method for computing annihilating differential operators to a zero dimensional algebraic local cohomology class, which is completely different from that in [3]. Actually, resulting algorithms do not need the computation of the b-function nor the use of the Gröbner basis computation in the Weyl algebra.

To be more precise, we introduce some notations. Let $F = \{f_1, \ldots, f_n\}$ be a regular sequence of polynomials $f_j \in \mathbb{Z}[z]$ of n variables $z = (z_1, \ldots, z_n)$ in $X = \mathbb{C}^n$. Denote by I the ideal of $\mathbb{Q}[z]$ generated by these n polynomials $f_1, \ldots, f_n$ and Z the variety $V(I) \subset X$ of the ideal I. Let $H^n_{[Z]}(\mathcal{O}_X) = \Gamma(X, \mathcal{H}^n_{[Z]}(\mathcal{O}_X))$ be the global section of the algebraic local cohomology group $\mathcal{H}^n_{[Z]}(\mathcal{O}_X)$ with support on Z where $\mathcal{O}_X$ is the sheaf of holomorphic functions on X. We consider the algebraic local cohomology class $\sigma_F = [1/f_1 \ldots f_n]$, the image by i of the element $\begin{bmatrix} 1 \\ f_1 \ldots f_n \end{bmatrix}$ of the extension group $Ext^n_{\mathcal{O}_X}(\mathcal{O}_X/I, \mathcal{O}_X)$ where i is the canonical mapping $i : Ext^n_{\mathcal{O}_X}(\mathcal{O}_X/I, \mathcal{O}_X) \to H^n_{[Z]}(\mathcal{O}_X)$. We give a method for constructing first and second order differential operators which annihilate the given algebraic local cohomology class σ_F and its *direct summands*. The main feature of this method is in its simplicity. Besides, the program is simple, and the operating speed is fast. The resulting algorithms thus provide practical tools for finding generators of the annihilating ideal in the Weyl algebra $A_n := \mathbb{Q}[z_1, \ldots, z_n]\langle \partial/\partial z_1, \ldots, \partial/\partial z_n \rangle$ of algebraic local cohomology class σ_F and its direct summands. In fact, it is known that "in many cases", the annihilating ideal can be generated by differential operators of order at most one and even the case of unimodal semiquasihomogeneous singularities, the annihilating ideal for the associated cohomology classes can be generated by differential operators of order at most two [2]. The present method has another applications to Grothendieck residue computations, Jacobi's interpolation formula [7] or, more generally, Grothendieck duality, which will be treated in the forthcoming papers [6].

In section 1, we derive a method for constructing first and second order linear partial differential operators which annihilate the algebraic local cohomology class σ_F. In section 2, we give a method for constructing first and second order linear partial differential operators which annihilate a direct summand of σ_F. In section 3, we present a multiplicity formula for the holonomic system corresponding to annihilating differential operators of order at most one, which serves to practical computations.

For convenience of explanations, we use the words "k-th order annihilators of the cohomology class σ" for expressing k-th order linear partial differential operators that annihilate the cohomology class σ. Throughout this paper, we fix a well-ordering $\succ$ on $\mathbb{N}^{2n}$ with $\mathbb{N} = \{0, 1, 2, \ldots\}$. We consider the case $I \neq \sqrt{I}$ because if $I = \sqrt{I}$ holds, the annihilating ideal of the algebraic local cohomology class σ_F is given by $A_n I$ [4,5].

1 Annihilators of the cohomology class σ_F

In this section, we give a method for constructing first order annihilators

$$p_1(z)\frac{\partial}{\partial z_1} + \cdots + p_n(z)\frac{\partial}{\partial z_n} + p(z) \tag{1}$$

and second order annihilators

$$\sum_{i=1}^{n}\sum_{j\geq i}^{n} p_{ij}(z)\frac{\partial^2}{\partial z_i \partial z_j} + \sum_{i=1}^{n} p_i(z)\frac{\partial}{\partial z_i} + p(z) \tag{2}$$

of the cohomology class σ_F.

Since, $g\sigma_F = 0$ holds for any $g = g(z) \in I$, it suffices to take $p(z)$ in (1) and (2) from the modulo class $\mathbb{Q}[z]/I$.

1.1 First order annihilators of σ_F

For two first order linear partial differential operators $Q = q_1(z)\frac{\partial}{\partial z_1} + \cdots + q_n(z)\frac{\partial}{\partial z_n} + q_0(z)$ and $R = r_1(z)\frac{\partial}{\partial z_1} + \cdots + r_n(z)\frac{\partial}{\partial z_n} + r_0(z)$ $(q_k(z), r_k(z) \in \mathbb{Q}[z])$, suppose that $q_k(z) = r_k(z) \mod I$, $k = 1, \ldots, n$ hold. Then, one has

$$\begin{aligned} Q - R &= \tfrac{\partial}{\partial z_1}(q_1(z) - r_1(z)) + \cdots + \tfrac{\partial}{\partial z_n}(q_n(z) - r_n(z)) \\ &\quad +(q_0(z) - r_0(z)) - (\tfrac{\partial q_1(z)}{\partial z_1} + \cdots + \tfrac{\partial q_n(z)}{\partial z_n} - \tfrac{\partial r_1(z)}{\partial z_1} - \cdots - \tfrac{\partial r_n(z)}{\partial z_n}) \\ &= (q_0(z) - r_0(z)) - (\tfrac{\partial q_1(z)}{\partial z_1} + \cdots + \tfrac{\partial q_n(z)}{\partial z_n} - \tfrac{\partial r_1(z)}{\partial z_1} - \cdots - \tfrac{\partial r_n(z)}{\partial z_n}) \end{aligned}$$

modulo $A_n I$. Thus it suffices to take coefficients $p_1(z), \ldots, p_n(z), p(z)$ in (1) from $\mathbb{Q}[z]/I$.

Let us consider a differential operator v of the form

$$v = p_1(z)\frac{\partial}{\partial z_1} + \cdots + p_n(z)\frac{\partial}{\partial z_n} \tag{3}$$

with $p_j(z) \in \mathbb{Q}[z]/I$, $j = 1, \ldots, n$.

Proposition 1 The following two conditions are equivalent.

(i) $vg(z) \in I$ for $\forall g(z) \in I$.

(ii) $\exists p(z) \in \mathbb{Q}[z]/I$ s.t. $(v + p(z))\sigma_F = 0$.

The above proposition yields a computation method for constructing first order annihilators of σ_F. The output of the following algorithm is a basis of the vector space

$$L^{(1)}(\sigma_F) = \{P = p_1(z)\tfrac{\partial}{\partial z_1} + \cdots + p_n(z)\tfrac{\partial}{\partial z_n} + p(z) \mid P\sigma_F = 0,\ p_j(z),\ p(z) \in \mathbb{Q}[z]/I\}$$

of first order differential operators.

Algorithm ($\mathfrak{A}$-1) (First order annihilators of σ_F)
Input : $F = \{f_1, \ldots, f_n\}$: a regular sequence of polynomials $f_j \in \mathbb{Z}[z]$.

1. Compute the canonical monomial basis $\{e_1(z), \ldots, e_d(z)\}$ of $\mathbb{Q}[z]/I$, where $d = \dim \mathbb{Q}[z]/I$.

2. Put $p(z) = \sum_{k=1}^{d} c_k e_k(z)$, $p_i(z) = \sum_{k=1}^{d} c_{ik} e_k(z)$ $(i = 1,\ldots,n)$ with indeterminate coefficients c_k, c_{ik}, $k = 1,\ldots,d$, $i = 1,\ldots,n$.

3. Determine the coefficients c_{ik}, $k = 1,\ldots,d$, $i = 1,\ldots,n$ so that $p_1(z)\frac{\partial g}{\partial z_1} + \cdots + p_n(z)\frac{\partial g}{\partial z_n} \in I$ for $\forall g = g(z) \in I$ holds.

4. Compute Gröbner basis of the ideal $\langle f_1^2,\ldots,f_n^2\rangle$. For n tuples $(p_1(z),\ldots,p_n(z))$ given by the above step, find $p(z)$ so that

$$-\sum_{i=1}^{n} p_i(z) \sum_{j=1}^{n} f_1 \cdots \frac{\partial f_j}{\partial z_i} \cdots f_n + p(z) f_1 \cdots f_n \in \langle f_1^2,\ldots,f_n^2\rangle$$

holds (by solving the membership problem).

5. Put $P := p_1(z)\frac{\partial}{\partial z_1} + \cdots + p_n(z)\frac{\partial}{\partial z_n} + p(z)$ where $p(z)$ and $p_i(z)$ are given by the above steps.

Output : A basis of $L^{(1)}(\sigma_F)$ of first order annihilators of σ_F.

Proposition 2 Let $Ann^{(1)}(\sigma_F)$ be the ideal generated by zeroth order and first order annihilators of the algebraic local cohomology class σ_F. Then $Ann^{(1)}(\sigma_F) = A_n(I \cup L^{(1)}(\sigma_F))$ holds.

1.2 Second order annihilators of σ_F

If $Ann^{(1)}(\sigma_F)$ does not define simple holonomic system, second order annihilators are candidates for generators of the annihilator $Ann(\sigma_F)$ in the Weyl algebra of σ_F. So, we give a method for computing generators of the ideal $Ann^{(2)}(\sigma_F)$ of annihilators of order at most 2. Especially, for efficiency of computations, we construct second order annihilators of σ_F that cannot be obtained by multiplying $\partial/\partial z_i$, $i = 1,\ldots,n$ to the first order annihilators.

We may assume all the coefficient $p(z), p_i(z), p_{ij}(z)$, $i = 1,\ldots,n$; $i \leq j \leq n$ of (2) are taken from $\mathbb{Q}[z]/I$. The reasoning is as follows. Assume that two second order differential operators

$$Q = \sum_{i=1}^{n}\sum_{j\geq i}^{n} q_{ij}(z)\frac{\partial^2}{\partial z_i \partial z_j} + \sum_{i=1}^{n} q_i(z)\frac{\partial}{\partial z_i} + q(z)$$

and

$$R = \sum_{i=1}^{n}\sum_{j\geq i}^{n} r_{ij}(z)\frac{\partial^2}{\partial z_i \partial z_j} + \sum_{i=1}^{n} r_i(z)\frac{\partial}{\partial z_i} + r(z)$$

are given. Suppose that $q_{ij}(z) = r_{ij}(z) \mod I$, $i = 1, \ldots, n$; $i \le j \le n$. Then the difference $Q - R$ modulo $A_n I$ becomes a first order differential operator.

We introduce a set V to be the set of first order differential operators given by Step 3 of Algorithm $\mathfrak{A}$-1 in §1.1, i.e.,

$$V = \{v = p_1(z)\frac{\partial}{\partial z_1} + \cdots + p_n(z)\frac{\partial}{\partial z_n} \mid vg(z) \in I, \forall g(z) \in I\}.$$

Denote by $N = \{\mathrm{hm}(v) \mid v \in V\}$ a set of the leading monomial $\mathrm{hm}(v)$ of the operator $v \in V$ with respect to the well-ordering $\succ$. Let M be the set of canonical monomial basis $\{e_1(z), \ldots, e_d(z)\}$ of $\mathbb{Q}[z]/I$ with respect to $\succ$. Denote by $N_i = \{e(z) \in M \mid e(z)\frac{\partial}{\partial z_i} \in N\}$ $(i = 1, \ldots, n)$ a set of coefficients of operators in N .

Lemma 3 Let $L_1, \ldots, L_s$ be a basis of the vector space $L^{(1)}(\sigma_F)$. Let R be a second linear partial differential operator with coefficients taken from $\mathbb{Q}[z]/I$. Then there exist first order differential operators $A_1, \ldots, A_s$ such that the difference $P = R - \sum_{j=1}^{s} A_j L_j$ is in the form

$$P = \sum_{i=1}^{n}\sum_{j\ge i}^{n} p_{ij}(z)\frac{\partial^2}{\partial z_i \partial z_j} + \sum_{i=1}^{n} p_i(z)\frac{\partial}{\partial z_i} + p(z) \tag{4}$$

with

$$\begin{cases} p_i(z) = \sum_{k=1, e_k \in M\setminus N_i}^{d} c_{ik} e_k(z), \\ p_{ij}(z) = \sum_{k=1, e_k \in M\setminus (N_i \cup N_j)}^{d} c_{ijk} e_k(z), \ i = 1, \ldots, n; i \le j \le n. \end{cases} \tag{5}$$

The above lemma tells us that, in order to construct an generator of $\mathrm{Ann}^{(2)}(\sigma_F)$, it suffices to consider differential operators of the form (4) with the condition (5).

Let $L^{(2)}(\sigma_F)$ be the set of second order annihilators of σ_F which has the form (4) with the condition (5). The output of the following procedure (Algorithm $\mathfrak{A}$-2) is a basis of $L^{(2)}(\sigma_F)$.

1. Take the canonical monomial basis $\{e_1(z), \ldots, e_d(z)\}$ of $\mathbb{Q}[z]/I$ where $d = \dim \mathbb{Q}[z]/I$.

2. Put $p(z) = \sum_{k=1}^{d} c_k e_k(z)$, $p_i(z) = \sum_{k=1, e_k \in M\setminus N_i}^{d} c_{ik} e_k(z)$, $p_{ij}(z) = \sum_{k=1, e_k \in M\setminus (N_i \cup N_j)}^{d} c_{ijk} e_k(z)$ $(i = 1, \ldots, n; i \le j \le n)$ with indeterminate coefficients c_k, c_{ik}, c_{ijk}, $i = 1, \ldots, n$; $i \le j \le n$; $k = 1, \ldots, d$.

3. Computing the image of $1/f_1 \dots f_n$ by the linear partial differential operator (4) with $p(z)$, $p_i(z)$, $p_{ij}(z)$ as above, find the numerator $u = u(z)$ in the right-hand side of the formula $P(1/f_1 \cdots f_n) = u/f_1^3 \cdots f_n^3$.

4. Determine coefficients c_k, c_{ik}, c_{ijk}, $i = 1, \dots, n$; $i \le j \le n$; $k = 1, \dots, d$ so that $u(z) \in \langle f_1^3, \dots, f_n^3 \rangle$ by solving a membership problem.

5. Put $P := \sum_{i=1}^n \sum_{j \ge i}^n p_{ij}(z) \frac{\partial^2}{\partial z_i \partial z_j} + \sum_{i=1}^n p_i(z) \frac{\partial}{\partial z_i} + p(z)$ with $p(z) = \sum_{k=1}^d c_k e_k(z)$, $p_i(z) = \sum_{k=1, e_k \in M \setminus N_i}^d c_{ik} e_k(z)$, $p_{ij}(z) = \sum_{k=1, e_k \in M \setminus (N_i \cup N_j)}^d c_{ijk} e_k(z)$, $i = 1, \dots, n$; $i \le j \le n$.

In the above procedure, we exploit the fact that $u(z) \in \langle f_1^3, \dots, f_n^3 \rangle$ is the necessary and sufficient condition for $P\sigma_F = 0$.

Proposition 4 Let $Ann^{(2)}(\sigma_F)$ be the ideal generated by annihilators of order at most two. Then $Ann^{(2)}(\sigma_F) = A_n(I \cup L^{(1)}(\sigma_F) \cup L^{(2)}(\sigma_F))$.

2 Semi local method

Assume that the primary decomposition of the ideal I is given by

$$I = I_1 \cap \cdots \cap I_m. \tag{6}$$

Denote by Z_ℓ the variety $V(I_\ell)$ of the ideal I_ℓ ($\ell = 1, \dots, m$). The algebraic local cohomology group $H^n_{[Z]}(\mathcal{O}_X)$ has the direct sum decomposition $H^n_{[Z]}(\mathcal{O}_X) = H^n_{[Z_1]}(\mathcal{O}_X) \oplus \cdots \oplus H^n_{[Z_m]}(\mathcal{O}_X)$ by algebraic local cohomology groups $H^n_{[Z_\ell]}(\mathcal{O}_X)$. Thus there are algebraic local cohomology classes $\sigma_\ell \in H^n_{[Z_\ell]}(\mathcal{O}_X)$ with support on Z_ℓ, $\ell = 1, \dots, m$ satisfying

$$\sigma_F = \sigma_1 + \cdots + \sigma_m. \tag{7}$$

Let us consider for instance the case where $Ann(\sigma_F) \ne Ann^{(1)}(\sigma_F)$. There exists at least one direct summand σ_i which satisfies $Ann(\sigma_i) \ne Ann^{(1)}(\sigma_i)$, whereas there is a possibility that $Ann(\sigma_j) = Ann^{(1)}(\sigma_j)$ for some other direct summand σ_j. Furthermore, the local behavior of the holonomic system $A_n/Ann^{(1)}(\sigma_F)$ at $\beta \in Z_\ell$ is completely equal to $A_n/Ann^{(1)}(\sigma_\ell)$. Thus, to understand explicitly local properties of holonomic systems attached to σ_F in question, it is desirable to compute annihilators of each direct summand.

In this section, we give a method for constructing first and second order annihilators of the direct summand σ_ℓ which serves to compute local behavior of σ_F.

Let $I^{(2)}$ and $I^{(3)}$ be the ideal given by squared and cubed of the functions $f_1, \ldots, f_n$, i.e., $I^{(2)} = \langle f_1^2, \ldots, f_n^2 \rangle$ and $I^{(3)} = \langle f_1^3, \ldots, f_n^3 \rangle$. Let $I^{(2)} = I_1^{(2)} \cap \cdots \cap I_m^{(2)}$ and $I^{(3)} = I_1^{(3)} \cap \cdots \cap I_m^{(3)}$ be the primary decomposition of the ideal $I^{(2)}$ and $I^{(3)}$ respectively, where the primary ideals are sorted as to satisfy $\sqrt{I_\ell} = \sqrt{I_\ell^{(2)}} = \sqrt{I_\ell^{(3)}}$.

2.1 A method for constructing first order annihilators

Recall that the annihilating ideal generated by zeroth order annihilators of σ_ℓ in the Weyl algebra A_n is given by $A_n I_\ell$. In order to construct the first order annihilators, it suffices to take the coefficients $p_1(z), \ldots, p_n(z), p(z)$ from $\mathbb{Q}[z]/I_\ell$.

Put

$$v = p_1(z)\frac{\partial}{\partial z_1} + \cdots + p_n(z)\frac{\partial}{\partial z_n} \tag{8}$$

for $p_j(z) \in \mathbb{Q}[z]/I_\ell$, $j = 1, \ldots, n$.

Proposition 5 The following two conditions are equivalent.

(i) $vg(z) \in I_\ell$ for $\forall g(z) \in I_\ell$.

(ii) $\exists p(z) \in \mathbb{Q}[z]/I_\ell$ s.t. $(v + p(z))\sigma_\ell = 0$.

The above Proposition provides a method for computing a basis of the vector space

$$L^{(1)}(\sigma_\ell) = \{P = p_1(z)\tfrac{\partial}{\partial z_1} + \cdots + p_n(z)\tfrac{\partial}{\partial z_n} + p(z) \\ \mid P\sigma_\ell = 0,\ p(z),\ p_i(z) \in \mathbb{Q}[z]/I_\ell\}$$

of first order annihilators of the cohomology class σ_ℓ.

Algorithm $\mathfrak{L}$-1

Input : $F = \{f_1, \ldots, f_n\}$: a regular sequence of polynomials $f_j \in \mathbb{Z}[z]$,
I_ℓ : one of primary ideals in (6)

1. Take a canonical monomial basis $\{e_1(z), \ldots, e_{d_\ell}(z)\}$ of $\mathbb{Q}[z]/I_\ell$ where $d_\ell = \dim \mathbb{Z}[z]/I_\ell$.

2. Put $p(z) = \sum_{k=1}^{d_\ell} c_k e_k(z)$, $p_i(z) = \sum_{k=1}^{d_\ell} c_{ik} e_k(z)$, $i = 1, \ldots, n$, with indeterminate coefficients c_k, c_{ik}, $k = 1, \ldots, n, i = 1, \ldots, n$.

3. Determine c_{ik}, $k = 1, \ldots, d_\ell$, $i = 1, \ldots, n$ for which $p_1(z)\dfrac{\partial g}{\partial z_1} + \cdots + p_n(z)\dfrac{\partial g}{\partial z_n} \in I_\ell$ holds for any $g = g(z) \in I_\ell$.

4. Determine c_k, $k = 1, \ldots, d_\ell$ so that $-\sum_{i=1}^{n} p_i(z) \sum_{j=1}^{n} f_1 \cdots \dfrac{\partial f_j}{\partial z_i} \cdots f_n + p(z) f_1 \cdots f_n \in I_\ell^{(2)}$ holds.

5. Put $P := p_1(z)\dfrac{\partial}{\partial z_1} + \cdots + p_n(z)\dfrac{\partial}{\partial z_n} + p(z)$ where $p(z)$ and $p_i(z)$, $j = 1, \ldots, n$ are given by the above steps.

Output : A basis of $L^{(1)}(\sigma_\ell)$ of first order annihilators of the algebraic local cohomology class σ_ℓ.

Proposition 6 Let $Ann^{(1)}(\sigma_\ell)$ be the ideal generated by zeroth and first order annihilators of the algebraic local cohomology class σ_ℓ. Then $Ann^{(1)}(\sigma_\ell) = A_n(I_\ell \cup L^{(1)}(\sigma_\ell))$ holds.

2.2 A method for constructing second order annihilators

In this section, for the algebraic local cohomology class σ_ℓ, we derive a method to construct second order annihilators

$$P = \sum_{i=1}^{n} \sum_{j \geq i}^{n} p_{ij}(z) \frac{\partial^2}{\partial z_i \partial z_j} + \sum_{i=1}^{n} p_i(z) \frac{\partial}{\partial z_i} + p(z). \tag{9}$$

Roughly speaking, this can be done by replacing the condition $u(z) \in I^{(3)}$ in Step 4 of the Algorithm $\mathfrak{A}$-2 in §1.2 to $u(z) \in I_\ell^{(3)}$. That is, the procedure below is based on the fact that $u(z) \in I_\ell^{(3)}$ is the necessary and sufficient conditions for $P\sigma_\ell = 0$.

Let

$$V_\ell = \{v = p_1(z)\frac{\partial}{\partial z_1} + \cdots + p_n(z)\frac{\partial}{\partial z_n} \mid vg(z) \in I_\ell, \ ^\forall g(z) \in I_\ell\}.$$

Denote by $N_\ell = \{\mathrm{hm}(v) \mid v \in V_\ell\}$ a set of the leading monomial $\mathrm{hm}(v)$ of the operator $v \in V_\ell$ with respect to the well-ordering $\succ$. Denote by $N_{\ell i} = \{e(z) \in M_\ell \mid e(z)\frac{\partial}{\partial z_i} \in N_\ell\}$ a set of coefficients of operators in V_ℓ where M_ℓ denotes the set of canonical monomial basis $\{e_1(z), \ldots, e_{d_\ell}(z)\}$ of $\mathbb{Q}[z]/I_\ell$ with respect to $\succ$. The following procedure (Algorithm $\mathfrak{L}$-2) gives us a basis of a vector space $L^{(2)}(\sigma_\ell)$ of second order annihilators of the algebraic local cohomology class σ_ℓ that are not obtained from first order annihilators belonging to $L^{(1)}(\sigma_\ell)$.

1. Take a canonical monomial basis $\{e_1(z), \ldots, e_{d_\ell}(z)\}$ of the quotient space $\mathbb{Q}[z]/I_\ell$ where $d_\ell = \dim \mathbb{Q}[z]/I_\ell$.

2. Put $p(z) = \sum_{k=1}^{d_\ell} c_k e_k(z)$, $p_i(z) = \sum_{k=1, e_k \in M_\ell \setminus N_{\ell i}}^{d_\ell} c_{ik} e_k(z)$, $p_{ij}(z) = \sum_{k=1, e_k \in M_\ell \setminus (N_{\ell i} \cup N_{\ell j})}^{d_\ell} c_{ijk} e_k(z)$ $(i = 1, \ldots, n; i \leq j \leq n)$ with indeterminate coefficients c_k, c_{ik}, c_{ijk}, $i = 1, \ldots, n$; $i \leq j \leq n$; $k = 1, \ldots, d_\ell$.

3. Computing the image of $1/f_1 \cdots f_n$ by the linear partial differential operator (9) with $p(z)$, $p_i(z)$, $p_{ij}(z)$ as above, determine the function $u = u(z)$ so that $P(1/f_1 \cdots f_n) = u/f_1^3 \cdots f_n^3$.

4. Determine c_k, c_{ik}, c_{ijk}, $i = 1, \ldots, n; i \leq j \leq n$; $k = 1, \ldots, d_\ell$ so that $u(z) \in I_\ell^{(3)}$.

5. Put

$$P = \sum_{i=1}^{n} \sum_{j \geq i}^{n} p_{ij}(z) \frac{\partial^2}{\partial z_i \partial z_j} + \sum_{i=1}^{n} p_i(z) \frac{\partial}{\partial z_i} + p(z)$$

   where c_k, c_{ik}, c_{ijk} are given by the above step.

We have the following.

Proposition 7 Let $Ann^{(2)}(\sigma_\ell)$ be the ideal generated by annihilators of order at most 2. Then $Ann^{(2)}(\sigma_\ell) = A_n(I \cup L^{(1)}(\sigma_\ell) \cup L^{(2)}(\sigma_\ell))$.

The use of the primary decomposition reduces the cost of computations considerably.

3 When does $A_n/Ann^{(1)}(\sigma_\ell)$ become simple?

It is known that if a given regular sequence F is "generic", the holonomic system $A_n/Ann^{(1)}(\sigma_F)$ (resp. $A_n/Ann^{(1)}(\sigma_\ell)$) is simple at each point of $V(I)$ (resp. $V(I_\ell)$). But we can not expect in general that $Ann^{(1)}(\sigma_\ell)$ is equal to the annihilating ideal $Ann(\sigma_\ell)$ in A_n of σ_ℓ. In this section, we give a practical criterion for $Ann^{(1)}(\sigma_\ell) = Ann(\sigma_\ell)$.

Here we introduce the following notation.

$$H_\ell = \{h(z) \in \mathbb{Q}[z]/I_\ell \mid v(h) \in I_\ell,\, v \in V_\ell\}.$$

H_ℓ becomes a finite dimensional vector space and can be easily computed after determining the space V_ℓ (Step 3 of Algorithm $\mathfrak{L}$-1). As multiplicities of $A_n/Ann^{(1)}(\sigma_\ell)$ at all points in V_ℓ are the same, denote it by $\mathrm{mult}_{V_\ell}(A_n/Ann^{(1)}(\sigma_\ell))$.

Theorem 8 The following formula holds.

$$\dim H_\ell = \dim \mathbb{Q}[z]/\sqrt{I_\ell} \cdot \mathrm{mult}_{V_\ell}(A_n/Ann^{(1)}(\sigma_\ell)).$$

Corollary 9 $Ann^{(1)}(\sigma_\ell) = Ann(\sigma_\ell)$ holds if and only if $\dim H_\ell = \dim \mathbb{Q}[z]/\sqrt{I_\ell}$ is satisfied.

Proof. We first recall that the holonomic system $A_n/Ann(\sigma_\ell)$ is simple at each point in Z_ℓ as a $\mathcal{D}_X$-module. If $Ann^{(1)}(\sigma_\ell) = Ann(\sigma_\ell)$, then $\mathrm{mult}_{V_\ell}(A_n/Ann^{(1)}(\sigma_\ell)) = 1$, which implies $\dim H_\ell = \dim \mathbb{Q}[z]/\sqrt{I_\ell}$. Conversely, from $\dim H_\ell = \dim \mathbb{Q}[z]/\sqrt{I_\ell}$, we have $\mathrm{mult}_\beta(A_n/Ann^{(1)}(\sigma_\ell)) = 1$. Thus $Ann^{(1)}(\sigma_\ell) \subseteq Ann(\sigma_\ell)$ yields $Ann^{(1)}(\sigma_\ell) = Ann(\sigma_\ell)$.

Thus, computing $\dim H_\ell$, we can find whether $Ann(\sigma_\ell)$ is generated by differential operators of order at most one or not.

Example Let $X = \mathbb{C}^2$ and $z = (x, y)$. The ordering $\succ$ is the lexicographic order $x \succ y$. Let F be a regular sequence $\{f_1, f_2\}$ given by functions $f_1 = 3x^4 - 6x^3 + 3x^2 + y^5$ and $f_2 = 5xy^4 + 7y^6$ and put $I = \langle f_1, f_2 \rangle$.

The output of Algorithm $\mathfrak{A}$-1 consists of 24 first order annihilators of σ_F. By computing $H = \{h(z) \in \mathbb{Q}[z]/I \mid v(h) \in I,\ v \in V\}$, we find $Ann(\sigma_F) \neq Ann^{(1)}(\sigma_F)$. Thus $Ann(\sigma) \neq Ann^{(1)}(\sigma)$ holds for some direct summand σ of σ_F. The Algorithm $\mathfrak{A}$-2 outputs three second order annihilators. Coefficients of these annihilators are taken from the 24 dimensional vector space $\mathbb{Q}[z]/I$.

The primary decomposition of the ideal $I = \langle f_1, f_2 \rangle$ is $I = I_1 \cap I_2 \cap I_3$ with $I_1 = \langle 7203y^4 + 10290y^2 + 625y + 3675, 5x + 7y^2 \rangle$, $I_2 = \langle y^4, x^2 - 2x + 1 \rangle$ and $I_3 = \langle y^8, 5y^4x + 7y^6, 15x^2 - 14y^7 + 5y^5 \rangle$. The algebraic local cohomology class σ_F has the decomposition $\sigma_F = \sigma_1 + \sigma_2 + \sigma_3$ with $\mathrm{supp}(\sigma_1) = Z_1$, $\mathrm{supp}(\sigma_2) = Z_2$, $\mathrm{supp}(\sigma_3) = Z_3$. For each local cohomology σ_j $(j = 1, 2, 3)$, we have the following.

(i) Since $I_1 = \sqrt{I_1}$, $Ann(\sigma_1) = A_n I_1$ holds.

(ii) The multiplicity of I_2 at V_2 is eight and thus $Ann(\sigma_2) \neq A_n I_2$. We have $H_2 = \mathrm{Span}\{1\}$ and thus $Ann^{(1)}(\sigma_2) = Ann(\sigma_2)$. In fact, we find that $Ann(\sigma_2)$ is generated by $x^2 - 2x + 1$, y^4 and two first order differential operators $y\frac{\partial}{\partial y} - \frac{14}{5}y^2 x + \frac{28}{5}y^2 + 4$ and $(-x+1)\frac{\partial}{\partial x} + (\frac{7}{5}y^2 - 3)x - \frac{7}{5}y^2 + 1$.

(iii) We have $\sqrt{I_3} = \langle x, y \rangle$ and $H_3 = \mathrm{Span}\{y^7, 1\}$. Since $\dim H_3 = 2$ is greater than one, $Ann^{(1)}(\sigma_3) \neq Ann(\sigma_3)$. Algorithm $\mathfrak{L}$-1 outputs 14 first order annihilators of σ_3 and Algorithm $\mathfrak{L}$-2 gives two second order annihilators. These operators are candidates of generators of the annihilator $Ann(\sigma_3)$. Actually, by calculating the dimension of the image space in $\mathbb{Q}[z]/I_3$ under the action of formal adjoint of operators P_1 and P_2 given below, we can verify that $y^8, 5y^4x + 7y^6, 15x^2 - 14y^7 + 5y^5$, P_1 and P_2 generate $Ann(\sigma_3)$.

$$P_1 = (84yx + 5y^4)\frac{\partial}{\partial x} + 42y^2\frac{\partial}{\partial y}$$
$$+(-\frac{187125625}{155649627}y^3 - \frac{78125}{1058841}y^2 - \frac{1206979}{7203}y - \frac{125}{49})x,$$

$$P_2 = (\frac{84}{5}yx - \frac{43}{18}y^4)\frac{\partial^2}{\partial x^2} + y\frac{\partial^2}{\partial y^2} + (\frac{50}{147}y + 9)\frac{\partial}{\partial y}$$
$$+((\frac{22107115}{3176523}y^2 - \frac{21741247}{324135}y - \frac{25}{441})x + \frac{16191391}{1361367}y^4 + \frac{125}{9261}y^3 - \frac{5}{63}y^2 + \frac{252}{5}y)\frac{\partial}{\partial x}$$
$$+(-\frac{1723699674605000}{211896265760109}y^3 + \frac{6871260672830}{1441471195647}y^2 + \frac{475739315798}{7004233215}y - \frac{11735350}{66706983})x$$
$$+\frac{4052165632630620775}{218041257467152161}y^7 + \frac{111788363134732224}{494424620106921}y^6 - \frac{303413966976175}{30270895108587}y^5$$
$$-\frac{1636548587117}{205924456521}y^4 + \frac{57504875}{1400846643}y^3 + \frac{18829750}{1361367}y^2 - \frac{32566558}{324135}y + 650441.$$

References

1. R. Hartshorne, Residues and Duality, Lecture Notes in Math. **20**, Springer, Berlin, 1966.
2. Y. Nakamura and S. Tajima, *The local cohomology classes attached to unimodal semiquasihomogeneous singularities*, (in preparation).
3. T.Oaku, *Algorithms for b-functions, restrictions, and algebraic local cohomology groups of D-modules*, Advances in Appplied Mathematics **19** (1997), 61–105.
4. F. Pham, Point de vue algébrique sur les systèmes différentiels lineaires., Progress in Math **2** Birkhäuser (1979), 25-170.
5. M.Sato, T. Kawai, M. Kashiwara, Microfunctions and pseudo-differential equations, Lecture Notes in Math. **287**, Springer, Berlin, 1973, pp.265–529.
6. S. Tajima and Y. Nakamura, *Computational aspects of Grothendieck local residues*, (in preparation).
7. S. Tajima, *Grothendieck duality and Hermite-Jacobi formulas*, Proceedings of International Conference on finite or infinite dimensional Complex Analysis, Dekker, 2000, pp.503–509.

D-MODULES FOR MACAULAY 2

ANTON LEYKIN

D-modules for *Macaulay 2* is a collection of the most recent algorithms that deal with various computational aspects of the theory of D-modules. This paper provides a brief guide, which gives examples of using the main functions of this package, as well as an overview of the core algorithms for D-modules and their applications.

1 Introduction

The *Macaulay 2* D-modules package is an implementation of the Weyl algebra and algorithms related to it in the computer algebra system *Macaulay 2*. Over the last decade there were many advances made in the computational theory of D-modules. Several newest algorithmic methods used in the analysis of hypergeometric differential equations are described in the recent book ([11]). Also we worked with a paper by Oaku and Takayama ([8]) providing, in particular, a detailed description of the restriction algorithm. The algorithms for computing localization, D-homomorphisms, and global b-functions of polynomials with parameters come from ([10]), ([13]), and ([5]) respectively. As to the applications that we describe here, we compute polynomial and rational solutions according to ([9]), local cohomology via Čech complex comes from a paper of Walther ([14]).

Macaulay 2, a noncommercial computer algebra system crafted by Grayson and Stillman, became one of the favorite tools for specialists in algebraic analysis. You are welcome to join the crowd by downloading the current distribution from the web (see ([4])). D-modules package comes as a part of it, for the most recent updates and online documentation see ([6]). If you would like to learn more about *Macaulay 2*, read a recently published book ([1]), which contains a set of very interesting examples of computations in algebraic geometry including a section by Walther featuring D-modules package.

At the end of the introduction, let us mention that there are several other systems that are capable of handling D-modules. First on the list is Takayama's system Kan ([12]), which is a specialized system for D-computations. There is an implementation of Weyl algebra in Maple by Chyzak ([3]). It is also implemented in Singular ([2]), although not included in the main package.

2 Making Weyl algebras

Throughout this paper k is a field of characteristic 0, $R_n(k) = k[x_1, ..., x_n]$ is the ring of polynomials in n variables and $A_n(k) = k\langle x_1, ..., x_n, \partial_1, ..., \partial_n\rangle$ is the corresponding Weyl algebra, i.e. an associative k-algebra generated by x's and ∂'s with the relations $\partial_i x_i = x_i \partial_i + 1$ for all i.

In this paper we refer to M as to a D-module if it is a finitely generated left module over a Weyl algebra $D = A_n$.

Weyl algebras are made in *Macaulay 2* by adding the option `WeylAlgebra` to a polynomial ring. For instance,

```
i1 : D = QQ[x_1,x_2,d_1,d_2,
              WeylAlgebra=>{x_1=>d_1,x_2=>d_2}]
o1 = D
o1 : PolynomialRing
```

makes the Weyl algebra with the commutation rules $d_i x_i = x_i d_i + 1$ for $i = 1, 2$.

Now we may do the usual things in *Macaulay 2* such as forming ideals and computing Gröbner bases:

```
i2 : I = ideal(x_1*d_1+2*x_2*d_2-5, d_1^2-d_2)
                                       2
o2 = ideal (x d  + 2x d  - 5, d  - d )
             1 1     2 2       1    2
o2 : Ideal of D
i3 : gb I
o3 = {0} | d_1^2-d_2 x_1d_1+2x_2d_2-5 x_2d_1d_2+...
o3 : GroebnerBasis
```

Weyl algebra construct belongs to the kernel of the system, to load the D-modules package, however, one has to type the following:

```
i4 : load "D-modules.m2"
```

3 Making D-modules

Construction of a D-module is similar to that of a module over a polynomial ring. For instance, it may be presented as a cokernel of a matrix with entries in a Weyl algebra:

```
i5 : A = matrix {{-x_1^3+x_2, 3*d_2*x_1^2+d_1, 0, 0},
                 {0, 0, -x_1^2+x_2, 2*d_2*x_1+d_1}};
                 2       4
o5 : Matrix D  <--- D
```

```
i6 : M = cokernel A;
i7 : isHolonomic M
o7 = true
i8 : makeCyclic A
                                    2 2                     2 2
o8 = HashTable{AnnG => ideal (x d  + 5x x d d  + 6x d  + ...
                                1 1        1 2 1 2       2 2
                  Generator => | x_2d_1 |
                               | 1      |

o8 : HashTable
```

Function isHolonomic checks whether a D-module is holonomic, more on this in the next section. Every holonomic D-module may be presented as a cyclic one: function makeCyclic finds such a presentation.

One can find the module D/I generated by a polynomial (rational function): here I is the annihilator ideal of the polynomial (rational function).

```
i9 : f = x_1^2-x_2^3;
i10 : PolyAnn f
                  3       2             3       2       2
o10 = ideal (- x d  + x d  - 2x , - x d  + x d  + 3x ,  ...
                 2 1     1 1     1     2 2     1 2     2
o10 : Ideal of D
i11 : g = 2*x_1*x_2;
i12 : RatAnn(g,f)
                     2        1   3 2     2 2
o12 = ideal (x d  + -*x d  + -, x d  - x d  - 6x x d  + ...
                1 1  3  2 2  3    2 2    1 2      1 2 1
o12 : Ideal of D
```

Also there are two functions gkz and AppellF1 that cook up ideals representing GKZ (Gelfand-Kapranov-Zelevinsky) systems and Appell F1 system respectively, which are discussed in ([11]).

4 Basic invariants

Let us now compute some basic invariants associated to a D-module D/I. First, we compute the dimension of I from the line i2 of the previous section and verify that I is indeed holonomic.

```
i13 : Ddim I
o13 = 2
```

Next, we compute its characteristic ideal, which is the initial ideal with respect to the differential order filtration.

```
i14 : charIdeal I

                2
o14 = ideal (d , x d  + 2x d )
              1   1 1     2 2

o14 : Ideal of QQ [x , x , d , d ]
                    1   2   1   2
```

Note that the initial ideal lives in a polynomial ring. Now, we may compute its singular locus, which is the projection of the characteristic variety minus the zero section on the cotangent bundle onto the base space.

```
i15 : singLocus I
o15 = ideal x
             2
o15 : Ideal of D
```

Finally, we compute the holonomic rank of the system, which tells us what the dimension of the space of solutions of the system is.

```
i16 : Drank I
o16 = 2
```

5 Main tools

In ([8]), Oaku and Takayama develop fundamental algorithms for functors in the category of D-modules. There are four main tools which are heavily utilized – b-functions, localization, resolutions, and restriction. Using them, one gets algorithms for Tor, Ext, local cohomology, deRham cohomology, and other functors.

5.1 b-functions

Given a weight vector $w = (-u, u)$ corresponding to a Gröbner deformation, we are able compute the b-function of D/I in the direction u as follows:

```
i17 : u = {1,3};
i18 : bFunction(I, u)
o18 = $s - 5
o18 : QQ [$s]
```

These types of b-functions with respect to the appropriate weight vectors are also used in Oaku's algorithm to compute global b-functions, a.k.a. Bernstein-Sato polynomials.

```
i19 : f = x_1^2+x_2^2;
i20 : globalBFunction f
         2
o20 = $s  + 2$s + 1
o20 : QQ [$s]
```

The function paramBpoly computes the list of possible global b-functions of a polymonial with parametric coefficients together with the corresponding conditions on the parameters.

```
i2 : D = QQ[a,b,c][x,y,dx,dy,WeylAlgebra=>{x=>dx,y=>dy}];
i3 : bList = paramBpoly(a*x^2+b*x*y+c*y^2,"quadratic");
i4 : bList/factorBFunction
                  2          1
o4 = {($s + 1) , ($s + -)($s + 1)}
                             2
o4 : List
```

Here we also use factorBFunction to factor the polynomials in the output. Factoring a b-function is a simple business due to the fact that the roots of a b-function are rational: this is why a separate function for this is provided.

If we consider a specification of parameters as a point of Proj $\mathbb{Q}[a, b, c]$ then it is proved in ([5]) the set corresponding to a global b-function in our list is constructible. The file named quadratic.tex, which is generated by this script, contains:

- $b(s) = (s+1)^2$ for $V(0) \setminus V(b^2 - 4 * a * c)$
- $b(s) = (s+1) * (s+1/2)$ for $V(b^2 - 4 * a * c)$

5.2 Localization

There is a function that computes the localization of a D-module by inverting a polynomial. We show how to compute $\mathbb{Q}[x, y, (x^2 - y^3)^{-1}]$:

```
i2 : D = QQ[x,y,dx,dy, WeylAlgebra=>{x=>dx,y=>dy}];
i3 : M = cokernel matrix{{dx,dy}};
i4 : f = x^2-y^3;
i5 : Dlocalize(M, f)
o5 = cokernel | xdx+2/3ydy+4 y2dx+2/3xdy y3dy-x2dy+6y2 |
                                      1
o5 : D-module, quotient of D
```

When calling **Dlocalize**, several variants of localization algorithm could be specified by adding a strategy option `Dlocalize(M,f,Strategy=>OTW)` to apply the algorithm from ([10]), or `Dlocalize(M,f,Strategy=>Oaku)` which is good for f-saturated modules and appears in ([14]). The latter usually works faster than the former.

5.3 Resolutions

To compute resolutions for D-modules, use the usual *Macaulay 2* command `res`:

```
i6 : I = gkz(matrix{{1,1,1},{0,1,3}}, {2,3})

                 3     2
o6 = ideal (D  - D D , x D  + x D  + x D  - 2, x D  + ...
             2     1 3   1 1    2 2    3 3       2 2

o6 : Ideal of QQ [x , x , x , D , D , D , WeylAlgebra => ...
                   1   2   3   1   2   3
i7 : D = ring I;
i8 : res I

      1      3      11      9
o8 = D  <-- D  <-- D   <-- D  <-- 0

     0      1      2       3      4

o8 : ChainComplex
```

To find a resolution by Schreyer method, we use the special command `Dres`:

```
i9 : Dres I

      1      8      16      12      3
o9 = D  <-- D  <-- D   <-- D   <-- D  <-- 0

     0      1      2       3       4      5

o9 : ChainComplex
```

Finally, we may compute V-strict resolutions, which are resolutions that respect a weight vector $w = (u, -u)$ associated to a Gröbner deformation. These resolutions are compatible with b-functions and thus become especially useful.

```
i10 : w = {1,3,5,-1,-3,-5};
i11 : Dres(I, w, Strategy => Vhomogenize)

       1      5      8      5      1
o11 = D  <-- D  <-- D  <-- D  <-- D  <-- 0

      0      1      2      3      4      5

o11 : ChainComplex
```

5.4 Restriction

Armed with b-functions and V-strict resolutions, we get an algorithm to compute the restriction functors, which are the Tor groups of a left D-module with the right D-module $D/\{x_1, \ldots, x_d\} \cdot D$. The following computes the derived restriction to the origin.

```
i12 : w = {1,3,5};
i13 : Drestrict(I, w, Strategy => Vhomogenize)

                       1
o13 = HashTable{0 => QQ }
                       2
                1 => QQ
                       1
                2 => QQ
                3 => 0

o13 : HashTable
```

By changing the weight vector, we may compute derived restriction to a subspace such as $\{x_1 = 0\}$.

```
i14 : w = {1,0,0};
i15 : Drestrict(I, w, Strategy => Vhomogenize)
o15 = HashTable{0 => cokernel | x_3D_3-1/2 0          ...
                              | 0          x_3D_3-1 ...
                1 => 0

o15 : HashTable
```

6 Applications

6.1 Solving holonomic systems

Polynomial solutions of I can be computed by duality or by Gröbner deformations.

```
i16 : PolySols I
o16 = {x x }
        1 3
o16 : List
```

More generally, the vector space $\mathrm{Hom}_D(\frac{D}{I}, N)$ corresponds to the N-valued solutions of I.

```
i17 : D = QQ[z,Dz, WeylAlgebra=>{z=>Dz}];
i18 : M = cokernel matrix{{(Dz-1)^2}};
i19 : N = cokernel matrix{{Dz*(Dz-1)}};
```

```
i20 : DHom(M,N)
o20 = {| zDz-2Dz |, | Dz |}
o20 : List
i21 : DHom(N,M)
o21 = {| -zDz+z-2Dz+3 |, | -Dz+1 |}
o21 : List
```

6.2 Local cohomology

One of the first algorithmic applications of D-modules to algebraic geometry was computing local cohomology. Let I be an ideal of $R_n = k[x_1, ..., x_n]$ and let M be an R_n-module, then the i-th local cohomology group $H_I^i(M)$ is defined as the i-th derived functor of the functor

$$\Gamma_Y(M) = \lim_{\to} Hom_{R_n}(R_n/I^m; M),$$

where the inductive limit is taken as m tends to infinity. We may generalize the definition by letting M be a D-module. D-structure proves to be useful, since whenever M is holonomic, so is $H_I^i(M)$ for every i. Hence, we can pass from viewing local cohomology groups as generally infinite R_n-modules to computing them as holonomic D-modules, which have finite description.

There are two algorithms available in the D-modules package: one due to Oaku and Takayama (uses restriction), another due to Walther (utilizes the Čech complex).

```
i2 : D = QQ[x_1..x_6, dx_1..dx_6,
              WeylAlgebra => toList(1..6)/(i->x_i=>dx_i)];
i3 : I = minors(2, matrix{{x_1, x_2, x_3}, {x_4, x_5, x_6}})
o3 = ideal (- x x  + x x , - x x  + x x , - x x  + x x )
               2 4    1 5     3 4    1 6     3 5    2 6
o3 : Ideal of D
i4 : H = localCohom ({0,1,3}, I,
          D^1/ideal{dx_1,dx_2,dx_3,dx_4,dx_5,dx_6})
o4 = HashTable{0 => subquotient ({0} | dx_6 dx_5 dx_4   ...
               1 => subquotient ({0} | 0     dx_5 0     ...
                                  {0} | 0     0    dx_4 ...
                                  {0} | -dx_6 0    0    ...
               3 => cokernel {0} | x_2^2x_4^2-2x_1x_2x_...
o4 : HashTable
i5 : pruneLocalCohom H
o5 = HashTable{0 => 0                                   ...
```

```
                             1 => 0
                             3 => {0} | x_4dx_4+x_5dx_5+x_6dx_6+6     ...
o5 : HashTable
```

In practice Walther's approach (option `Strategy=>Walther`) shows to be faster than the Oaku-Takayama method (option `Strategy=>OaTa`).

6.3 DeRham cohomology

Another exciting application of D-modules is for computing the deRham cohomology groups. Using the integration and localization functors, Oaku and Takayama showed how to compute these groups for the complement of an affine complex hypersurface. In ([15]) Walther generalized this algorithm to complements of affine complex varieties, and also showed how to compute the cup product structures.

We have implemented an algorithm for the hypersurface case:

```
i6 : R = QQ[y,z];
i7 : f = y^2-z^3;
i8 : deRham f

                             1
o8 = HashTable{0 => QQ }
                             1
                   1 => QQ
                   2 => 0

o8 : HashTable
```

7 Example: rank jumps in A-hypergeometric systems

In this section we shall give a practical example of employing the D-modules package. This example is borrowed from the work of Matusevich ([7]).

For an integer $d \times n$ matrix $A = (a_{ij})$ with the first row entries equal to 1 and a vector $\beta \in \mathbb{C}^d$ we define the GKZ(Gelfand-Karpanov-Zelevinsky) *A-hypergeometric system with parameter β* to be the D-ideal $H_A(\beta)$ generated by

$$\partial^u - \partial^v, \text{ where } u, v \in \mathbb{N}^n \text{ such that } A \cdot u = A \cdot v, \tag{1}$$

$$\sum_{j=1}^{n} a_{ij} x_j \partial_j - \beta_i,\ i = 1, ..., d. \tag{2}$$

The commutative ideal of $\mathbb{C}[\partial_1, ..., \partial_n]$ generated by operators (1) is denoted by I_A and referred to as *underlying toric ideal.*

It is a fact that if I_A is Cohen-Macaulay then the D-rank of $H_A(\beta)$ is equal to vol(A), the normalized volume of the convex hull of A viewed as an n-point configuration in $\mathbb{Z}^d$.

However, if I_A is non-Cohen-Macaulay, it is possible that rank$(H_A(\beta)) >$ vol(A). For codimension 2 case it was proved that the *exceptional set* of parameters β for which the inequality holds is a nonempty constructible set.

Example. Using *Macaulay 2*, for

$$A = \begin{pmatrix} 1 & 1 & 1 & 1 & 1 \\ 1 & 0 & -1 & 0 & 3 \\ 0 & 1 & -2 & 0 & 3 \end{pmatrix}, \ \beta = \begin{pmatrix} 3 \\ 3 \\ 2 \end{pmatrix}$$

we are going to show that β is an exceptional parameter for A.

```
i2 : A = matrix{{1,1,1,1,1},
        {1,0,-1,0,3},
        {0,1,-2,0,3}};

                 3        5
o2 : Matrix ZZ  <--- ZZ
i3 : b = {3,3,2};
i4 : H = gkz(A,b)
             2                3         2
o4 = ideal (D D   - D D D  , D D  - D D D  , ...
             1 2     3 4 5    1 4    2 3 5
o4 : Ideal of QQ [x , x , x , x , x , D , ...
                   1   2   3   4   5   1
i5 : time Drank H
     -- used 89.75 seconds
o6 = 10
```

Here we used function **gkz** to construct the GKZ A-hypergeometric system with parameter β, and then computed the D-rank of the resulting D-ideal. (Operator **time** put in front of any *Macaulay 2* command prints out the computation time.)

The rank is 10, however, an easy computation shows that the normalized volume of A is only 9. Thus, we conclude that the system experiences a rank jump of 1 for the parameter β.

References

1. *Computations in Algebraic Geometry with Macaulay 2.* Algorithms and Computation in Mathematics, Vol. 8, 2001
2. *Singular: a computer algebra system for polynomial computations.* `http://www.singular.uni-kl.de/`

3. Chyzak, Frédéric. *Mgfun Project.* `http://pauillac.inria.fr/algo/chyzak/mgfun.html`
4. Grayson, Daniel; Stillman, Michael. *Computer algebra system Macaulay 2.* `http://www.math.uiuc.edu/Macaulay2`
5. Leykin, Anton. *Constructibility of the set of polynomials with a fixed Bernstein-Sato polynomial: an algorithmic approach.* Effective methods in rings of differential operators. J. Symbolic Comput. 32 (2001), no. 6, 663–675.
6. Leykin, Anton; Tsai, Harrison. *D-modules for Macaulay 2.* `http://www.math.umn.edu/ leykin/Dmodules`
7. Matusevich, Laura. *Rank jumps in codimension 2 A-hypergeometric systems.* Effective methods in rings of differential operators. J. Symbolic Comput. 32 (2001), no. 6, 663–675.
8. Oaku, Toshinori; Takayama, Nobuki. *Algorithms for D-modules – restriction, tensor product, localization, and local cohomology groups.* J. Pure Appl. Algebra 156 (2001), no. 2-3, 267–308.
9. Oaku, Toshinori; Takayama, Nobuki; Tsai, Harrison. *Polynomial and rational solutions of holonomic systems.* Effective methods in algebraic geometry (Bath, 2000). J. Pure Appl. Algebra 164 (2001), no. 1-2, 199–220.
10. Oaku, Toshinori; Takayama, Nobuki; Walther, Uli. *A localization algorithm for D-modules.* Symbolic computation in algebra, analysis, and geometry (Berkeley, CA, 1998). J. Symbolic Comput. 29 (2000), no. 4-5, 721–728.
11. Saito, Mutsumi; Sturmfels, Bernd; Takayama, Nobuki. *Gröbner deformations of hypergeometric differential equations.* Algorithms and Computation in Mathematics, 6. Springer-Verlag, Berlin, 2000.
12. Takayama, Nobuki. *kan/sm1: a computer algebra system for algebraic analysis.* `http://www.math.sci.kobe-u.ac.jp/KAN/`
13. Tsai, Harrison; Walther, Uli. *Computing homomorphisms between holonomic D-modules.* Effective methods in rings of differential operators. J. Symbolic Comput. 32 (2001), no. 6, 597–617.
14. Walther, Uli. *Algorithmic computation of local cohomology modules and the local cohomological dimension of algebraic varieties.* Effective methods in algebraic geometry (Saint-Malo, 1998). J. Pure Appl. Algebra 139 (1999), no. 1-3, 303–321.
15. Walther, Uli. *Computing the cup product structure for complements of complex affine varieties.* Effective methods in algebraic geometry (Bath, 2000). J. Pure Appl. Algebra 164 (2001), no. 1-2

COMPUTER GRAPHICS IN MATHEMATICAL RESEARCH, FROM ICM 1978 TO ICM 2002: A PERSONAL REFLECTION

THOMAS F. BANCHOFF

Brown University, Providence, RI 02912, USA

Six International Congresses ago, a 45-minute invited address in Helsinki introduced the use of computer animation in mathematical research and teaching. Now, twenty-four years later, we can return to the same themes and see how the advent of raster graphics, real-time display, and interactivity on the Internet can enrich our understanding of the geometry of surfaces, both for research and teaching. In the sections below, we follow the outline of the article in the Proceedings of the 1978 ICM, comparing the original text (in small type) to some current developments of the ahuthor and his associates. Our illustrations recreate the wire-frame images from the original article and juxtapose them with modern images using software developed by Davide Cervone. The appendix is a narrative of the circumstances leading up to the presentation in the Helsinki ICM.

Introduction

In the Helsinki ICM, the title of the talk was "Computer Animation and the Geometry of Surfaces in 3- and 4-Space" and that remains the focus of the research and teaching of the author to this day.

Geometers have always used any available media to help them illustrate their work with diagrams, pictures, and models. Modern computer graphics provides a new medium with great potential both for teaching and research. Older methods of representing curves and surfaces by drawings on a blackboard or models in wire or plaster are frequently found to be inadequate in many important geometric problems, specifically those which involve objects undergoing transformations or objects which exist properly in the fourth dimension or higher. A high-speed graphics computer makes it possible to approach and solve such problems by methods that were unavailable only a few years ago.

At the time of the writing of the paper, the "high-speed graphics computer" was a device that would rotate a not very complicated wire-frame model in 3-space in real time. Our system included two improvements that were especially useful for our research—a hardware matrix multiplier designed by Harold Webber that made it possible for us to rotate figures in real time in 4-space, and a routine that could interpolate linearly between wire-frame surfaces with the same parametrization.

The next paragraph of the original article gave an introduction to computer graphics imagery that we take for granted today. There was a short discussion of motion clues versus stereoscopic pairs as the best way to view complicated geometric objects. Then, as now, the impressions created by objects in motion seem to work best for most viewers, especially in large audiences.

We provide a background by rotating a wire-frame model slowly about a vertical axis in space. Subsequent deformations take place within this background. For example the slices of a surface made by planes parallel to a fixed direction appear to the viewer to be planar slices of the rotating figure. Although slicing by a linear function, or more generally by some other function, does require some computational ability, the machine operates quickly enough that it is possible to view a sequence of slices in "real time", as if observing an object through a window as it rotated in the next room.

Modern rendering techniques have largely made the background rotation unnecessary, although we ordinarily provide rotation about a vertical axis as an aid in viewing complicated surfaces. The slicing techniques referred to used a standard "marching squares" algorithm to identifiy intersections of edges with the slicing plane and to connect points on the same rectangle in the domain by segments. Since the parameter domains only included a few hundred rectangular cells, the computer could produce moving slices of a rotating object at a rate fast enough for previewing scenes.

The effect, however, is greater if in addition to the slice, the film displays as well the part of the surface lying below or above the slice—the technique of "water-level slicing". Since this technique requires more time per picture and since it is especially well suited to representations using color, this technique is used primarily in the animation mode, where one picture is made at a time, and then the finished film is projected at 24 frames per second. For color, filters are used and each frame is exposed several times for the different portions of the picture.

It is hard to imagine computer workstations without color these days, but in the seventies they were rare. Multiply exposing each frame of a film to achieve a short sequence in color was intensely laborious, but the results were a striking preview of what we have today.

In addition to projection, rotation, and slicing it is possible to use linear interpolation between figures with the same parametrization. Again this is fast enough that the technique can be employed for real time manipulation of the figures for videotapes or for on-line research.

Video technology was extremely primitive at this time. There were competing standards and sizes of videotape, and it was nearly impossible to show a tape unless you brought the equipment with you, and even then it was impossible to project them for a large audience. Although standards took place very quickly in the United States, it is still complicated to show a videotape in other countries.

The Hypercube: Projections and Slicing

This film treats the convex hull of the sixteen points $(\pm 1, \pm 1, \pm 1, \pm 1)$ in 4-space, first by orthogonal projection then by central projection from 4-space to 3-space. In each case we rotate in the coordinate planes xy, yu, xw, yw, and zw, ending at the original position. We then slice each figure by hyperplanes perpendicular to the vectors $(1,0,0,0)$ then $(1,1,0,0)$ then $(1,1,1,0)$ and finally $(1,1,1,1)$. For a more thorough description of this film, see Banchoff.[4]

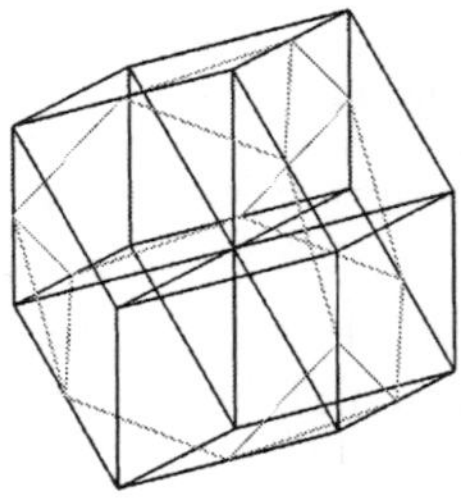

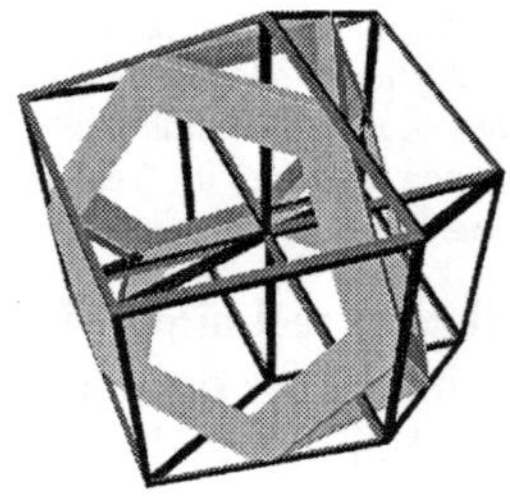

Figure 1. A frame from "The hypercube: projections and slices", and a modern version from the artwork "Iced cubes".

In these days, programming images of hypercubes is a beginning exercise in introductory courses in computer graphics. Instead of demanding a workstation, it is possible to realize scenes on a laptop computer. Even then the topic had been treated by several researchers, most notably A. Michael Noll at Bell Laboratories.[24] Our main contribution was a scripted tour of the 4-dimensional cube with three movements, orthographic projections, central projections, and slicing by planes and hyperplanes. After a quarter of a century, this film, now available in video, is still in demand, especially in schools and colleges. For modern interactive versions of the object, see for example the two cube sequences in the interactive art exhibit site "Para Álem da Terceira Dimenšao".[16,17] The reference[4] is a short description in an article where the illustrations came from Polaroid pictures taken directly off the computer screen!

Complex Function Graphs

This film treats graphs of complex functions $w = f(z)$ considered as parametric surfaces (x, y, u, v) in 4-space, where $z = x + iy$ and $w = u + iv$. In each case orthographic projection into (x, y, u) is used to get the graph of the real part of w, then rotation in the uv plane gives (x, y, v), the graph of the imaginary part of w. Rotating the original graph in the xv plane leads to (y, u, v), the graph of the imaginary part of the inverse function of f. Finally, projection to (x, u, v) gives the graph of the real part of the inverse function (Figure 2).

A particularly interesting example is the exponential function $w = e^z$ with the inverse relation $z = \log(w)$. The graph is given by $(x, y, e^x \cos(y), e^x \sin(y))$ in 4-space. Projection to (x, y, u) gives the real part of the exponential (Figure 2, left). The projection (y, u, v) gives a right helicoid which represents the imaginary part of the Riemann surface for the logarithm (Figure 2, center). The projection (x, u, v) gives a surface of revolution of a real exponential function as the real part of the logarithm.[3]

Color certainly is a considerable help in analyzing these objects. Coloring the axes and displaying a coordinate frame during rotations is a very effective way of keeping track of positions in real four-dimensional space. The authors have devised other methods for displaying complex function graphs, both of which have played central roles in expositions of art and mathematics, most recently in Lisbon.[16]

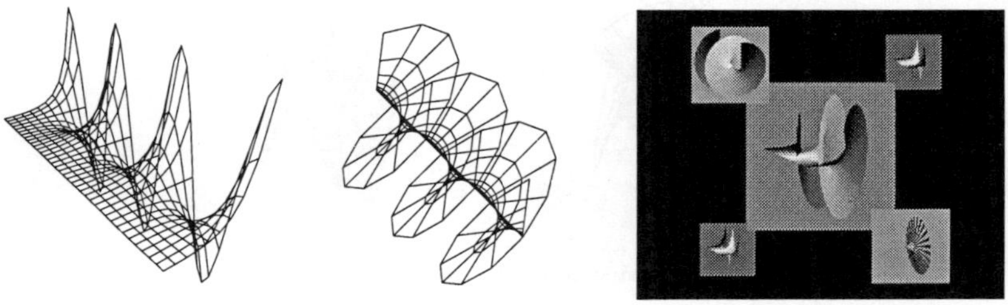

Figure 2. Graphs of the real part of the exponential function and the imaginary part of the logarithm, together with a "tetraview" for the exponential function.

A method with promise is the "tetraview", a way of displaying the real and imaginary parts of a function and its inverse relation in such a way that we can see the deformation from any position to any other by selecting edges in the one-skeleton of a tetrahedron whose vertices represent the four main projections from 4-space. We can also consider the intermediate position at the barycenter of the tetrahedron, the average of the four extreme views.

The Gauss Map, A Dynamical Approach

In the Helsinki ICM, the film "The Gauss Map: A Dynamical Approach" was a preliminary version that led to the 1982 monograph "Cusps of Gauss Mappings"[7] with Clint McCrory and Terence Gaffney. The illustrations in that volume were wire-frame images similar to those in the Proceedings article. Subsequently that monograph has appeared in an online updated version[8] with highly rendered images produced by Dan Dreibelbis, using software developed by Davide Cervone.[20]

Both the original article and the monograph presented of one-parameter families of surfaces involving the appearance and disappearance of cusps on the Gauss mapping. One of them, a perturbation of the "monkey saddle", exhibits a disc on the surface where the Gaussian curvature is positive such that the spherical image map has three cusps on the boundary curve. This provided an example that clarified a construction of Réné Thom.[6]

We consider the monkey saddle, with an isolated point of zero Gaussian curvature, and perturb to get the graph of $(x, y, x^3 - 3xy^2 + k(x^2 + y^2))$. For $k = 0$, this surface has a Gauss mapping with a ramification point of order 2, and for k not zero, the image of the parabolic curve will have three cusps (Figure 3).

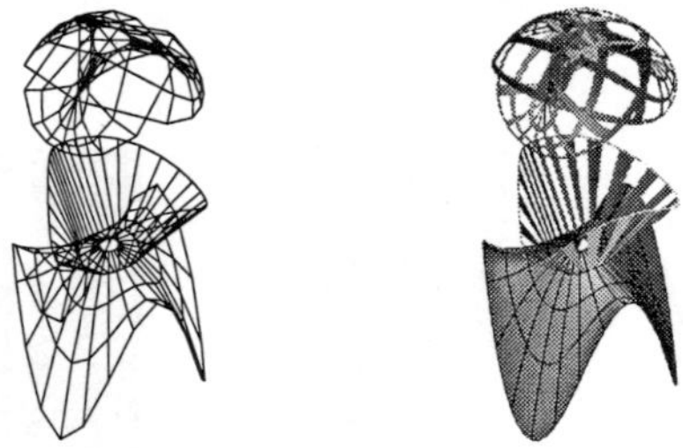

Figure 3. The Gauss map of a perturbed monkey saddle.

We show the spherical image of a circle $x^2 + y^2 = r^2$ as r changes. We show the linear interpolation between the surface and its Gauss spherical image so that the singularities of the Gauss map are expressed as limits of singularities of homothetic images of parallel surfaces of the original surface. We then show the spherical image of a test curve centered on the curve $r =$ constant and indicate the behavior of the asymptotic vectors in a neighborhood of a cusp of the Gauss mapping.

Included in the previous paragraph is one of the most striking applications of linear interpolation between surfaces with the same parametrization. If we consider the family $uX + (1 - u)N$ interpolating between a surface X and its Gauss mapping, then this can be viewed as a scaled version $u(X + (1-u)/uN)$ of the parallel surface at distance $r = 1/u - 1$. Thus the Gauss mapping can be thought of as "the parallel surface at infinity".

Various characterizations of the singularities of the Gauss map in terms of lines of curvature, ridges, and double tangencies are included in "Cusps of Gauss Mapping". Using the linear interpolation between a surface and its spherical image gives an interpretation of a cusp of the Gauss mapping as a limit of swallowtail points on the scaled parallel surfaces.

The Veronese Surface

The Veronese surface is an embedding of the real projective plane which starts with the hemisphere $x^2 + y^2 + z^2 = 1$, $z \leq 0$ and maps each point (x, y, z) to 6-space. The projection of this surface $(x^2, y^2, z^2, \sqrt{2}xy, \sqrt{2}yz, \sqrt{2}zx)$ from 6-space into 4-dimensional space given by $(\sqrt{2}xz, \sqrt{2}yz, (1/\sqrt{2})(x^2 - y^2), \sqrt{2}xy)$ is again an embedding and we examine a family of projections of this surface into 3-dimensional subspaces (all of which must have local singularities). This will appear in a paper on normal Euler classes by the author.

The paper referred to was published, in expanded form, as a collaboration with Ockle Johnson[11] twenty years after the Helsinki Congress! In the meantime, several other papers of the author have used the fact that the normal Euler class of a surface embedded in four-space can be obtained as an indexed sum of singularities of any generic orthogonal projection into a hyperplane. Since the normal Euler class of an embedded real projective plane is non-zero, there must be singular points for almost any such projection.

The projection into the last three coordinates gives a cross-cap with two pinch points (Whitney umbrella points). The linear interpolation of the left hemisphere into the cross-cap is a regular homotopy right up to the last instant when opposite points on the equator are identified, forming a segment of double points.

Deforming a hemisphere into a cross-cap is a another remarkable use of linear interpolation between surfaces with the same parametrization. It is a challenge, however, to position the surfaces so that the intermediate stages are all embedded with two-fold symmetry. One way is to interpolate between the cross-cap

$$\tfrac{1}{\sqrt{2}}(\sin u \sin v,\ \tfrac{1}{2}(\cos 2u)(1+\cos 2v),\ \tfrac{1}{2}(\sin 2u)(1+\cos 2v))$$

and the hemisphere given by $(\sin v, -\sin u \cos v, \cos u \cos v)$ where u goes form 0 to π and v goes from -$\pi/2$ to $\pi/2$.

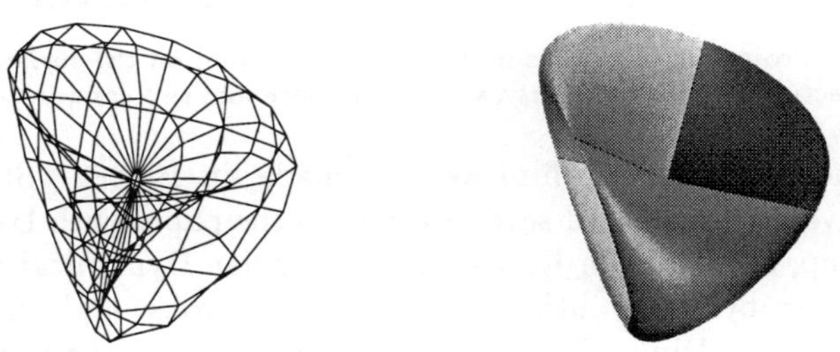

Figure 4. The projection of the Veronese surface as Steiner's Roman surface.

Rotating in the plane of the first and third coordinates gives a deformation from the cross-cap to Steiner's Roman surface $(\sqrt{2}xz, \sqrt{2}yz, \sqrt{2}xy)$ (Figure 4) with tetrahedral symmetry. This projection has six pinch points that are the end-points of three double point segments intersecting in a triple point. These examples are described in the classical book, "Geometry and the Imagination" by Hilbert and Cohn-Vossen.[23]

The embedding in 4-space is tight (i.e. almost every height function when restricted to the surface has exactly one maximum and one minimum) and this property is shared by the images in 3-dimensional subspaces. These examples lead to the conjecture that any stable tight mapping of the real projective plane into 3-space must have either two pinch points or six pinch points.

This conjecture was established in the Ph.D. thesis of Leslie Coglan, under the author's direction.[21,22]

The Flat Torus in the 3-Sphere

The flat torus is an embedding as a product of two circles in 4-space considered as the product of two planes, i.e. $(\cos u, \sin u, \cos v, \sin v)$. This torus is a surface on the 3-sphere of radius $\sqrt{2}$. We may project stereographically from $(0,0,0,\sqrt{2})$ to obtain a torus of revolution in 3-space.

Rotating the flat torus in the plane of the first and fourth coordinates produces a one-parameter family

$$(\cos a \cos u + \sin a \sin v, \sin u, \cos v, -\sin a \cos u + \cos a \sin v).$$

These rotated images of the flat torus project to a family of cyclides of Dupin, all conformally equivalent to the original torus. In particular when $a = \pi/2$ and the point $(0,0,0,\sqrt{2})$ is on the rotated torus, the projected surface is a non-compact cyclide which separates all of 3-space into two congruent parts.

Figure 5. Projection of a torus in 4-space from a point on the surface, so that the projection extends to infinity, separating space into two congruent regions.

This sequence was the first film we ever made, in 1968–9. Stills from the sequence have been recreated several times, in the paper with David Laidlaw, Fred Bisshopp, and Hüseyin Koçak on Hamiltonian Dynamical Systems,[18] in the related film by these authors and David Margolis, and on the cover of "Beyond the Third Dimension".[13] The flat torus was the first example to be developed by Davide Cervone and the author at the Geometry Center of the University of Minnesota in 1994, and it is one of the featured items in both our Providence Art Club and Lisbon exhibits.[15,16]

The cyclides of Dupin and spheres are the only closed surfaces in 3-space that have the spherical two-piece property, i.e. any sphere separates them into at most two pieces. Their inverse stereographic projections are the only surfaces on the 3-sphere that are tight, i.e. every hyperplane separates them into at most two pieces.[1]

Tight and taut surfaces and their generalization have undergone extensive development over the past twenty-five years. A full account of the subject can be found in the volume edited by Shiing-Shen Chern and Thomas Cecil, including a posthumous article by Nicolaas Kuiper and an extensive survey by Wolfgang Kühnel and the author.[12]

Appendix

How did it come about that an invited 45-minute address on computer graphics in mathematical research occurred at the International Congress of Mathematicians in Helsinki in 1978? What did we know at that time and what did we expect? What has actually happened over the past 24 years with respect to the topics featured in that first computer-illustrated invited talk at an ICM? This report at the ICMS will address the changes that have taken place in the author's collaborations with colleagues and students, particularly concerning advances in software for research and teaching in the areas of geometry and topology.

In the early 1970s, the only medium for presenting computer animation to an audience was film. In 1967, when I arrived at Brown University, I was fortunate to meet an applied mathematician, Charles Strauss, just finishing his Ph.D. in three-dimensional computer graphics, the first doctoral thesis supervised by the world expert on the subject, Andries van Dam. We began using the techniques from his thesis to analyze surfaces in four-dimensional space, projected orthographically or stereographically into three-space, and within one year, we had produced our first film, "The Hypertorus" (described above). It was a laborious process, waiting the better part of a minute for each image of a wire-frame model with 400 vertices, then taking two pictures and instructing the computer to produce the next image. It took all night in a darkened room before a storage tube to generate a few seconds worth of film. What we saw convinced us that not only was it worth the effort. We would never be satisfied again with still images.

It was nearly five years before we succeeded in producing our next films, preliminary versions of "The Hypercube: Projections and Slicing" and "Complex Function Graphs". I showed these in a two-part series at Berkeley during my first sabbatical, 1973–4. The first talk was titled "*Neue Polyhedralische Methoden in der Differentialgeometrie*" and the second was "*Disquisitiones Generales super Superficies Polyhedrales*". My Ph.D. advisor, Professor S.-S. Chern, showed interest in the project.

Two years later I gave a presentation at a symposium in honor of Prof. Chern. Afterwards he said to me that he thought it was time to present this to the world. He would put in a word, he said, with the selection committee for the ICM coming up in Helsinki. In late summer, 1977, I received a call from Prof. Armand Borel at the Institute for Advanced Study in Princeton. Could I tell him something about these films that had been mentioned? I said I would be coming to New Jersey in two weeks for a talk and I could show some films to him then. He said he would have to know about them before that, and perhaps I could describe them on the phone. "I'll be there tomorrow," I said. Prof. Borel put up an announcement, and over forty people arrived for a presentation in the IAS theater. I narrated films on the hypercube, the Gauss mapping, and an early version of the Veronese surface, and presented two films on complex functions with their own soundtracks. The audience was appreciative.

Afterwards in his office, Prof. Borel said he thought that it would be good to have a showing of these films at the ICM. He indicated that there could be a side event some evening, or there could be an invited talk in Section XIX, on History and Pedagogy. "You would probably prefer the latter?" he asked. I agreed, and that was the invitation.

Charles Strauss and I worked hard that next year, producing the version of "The Hypercube: Projections and Slicing" that is still used to this day. Since there were no color monitors available at that time, we achieved color by using color filters and a quadruple exposure for each frame, a semi-automated process that was still quite labor-intensive. The other films were all black-and-white, using vector graphics images with no filled surfaces. We worked especially on the Veronese surface, a series of images of the real projective plane which some exceptional singular projections.

At the ICM in Helsinki in 1978, the invited presentation was early in the meeting and over four hundred mathematicians attended. A few days later, Prof. Michael Atiyah asked if I would give another showing since many people had heard about the films only after the address. Another full audience watched a second showing, in which I included the original "Hypertorus" film. The article for the Proceedings used screen photographs of stills from the films as illustrations. Most of the article is reproduced above, along with annotations and new renditions of many of the images.

The purpose of the talk in Helsinki was to show how this new medium could offer new insights and new possibilities for teaching and research. In this article, we make particular mention of advances that have taken place using computer graphics in various aspects of mathematics, especially when visualization contributes new insights into geometric phenomena, new conjectures, and new methods of proof.

References

1. Banchoff, Thomas F., The spherical two-piece property and tight surfaces in spheres. *J. Differential Geometry* **4** (1970), 193–205.
2. —, with Charles Strauss, Computer-animated four dimensional geometry, *Proc. Amer. Assoc. Advancement of Science*, Washington, 1978.
3. —, with Charles Strauss, Real-time computer graphics techniques in geometry, The Influence of Computing on Mathematical Research and Education, *Proc. Sympos. Appl. Math.*, vol. 20, Amer. Math. Soc., Providence, R. I., 1974, pp. 105–111.
4. —, with Charles Strauss, Real-Time Computer Graphics Analysis of Figures in Four-Space, *American Association of the Advancement of Science Selected Symposium* **24** (1978), Westview Press, Colo., 159–168.
5. —, Computer animation and the geometry of surfaces in 3- and 4-space, Proceedings of the International Congress of Mathematicians (Helsinki, 1978), Acad. Sci. Fennica, Helsinki, 1980, 1005-1013.
6. —, with Réné Thom, Sur les points paraboliques des surfaces: erratum et complements, *C. R. Acad. Sc. Paris* **291** (27 Octobre 1980), 503–505.
7. —, with Terence Gaffney and Clint McCrory, Cusps of Gauss mappings, *Research Notes in Mathematics* **55**. Pitman (Advanced Publishing Program), Boston, Mass., London, 1982.
8. —, with Terence Gaffney, Clint McCrory and Dan Dreibelbis, Cusps of Gauss mappings, the Electronic Version, European Mathematical Information Service, ⟨`http://www.emi.de/monographs/CGM`⟩, 2000.
9. —, Normal Curvatures and Euler Classes for Polyhedral Surfaces in 4-Space, *Proc. A.M.S.*, 92 (1984), 593–596.
10. —, Differential Geometry and Computer Graphics, *Perspectives in Mathematics*, Anniversary of Oberwolfach (1984) Birkhäuser-Verlag,

Basel, 43–60.

11. —, with Ockle Johnson, The normal Euler class and singularities of projections for polyhedral surfaces in 4-space, *Topology* **37** (1998) no. 2, 419–439.
12. —, with Wolfgang Kühnel, Tight submanifolds, smooth and polyhedral, *Tight and taut submanifolds* (Berkeley, CA, 1994), 51–118, *Math. Sci. Res. Inst. Publ.*, **32**, Cambridge Univ. Press, Cambridge, 1997.
13. —, *Beyond the third dimension: Geometry, computer graphics, and higher dimensions.* Amended and enhanced version of the 1990 original. Scientific American Library Paperback, 33. *Scientific American Library*, New York, 1996.
14. —, with Davide P. Cervone, Illustrating "Beyond the Third Dimension", *The visual mind*, 85–92, *Leonardo* Book Ser., MIT Press, Cambridge, MA, 1993.
15. —, with Davide P. Cervone, *Surfaces Beyond the Third Dimension*, virtual art gallery, 1996; ⟨`http://www.math.brown.edu/~banchoff/art/PAC-9603/`⟩.
16. —, with Davide P. Cervone, *Para Álem da Terceira Dimensão*, physical and virtual art exhibit, Lisbon, 2000; ⟨`http://alem3d.obidos.org/`⟩.
17. —, with Davide P. Cervone, A virtual reconstruction of a virtual exhibit, in *Multimedia Tools for Communication Mathematics*, Springer-Verlag, Berlin, Heidelberg, 2002, 29–38.
18. —, with H. Koçak, F. Bisshopp and D. Laidlaw, Topology and Mechanics with Computer Graphics: Linear Hamiltonian Systems in Four Dimensions, *Advances in Applied Mathematics* (1986), 282–308.
19. —, *Beyond the Third Dimension*, (1990) New York: W. H. Freeman & Co., Scientific American Library, 1–210.
20. Cervone, Davide P., `StageTools` Geometry Package, ⟨`http://www.math.union.edu/locate/StageTools`⟩.
21. Coghlan, Leslie, Tight stable surfaces, I, *Proc. Roy. Soc. Edinburgh Sect. A* **107** (1987) no. 3–4, 213–232.
22. Coghlan, Leslie, Tight stable surfaces, II, *Proc. Roy. Soc. Edinburgh Sect. A* **111** (1989) no. 3–4, 213–229.
23. Hilbert, D. and Cohn-Vossen, S., *Geometry and the imagination*, Chelsea, New York, 1952.
24. Noll, A. Michael, A Computer Technique for Displaying n-Dimensional Hyperobjects, *Communications of the ACM*, **10** (1967) no. 8, 469–473; Reprinted in *Hypergraphics*, D. Brisson, ed., American Association for the Advancement of Science Selected Symposium 24 (1978) Westview Press, Colo., 159–168.

GENERATING OPTIMAL CURVES VIA THE C^{++} STANDARD LIBRARY

ANDERS LINNÉR
Department of Mathematical Sciences, Northern Illinois University, DeKalb, IL 60115, USA
E-mail: alinner@math.niu.edu

Recent computational breakthroughs in infinite-dimensional optimization have produced explicit, as well as numerically advantageous, gradient formulas. By following the negative gradient trajectories, along so-called steepest descent, the minimizing curve is sought. The C^{++} standard library supports the assembly of a user-defined type based on the standard cubic splines, and a corresponding operator algebra that is capable of representing all the gradient expressions while preserving the conventional notation. It is crucial that the algebra implements, and is 'closed' under, composition with both user-defined and built-in functions, as well as the calculus operations integration and differentiation. Examples illustrate both fixed and free endpoints, and particular attention is paid to the notoriously difficult isoperimetric constraints. In the spirit of 'component technology', the implementation uses the 'boost extension' of the standard library.

1 Introduction

1.1 Objective

This article concerns the combined use of object-oriented methods and generic programming to generate optimal curves in a wide range of variational problems. The ideas behind the algorithm have their origin in infinite dimensional differential geometry. It is reasonable to select the largest possible function space permitted by the nonlinear functional one seeks to minimize. The geometry is subsequently determined by the choice of a Sobolev metric. To simplify matters, this article will stick to one such choice that leads to the least complicated set of gradient formulas. As exposed in [1], there are alternative metrics with geometrically distinct behavior, so the choice should generally speaking not be capricious. Once the metric is in place, the notion of steepest descent along negative gradient trajectories in the manifold of constrained curves is well defined. The idea is to select some initial curve that satisfies all constraints and then reduce the value of the functional using steepest descent inside the manifold of constrained curves. It must be understood that this manifold is not expected to be flat in the presence of isoperimetric constraints. This essential nonlinearity is the source of much computational trouble. Periodic boundary conditions also fit well into the present framework, and arguably is a major reason behind the interest in this kind of approach. The present article puts forth an 'Euler operator' and a 'Work integral' that lead to computationally very favorable formulas. This is supported by showing that the most elementary numerical methods suffice as building blocks of the algorithm that generates the solutions.

1.2 Interpolation tool

Since the steepest descent cannot be given explicitly, any discrete counter part must deal with a serious loss of information at the first step of the descent. Specifically, the initial curve is known at all points, but this is not the case for the curve after one step of the descent. It follows that the numerical methods employed that require information away from the known discrete representative must be supported by some interpolation scheme. The implementation presented here is based on cubic spline interpolation. A **C++ class** called S is developed as a means to represent intermediate curves and at the same time have complete computational flexibility. This spline class is equipped with both integral and differential operators as well as support of composition with arbitrary functions. Once all is put together, formulas are expressed in the computer code in a form that makes the translation from their matematical expressions completely transparent.

1.3 Organization

Section 2 gives a brief summary of the gradient formulas and how they are translated into computer code. The resolution of the complications due to isoperimetric constraints is also indicated. Section 3 presents the core parts of the implementation. Special attention is of course paid to the S class. Section 4 illustrates the algorithm using eight examples. There is also a discussion about convergence and timing. Section 5 gives a summary and references.

2 Gradients

2.1 From gradient formula to computer code

Let the inner product be given by $\langle w, v\rangle = w(0)v(0) + \int_0^1 \dot{w}(t)\dot{v}(t)dt$. Suppose the functional is given by $F(x) = \int_0^1 f(t, x(t), \dot{x}(t))\, dt$. Define the Euler operator by $E_x^f(t) = \frac{\partial f}{\partial \dot{x}}(t, x(t), \dot{x}(t)) - \int_0^t \frac{\partial f}{\partial x}(s, x(s), \dot{x}(s))\, ds$.

In the case of fixed endpoints, the gradient formula is given by [2]

$\nabla F(x) = \int_0^t E_x^f(s)ds - t\int_0^1 E_x^f(s)ds$. The gradient clearly vanishes at both endpoints. Observe that the gradient is a function that can be expressed in terms of x and $\dot{x}$ using operations provided by the class S. Assume that x is an S-object representing the current curve. The $\mathbf{C}^{++}$ code corresponding to the gradient formula is given by

```
S Euler = fxDot - Integral(fx);
Gradient = Integral(Euler) - Identity*CompleteIntegral(Euler);
```

Ultimately, the definitions of the spline class should be placed in their own namespace (Spline say). This protects against name clashes, since S will be replaced by Spline::S, much the same way names from the standard library are prefixed by *std::*. The values of f, and the partial derivatives fx and fxDot are supplied by the user. The coded gradient formula exposes the central idea of the implementation: *Everything representing a point in the infinite-dimensional space must be an S-object.* The S-object Identity has the property that Identity(t) = t for all t. It is incorrect to replace Identity by a **double** t since this creates a gradient where the t in front of the second integral is the current value of t as the statement is executed.

2.2 Different gradients due to side conditions

The implementation supports gradient formulas for free endpoints, free left-hand endpoint, free right-hand endpoint, and fixed endpoints. The inner product induces a metric in the space of curves. A change in the metric alters the gradients. The steepest descent trajectory is different. Geometric properties, such as symmetries, are preserved in some metrics but not in others. Observe that, although the gradient depends on the metric, the collection of curves where the gradient vanishes is fixed. There are also differences in the level of smoothness in the gradient formulas as the metric changes. All gradient formulas are expressed in terms of the Euler operator and the Work integral $W_x^f = \int_0^1 \frac{\partial f}{\partial x}(t, x(t), \dot{x}(t))\, dt$.

2.3 Examples with free right-hand endpoint

To see examples where the work integral is present, consider the case of a free right-hand endpoint so the projection has changed. The gradient is given by $\nabla^{\pi,0} F(x) = \int_0^t E_x^f(s)ds + W_x^f t$. It is clear that the gradient vanishes at the left-hand endpoint, but not necessarily at the right-hand endpoint. Here a second superscript indicates that only the left-hand endpoint is fixed.

2.4 Isoperimetric Constraints

Recall that the case of fixed endpoints generated translated linear subspaces and the need for an orthogonal projection to compute the gradient vector field. The tangent space to a translated linear subspace is the linear subspace itself. Isoperimetric constraints, $G : H \to \mathrm{R}$, defined in a Hilbert space H, are more complicated since they generate tangent spaces that vary from point to point. Since $G(x) = 0$ for each x that satisfies the isoperimetric constraint G, it must be that the directional derivative satisfies $DG(x)\, v = 0$ for all tangential v, and the familiar $\langle \nabla G(x), v \rangle = 0$ holds. Now consider $F : H \to \mathrm{R}$ and its gradient vector field. The idea is to introduce a scalar field $\lambda(x)$ and express the tangential component $\nabla^{\pi} F$ as $\nabla^{\pi} F(x) = \nabla F(x) - \lambda(x) \nabla G(x)$. The condition $DG(x)\, v = 0$ leads to

$$\lambda(x) = \frac{DG(x) \nabla F(x)}{DG(x) \nabla G(x)}.$$

For this approach to work, it is necessary to assume that $\nabla G(x) \neq 0$ for all x such that $G(x) = c$. This assumption is referred to as regularity.

3 Implementation

3.1 The spline class

In this section the code is dissected and the different techniques used in the implementation are highlighted. Objects of the class S represent the curves and the gradients. S is given by the following **C++** code. A reader unfamiliar with **C++** may think of a S as an example of a user-defined type like the built-in **double**. Unlike built-in types, a user-defined type must supply functions to control creation and manipulation of its objects. The boost extension [6] to the standard library will leverage these functions and restore some of the functionality available to built-in types. Reserved words are in bold. Names from either the standard library or the boost extension are in italic.

```
#include "operators.hpp" // boost extension
#include <vector>
#include <string>
```

```
#include <functional>
class FunctionObject; // Forward declaration
class S : private boost::addable<S>, boost::addable<S,double>,
boost::subtractable<S>, boost::subtractable<S,double>,
boost::multipliable<S>, boost::multipliable<S,double>,
boost::dividable<S>, boost::dividable<S,double>
{public: friend FunctionObject;
S(std::vector<double>, std::vector<double>);
double operator()(double);
void Show() const;
void Save(const std::string&) const;
S& operator+=(const S&); S& operator+=(const double&);
S& operator-=(const S&); S& operator-=(const double&);
S& operator*=(const S&); S& operator*=(const double&);
S& operator/=(const S&); S& operator/=(const double&);
S Integral(double) const; S Derivative()const;
private: std::vector<double> m_p, m_v, m_vdotdot;};

class FunctionObject : public std::unary_function<double,double>
{public: FunctionObject(double (*)(double));
double operator()(double) const;
S operator()(const S&) const;
private: double (*m_f)(double);};
```

The extension uses the eight user-supplied member operators to define almost all the standard algebraic operations. As an example look at the boost extension's addable class which permits the compiler to properly interpret the plus in expressions like `x*x + xDot*xDot`:

```
template<class T> class addable
{ friend T operator+(T x, const T& y) {return x += y;} };
```

The class S has three private data members stored as dynamically allocated vectors based on the built-in type double. To create an object of class S representing a function x, it suffices to supply a discrete set of parameters p_i and the corresponding values $v_i = x(p_i)$. The second derivative of the cubic spline, passing through the points (p_i, v_i) in order, is automatically computed and saved in the third data member. To illustrate this consider how the Identity object is created and the initial x.

```
double* p; //Sample points
p = new double[NumberOfCubicPolynomials + 1]; //Uniform sample
```

```
for(unsigned int i = 0; i <=NumberOfCubicPolynomials; ++i)
p[i] = double(i)/NumberOfCubicPolynomials;
//Initialize a standard vector
const std::vector Points(p,p + NumberOfCubicPolynomials + 1);
delete[ ] p; //Clean-up
S Identity = S(Points,Points); //Base S-object
FunctionObject Initial(initial); //FunctionObject from initial curve
S x = Initial(Identity); //Corresponding S-object
S xDot = Derivative(x);
```

Armed with its three data members, S has member functions to compute objects of class S corresponding to the derivative as well as the integral operation. The S class is also equipped with an evaluation operator. This method calls a cubic spline routine to calculate the value. There is a friend class FunctionObject to facilitate composition with functions. Specifically, if s is an object of class S, then s(0.3) returns the value of s at 0.3. Moreover, if f is an object of the FunctionObject class, then f(s) is an object of the S class so in terms of the composition operation $\circ$, $f(s)(0.3) = f \circ s(0.3)$.

3.2 Cubic splines

The cubic spline routines are based on the code given in Numerical Recipes in C [4]. The main difference is the use of vectors from the standard library rather than the built-in arrays. This renders the use of memory allocation routines in the Recipes obsolete. The calling interface is simplified since there is no need to pass the dimension of the vectors to each routine. To this end, the implementation uses the *size*() member function from <*vector*>. In addition, the implementation supplies two extensions to the cubic spline recipes. The first computes the value of the derivative of the spline. The second computes the value of the integral with a supplied initial value to facilitate piecewise integration. To retain notational compatibility with the code in the recipes, this part of the code uses indices and array notation.

```
double SplineDerivativeValue(std::vector<double>& x, std::vector<double>& y,
std::vector<double> yDotDot, double t)
{int klo=0, khi = x.size() - 1,k;
while( khi-klo > 1 )
{k=(khi+klo) >> 1; if(x[k] > t) khi=k; else klo=k;}
double h,a,b; h = x[khi]-x[klo]; a = (x[khi]-t)/h; b = (t-x[klo])/h;
return (y[khi]-y[klo])/h + ((1.0 - 3.0*a*a)*yDotDot[klo] + (3.0*b*b -
```

```
1.0)*yDotDot[khi])*h/6.0;}

double SplineIntegralValue(std::vector<double>& x, std::vector<double>& y,
std::vector<double> yDotDot, double t)
{int klo=0, khi = x.size() - 1, k;
while( khi-klo > 1 )
{k=(khi+klo) >> 1; if(x[k] >= t) khi=k; else klo=k;}
double h,a,b; h = x[khi]-x[klo]; a = (x[khi]-t)/h; b = (t-x[klo])/h;
return (b*b*y[khi] + (1.0 - a*a)*y[klo])*h/2.0 + ((2.0*a*a - a*a*a*a -
1.0)*yDotDot[klo] + (b*b*b*b - 2.0*b*b)*yDotDot[khi])*h*h*h/24.0;}
```

The Spline routine in the recipes updates the second derivative vector. The derivative and integral S member functions are given by the following code.

```
S S::Derivative() const {S New(*this), Old(*this);
std::vector<double>::const_iterator pIterator = Old.m_p.begin();
std::vector<double>::iterator vIterator = New.m_v.begin();
while( pIterator != Old.m_p.end() )
*vIterator++ = SplineDerivativeValue(Old.m_p,Old.m_v,Old.m_vdotdot,*pIterator++);
Spline(New.m_p,New.m_v,New.m_vdotdot);return New;}

S S::Integral(double InitialValue) const {S New(*this), Old(*this);
std::vector<double>::const_iterator pIterator = Old.m_p.begin();
std::vector<double>::iterator vIterator = New.m_v.begin(); *vIterator++ =
InitialValue; ++pIterator;
while( pIterator != Old.m_p.end() )
*vIterator++ = SplineIntegralValue(Old.m_p,Old.m_v,Old.m_vdotdot,*pIterator++);
std::partial_sum(New.m_v.begin(),New.m_v.end(),New.m_v.begin());
Spline(New.m_p,New.m_v,New.m_vdotdot); return New;}
```

Note how the standard library's generic algorithm *std::partial_sum* in *<numeric>* is very helpful here. It is interesting to compare the index-free coding style of the last two functions and the pre-standard-template-library style of the Recipes code. The C++-standards committee exhibited considerable leadership when it decided to incorporate most of the standard template library [5] into the standard library. To further conform to standard mathematical notation it is useful to define four global functions.

```
S operator-const double& x, const S& y) {return (-1.0)*(y-x);}
```

```
S Integral(const S& x){return x.Integral(0.0);}
double CompleteIntegral(const S& x){return x.Integral(0.0)(1.0);}
S Derivativeconst S& x){return x.Derivative();}
```

In what follows the cubic splines are shielded from the rest of the implementation, and any collection that provides the same functionality could take the place of the cubic splines.

3.3 FunctionObject

The implementation of composition in the FunctionObject class is given by the following code.

```
doubleFunctionObject::operator()(double t) const {return m_f(t);} S
FunctionObject::operator()(const S& x) const {S result(x);
std::transform(x.m_v.begin(),x.m_v.end(),result.m_v.begin(),*this);
Spline(result.m_p, result.m_v, result.m_vdotdot); return result;}
```

A FunctionObject is equipped with one 'evaluation operator' and one 'composition operator'. The composition operator uses the evaluation operator through self-reference inside the generic algorithm *std::transform*. The abstraction built-in to the standard template library is here paying great dividend.

4 Examples

4.1 Eight Examples

Here is a series of representative functionals F and constraints G:

$$F_1(x) = G_{6,7}(x) = \int_0^1 \sqrt{1+\dot{x}^2(t)}\,dt, \qquad F_2(x) = \int_0^1 x^2(t) + \dot{x}^2(t)\,dt,$$

$$F_3(x) = \int_0^1 \left(1-\dot{x}^2(t)\right)^2 dt, \quad F_{4,5,6,7}(x) = \int_0^1 \frac{\sqrt{1+\dot{x}^2(t)}}{\sqrt{v_0^2 - x(t)}}dt, \quad F_8(x) = \int_0^1 \dot{x}^2(t)\,dt,$$

$$G_8(x) = \int_0^1 (\cos(x(t)), \sin(x(t)))dt.$$

The solutions in the eight examples are: 1) Straight line. 2) $\sinh(t)/\sinh(1)$. 3) Derivative ± 1. 4,5) Cycloid with or without fixed end. 6,7) As far as I know, the Brachistochrone of fixed length is not in general known. 8) Elastic curve of fixed length, which can be expressed in terms of Jacobi's Elliptic functions. Observe that only the solution is plotted in black. As the steepest

Figure 1

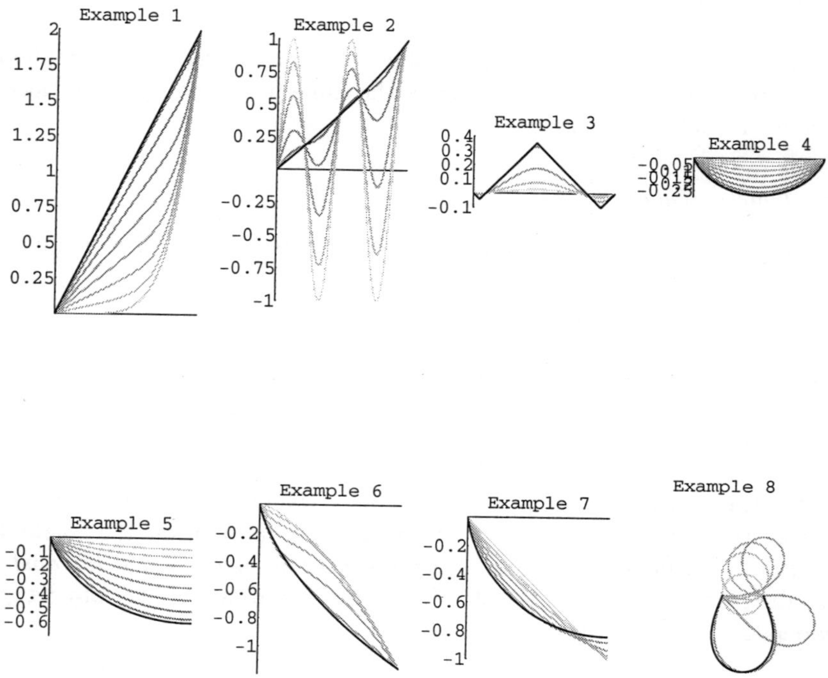

descent proceeds, the curves are shaded progressively darker.

4.2 Convergence criterion and computational speed

In all the examples the limit curve drawn in black corresponds to a curve where the norm of the projected gradient vector field is less than 10^{-6}. The user supplies this value along with other parameters that control the behavior of the algorithm. The number n of cubic splines carried by each S-object is one such important parameter. To measure the speed of the algorithm a method for timing generic algorithms is employed, see Chapter 19 in [3].
The results are split as baseline time b_i(preparation of data) and computa-

n	b_0	t_0	f_0	b_1	t_1	f_1	b_2	t_2	f_2
4	0.46	0.27		1.30	0.99		0.53	1.60	
8	0.56	0.36	1.33	1.70	1.33	1.34	0.65	2.14	1.34
16	1.20	0.76	2.13	3.31	2.71	2.04	1.36	4.60	2.14
32	1.66	1.15	1.50	4.88	4.13	1.52	1.84	6.93	1.51
64	2.78	2.01	1.75	8.47	7.40	1.79	3.08	12.3	1.76
128	4.81	3.60	1.79	14.9	13.6	1.84	5.21	22.6	1.83
256	8.85	7.02	1.95	28.6	27.0	1.98	9.52	43.1	1.91
512	16.9	13.4	1.90	55.6	55.6	2.06	18.2	81.8	1.90
1024	33.3	29.2	2.18	111	88.9	1.60	37.0	163	1.99
2048	66.7	58.3	2.00	250	250	2.81	71.4	429	2.63
4096	143	107	1.84	500	500	2.00	143	857	2.00

Example	1	2	3	4	5	6	7	8
n	32	64	256	128	128	512	512	128
h	0.5	0.1	0.001	0.001	0.001	0.01	0.01	0.01
steps	256	64	16384	20480	40960	4096	8192	512

tional time t_i. The values are given in milliseconds and the index i corresponds to the number of isoperimetric constraints. The timing is for a Sun Ultra 5 running Solaris 7 using GNU g++ version 2.95 compiled with speed optimization. The ratio $f_i(n) = t_i(n)/t_i(n/2)$ is listed and Examples 1, 6, and 8 are tabulated as index 0, 1, and 2 respectively. The next table lists, by example, the step sizes h and the number of steps taken before the norm of the projected gradient vector is zero within the acceptable tolerance of 10^{-6}. All constraints are satisfied to at least three digits.

By combining the information in the two tables one estimates example 6 to take between $4096 \cdot t_1(512)$ and $4096 \cdot (b_1(512) + t_1(512))$ milliseconds. With $b_1(512) = t_1(512) = 55.6$ milliseconds the estimate is 4 to 8 minutes. A stopwatch trial yielded 5 minutes and 40 seconds. It is increasingly difficult to visually distinguish between curves of tolerances less than 10^{-4}, so in practice these times can often be reduced by a factor of four without a loss of geometric insight.

5 Conclusions

About a decade ago, while working on implementations of infinite-dimensional problems, it became clear to the author that it is necessary to bridge the notational gap between mathematical formulas and the corresponding computer code. Back then, the coding of complicated mathematical formulas, and the demand for performance, yielded very frail and difficult to extend programs. The emergence of **C**$^{++}$ and the standard template library has

changed all of this. As this article demonstrates, it is now possible to write code where the notational gap between the mathematical formulas and the computer code has all but disappeared. Significantly, this does not come with severe, or even noticeable, performance degradation. Moreover, the code here is easily extended to cover more than two isoperimetric constraints, or more than one metric. It is also designed to be user-friendly in the sense that next to no knowledge of **C**$^{++}$ is required to add code for new examples. All that is needed is a straightforward translation of the mathematical expressions corresponding to the integrand of each functional involved and its partial derivatives. The complete software is available at: http://www.math.niu.edu/~alinner/Software/CoV00/
The reader seeking more background material and detailed explanations should consult the site www.math.niu.edu/~alinner and the link to the unabridged version of this article. The software has been tested under UNIX Solaris 7 and Windows 98 Second edition using GNU g++ version 2.95 and Borland C++ Builder 4.0, respectively. One reasonable approach to improve the performance of the software is to combine the Euler step with Adams-Bashforth's method and invoke higher order formulas as soon as enough points are calculated. The subsequent freedom to aggressively increase the step size is desirable. In the context of 'critical curves at infinity', this feature is essential. Examples of this, using a completely different implementation, are given in [1]. The current implementation avoids all but the most basic numerical methods to showcase the strengths of the underlying gradient formulas as well as the power of **C**$^{++}$ and the standard library.

References

1. A. Linnér in *Symmetrized Curve-Straightening*, To appear in Differential Geometry and its Applications, 2002
2. A. Linnér, *Gradients, preferred metrics and asymmetries*, Preprint 2000, *http://www.math.niu.edu/~alinner/*
3. D.R. Musser and G.J. Derge *et al STL Tutorial and Reference Guide, C++ Programming with the Standard Template Library* 2nd ed., (Addison-Wesley, 2001) 279-289.
4. W.H. Press and S.A. Teukolsky *et al Numerical Recipes in C*, The Art of Scientific Computing 2nd ed.,(Cambridge University Press, 1992) 113-116.
5. A.A. Stepanov and M. Lee, *The standard template library*, Technical Report HP-94-34, Hewlett- Packard, April, 1994. Revised July 7, 1995 .
6. Public domain, *http://www.boost.org/*

A MATHEMATICAL WEBSERVICE FOR RECOGNIZING THE UNKNOT

M. V. ANDREEVA [*], I. A. DYNNIKOV [†] AND K. POLTHIER [‡]

In this paper we present the implementation of a partial knot recognition algorithm as a mathematical web service on the internet. Knots may interactively be loaded and edited, and then checked for being unknottet. We use the approach by Dynnikov, which is based on a combinatorial represenation of knots in a three-page book, and we use JavaView as visualization environment.

1 Introduction

Among the most interesting algorithmic problems of topology is recognition of the unknot. Colloquially speaking this means to decide if a given knot is really knotted. Being interesting by their own, a strong mathematical application of knots is their usage in different representations of three-manifolds. For instance, any compact three-manifold without boundary can be represented as a covering of the three-sphere branched along some knot. In recent years knot theory has found interesting applications outside mathematics, for example, in biochemistry where the topological shape of knotted molecules sometimes contributes to their chemical properties.

In this paper we present the implementation of a knot recognition algorithm as a mathematical web service on the internet. The overall purpose of this activity is motivated by the following two goals:

1. We would like to use an experimental laboratory for the recognition of knots which is maintained on a central server but has world-wide access through the internet. This is essential for our own experiments since our co-authors happen to travel around and would hardly be able to synchronize their local computer or notebook with the latest version of the laboratory.

2. We would like to offer other researchers a simple access to perform own experiments without the need to run through software installation, or

[*]INSTITUTE OF INFORMATION SYSTEMS, 630090 NOVOSIBIRSK, RUSSIA, E-MAIL: MARY@IIS.NSK.SU

[†]MECHANICS AND MATHEMATICS DEPARTMENT, MOSCOW STATE UNIVERSITY, VOROB'EVY GORY, 119899 MOSCOW, RUSSIA, E-MAIL: DYNNIKOV@MECH.MATH.MSU.SU

[‡]INSTITUT FÜR MATHEMATIK, TECHNISCHE UNIVERSITÄT BERLIN, STRASSE DES 17. JUNI 136, 10623 BERLIN, GERMANY, E-MAIL: POLTHIER@MATH.TU-BERLIN.DE

even worse, to repeated software installations when the laboratory is upgraded.

In section 2 we recall the principles behind the recognition algorithm, in section 3 we describe the basics of the web service and its components, and in section 4 we discuss in detail the functionality of the user interface and how to perform practical experiments.

2 Review of the Recognition Algorithm

Here a *knot* is a simple polygonal line in the Euclidean three-space $\mathbb{R}^3$. Two knots Γ_1 and Γ_2 are called *isotopic* if there exists an orientation preserving piecewise linear homeomorphism of $\mathbb{R}^3$ to itself that sends Γ_1 to Γ_2. In topology, knots are considered up to isotopy. The union of a few non-intersecting knots in the three-space is called a *link*. The notion of isotopy naturally extends to links.

A knot is said to be an *unknot* or a *trivial knot* if it is isotopic to a knot that lies in a plane. Otherwise we say that a knot is non-trivial. The simplest example of a non-trivial knot is the trefoil knot shown in Figure 1.

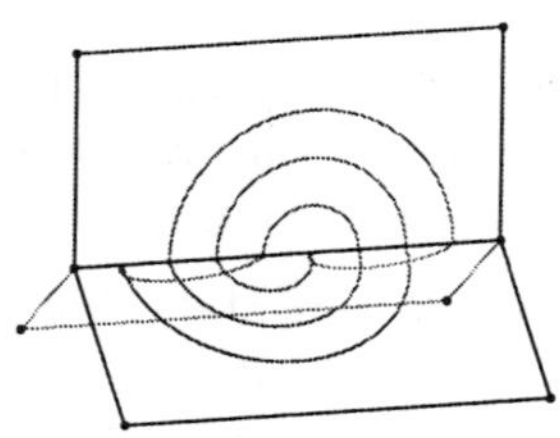

Figure 1. The trefoil knot

The problem of recognizing the unknot was theoretically solved by W. Haken who introduced an algorithm that has at least exponential complexity [8]. It was used by J. Birman and H. Hilden for recognizing the three-sphere among three-manifolds represented by genus two Heegaard splittings or, more general, by two-sheeted coverings of the three-sphere branched along links [3]. Some other topological problems need an effective algorithm for recognizing the unknot. Up to now no algorithm of polynomial complexity for solving this problem is known. The question of the existence of such an algorithm is an important and well-known unsolved problem of algorithmic topology.

One of the possible approaches to the problem consists in finding a *partial* algorithm for recognizing the unknot with polynomial running time and then see if one can improve it to be a real algorithm. 'Partial' formally means that the algorithm checks some sufficient condition for knot triviality. If the condition is satisfied, the positive answer is given. If not, the algorithm ends up with an uncertain answer.

One of the most natural ways to construct a partial algorithm is by using some simplification of the knot presentation. The most naive approach, using simplification by Reidemeister moves, does not work already for knots with twenty crossings.

An partial algorithm based on simplification of a combinatorial presentation of the knot complement was proposed by S. V. Matveev.

Recently one of us (I.D.) proposed another partial algorithm for recognizing the unknot. It is based on a simplification of the knot with respect to a complexity different from the number of crossings. An advantage is the small running time for the algorithm, $O(\text{complexity}^3)$, which allows to run it for monster knots (having, say, 1000 crossings on a planar diagram). The result of the algorithm is a knot isotopic to the initial one that cannot be simplified any further by certain type transformations.

The estimation to what complexity the algorithm will distinguish between trivial and non-trivial knots is a hard problem. We tried to run it for a few monster unknots (see [4,5] for an example) with hundreds of crossings and they have been successfully untangled. However, some unknots are not simplified completely by the algorithm. So far, we do not have such an example with crossing number < 50. One of the purposes of our project is to visualize the simplifying process and possibly find new methods to improve the algorithm.

See the example of such procedure applied to one of the Goeritz diagrams on the figure 2.

The subject of our talk is exposition of its computer realization [1] which may serve as a good tool for working topologists. This package is based on the JavaView library developed by one of us (K.P.) [2].

The approach is based on a simple and very well known observation that every link in the three-space is topologically equivalent to a link that lies entirely in a three-page book, i.e. the union of three half-planes with common boundary. Though being not convenient for human perception, this way of presenting links seems to be very efficient for handling knots by computer. It provides a very quick way from a combinatorial description of a link to its three-dimensional presentation. A three-page link admits a lot of transformations that preserve its isotopy class and can be easily found. This fact is used in the knot simplification tool which we are developing now. For algebraic

Figure 2. Untangling Goeritz 13-21

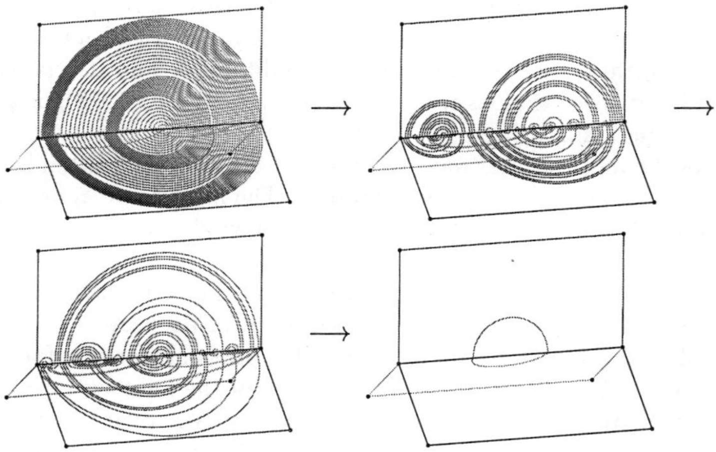

aspects of the three-page representation of links we refer to [6,7].

3 The Web Service

The web service at [1] consists of a Java applet embedded in a HTML web page. The Java applet is based on the Javaview software library plus an additional library which contains the knot recognition and editing functionality.

In contrast to other mathematical web services the software runs on the local computer of each user, and not on our server. This distinguishes our Java based approach from web services which are, for example, built with webMathematica. The advantage for us is a reduced requirement on the computation power of our web server, especially, if multiple users are accessing web services. The user has the benefit that he can choose what ever computational resources he needs by selecting an appropriate computer in his environment. But usually there is no big need to care about this issue on the user side. For smaller knots the simplification algorithms runs fast even on laptops, that means, basically the service is available everywhere on world.

4 Components of the User Interface

The user interface of the web services at [1] shows a 3D geometry viewer for interactive modifications on the currently investigated knot and a control panel to enable different functionalities:

1. view knots available in the database of the server
2. create your own knot by entering a 3D curve or a Gauss code
3. simplify or perform partial transformation a knot
4. interactively edit a knot.

We shall now explain them very briefly.

1. View a knot.

To download a knot from the server click 'Choose knot' button. At the present time, we have a collection of knots presented in jkb format, as 3D curve (∗.txt files), and as a Gauss code (∗.gc files). Files 31.jkb, 141.jkb, 322.jkb, and 539 have been generated automatically from hand-made (I.D.) planar diagrams of unknots having 31, 141, 322, and 539 crossings, respectively. These unknots were used to test the simplifier.

The unknot with 539 crossings is a satellite of this unknot (see [4] for a picture). The source of the knots in txt format is the Knot-Plot web site, which is created by Robert Scharein [9] and contains a collection of knots and links with small crossing number (less than or equal to 10). The files with gc extension contain Gauss codes of the same links. These codes were generated automatically from the ∗.txt files by a program written by S. E. Koval. For converting a planar diagram to a three-page diagram the algorithm described in [4] is used.

2. Input a knot.

To input your own knot click 'Input knot' button and choose a method of input. Now available methods are using a 3D presentation and the Gauss code.

2.a. Input a knot as a 3D curve. To input a knot presented by a 3D polygon just type coordinates of all vertices in the order they follow on the knot. Coordinates must be separated by blank spaces or new lines. For example the trefoil knot is presented by a polygon with vertices $(0,0,0)$, $(3,0,1)$, $(0,2,0)$, $(1,0,1)$, $(0,4,0)$, $(4,0,1)$, $(0,1,0)$, $(2,0,1)$, $(0,3,0)$, $(0,0,1)$. To input it, type

0 0 0 3 0 1 0 2 0 1 0 1 0 4 0 4 0 1 0 1 0 2 0 1 0 3 0 0 0 1

and click 'ok' button. The Gauss code of the knot appears. Click 'ok' button again to see the three-page picture of the knot. If your link consists of more than one connected component, separate records for different components by a blank line. At the present state, for converting a 3D polygon to a planar diagram of a knot, projection along the z-axis is used. we would like to warn that if the projection of your knot along the z-axis is not generic (i.e. two vertices are projected to the same point, a vertex is projected to an internal point of an edge, or a triple crossing occurs), then the conversion may fail.

2.b. Input the Gauss code of a knot.

If you have planar diagram of a knot or link you may input it by typing its Gauss code. The diagram must be connected and have at least one crossing. We use an encoding a little different from what is usually called Gauss code, providing additional information that allows to distinguish knots from their mirror images. To input the link do the following. For each component, choose an orientation and a starting point. Assign different indices, which must be natural numbers, to all the crossings. For each connected component, go along it starting from the marked point in the chosen direction, and each time you come through a crossing, write down the index of the crossing preceded by two symbols: the first one must be + or − indicating whether you go over or under the crossing, and the second must be > or < sign indicating whether you see the arc that you intersect going from left two the right or from right to the left, respectively. Entries for different crossings must be separated by spaces. When you come back to the starting point of the chosen component, proceed with the next one. Separate entries for different components by the keyword 'component'.

3. Simplify a knot.

If you press the 'Simplify' button, the knot simplifier starts. It implements an algorithm described in [4]. The algorithm is based on a monotonic simplification of the three-page diagram by transformations that preserve the isotopy class of the knot (or link). Most frequently used transformations are of the following type. A connected component of the intersection of the link with a plane formed by two of the three pages is removed. Then the rest of the link is deformed within the three-page book so that the endpoints are kept fixed. Finally, the endpoints are connected by another curve in the same plane where the removed part of the link lied. At each step, we chose a transformation of this kind that leads to the simplest possible diagram, i.e. having the least number of points at the triple line.

At the present state, the simplifier does not search among all three-page diagrams that can be obtain from the current one by available transforms and does not make a "canonical" choice. So, if you click the 'Simplify' button again, the picture may change.

The running time for simplification of the knot of complexity n is bounded from above by $O(n^3)$.

4. Edit a knot.

The only available editing tool for now is a cyclic permutation of vertices on the binding line, which is achieved by pressing the 'Vertex permutation' button. Other editing tools are coming soon.

Of course, the developing of the Knot Simplifier is still in progress and more things have to be done such as:

1. introducing more editing tools;
2. converting to/from other formats;
3. comparing and recognizing some simple knots which are not unknots.

Acknowledgement 1. *This project was supported by Sonderforschungsbereich 288 "Differential Geometry and Quantum Physics" at Technische Universität in Berlin. We also would like to thank S. E. Koval and I. A. Taimanov who collaborated in this project at its different stages.*

References

1. The Knot Simplifier web service is available at `http://www.javaview.de/services/knots`.
2. K. Polthier, S. Khadem-Al-Charieh, E. Preuß, and U. Reitebuch. Publication of interactive visualizations with JavaView. In J. Borwein, M. H. Morales, K. Polthier, and J. F. Rodrigues, editors, *Multimedia Tools for Communicating Mathematics.* Springer Verlag, 2002. `http://www.javaview.de`.
3. J. Birman, H. Hilden. Heegaard splittings of branched coverings of S^3. *Transactions of the Amer. Math. Soc.* **213** (1975), 315–352.
4. I. A. Dynnikov. Three-page link presentation and an untangling algorithm. In: Proc. of the International Conference *Low-Dimensional Topology and Combinatorial Group Theory*, Chelyabinsk, July 31–August 7, 1999; Kiev, 2000, pp. 112–130.
5. I. A. Dynnikov. A new way to represent links. One-dimensional formalism and an untangling technology, *Acta Applicanda Mathematicae*, to appear.
6. I. A. Dynnikov. Three-page approach to knot theory. Encoding and local moves. *Functional Analysis and Its Appl.* **33:4** (1999), 260–269.
7. I. A. Dynnikov. Three-page approach to knot theory. Universal semigroup. *Functional Analysis and Its Appl.* **34:1** (2000), 24–32.
8. W. Haken. Theorie der Normalflächen. Ein Isotopiekriterium für der Kreisknoten. *Acta Math.* **105** (1961), 245–275.
9. R. Scharein. The Knot-Plot Homepage `http://www.pims.math.ca/knotplot`.

MAKING THE MOVE: THE NEXT VERSION OF *CINDERELLA*

ULRICH KORTENKAMP

Institut für Informatik, Freie Universität Berlin, Takustr. 9, 14195 Berlin, Germany

E-mail: kortenkamp@inf.fu-berlin.de

JÜRGEN RICHTER-GEBERT

Lehrstuhl Geometrie und Visualisierung, Technische Universität München, Germany

E-mail: richter@ma.tum.de

Cinderella is a software package for interactive or *dynamic* geometry. Its first version was published in 1999 and was the first geometry software to be based on the theory of complex tracing[3], thus avoiding mathematical inconsistencies and unmotivated discontinuities (modulo numerical errors). At the ICMS 2002 we will introduce the next major version of Cinderella and highlight its new features and new concepts.

1 The Interactive Geometry Software Cinderella

Let us briefly discuss the main features of the first version of Cinderella[12]. With Cinderella, you can easily create *constructions*, figures consisting of points, lines, circles, conics and other geometric elements, together with relations that describe the mathematical connections between the elements.[5] A standard example is a triangle, consisting of three points and three segments connecting each pair of points, and the altitudes in that triangle. The representation of the construction within the software makes it possible to move the vertices of the triangle (using the mouse) and get a new drawing for each position while the segments and the altitudes are updated accordingly.

The mathematical core of Cinderella is based on projective geometry and Cayley-Klein geometries for maximal generality. All geometric constructions are designed to work in all meaningful situations and to have as few special cases as possible. At the same time it is possible to have different, simultaneous, interactive views of a geometric construction, like its 2D representation, a spherical projection, polarized versions of this or special views like the Poincaré disk for hyperbolic geometry.

As soon as multi-valued geometric constructions, like the intersections of two circles or conics, or angle bisectors, are used, it is necessary to handle these with special care to avoid mathematical inconsistencies. Using a continuation technique on associated Riemann surfaces[4] Cinderella can guarantee – up to

```
A = FreePoint;
B = FreePoint;
C = FreePoint;
a = Line(B,C);
b = Line(A,C);
c = Line(A,B);
d = AngleBisector(b,c);
e = AngleBisector(a,c);
f = AngleBisector(a,b);
D = Intersection(d,e);
```

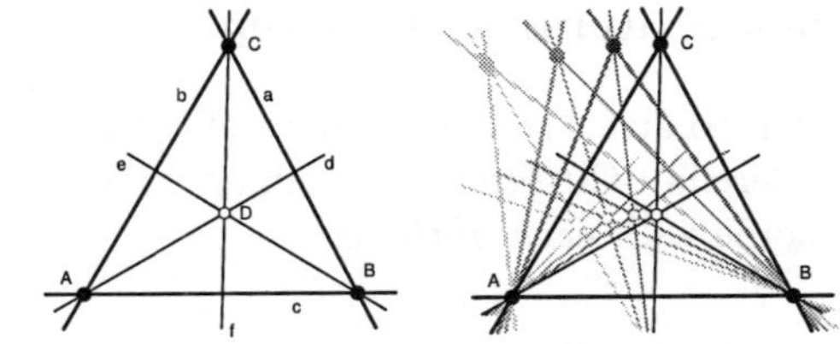

internal representation of the construction sequence — static drawing — idea of the dynamic construction

Figure 1. The basic principle of dynamic geometry software: A construction consists of movable, independent objects (here: the free points A, B, and C, and dependent objects, that are defined using already existing elements. Whenever the position of a free element is changed, the dependent elements are updated accordingly.

numerical errors – that continuous changes of parameters (i.e. free points) will cause continuous changes of the construction (within its realization space). As a consequence we do have a notion of "dynamic theorems", consisting of a sequence of geometric construction steps and an element of the "fiber" over a given assignment of free values as a "germ." The theory behind this gives rise to several mathematical questions that are interesting on their own,[6] but are not covered here.

The intuitive user interface of Cinderella hides most of the mathematical details but exposes the mathematical content of geometry and makes a proper handling of geometry available not only for academic research, but also for teaching on all levels, from K-12 to university. An integrated randomized theorem checking engine is used for computer guided exercises which can be authored with Cinderella. As Cinderella is written in Java, these can be integrated in web pages to create interactive learning environments.

Although Cinderella is a fairly large software package with a lot of functionality, it lacks some essential features that other programs[2,?] offer. Some of them were not implemented because the authors did not have the time or just did not think of them as being necessary, some others are missing because they did not fit into the mathematical framework. For example: As Cinderella is based on complex numbers internally, and even paths of free points within the real numbers are deformed into complex paths, there is no easy way to introduce orientations – it is not possible to order the complex numbers. It was clear that a next version will not be done by just adding missing features, but there had to be new concepts, both mathematically and from a software development point of view.

2 New Features And Concepts

We want to give a rough overview of the new features in the upcoming version of Cinderella – Cinderella 2.0 – and point out some of the difficulties or benefits that come with each of these.

2.1 Transformations

Transformations are at the core of geometry. In fact, the very concept of geometric invariant theory in the sense of Felix Klein is based essentially on the concept of transformations. Cinderella 2.0 introduces transformations not only as operations on geometric objects, but also as geometric objects themselves. It is possible to define translations (with respect to different geometries, e.g. hyperbolic translations), similarities, affine and projective transformations, rotations and others by associating points to their images. These transformations can be composed or inverted – it is possible to work within the full transformation group. All interaction with the transformations is done visually on screen. Any of these can then be easily applied to other geometric objects, e.g. circles. Of course the transformations are themselves dynamic: If their defining objects change, they change, too, as do the images from their application. Even more: the transformations are fully integrated in the automatic proving engine that recognizes whether the same transformation is created in two different ways. This makes it possible to perform calculations in transformation groups.

For an even deeper understanding of the structural behavior of transformations it is possible to create non-deterministic iterated function systems (IFS) by assigning probabilities to them. Using different colors schemes one can visualize the self-similarities and fixpoints.

2.2 Logical Operations

Over the complex numbers two circles always have two intersection points (counted with multiplicity) or they coincide. This is a great way to construct the perpendicular bisector of a segment: take any two circles of the same radius around the endpoints of the segment and draw the line through their intersections. This line will always be a real line, because the two points of intersection are complex conjugates in case they are complex solutions. Thus even if the two circles are "too small" and do not intersect we have a valid construction of the perpendicular bisector (the radical axis). Unfortunately, this is not a desired behavior in K-12 classroom use, as the students rarely know complex numbers, and it might cause more confusion than delight.

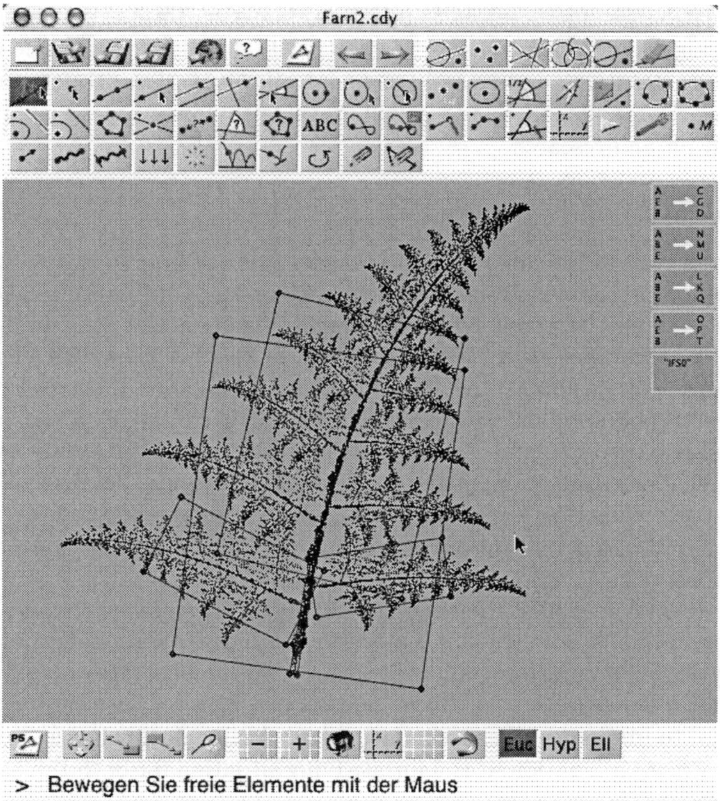

Figure 2. Iterated function systems (IFS) create self-similar structures. Many processes found in nature can be created with IFS, like this fern.

Another situation where elements should vanish although they are valid and real has to be considered with the intersection of line segments. In the former release of Cinderella line segments are "clipped" lines, and as any two distinct lines do have a point of intersection in the projective plane. In order to introduce conditionally vanishing points we had to think of a way to decide whether a point of intersection should appear on the display or not, while still having it accessible for the mathematical kernel – otherwise our concept of theorem could not work.

We ended up with introducing the possibility of showing an element based on a given condition, like "lies on segment S" or "is right of ℓ". The natural generalization was furthermore to introduce boolean objects that can be composed using logical operations (AND, OR, NOT), making it possible to show parts of construction conditionally in a controlled way. The arbitrary boolean formulas are very flexible and they will prove useful for complex visualizations in education and research.

2.3 Functions

A common request for enhancement is the addition of functions to Cinderella: Based on values gained from measurements or as coordinates of geometric elements it should be possible to calculate new values and feed them back into Cinderella as number objects or as coordinates of other geometric elements.

Here it is very important which functions are asked for. If we could restrict ourselves to polynomials, rational functions, or even single-valued complex functions like sin, cos, exp, this were very easy to add. If we also want to allow multi-valued functions – and we do want to do this, because $\sqrt{\cdot}$ is such a multi-valued function – we can do the same complex tracing as we do with the other multi-valued geometric constructions. This does not cause a major problem, as well.

The theory on which the tracing strategy of Cinderella is based implies that every value is not only continuous in the input parameters, but has to be analytic. So we must not allow functions like the absolute value function, as it is not differentiable in 0. Even complex conjugation, seemingly harmless and trivial to implement, has to be forbidden. Although this has a very sound mathematical explanation, it is a unsatisfying situation for both the users and the authors of the software.

The idea to circumvent this restriction may sound trivial, but it is the key to many other improvements: Instead of doing the calculation for a non-analytically dependent element directly, we move the element to its new position *as if it were done by the user interactively.* By splitting up the chain of dependencies we can allow any – even discontinuous – function to be applied to dependent elements. The only negative effect is that the theorem checking engine will not be able to find theorems that stretch across non-analytic functions. Note however that our notion of dynamic theorem does not even include any statement about such situations; the theory does not cover this at all. So it is quite comforting that we do not find theorems.

2.4 Communication

An issue that has gained attention among math software developers[7] is the ability to exchange mathematical content. There are ways to communicate mathematical content based on its representation; TeX and LaTeX are de facto standards when it comes to encoding math on a computer such that another mathematician can read it. MathML[14] by now has become an accepted standard format for mathematical material, as it is a well-structured format (being based on XML), and its industry support. The main advantage of MathML is that is can be parsed easily to recreate and thus can be used for computer-computer communication protocols. However, MathML is still and will be forever a format for the representation of math, the typesetting, the syntax. Currently, the only way to capture semantic information is the upcoming OpenMath standard as defined by the OpenMath consortium[11].

Cinderella 2.0 does support exchanging OpenMath content, both for functions and also for the upcoming new content dictionaries for plane geometry[1]. This makes it also possible to use Cinderella as a server side extension for web servers, rendering OpenMath to 2D constructions.

Two other built in ways of communication should be mentioned: Cinderella offers a TCP/IP based communication model that can be used to synchronize two constructions via the internet. Two or more people can work on different machines and manipulate the same construction simultaneously. This creates exciting new perspectives for distant teaching and remote research. Also new is a Java API that can be used for custom extensions. Using Cinderella as a visualization and manipulation device it is easy to write Java code demonstrating for example certain algorithms from computational geometry or computer graphics.

The Java API can also be used to interface directly with Mathematica[10] using J/Link or other computer algebra software that is able to execute Java code. The communication is not bound to have Cinderella as visualization component only, but the mathematical kernel and the complex tracing strategy can also be used from the outside, for example for the visualization of Riemann surfaces[4].

2.5 Scripting and Macros

With the former versions of Cinderella there is no easy way to create "macros" or "scripts" – subroutines that automate repeating tasks. This is not only annoying if one wants to create complex constructions consisting of many similar subconstruction, but also from an educational point of view: The proper identification of substructures is a key skill to be learned in mathematics. The next version of Cinderella supports the automatic creation of scripts from a selected part of a construction.

There is also an alternative way of building scripts: A contributed parser[9]

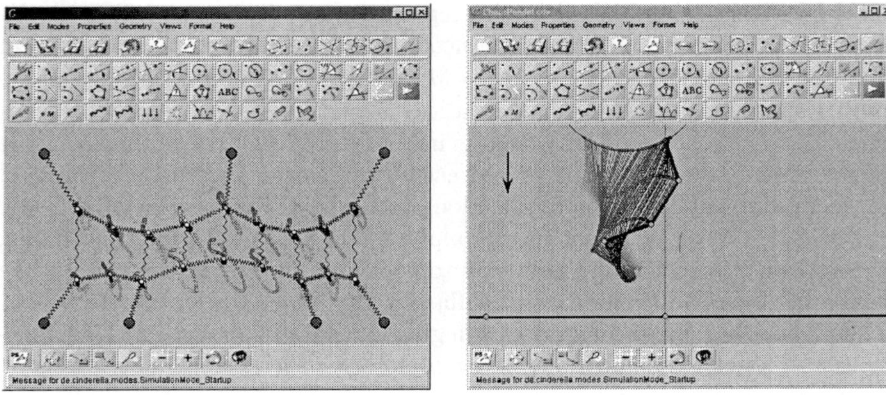

Figure 3. The new physics simulation features of Cinderella enable users to work with springs and other physical objects. On the left you see a network of springs attached to some fixed points, on the right there is a snapshot of a triple pendulum showing chaotic behavior.

is able to "understand" natural language descriptions of geometric constructions and transforms them into dynamic constructions within Cinderella. Currently this works for German language only, but probably the concepts can be extended to other languages as well, with varying success.

2.6 Physics Simulation

During the development of the Mathematica integration (see Sec. 2.4) all examples we did came from physics. This raised the question whether the ability to have basic interaction simulating physical effects should not be included natively into Cinderella. It turned out to be relatively easy to have a qualitatively correct simulation of masses, springs, and gravity, in many cases, but it is hard to have quantitatively correct and a numerically stable simulation, especially for very stiff springs and other extreme values. However, the current system can be used for explorations in physics. It is also possible to use the physical elements for special effects, like creating projectively correct drawings of Schlegel diagrams for 3-dimensional polytopes.

As springs can change their length periodically and masses can rotate around other masses there is even more dynamism and more possibilities for animations than before. You can build apparatus that actually do things, or plug together "animals" like you do with sodaplay[?].

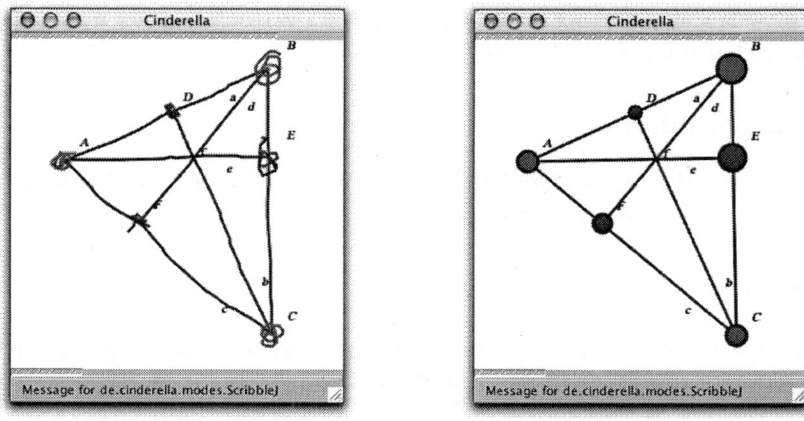

A hand-drawn sketch... ...and its automatic recognition

Figure 4. The new sketching recognition is useful with special hardware devices, like handhelds with touch screen or interactive whiteboards. Relations like perpendicularity or parallelism can be expressed using extra gestures.

2.7 Sketching

During the development of Cinderella we had the opportunity to test new hardware. First of all, we had access to an interactive whiteboard, a projection screen that is able to track movements of a special pen. It is possible to work on this screen with a computer display projection in a very natural way, the interaction is immediate and accurate. Still, we wanted to make working with Cinderella on this whiteboard be as easy as working with a conventional blackboard. This was our starting point for automatic sketch detection: A hand drawn sketch is interpreted and automatically converted into a construction, i.e. geometric elements that satisfy certain relations. Gestures are used for adding non-trivial content, for example, one can indicate that two lines are parallel by giving both of them a small mark, or two lines "are made" orthogonal by placing an orthogonality-mark at their intersection.

This gesture-based constructing is also very useful on another device, located at the other end of the scale. We were able to create versions of Cinderella that run on handheld devices, e.g. the Casio Cassiopeia E-125 or the SHARP Zaurus SL-5000 series. These handhelds typically have an approx. 3×4 inch touchscreen with a resolution of 240×320 pixels, operated with a pen. The small screen makes it necessary to have as few as possible user interface elements on screen. This makes gesture-based constructing the primary choice for these devices.

3 Further Information

More information about Cinderella and a demo version is available on the WWW at http://www.cinderella-geometry.com.

References

1. A. Cohen, H. Cuypers, U. Kortenkamp, *A content dictionary for plane geometry (tentative title)*, in preparation.
2. N. Jackiw, *The Geometer's Sketchpad*, Key Curriculum Press, Berkeley (1991–2002).
3. U. Kortenkamp, *Foundations of Dynamic Geometry*, Ph.D. thesis, ETH Zürich (1999).
4. U. Kortenkamp, *Visualizing Riemann Surfaces Using Cinderella*, in preparation for VisMath 2002.
5. U. Kortenkamp, J. Richter-Gebert, *A dynamic setup for elementary geometry*, Proceedings of MTCM 2000, Multimedia Tools for Communicating Mathematics, Lisboa, Springer-Verlag, Heidelberg (2002).
6. U. Kortenkamp, J. Richter-Gebert, *Complexity issues in Dynamic Geometry*, Proceedings of the Smale Fest 2000, Hongkong (accepted).
7. U. Kortenkamp, *The Future of Mathematical Software*, Proceedings of MTCM 2000, Multimedia Tools for Communicating Mathematics, Lisboa, Springer-Verlag, Heidelberg (2002).
8. J.-M. Laborde and F. Bellemain, *Cabri Geometry II*, Texas Instruments, (1993–1998).
9. D. Materlik and U. Kortenkamp *A Parser For Natural Language Descriptions of Geometric Constructions*, Technical Report, Freie Universität Berlin, to appear (2002).
10. Mathematica, Wolfram Research Inc., http://www.wolfram.com.
11. The OpenMath Consortium, http://www.openmath.org.
12. J. Richter-Gebert and U. Kortenkamp, *The Interactive Geometry Software Cinderella*, Springer-Verlag, Heidelberg (1999).
13. Sodaplay, http://www.sodaplay.com.
14. W3C Math Home, http://www.w3.org/Math.

PRIME DECOMPOSITION OF POLYNOMIAL IDEALS OVER FINITE FIELDS

KAZUHIRO YOKOYAMA
Graduate School of Mathematics, Kyushu University
6-10-1 Hakozaki, Higashi-ku, Fukuoka, 812-8581, Japan
yokoyama@math.kyushu-u.ac.jp

In this paper we propose new algorithm for prime decomposition of polynomial ideals over finite fields. Aiming for practical computation, we employ the strategy by Gianni *et al.*, and modify "decomposition using generic position" with special treatment for positive characteristic.

1 Introduction

The theory of primary decomposition of ideals in noetherian rings is very classical and many works [4] were done for its computation over fields of characteristic 0. In positive characteristic, existing works [8,6,9,5] on the subject or its related topic were aimed mainly for theoretical study with general setting. However, to realize practical algorithm for primary decomposition of polynomial ideals over finite fields is not only very interesting as a computational problem, but also very useful for studies on pure mathematics and those on engineering problems. Thus, our goal is to obtain a practical algorithm for primary decomposition of polynomial ideals over finite fields. To do this, we can apply "localization technique" [12], where we can extract primary components from prime divisors. This technique is based on Gröbner basis computation and does not depend on the characteristic of coefficient fields. Therefore primary decomposition computation can be reduced efficiently to prime decomposition computation. But, for prime decomposition, there are much differences between cases in characteristic 0 and those in positive characteristic, and we cannot apply methods used for polynomial ideals over the rational number field directly to those over finite fields.

In this paper we propose new algorithm for prime decomposition as the first step for practical computation. Aiming for practical computation, we employ the well-established strategy by Gianni *et al.* [7] and modify "decomposition using generic position", which works very successfully in case of characteristic 0. To solve defects derived from positive characteristic, we introduce notions "separable ideals" and "separable closure of ideals". As the whole procedure consists of all basic computations, Gröbner basis computation and polynomial factorization over finite fields, its efficiency does depend on those of such computations. Unfortunately, as the whole procedure on real computer has not been implemented yet, the practicality is not examined. (This will be done in our next study.) Instead, we discuss possible improvements for practical implementation. We note that the problem here is much related to *radical ideal computation in positive characteristic* [9,5].

2 Mathematical Background

We give key points on computation of prime decomposition in positive characteristic. Throughout this paper, we consider a polynomial ring $K[x_1, \ldots, x_n]$, where K is a finite field $GF(q)$ of order q and of characteristic p, and we denote the set of variables $\{x_1, \ldots, x_n\}$ by X. For a noetherian commutative ring R, we write $Id_R(f_1, \ldots, f_t)$ for the ideal generated by elements $f_1, \ldots, f_t$ in R, and $(I : f)$ for the quotient ideal of an ideal I by an element f in R. For an ideal I of R, we denote its radical by $\sqrt{I}$ and denote by $\mathrm{Ass}(I)$ and $\mathrm{Ass}_{iso}(I)$ the set of all prime divisors of I and that of all isolated prime divisors of I, respectively. Then, $\sqrt{I} = \cap_{P \in Ass_{iso}(I)} P$, and $\mathrm{Ass}(\sqrt{I})$ coincides with $\mathrm{Ass}_{iso}(I)$. Moreover, for a polynomial ideal J of $L[Z]$, where L is an extension field of K and Z is a set of variables, we denote the algebraic variety of J, i.e. the set of all zeros of J, by $V_{\tilde{L}}(J)$, where we consider zeros in the algebraic closure $\tilde{L}$ of L. Conversely, for an algebraic variety W, we denote its corresponding ideal $\{f \in L[Z] \mid f(\alpha) = 0 \text{ for any } \alpha \in W\}$ by $I_{L[Z]}(W)$. Here, by the prime decomposition of I, we mean to compute $\mathrm{Ass}_{iso}(I)$.

2.1 Successive and Simultaneous Approaches

For prime decomposition, there are two approaches; *successive* one and *simultaneous* one. Kalkbrener [8] discussed the executability of the prime decomposition of ideals of $R[x]$ by inductive arguments for a noetherian commutative ring R (with identity) under the assumption that one can compute the prime decomposition of ideals of R and one can compute factorization of univariate polynomials over the quotient field $Q(R/P)$ for a prime ideal P. In our case, by computing pure-dimensional components [7,12,3], we can reduce the problem to that of 0-dimensional ideals over rational function fields. Letting $L = K(Y)$ for some $Y \subset X$, one can compute the prime decomposition of the elimination ideal $I \cap L[Z \cup \{z\}]$ for $Z \subset X \setminus Y$ and $z \in X \setminus (Z \cup Y)$ from that of $I \cap L[Z]$, if one can compute the factorization of univariate polynomials over any algebraic extension field of L. This corresponds exactly to "construction of successive extension fields over rational function fields", and we call it *successive approach*.

Thus, if we employ *successive approach*, we have to argue efficient factorization of univariate polynomials over algebraic extension fields of rational function fields. However, since this factorization shall be reduced to factorization of polynomials over the ground field, it seems much better, for practical realization of prime decomposition, to employ another approach *simultaneous strategy*, where the prime decomposition will be executed all at once. Thus, we adopt *simultaneous approach*, where we employ the strategy [7] and *decomposition using generic position*, and it has the following feature.

- We introduce a new notion *separable closure* to overcome certain difficulty appearing in positive characteristic case.
- We always consider factorization of polynomials only over the ground (perfect) field $K = GF(q)$, which makes the procedure much simpler.

2.2 Separability and Decomposition Using Generic Position

Now we provide important notions for the computation. Let Y be a proper subset of X, $Z = X \setminus Y$ and $L = K(Y)$. For simplicity, we write $Z = \{x_1, \ldots, x_s\}$ and $Y = \{x_{s+1}, \ldots, x_n\}$. Moreover, we use t as a new variable.
Definition 2.1 *For an ideal J of $L[Z]$, J is said to be* separable *if J satisfies the following:*
(1) J is a 0-dimensional radical ideal.
(2) For every prime divisor P of J, the residue class ring $L[Z]/P$ is a separable extension field of L.

The separability here corresponds to *that in successive approach* [5].
Definition 2.2 *For a univariate polynomial $f(x)$ over L, f is said to be* separable *if f has no multiple root in the algebraic closure $\tilde{L}$ of L. Moreover, if there is a separable polynomial h such that $f(x) = h(x^{p^e})$ for some non-negative integer e, h is called the* separable closure *of f and denoted by* $\mathrm{sc}(f)$.
Definition 2.3 *Let J be a 0-dimensional ideal of $L[Z]$. For a polynomial $f(Z)$ in $L[Z]$, its* minimal polynomial m_f with respect to J *is defined as the (monic) univariate polynomial over L having the smallest degree among all univariate polynomials h such that $h(f)$ belongs to J.*

For each variable x in Z, its minimal polynomial m_x with respect to J is a generator of the elimination ideal $J \cap L[x]$.
Proposition 2.4 *Let J be a 0-dimensional ideal of $L[Z]$. If, for each x in Z, its minimal polynomial m_x with respect to J is separable, then J is separable.*
Proof. By the definition of separability, $\gcd(m_x, dm_x/dx) = 1$ for every x in Z. By Seidenberg's Lemma 92 [11] (or Lemma 8.13 [2]), J is a radical ideal.

Next consider $Ass(J)$. Then, for each $P \in Ass(J)$, $L' = L[Z]/P$ is an extension field of L and $L' \cong L(\alpha_1, \ldots, \alpha_s)$ for any $\alpha = (\alpha_1, \ldots, \alpha_s)$ in $V_{\bar{L}}(P)$. Since each α_i is a root of a separable m_{x_i}, each α_i is a separable element over L and thus, $L(\alpha_1, \ldots, \alpha_s)$ is a separable extension field of L. ∎

Definition 2.5 *Let J be a separable ideal of $L[Z]$. A polynomial $g(Z)$ in $L[Z]$ is said to be* in generic position with respect to J *if, for each prime divisor P of J, $g(Z)$ is a primitive element for the separable extension $L[Z]/P$.*

If K has enough elements, a polynomial $g(Z)$ in generic position can be found among linear polynomials $\sum_{x_i \in Z} a_i x_i$, $a_i \in K$, by using Proposition 2.6 as an efficiently computable criteria. (Proposition 2.6 can be considered as a special case of Proposition 8.69 [2], which can be shown by the co-maximality of prime divisors.) In this case the minimal polynomial m_g can be considered as a polynomial in a new variable t and Y over K by removing its denominator.

Proposition 2.6 *Let J be a separable ideal of $L[Z]$. A polynomial $g(Z)$ is in generic position with respect to J if and only if* $\deg(m_g) = \dim_L(L[Z]/J)$ *for its minimal polynomial m_g with respect to J. Moreover, if $m_g = m_1 \cdots m_r$ is the irreducible factorization of m_g over L, then $P_i = Id_{L[Z]}(J, m_i(g))$ is a prime divisor for each m_i, and $J = \cap_{i=1}^r P_i$ is the prime decomposition of J.*

Thus, for a separable ideal, once we find a polynomial in generic position, one can compute its prime decomposition by using factorization of the minimal polynomial. Here, we call this procedure *decomposition using generic position.* When K is of characteristic 0, each radical ideal is separable and almost every linear polynomials are in generic position. However, in our case, there are certain computational problems for applying it.

- Even if J is a radical ideal, J may not be separable, which implies that the existence of a polynomial in generic position is not guaranteed. Moreover, even if J is separable, there may not be a polynomial in generic position among all linear polynomials over K.

- For radical ideal computation, we cannot apply Seidenberg's theorem [11]. Instead, we have to use other existing algorithms [9,5]. But, if the ground field is not perfect, radical computation turns very complicated and hard. Moreover, we also have to make special treatment on square-free decomposition of polynomials if the ground field is not perfect.

Example 2.7 *In $GF(p)(u,v)[x,y]$, $Id(x^p - u, y^p - v)$ is prime but not separable. Thus, there is no polynomial in generic position. In $GF(p)(z)[x,y]$, $Id(x^p - z, y^p - z)$ is a primary ideal associated with a prime ideal $Id(x^p - z, x - y)$. For the ideal, the minimal polynomial of each variable is irreducible but not separable.*

2.3 Separable Closure of Ideal

Definition 2.8 *Let J be a 0-dimensional ideal of $L[Z]$. If an ideal J' of $L[Z]$ satisfies the following conditions, then we call it the* separable closure of J *and denote it by* $\mathrm{sc}(J)$.
(1) *J' is a separable ideal of $L[Z]$.*
(2) *There is a correspondence between zeros of J (in $V_{\tilde{L}}(J)$) and those of J' (in $V_{\tilde{L}}(J')$) as follows: For each zero $\alpha = (\alpha_1, ..., \alpha_s)$ of J there exists a zero $\beta = (\beta_1, ..., \beta_s)$ of J' uniquely such that $\beta_i = \alpha_i^{p^{e_i}}$ for each i, where e_i is a non-negative integer determined by α.*

Theorem 2.9 *For a 0-dimensional ideal J of $L[Z]$, there exists* $\mathrm{sc}(J)$ *uniquely. Moreover, there is a correspondence between prime divisors of J and those of* $\mathrm{sc}(J)$*: Suppose a prime divisor P of J corresponds to a prime divisor Q of* $\mathrm{sc}(J)$*. Then, there exist non-negative integers $e_1, \ldots, e_s$ such that each zero $(\alpha_1, ..., \alpha_s)$ of P corresponds uniquely to a zero $(\alpha_1^{p^{e_1}}, ..., \alpha_s^{p^{e_s}})$ of Q.*

Proof. For each prime divisor P of J, we fix its zero $\alpha^{(0)} = (\alpha_1^{(0)}, \ldots, \alpha_s^{(0)})$. Then $L[Z]/P \cong L(\alpha_1^{(0)}, \ldots, \alpha_s^{(0)})$. Now consider the minimal polynomial m_{x_i} of x_i with respect to P for each x_i in Z. As P is a maximal ideal, m_{x_i} is irreducible and we can write it uniquely as $m_{x_i}(t) = \mathrm{sc}(m_{x_i})(t^{p^{e_i}})$, where t is a new variable, $\mathrm{sc}(m_{x_i})$ is the separable closure of m_{x_i} and e_i is some non-negative integer. (m_{x_i}, e_i do not depend on the choice of $\alpha^{(0)}$.) Letting $\beta_i^{(0)} = (\alpha_i^{(0)})^{p^{e_i}}$ for $i = 1, \ldots, s$, $L(\beta_1^{(0)}, \ldots, \beta_s^{(0)})$ is a separable extension of L, since every $\beta_i^{(0)}$ is a separable element over L.

Set $W_P = \{(\beta_1, \ldots, \beta_s) \mid$ there is a zero $\alpha = (\alpha_1, \ldots, \alpha_s)$ of P such that $\beta_i = \alpha_i^{p^{e_i}}$ for $i = 1, \ldots, s\}$. Since P is a maximal ideal in $L[Z]$, $V_{\tilde{L}}(P)$ consists of all algebraic conjugates of $\alpha^{(0)}$. Thus, W_P also consists of all algebraic conjugates of $\beta^{(0)} = (\beta_1^{(0)}, \ldots, \beta_s^{(0)})$ and it forms a *minimal invariant* set in $(\tilde{L})^s$ for all L-isomorphisms of $\tilde{L}$. Then, letting $Q_P = I_{L[Z]}(W_P) = \{f(Z) \in L[Z] \mid f(\beta) = 0$ for every $\beta \in W_P\}$, it follows that Q_P is a prime ideal of $L[Z]$ and $L(\beta_1^{(0)}, \ldots, \beta_s^{(0)}) \cong L[Z]/Q_P$ by considering the corresponding algebraic extension fields $L[Z]/P$ and $L[Z]/Q_P$ as successive extension fields $(\cdots((L(\alpha_1^{(0)})(\alpha_2^{(0)})\cdots)(\alpha_s^{(0)})$ and $(\cdots((L(\beta_1^{(0)})(\beta_2^{(0)})\cdots)(\beta_s^{(0)})$. Then, $J' = \bigcap_{P \in \mathrm{Ass}(J)} Q_P$ satisfies the conditions in Definition 2.8. ∎

We call $E = (e_1, \ldots, e_s)$ the *exponent vector* of P. The correspondence may not be one to one, i.e. distinct prime divisors of J may correspond to the same prime divisor of $\mathrm{sc}(J)$. In this case those are distinguished by their exponent vectors. If J is of *special type* defined below, every prime divisor of J has the same exponent vector, and the correspondence is one to one.

Definition 2.10 *Let J be a 0-dimensional ideal of $L[Z]$. J is said to be* of special type, *if the minimal polynomial* m_{x_i} *of* x_i *with respect to* J *is irreducible for every* x_i *in* Z.

Example 2.11 *Consider the second example in Example 2.7.*

$$J = < x^p - z, y^p - z > \leftrightarrow \mathrm{sc}(J) = < x - z, y - z >$$
$$V(J) \ni (\sqrt[p]{z}, \sqrt[p]{z}) \leftrightarrow (z, z) \in V(\mathrm{sc}(J))$$
$$P = < x^p - z, x - y > \leftrightarrow \mathrm{sc}(J) = Q = < x - z, y - z >$$

Once one compute the separable closure $\mathrm{sc}(J)$, one can compute the prime decomposition of $\mathrm{sc}(J)$ by *decomposition using generic position*, from which one can extract the prime divisors of J.

2.4 Construction of sc(J) *and Construction of* P *from* Q

In our algorithm, we do not compute $\mathrm{sc}(J)$ directly from J, but we compute ideals J_j such that $\sqrt{J} = \cap_{j=1}^r \sqrt{J_j}$ and every J_j is of special type. (Here we call each J_j' *an intermediate ideal* of J.) For each J_j, we compute $\mathrm{sc}(J_j)$.

Take an intermediate ideal J_j and we write it simply by H. Then H is a 0-dimensional ideal of special type in $L[Z]$. By the definition, the minimal polynomial m_{x_i} of x_i with respect to H is irreducible over L for every x_i in Z. Then there is the separable closure $\mathrm{sc}(m_{x_i})$ for each m_{x_i}. And $m_{x_i}(t) = \mathrm{sc}(m_{x_i})(t^{q_i})$ and $q_i = p^{e_i}$ for each $x_i \in Z$. Now we define the Frobenius map:

$$\phi_E : L[Z] \ni f(x_1, \ldots, x_s) \to f(x_1^{q_1}, \ldots, x_s^{q_s}) \in L[Z],$$

where $E = (e_1, \ldots, e_s)$. Then, we can compute $\mathrm{sc}(H)$ as follows.

Theorem 2.12 *For the separable closure* $\mathrm{sc}(H)$ *of* H, *we have*

$$\mathrm{sc}(H) = \phi_E^{-1}(H) = \{f \in L[Z] \mid \phi_E(f) \in H\}.$$

Moreover, there is a one to one correspondence between prime divisors P *of* H *and those* Q *of* $\mathrm{sc}(H)$ *such that*

$$Q = \mathrm{sc}(P) = \phi_E^{-1}(P).$$

Proof. Let $H' = \phi_E^{-1}(H)$. For each x_i in Z, as $\mathrm{sc}(m_{x_i})(x_i^{q_i}) = m_{x_i}$ belongs to H, $\mathrm{sc}(m_{x_i})(x_i)$ also belongs to H'. Then, H' is separable, since its minimal polynomial $\mathrm{sc}(m_{x_i})$ is a separable polynomial for every variable x_i in Z.

On the other hand, there is a one to one correspondence between $V_{\tilde{L}}(H)$ and $V_{\tilde{L}}(\phi_E^{-1}(H)) = V_{\tilde{L}}(H')$. Because, for each zero $\alpha = (\alpha_1, \ldots, \alpha_s)$ of H, $\beta = (\alpha_1^{q_1}, \ldots, \alpha_s^{q_s})$ is a zero of H', and conversely, for each zero $\beta = (\beta_1, \ldots, \beta_s)$ of H', $\alpha = (\sqrt[q_1]{\beta_1}, \ldots, \sqrt[q_s]{\beta_s})$ is a zero of H. By Definition 2.8, we have $H' = \mathrm{sc}(H)$ and the correspondence between $\mathrm{Ass}(H)$ and $\mathrm{Ass}(\mathrm{sc}(H))$. ∎

Now suppose that we have obtained all prime divisors of $\mathrm{sc}(H)$. Then, we can recover prime divisors or their associated primary ideals as follows.

Proposition 2.13 *Let Q be a prime divisor of* $\mathrm{sc}(H)$, *P its corresponding prime divisor of H and $P_0 = Id(\phi_E(Q))$. Then, $\sqrt{P_0} = P$, that is, P_0 is either the corresponding prime divisor or its associated primary ideal.*

Proof. Consider each zero $\alpha = (\alpha_1, \ldots, \alpha_s)$ of P_0. As $P_0 = Id(\phi_E(Q))$, $(\alpha_1^{q_1}, \ldots, \alpha_s^{q_s})$ must be a zero of Q, and hence α is a zero of the corresponding prime divisor P of Q. Thus, $V_{\tilde{L}}(P_0) \subset V_{\tilde{L}}(P)$. But, as P is a maximal ideal, we have $V_{\tilde{L}}(P_0) = V_{\tilde{L}}(P)$ and $\sqrt{P_0} = P$. ∎

Frobenius Map Computation: Both of *inverse Frobenius map computation* $\phi_E^{-1}(H)$ and *Frobenius map computation* $Id(\phi_E(Q))$ can be done by elimination ideal computation. (See Chapter 2 [1].)

As to *inverse Frobenius map*, we introduce an elimination ordering $x_i >> y_j$ and compute a Gröbner basis G_0 of $Id(H \cup \{x_i^{p^{e_i}} - y_i | 1 \leq i \leq s\})$ in $L[x_1, \ldots, x_s, y_1, \ldots, y_s]$. Then, $G_0 \cap L[y_1, \ldots, y_s]$ with y_i replaced by x_i for each i is a Gröbner basis of $\phi_E^{-1}(H)$. (See Proposition 2.5 [9].)

As to *Frobenius map*, we introduce an elimination ordering $x_i >> y_j$ and compute a Gröbner basis G_1 of $Id(Q \cup \{y_i^{p^{e_i}} - x_i | 1 \leq i \leq s\})$ in $L[x_1, \ldots, x_s, y_1, \ldots, y_s]$. Then, $G_1 \cap L[y_1, \ldots, y_s]$ with y_i replaced by x_i for each i is a Gröbner basis of $Id(\phi_E(Q))$. This can be shown by using the property that ϕ_E is a ring endomorphism.

Radical Ideal Computation: When $J \neq \sqrt{J}$, the ideal P obtained by Frobenius map computation may not be a prime ideal but a primary ideal. Thus, we have to compute $\sqrt{P}$. However, as our goal is to compute the prime divisors of I and $(\sqrt{P})^c = \sqrt{P} \cap K[X]$ is the wanted prime divisor of I, we compute $(\sqrt{P})^c$ directly by using $(\sqrt{P})^c = \sqrt{P^c} = \sqrt{P \cap K[X]}$, where P^c denotes the *contraction* of P. In this case, *as the ground field K is $GF(q)$, we can compute the radical $\sqrt{P^c}$ efficiently by Matsumoto's algorithm* [9], which consists of inverse Frobenius map computation and p-th root computation of field elements. As P contains J, the radical ideal computation of P^c seems much efficient than that of J^c.

3 Computation of Prime Decomposition

Here we give the whole algorithm proposed here. Let I be an ideal of $K[X]$.

3.1 Reduction to 0-dimensional Ideals

First we compute pure-dimensional components from I by techniques using independent sets modulo I. (See Chapter 8 [2] for the detail.) Using Gröbner basis, we compute a maximal strongly independent set Y modulo $\sqrt{I}$, and lift I to its *extension ideal* J of $K(Y)[Z]$, where $Z = X \setminus Y$. Then, for each prime divisor P of J, we extract its corresponding prime divisor $P \cap K[X]$ by *contraction* computation, and we have the following prime decompositions:

$$\sqrt{J} = \cap_{i=1}^{r} \sqrt{Q_i}, \quad \sqrt{I} = \cap_{i=1}^{r} (\sqrt{Q_i} \cap K[X]) \cap \sqrt{I'},$$

where $I' = Id_{K[X]}(I, f)$ for some polynomial f computed from J. As I' contains I properly, we can compute all prime divisors of I in finitely many steps by applying the above computation recursively to I'. As we want to compute prime divisors of I, we can incorporate any of useful decompositions [3,12].

3.2 Intermediate Decomposition

We consider a 0-dimensional ideal J of $L[Z]$, where $Y \subset X$, $L = K(Y)$ and $Z = X \setminus Y$. For each variable $x_i \in Z$, we compute its minimal polynomial $m_{x_i}(t)$ with respect to J, which is considered as a polynomial in t and Y over K by removing its denominator. Then, we factorize it over K as $m_{x_i}(t) = \prod_j m_{i,j}(t)^{e_{i,j}}$, where each $m_{i,j}$ is irreducible over K and so over $K[Y]$. By Gauss' lemma, $m_{i,j}$ is also irreducible over L. Adjoining $m_{i,j}$'s to J, we have the following intermediate decomposition, where each J_k is of special type:

$$\sqrt{J} = \cap_{k=1}^{r} \sqrt{J_k},$$

Let $\mathcal{F}_i$ be the set of all distinct irreducible factors of m_{x_i} over K, and $n_i = \#\mathcal{F}_i$ for each i. Since intermediate ideals are of form $Id_{L[Z]}(J, g_1, \ldots, g_s)$, where each g_i is chosen from $\mathcal{F}_i$, we have to deal with $n_1 \cdots n_s$ combinations of $(g_1, \ldots, g_s)$, which will require many computations. However, as many of those tend to coincide with $L[Z]$, it is better to provide an efficient check for finding unnecessary combinations. To do this, *successive adjoining g_i to ideals* seems work well, since we can omit unnecessary pairs in earlier stage.

INCREMENTAL SEARCH FOR INTERMEDIATE DECOMPOSITION:

1. Set $\mathcal{J} = \{J\}$.
2. For $i = 1$ to s, do the following.
    2.1. Set $\mathcal{J}_0 = \mathcal{J}$ and $\mathcal{J} = \emptyset$.
    2.1. For each H in $\mathcal{J}_0$, do the following.
        2.1.1. Take one factor h from $\mathcal{F}_i$.
        2.1.2. Compute the Gröbner basis of $Id_{L[Z]}(H, h)$.
        2.1.3. If $Id_{L[Z]}(H, h) \neq L[Z]$, then $\mathcal{J} = \mathcal{J} \cup \{Id_{L[Z]}(H, h)\}$.
        2.1.4. Go to Step 2.1.1.
3. Return $\mathcal{J}$.

3.3 Prime Decomposition of Ideals

Let $\mathcal{J}$ be the intermediate decomposition of J. For each H in $\mathcal{J}$, we can compute sc(H) by inverse Frobenius map computation. And from the prime divisors of $\mathrm{sc}(H)$, we can reconstruct their corresponding prime divisors of H. (See Section 2.4). As to prime decomposition of $\mathrm{sc}(H)$, since $\mathrm{sc}(H)$ is a separable ideal of $L[Z]$, we can compute its prime decomposition by *decomposition using generic position.*

By Proposition 2.6, we can find a polynomial g in generic position by checking if $\deg(m_g) = \dim_L(L[Z]/\mathrm{sc}(H))$ for its minimal polynomial m_g. For the efficiency, we want to find a linear polynomial $g(Z)$ in generic position. For this, we can use the following.

Lemma 3.1 (Theorem 4.2 [13]**)** *Let $T = s \times \ell \times \dim_L(L[Z]/\mathrm{sc}(H))$, where $s = \#Z$ and $\ell = \#\mathrm{Ass}(\mathrm{sc}(H))$. Then, if $\#K > T$, there exists a polynomial g in generic position among all linear polynomials in Z over K.*

The bound in Lemma 3.1 is merely theoretical and it will be more likely that we will find a linear polynomial in generic position even if K does not satisfies the bound. This will be examined in our next work by experiments.

Now we consider the case where we must replace K with its extension field K_1. In this case first we deal with the ideal $J_1 = K_1 \otimes J$ of $K_1(Y)[Z]$. Then, we apply *decomposition using generic position* to obtain the set $\mathcal{P}'$ of all the prime divisors of J_1 or their associated primary ideals. And, by contraction, we obtain the set $\mathcal{P}_{K_1}$ of all the prime divisors of $(J_1)^c$ or their associated primary ideals. By the action of the Galois group $\mathcal{G} = Galois(K_1/K)$, $\mathcal{P}_{K_1}$ will be divided to $\mathcal{G}$-orbits. Suppose that $P_{K_1,1}, P_{K_1,2}, \ldots, P_{K_1,r}$ form one $\mathcal{G}$-orbit. Then, $W = V_{\tilde{K}}(P_{K_1,1}) \cup \cdots \cup V_{\tilde{K}}(P_{K_1,r})$ forms a *minimal invariant* set among unions of sets in $\mathcal{P}_{K_1}$ for $\mathcal{G}$. (We note $\tilde{K} = \tilde{K}_1$.) Thus, it follows that there exists a prime divisor P of J^c uniquely such that $V_{\tilde{K}}(P) = W$.

Now consider P_{K_1} in $\mathcal{P}_{K_1}$, again. As K_1 is a finite extension field of K, it can be considered as $K[T]/P_0$, where T is a set of new variables and P_0 is a maximal ideal of $K[T]$. Consider a Gröbner basis G_{K_1} of P_{K_1} as a set of $K[T, X]$ and let $P' = Id_{K[T,X]}(P_0, G_{K_1})$. Then, P_{K_1} contains $P' \cap K[X]$ as subsets of $K_1[X]$, and $\mathcal{G}$-conjugates of P_{K_1} also contain $P' \cap K[X]$. From this fact, it can be shown that $P' \cap K[X]$ is the prime divisor of J^c corresponding to P_{K_1}. This divisor can be computed by Gröbner basis computation with an elimination ordering $T >> X$,

Main Procedure: As a summary, we outline the whole procedure. (If $\#K$ is too small to find a linear polynomial in generic position, we use certain extension field K_1 over K. Afterward, we recover the prime divisors of I.)

PROCEDURE OF PRIME DECOMPOSITION:

1. Compute the dimension of I and a maximal strongly independent set Y.
   Let $Z = X \setminus Y$ and $L = K(Y)$.
2. Compute the *extension* $J = I^e = IL[Z]$ of I.
   (We compute a Gröbner basis G^e with respect to some efficient ordering, and

compute a polynomial f from all leading coefficients of elements of G^e. Then, $I = I^{ec} \cap (I : f^\infty)$, where $I^{ec} = J \cap K[X]$.)

3. Compute prime divisors of J. (Sometimes, we compute primary ideals associated with prime divisors of J.)
   3.1. Compute the minimal polynomial m_{x_i} of x_i w.r.t. J for each x_i in Z.
   3.2. Factorize each m_{x_i} as a polynomial over K.
   3.3. Compute the intermediate decomposition $\mathcal{J}$ by INCREMENTAL SEARCH.
   3.4. For each $J_i \in \mathcal{J}$, compute $\mathrm{sc}(J_i)$ by *inverse Frobenius map computation.*
   3.5. Compute the prime decomposition of $\mathrm{sc}(J_i)$ by *decomposition using generic position*, similarly as in the characteristic 0 case.
   3.6. Compute the prime divisors or its associated primary ideals of J from the prime divisors of $\mathrm{sc}(J_i)$ by *Frobenius map computation.*
4. Compute the prime divisors of I^{ec} from prime divisors or associated primary ideals of J_i's by *contraction* and by *radical ideal computation over K using Matsumoto's algorithm* [9].
5. Unless $(I : f^\infty) = K[X]$, go back to Step 1, where we apply $(I : f^\infty)$ instead of I.

Remark 3.2 *As we want to compute the prime divisors, we can use $(I : f)$ instead of $(I : f^\infty)$ or any other decomposition mentioned in Section 3.1.*

4 Concluding Remarks

We proposed a concrete algorithm for prime decomposition of polynomial ideals over finite fields. Since the algorithm consists of practical procedures based on Gröbner basis computation and polynomial factorization over finite fields, it shall have certain practical behavior. Its practicality will be examined by experiments on real computer in our next study. (With Prof. Noro, the author will report its experiments in the subsequent full paper.) Instead, we list possible improvements for practical realization of the new algorithm.

- As we want to compute prime divisors of I, we can incorporate any of useful decompositions [3,12].

- Since polynomials to be factorized tend to have larger degrees, efficient polynomial factorization is required. (See experiments [10] for its practical implementation.) Especially, we can use efficient field arithmetics for small fields, which shall improve the total efficiency.

- The special procedure aimed for non separable case seems to work mainly for small characteristic cases like $p = 2, 3, 5$, because it will be hard to handle ideals J with larger $\dim_L(L[Z]/J)$ in practice. (If the characteristic p is not small, ideals we can handle shall be separable and we simply apply ordinary *decomposition using generic position.*) Thus, the specialty of such small characteristics must be used much more.

- The choice of admissible ordering shall effect the total efficiency. The practical effect of ordering will be examined by experiments. If we use a lexicographical ordering, the total behavior seems very close to that of *successive approach*, where we compute a simple extension field successively based on polynomial factorization over extension fields.

Acknowledgments

The author would like to thank Dr. Ryutaroh Matsumoto for helpful discussions and Prof. Masayuki Noro for his supports on implementation.

References

1. Adams, W.W., Loustaunau, P. (1994) *An Introduction to Gröbner Bases.* Graduate Studies in Mathematics **3**, American Mathematical Society.
2. Becker, T., Weispfenning, V. (1993). *Gröbner Bases.* Springer-Verlag, New York.
3. Caboara, M., Conti, P., Traverso, C. (1997) Yet another ideal decomposition algorithm. Proceedings of AAECC 12, L.N.C.S. **1255**, Springer, 39-54.
4. Decker, D., Greuel, G.-M., Pfister, P. (1999) Primary decomposition: algorithms and comparisons. *Algorithmic Algebra and Number Theory*, Springer, 187-220.
5. Fortuna E., Gianni P., Trager B. (2001) Computation of the radical of polynomial ideals over fields of arbitrary characteristic. Proceedings of ISSAC'01, ACM Press.
6. Gianni, P., Trager, B. (1996) Square-free algorithms in positive characteristic. *Appl. Alg. in Eng. Comm. and Comp*, **7**, 1-14.
7. Gianni, P., Trager, B., Zacharias, G. (1988). Gröbner bases and primary decomposition of polynomial ideals. *J. Symb. Comp.* **6**, 149-167.
8. Kalkbrener, M. (1994) Prime decomposition of Radicals in Polynomial Rings. *J. Symb. Comp.* **18**, 365-372.
9. Matsumoto R. (2001) Computing the radical of an ideal in positive characteristic. *J. Symb. Comp.* **32**, 263-271.
10. Noro, M., Yokoyama, K. (2002). Yet another practical implementation of polynomial factorization over finite fields. to appear in the proceedings of ISSAC'02.
11. Seidenberg, A. (1974) Constructions in algebra. *Trans. Amer. Math. Soc.* **197**, 272-313.
12. Shimoyama, T., Yokoyama, K. (1996). Localization and primary decomposition of polynomial ideals. *J. Symb. Comp.* **22**, 247-277.
13. Yokoyama, K., Noro, M., Takeshima, T. (1992). Solution of systems of algebraic equations and linear maps on residue class rings. *J. Symb. Comp.* **14**, 399-417.

PARTIAL STANDARD BASES FOR FAMILIES

ANNE FRÜHBIS-KRÜGER

FB Mathematik, Universität Kaiserslautern, 67653 Kaiserslautern, Germany

When considering a family $(X,0) \longrightarrow (T,0)$ of singularities over a base space $(T,0)$, it is often necessary to study the behaviour of invariants in family, that is simultaneously for all fibres. This is the situation for which the algorithm presented in this papers is designed. It is a variant of a standard bases algorithm which takes into account the fact that we want to view certain variables as parameters and that we want to do calculations for all fibres simultaneously. We only need an answer in a small neighbourhood of $0 \subset T$ and can hence compute effectively in "real life" examples. To illustrate this, the algorithm is applied to 2 problems from singularity theory: in the first one it is used as a tool to find an example showing that the module of K_V-liftable vector fields is not necessarily free for the versal unfolding of a section of a free complete intersection. The second application is to explicitly determine the complete list of adjacencies for simple complete intersection singularities in $(\mathbb{C}^3, 0)$.

1 Motivation

How does a numerical invariant change in a given family of singularities? Questions of this kind can in some cases be answered very easily, as for example in the case of the Milnor number μ or the δ-invariant by the principle of conservation of number. This principle states that if in a family a singularity splits up into several singular points by varying the parameter, the sum over the values of the invariant at these singular points is again the one of the original singularity. In some other cases like the Tjurina number, on the other hand, considering the fibre at zero only provides a partial answer, a bound for the invariant. In this situation, we are interested in conditions on the parameters of the family which ensure that the invariant has a desired value. In this section, we describe the situation in detail before proceeding to algorithmic considerations in the subsequent sections.

Let $(X,0) \longrightarrow (T,0)$ be a family of singularities [a] with base space $(T,0) \subset (\mathbb{C}^m, 0)$ and with fibre $(X_0, 0) \subset (\mathbb{C}^n, 0)$ at 0. Let U and V be sufficiently small open neighbourhoods of 0 in $(\mathbb{C}^n, 0)$ and in $(\mathbb{C}^m, 0)$, respectively, and pass to suitable representatives X_0, X and T such that $X_0 = X \cap \pi^{-1}(0)$ where $\pi : U \times V \longrightarrow V$ is the canonical projection. For $t \in T$, we put $X_t = X \cap \pi^{-1}(t)$. Using this notation, the behaviour of the type of numerical invariants, which we are interested in, is described by the following lemma:

[a] Non-isolated singularities are also allowed, using appropriate invariants.

Lemma 1.1 *Let $\mathcal{M}$ be a sheaf on X with the following properties:*

- *$\mathcal{M}$ is a coherent $\mathcal{O}_X$-module.*
- *$\mathcal{M}_0$, the stalk of $\mathcal{M}$ at 0, is an $\mathcal{O}_{T,0}$ module of finite rank.*

For all sufficiently small neighbourhoods $U' \subset U$ of 0 in $\mathbb{C}^n$, there exists an open neighbourhood $V' \subset V$ of 0 in $\mathbb{C}^p$ such that for all $t \in V' \cap T$:

$$\dim_{\mathbb{C}}((\mathcal{M}|_{X_0})_0) \geq \sum_{p \in X'_t} \dim_{\mathbb{C}}((\mathcal{M}|_{X_t})_p),$$

where $X'_t = X_t \cap (U' \times \{t\})$.

This lemma is, of course, a weakened form of the principle of conservation of number and can be proved using the same type of arguments. For our further considerations, however, it is helpful to have a closer look at the second part of the proof again:

Proof (sketch, emphasizing the steps we need later on): Let us consider the analytic space $Z = \mathrm{Supp}(\mathcal{M})$. In the same way as in the proof of the principle of conservation of number (see e.g. [9]), it can be shown that the natural projection map $f : (Z, 0) \longrightarrow (T, 0)$ is finite.
We take a representative $f : Z \longrightarrow T$ such that f is finite and $f^{-1}(0) = \{0\}$. Then $f_*\mathcal{M}$ is a coherent sheaf, whose stalk at 0 is by assumption $\mathcal{O}^k_{T,0}/K$ for suitable $k \in \mathbb{N}$ and a suitable submodule K. Coherence of $f_*\mathcal{M}$ now implies that there exists an open neighborhood V' of 0 in T such that $f_*\mathcal{M}|_{V'}$ has a presentation of the structure

$$\mathcal{O}^l_T|_{V'} \xrightarrow{M} \mathcal{O}^k_T|_{V'} \longrightarrow f_*\mathcal{M}|_{V'} \longrightarrow 0.$$

Considering a point $t_0 \in V'$ and the points $q_1, \ldots, q_s \in Z$ in the fibre of f at t_0, we then know

$$\oplus_{i=1}^s \mathcal{M}_{q_i} = (f_*\mathcal{M})_{t_0} \cong \mathcal{O}^k_{T,t_0}/Im(M)$$

because f is finite. This implies that

$$\sum_{i=1}^s \dim_{\mathbb{C}}(\mathcal{M}|_{X_{t_0}})_{q_i} = k - rank(M_{t=t_0}). \qquad \text{q.e.d.}$$

This shows that the object we need to compute is an $\mathcal{O}_T|_{V'}$ presentation matrix of $\mathcal{M}|_{V'}$. The method to do this will be a slightly modified standard basis algorithm which will be introduced in detail in the next section.

2 Partial Standard Bases

In order to solve the problem formulated in the previous section, namely to compute a presentation matrix over the ring of the base of the family, we will follow the steps of the standard basis algorithm applied to the fibre at zero[b]. This will provide us with a set of generators for our module from which we can easily determine the columns of the desired matrix - up to a certain order in the perturbation variables.

Let $I_0 = (f_1, \ldots, f_r) \subset F_0 = R^k$ be a submodule of finite colength[c]; let $>$ be a fixed monomial ordering on F_0 for which the underlying monomial ordering on R is a degree ordering. Let us denote by $g_1, \ldots, g_d$ a $\mathbb{C}$-vector space basis of F_0/I_0 (resulting from a standard basis computation of I w.r.t. the given ordering). For studying families of objects over R, we have to pass to the ring $R_{\underline{t}} = R \otimes_K (K[[\underline{t}]]/J)$ where $t_1, \ldots, t_s$ denote the parameters of the family. Let the module $I_{\underline{t}} \subset F_{\underline{t}} = R_{\underline{t}}^k$ be generated by $\tilde{f}_i = f_i + \sum_{j \in J_i} t_j h_{ij}$ for $1 \leq i \leq r$ where $J_i \subset \{1, \ldots, s\}$ and $h_{ij} \in F_{\underline{t}}$ for $1 \leq i \leq r$, $j \in J_i$.
As an ordering on our new ring $R_{\underline{t}}$, we choose a local degree ordering on the $\underline{t}$ as the first block and the original ordering $>$ as the second. Note that this ordering makes sure that a leading term can only be divisible by a t_i if there is no other term which is not element of $(\underline{t})R_{\underline{t}}$.

Definition 2.1 *A partial standard basis S for the family of modules $I_{\underline{t}}$ w.r.t. this ordering is a generating system of this $R_{\underline{t}}$-module which is the union of the following three subsets:*

- S_J *a standard basis of the ideal $J \subset K[[\underline{t}]]$*
- $S_{\underline{x}}$ *a set of elements of $I_{\underline{t}}$ whose restriction to $\underline{t} = 0$ is non-zero*
- $S_{\underline{t}}$ *a set of elements of $I_{\underline{t}}$ which are not in the subring $K[[\underline{t}]]$ and whose restriction to $\underline{t} = 0$ is zero*

such that

1. *$S_{\underline{x}}|_{\underline{t}=0}$ is a standard basis of I_0 with respect to the given ordering on F_0*
2. *$S_{\underline{t}}$ contains those spolys of elements of $S_{\underline{x}}$ whose part of degree zero in $\underline{t}$ reduces to 0 by elements of $S_{\underline{x}}$*

Allowing in our theoretical considerations an infinite number of steps in the normal form calculations, we can now assume that all elements of $S_{\underline{t}}$ are in normal form w.r.t. $S_{\underline{x}} \cup S_J$.

Lemma 2.2 *By adding to $S_{\underline{t}}$ elements of the form $NormalForm(x^\alpha \cdot f | S_{\underline{x}} \cup S_J)$, where $f \in S_{\underline{t}}$ and $\alpha \in \mathbb{N}^n$, we obtain a set of generators for the module of $K[[\underline{t}]]/J$-relations of $F_{\underline{t}}/I_{\underline{t}}$.*

Proof: The $K[[\underline{t}]]/J$-relations of $F_{\underline{t}}/I_{\underline{t}}$ are given by $I_{\underline{t}} \cap \sum_{i=1}^{s} (K[[\underline{t}]]/J) g_i$ (considered as an intersection of $K[[\underline{t}]]/J$-submodules of $R_{\underline{t}}$). Because the partial standard basis S is generating $I_{\underline{t}}$ and because the elements of $S_{\underline{x}}$ do

[b]For a description of the standard basis algorithm see for example [1], for a description with respect to local or mixed orderings [7]

[c]$f_1, \ldots, f_r$ a set of generators for I_0, not necessarily minimal

not lie in that intersection, the desired generators have to be of one of the following forms:

- an element of $S_{\underline{t}}$
- a normal form of an spoly of an element of $S_{\underline{x}}$ and $S_{\underline{t}}$.

In the second case, the element of $S_{\underline{t}}$ is multiplied by a monomial not involving any t_i, when forming the spoly. Up to a $K[[\underline{t}]]/J$-linear combination of elements of $S_{\underline{t}}$, this is the same as multiplying the element of $S_{\underline{t}}$ with this suitable monomial and calculating the normal form. This shows that the elements we have to add to $S_{\underline{t}}$ are of the claimed structure. q.e.d.

This lemma also explains why we did not compute a standard basis, but stopped halfway through the calculations: the remaining steps would have created additional elements of $S_{\underline{t}}$ which are either redundant or obtained in the subsequent step of computing the presentation matrix anyway. The following small example illustrates this:

Example 2.3 *Let us consider $\mathbb{Q}[[t_1, t_2, t_3, x, y, z]]$ with a block ordering such that the first block is a local weighted degree reverse lexicographic ordering on the $\underline{t}$ using weights -6, -7, -1 and the second block is a local degree reverse lexicographic ordering on x, y and z. Let $I_{\underline{t}} = (x+y^3, xy-t_1t_3+t_2^2, xz+t_3^2-t_1^3, x^2-y^3)$. Then a standard basis of $I_{\underline{t}}$ is given by the six polynomials*

$$x, y^3, t_3^2 - t_1^3, -t_1t_3 + t_2^2, t_2^2t_3 - t_1^4, t_1^5 - t_2^4,$$

but a partial standard basis is

$$S_{\underline{x}} = x, y^3; S_{\underline{t}} = t_3^2 - t_1^3, -t_1t_3 + t_2^2.$$

In particular, we see that the standard basis computation has created two more polynomials which result from spolys not involving any x, y or z. These would then give rise to 6 extra rows in the presentation matrix.

¿From the algorithmic point of view, it is appropriate to split the calculation of the presentation matrix into two parts, the algorithm to determine a partial standard basis of $I_{\underline{t}}$ and the computation of the generators of the $K[[\underline{t}]]/J$-relations of the g_i. In addition to the partial standard basis, the first algorithm also computes the K-basis $\{g_i\}$ of F_0/I_0 (w.r.t. the given monomial ordering) and a bound b such that all monomials of weighted degree greater than b are in the initial module of $S_{\underline{x}}|_{\underline{t}=0}$. If R is local, then this is the highest corner of $S_{\underline{x}}|_{\underline{t}=0}$.

Algorithm 1: partial standard basis
<u>Input</u>: module $I_{\underline{t}}$ and ideal J in $R \otimes_K K[[\underline{t}]]$
<u>Output</u>: partial standard basis S_J, $S_{\underline{x}}$, $S_{\underline{t}}$ of $I_{\underline{t}}$;

monomial basis $\underline{g}$ of F_0/I_0 and bound b

- compute a standard basis S_J of J
- follow the steps of the standard basis algorithm for the set $\{\tilde{f}_i | 1 \leq i \leq r\}$, but dropping all pairs where one of the two polynomials has a leading monomial in $(\underline{t})$; call the resulting set S_{temp}. (In this step, no tail reduction and no reduction of elements with leading monomials in $(\underline{t})$ should be done!)
- $S_{\underline{x}} = \{f \in S_{temp} | L(f) \notin (\underline{t})\}$, $S_{\underline{t}} = S_{temp} \setminus S_{\underline{x}}$
- over R compute the monomials g_i not in the ideal generated by the leading monomials of $S_{\underline{x}}$ and the bound b
- return(S_J,$S_{\underline{x}}$, $S_{\underline{t}}$, $\underline{g}$, b)

This algorithm terminates, because the steps performed in it are exactly the ones of a usual standard basis algorithm for J and for I_0.
Concerning the second algorithm, we are facing the difficulty that we might be forced to do power series reductions, that is there might be infinitely many steps, when trying to express the elements of $S_{\underline{t}}$ as a $K[[\underline{t}]]/J$-linear combinations of the g_i. To avoid this, we have to introduce a bound, say c, on the degree in $\underline{t}$ up to which computations should be exact. Considering more closely a step that introduces new terms, which are greater than the leading term w.r.t. the original monomial ordering in the variables $\underline{x}$, we see that the degree in $\underline{t}$ of this new monomial is greater (at least by one) than the one of the monomial that has been reduced. This and the chosen bound up to which we compute, ensure that the following algorithm terminates:

Algorithm 2: computing the presentation
Input: Output of Algorithm 1 and bound c
Output: matrix M such that $g \cdot M$ generate $I_{\underline{t}} \cap \sum_i (K[[\underline{t}]]/J)g_i$ as a $K[[\underline{t}]]/J$ module

- $B = b \cdot c + 1$
- $S_{temp} = \{a \cdot m | a \in R \text{ monomial}, wdeg(a) < B, m \in S_{\underline{t}}\}$
- S_{temp} =NormalForm($S_{temp}|S_{\underline{x}} \cup S_J$) cutting off terms of weighted degree greater than $B+b$ where to each t_i the weight b is assigned
- for the j-th element of S_{temp} collect the coefficient of g_i into the i-th entry of the j-th column of a $d \times \#S_{temp}$ matrix M
- optionally compute a minimal generating set of the $K[[\underline{t}]]/J$-module generated by the columns of M and call it M again
- return(M)

If R is local, I_0 is quasihomogeneous and all perturbations are of strictly positive weight, the algorithm terminates without use of the bound c in the reduction step. In this special situation, monomials a in the second step can be chosen of weighted degree smaller than b.

3 Applications

3.1 $\mathcal{K}_V$-liftable vector fields

In this first application, our goal is to compute an example where the module of $\mathcal{K}_V$-liftable vector fields is not free for the versal unfolding of a section of a free complete intersection. Computing liftable vector fields is a rather difficult task; successful computations have been made in the case of hypersurface singularities ([15] and algorithmically [3], [14]) and by using duality results ([11], [2]). ¿From a different point of view, the one of deformation theory, calculations of the same flavour appear when dealing with the kernel of the Kodaira-Spencer map ([10], [12] for hypersurface singularities and [4] for Cohen-Macaulay codimension 2 singularities). In fact, dealing with this question was the starting point for developing the algorithm introduced in section 2, and I should like to thank J. Damon very much for asking me for an algorithmic way to do these calculations.

We want to give an example where genericity of Morse type singularities does not hold for $\mathcal{K}_V$ in the case of free complete intersections. Before considering the example, let us briefly recall the definition of $\mathcal{K}_V$-equivalence: Let $V,0 \subset \mathbb{C}^p,0$ be a complete intersection singularity and $f_0 : \mathbb{C}^n,0 \longrightarrow \mathbb{C}^p,0$ a holomorphic map germ. This map germ is viewed as a section of V so that the singularity $V_0 = f_0^{-1}(V)$ is a pullback given by the following diagram:

$$\begin{array}{ccc} \mathbb{C}^n,0 & \xrightarrow{f_0} & \mathbb{C}^p,0 \\ \uparrow & & \uparrow \\ f_0^{-1}(V) = V_0,0 & \longrightarrow & V,0 \end{array}$$

$\mathcal{K}_V$-equivalence is then given by the group

$$\mathcal{K}_V = \{\Psi \in \mathcal{K} | \Psi(\mathbb{C}^n \times V) \subseteq \mathbb{C}^n \times V\},$$

where $\mathcal{K}$ denotes the contact group introduced by Mather [13].

The example where genericity of Morse type singularities does not hold is the following: We consider a singularity $V,0 \subset \mathbb{C}^5,0$ given by the ideal $I = (abc, de) \subset \mathbb{C}\{a,b,c,d,e\}$ and a linear section $f_0 : \mathbb{C}^2 \longmapsto \mathbb{C}^5$ defined by the tuple $x+2y, x+y, y-x, x-3y, y-5x$.
To determine Derlog(V) $= \{\zeta \in \theta_5 | \zeta(I) \subseteq I\}$, we first compute the syzygy matrix of the module $J(I)+I\cdot\mathcal{O}^2_{\mathbb{C}^5,0}$, where $J(I)$ denotes the module generated by the columns of the Jacobian matrix. Dropping the rows of the syzygy

matrix which do not correspond to generators of $J(I)$, we obtain the following matrix

$$\begin{pmatrix} a & de & 0 & 0 & 0 & 0 & 0 & 0 & 0 & 0 \\ 0 & 0 & b & de & 0 & 0 & 0 & 0 & 0 & 0 \\ 0 & 0 & 0 & 0 & c & de & 0 & 0 & 0 & 0 \\ 0 & 0 & 0 & 0 & 0 & 0 & d & abc & 0 & 0 \\ 0 & 0 & 0 & 0 & 0 & 0 & 0 & 0 & e & abc \end{pmatrix}$$

whose columns generate Derlog(V). Knowing Derlog(V), we then determine a monomial $\mathbb{C}$-vector space basis of the $\mathcal{K}_V$-normal space $N\mathcal{K}_V \cdot f_0$ (by a standard basis compuation of $T\mathcal{K}_V \cdot f_0$ with respect to an ordering which first considers the module components and then a degree reverse lexicographical ordering). Using new variables $t_1, \ldots, t_5$ as unfolding parameters, the versal unfolding is then given by

$$((x+2y)\cdot(x+y)\cdot(y-x+t_1),(x-3y+t_2y+t_3)\cdot(y-5x+t_4y+t_5)).$$

Now we apply algorithm 1 to the computation of the standard basis of $T\mathcal{K}_{V,un,e}$ which provides us with a set of $\mathbb{C}\{x,y,\underline{t}\}$ generators of the relations among the $\mathbb{C}\{\underline{t}\}$-generators of $N\mathcal{K}_{V,un,e}$. In order to apply algorithm 2, we, now, choose 4 as the degree in $\underline{t}$ up to which we want to compute. (This turned out to be a good choice for our example.). Dropping those elements in the result which are obviously in the $\mathbb{C}\{\underline{t}\}$-module generated by the other ones, we obtain the 10 generators from the algorithm [d]:

$$f_1 = \begin{pmatrix} 0 \\ -5t_3t_5 \\ 0 \\ 2t_5^2 \\ 0 \end{pmatrix}, f_2 = \begin{pmatrix} 0 \\ -50t_1t_5 \\ 0 \\ 33t_1t_5 - 21t_5^2 \\ 0 \end{pmatrix}, f_3 = \begin{pmatrix} 0 \\ 50t_1t_4 + 15t_3t_4 \\ 0 \\ 15t_4t_5 - 33t_1t_4 \\ 0 \end{pmatrix}, f_4 = \begin{pmatrix} 0 \\ 550t_1^2 + 50t_5^2 \\ 0 \\ 107t_5^2 - 363t_1^2 \\ 0 \end{pmatrix},$$

$$f_5 = \begin{pmatrix} 0 \\ 1155t_1t_3 - 100t_5^2 \\ 0 \\ -214t_5^2 \\ 0 \end{pmatrix}, f_6 = \begin{pmatrix} 0 \\ 100t_1t_2 + 105t_2t_3 \\ 0 \\ -66t_1t_2 \\ 0 \end{pmatrix}, f_7 = \begin{pmatrix} 0 \\ 3465t_3^2 + 500t_5^2 \\ 0 \\ 3465t_1t_3 + 2079t_3^2 + 1070t_5^2 \\ 0 \end{pmatrix},$$

$$f_8 = \begin{pmatrix} 0 \\ -2000t_1 - 2100t_3 - 195t_2t_3 \\ 950t_1t_3 + 750t_3^2 \\ 1320t_1 + 594t_1t_2 + 340t_1t_4 - 150t_4t_5 \\ 670t_1t_5 - 150t_5^2 \end{pmatrix}, f_9 = \begin{pmatrix} 0 \\ -40t_3 + 20t_2t_3 \\ 20t_1t_3 + 20t_3^2 \\ 16t_5 - 4t_4t_5 \\ 20t_1t_5 - 4t_5^2 \end{pmatrix} f_{10} = \begin{pmatrix} t_1 \\ 0 \\ t_3 \\ 0 \\ t_5 \end{pmatrix}$$

It is then easy to check that we need $f_8, \ldots, f_{10}$ in a minimal set of generators, since they are the only generators of order 1. Computing normal forms with respect to a standard basis of this set, only f_7 still has a non-zero 4-th component, its leading monomial is t_3^2 (as before computed modulo

[d]We have chosen an ordering on the module components which considers them in the following order: 4,2,5,3,1.

$(\underline{t})^4)$. Reducing by a standard basis of $f_7, \ldots, f_{10}$, we then see that we need additional generators: The leading monomials of the normal forms of f_1 and f_5 are $t_2t_3t_5$ resp. t_1t_5 in the second component. Hence the module of $\mathcal{K}_V$ liftable vector fields cannot be generated by 5 elements and it can thus not be a free submodule of $\mathbb{C}\{\underline{t}\}^5$.

3.2 Adjacencies for simple isolated space curve singularities

In the previous application, we were interested in a set of generators for the module of relations. Here, we are interested in something different: the module we are considering is the relative T^1 of a family and we want to determine conditions which ensure that for fibers over points in the base space satisfying these conditions the Tjurina number has a certain value. We will use this to determine a complete list of adjacencies for the simple isolated complete intersection space curve singularities. The first contribution to this list was given by Giusti in the article classifying the simple isolated complete intersection singularities (= simple ICIS) ([6]), 2 further adjacencies were found by Goryunov ([8]) and it was shown in [16] that this completes the list of adjacencies among the simple space curve ICIS. But it was not known whether there would be further unknown adjacencies to hypersurface singularities for the exceptional simple space curve ICIS.

Figure 1. Adjacency diagram (= diagram of deformations) for simple isolated complete intersection space curve singularities

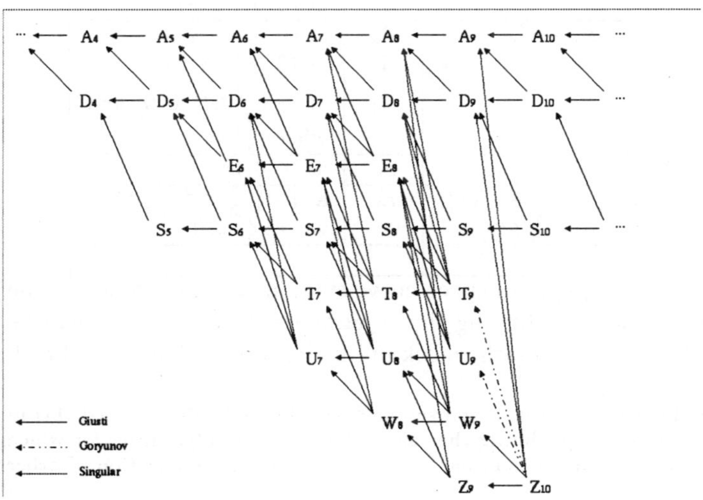

Starting from the versal deformation of the given singularity, $I_{\underline{t}} = F_1(\underline{x},\underline{t}),\ldots,F_k(\underline{x},\underline{t})$, $F_i(\underline{x},\underline{0}) = f_i$ for $1 \leq i \leq k$, we compute a partial standard basis and an $\mathcal{O}_T$ presentation matrix of the relative T^1:

$$T^1_{\underline{t}} = \mathbb{C}\{\underline{x},\underline{t}\}^k/((\frac{\partial \underline{F}}{\partial x_1},\ldots,\frac{\partial \underline{F}}{\partial x_n}) + I_{\underline{t}}\cdot \mathbb{C}\{\underline{x},\underline{t}\}^k).$$

Since the singularities of types $T_7, T_8, T_9, U_7, U_8, U_9, W_8, W_9, Z_9$ and Z_{10} are quasihomogeneous, the presentation matrix will contain a row with the Euler relation in each of these cases. Hence the Tjurina number drops by at least one as soon as one of the perturbation parameters is non-zero. Asking that it drops by at most one is the same as considering the locus where the 2-minors of the presentation matrix vanish simultaneously.[e]

The following adjacencies are computed by primary decomposition of the ideal generated by the 2-minors of the presentation matrix. For each component the singular locus in the fibre is then computed and, if it is just one point, the type of the singularity is determined. In particular, adjacencies to singularities of Tjurina number $\tau_0 - 1$ which are not found in this way do not exist, since there is no corresponding stratum in the base space.
For the singularity T_7 the versal family is given by $(yz + t_1 + t_2x, x^2 + y^3 + z^3 + t_3 + t_4y + t_5z + t_6y^2 + t_7z^2)$. The following table shows a part of the list of all associated primes of the ideal of 2-minors of the presentation matrix. For each associated prime the type of singularity is stated as well:

stratum	type	
$D(t_7)\cap V(t_1,\ldots,t_5,4t_6 - t_7^2)$	D_5 A_1	at $(0,0,0)$ and at $(0,0,-\frac{1}{2}t_7)$
$D(t_7)\cap V(t_1,t_2,27t_3 - t_7^3, 3t_4 - t_5^2,$ $3t_6 - t_7^2, \zeta t_5 - t_7)$	A_2 A_2 A_1	at $(0,0,-\frac{1}{3}t_7)$, at $(0,-\frac{1}{3}\zeta t_7,0)$, at $(\pm i\sqrt{\frac{1}{27}t_7^3},0,0)$ and where $\zeta^3 = 1$
$D(t_7)\cap V(t_1,\ldots,t_6)$	S_6	at $(0,0,0)$
$D(t_2)\cap V(t_1 + \frac{9}{2}t_2^4, t_3 + \frac{675}{16}t_2^6, t_4 - \frac{81}{4}\zeta t_2^4,$ $t_5 + \frac{27}{4}\zeta^2 t_2^2, t_6 - \frac{81}{4}\zeta^2 t_2^4, t_7 + \frac{27}{4}\zeta t_2^2)$	E_6	at $(\frac{9}{2}t_2^3, \frac{3}{2}\zeta^2 t_2^2, \frac{3}{2}\zeta t_2^2)$ where $\zeta^3 = 1$
...		

The adjacencies of simple ICIS space curve singularities are shown in the figure 1. The computations for finding the adjacencies in the diagram can be found at the URL http://www.mathematik.uni-kl.de/ anne/adjacencies.html.

[e]The meaning of the higher minors is of course analogous, but we will not need this in the following, because it turns out during the calculations that all adjacencies for which the Tjurina number drops by more than one can be obtained by composition of other adjacencies.

4 Implementation and Timings

When implementing algorithm 1 in SINGULAR, it turned out that using the degree bound 1 on the perturbation parameters $\underline{t}$ was not possible, since the chosen ordering is a block ordering for which degree bounds are not supported; algorithm 1' avoids this problem:[f]

Algorithm 1': partial standard basis (practical approach)

Input: ideals $I_{\underline{t}}$ and J in $R \otimes_K K[[\underline{t}]]$

Output: partial standard basis S_J, $S_{\underline{x}}$, $S_{\underline{t}}$ of $I_{\underline{t}}$;
monomial basis $\underline{g}$ of F_0/I_0 and bound b

- compute a standard basis S_J of J
- Define a module $\tilde{I} \subset (R \otimes_K K[[\underline{t}]])^2$ generated by the vectors $(f_i, \tilde{f}_i)$, $\forall 1 \leq i \leq r$.
- Choose a monomial ordering which first considers the module components, compatible to the original monomial ordering.
- Perform a standard basis computation of $\tilde{I}$ which only involves the first component, that is in which the condition that an spoly reduces to zero is replaced by the condition that the leading term is not in the first component. Call the resulting set S_{temp}.
- $S_{\underline{t}} = \{\text{second comp. of } v \mid v \in S_{temp},\ Lm(v) \text{ not in first comp.}\}$,
  $S_{\underline{x}} = \{\text{second comp. of } v \mid v \in S_{temp}, Lm(v) \text{ in first comp.}\}$,
- From this point on, proceed as in algorithm 1.

In algorithm 2, it is useful to first reduce the original generators of $S_{\underline{t}}$ with $S_{\underline{x}} \cup S_J$ before multiplying with the monomials in $\underline{x}$ for avoiding redundant columns in the presentation matrix. [g]

The algorithms 1' and 2 are implemented in the SINGULAR library `partsb.lib` which also uses the dynamical module `kstd.so` [5]. Some timings for the 2nd application (AMD Athlon 1000MHz, 1GB RAM):

singularity	parameters in versal family	time(s)
S_6	6	1
T_7	7	4
U_8	8	31
Z_9	9	74

Acknowledgements

I should like to thank the organizers of the special semester on Singularities at

[f] Here, it is only stated for ideals, but it can also be used for modules by introducing k new variables, say $e_1, \ldots, e_k$, writing e_i for the i-th unit vector of the free module F and computing modulo $(e_1, \ldots, e_k)^2$.

[g] If time is not the crucial issue, but memory usage, it is also possible to interreduce the set S_{temp} several times while constructing it, e.g. each time after adding a certain number of new generators.

the Newton Institute for their hospitality and the great atmosphere in which the first ideas leading to this article were born. There the problems treated in the two applications arose from discussions with J. Damon and C.T.C. Wall. I would also like to thank the development team of SINGULAR [17] and, in particular, Hans Schönemann and Kai Krüger for many fruitful discussions.

References

1. Cox,D., Little,J., O'Shea,D.: *Ideals, Varieties and Algorithms*, Springer-Verlag (1996)
2. Damon,J.: *The unfolding and determinacy theorems for subgroups of A and K*, Memoirs of the AMS 50 no. 306 (1984)
3. Damon,J., Galligo,A.: *Universal Topological Stratification for the Pham Example*, Bull.Soc.math.France 121 (1993), pp. 153–181
4. Frühbis-Krüger,A.: *Computing Moduli Spaces of Space Curve Singularities*, in Proceedings of MEGA 2000, JPAA 164 (2001), pp. 165–178
5. Frühbis-Krüger,A., Krüger,K.: *Dynamical Modules in Singular*, preprint in preparation
6. Giusti,M.: *Classification des singularités isolées simples d'intersections complétes*, Proc.Symp.Pure Math. **40**, 457–494 (1983)
7. Greuel,G., Pfister,G.: *Advances and Improvements in the Theory of Standard Bases and Syzygies*, Arch. Math. **66**, 163–179, (1996)
8. Goryunov,V.: *Singularities of projections of full intersections*, Journal of Soviet Math. 27(3) (1984), pp. 2785–2811
9. de Jong,T., Pfister,G.: *Local Analytic Geometry*, Vieweg, (2000)
10. Laudal,O.A., Pfister,G.: *Local Moduli and Singularities*, Lect.Notes Math. 1310, Springer-Verlag, (1988)
11. Looijenga,E.: *Semi-universal deformation of a simple elliptic hypersurface singularity I: unimodularity*, Topology 16 (1977), pp. 257–262
12. Martin,B., Pfister,G.: *The Kernel of the Kodaira-Spencer map ...*, J.Symb.Comp. **7**, 527–531 (1989)
13. Mather,J.: *Finitely determined map germs*, Publ. I.H.E.S. 36 (1968), pp. 127–156
14. du Plessis,A., Wall,C.T.C.: *The geometry of Topological Stability*, LMS Monographs, New Series 9 (1995)
15. Saito,K.: *Theory of logarithmic differential forms and logarithmic vector fields*, J.Fac.Sci.Univ.Tokyo Sect.Math. 27 (1980), pp. 265–291
16. Schröter,C., Schuba,S.: *The deformations of the simple space curve singularities*, Math. Z. 229(1998), pp. 607–614
17. SINGULAR *2.2*, to appear in 2002, http://www.singular.uni-kl.de/

AN ENVIRONMENT FOR SYMBOLIC AND NUMERIC COMPUTATION

G. DOS REIS
CMLA, ENS Cachan - CNRS (UMR 8536)
61, Avenue du Président Wilson, 94235 Cachan - Cedex, France
dosreis@cmla.ens-cachan.fr

B. MOURRAIN* & PH. TRÉBUCHET*
GALAAD, INRIA,
BP 93, 06902 Sophia Antipolis, France
{mourrain,ptrebuc}@sophia.inria.fr

F. ROUILLIER
SPACES, INRIA,
615, rue du Jardin Botanique, B.P. 101, 54602 Villers-lès-Nancy Cedex France
Fabrice.Rouillier@loria.fr

We describe the environment for symbolic and numeric computations, called SYNAPS (Symbolic and Numeric APplicationS) and developed in C++. Its aim is to provide a coherent platform integrating many of the nowadays freely available software in scientific computing. The approach taken here is inspired by the recent paradigm of software developments called active library. In this paper, we explain the design choices of the kernel and their impact on the development of generic and efficient codes for the treatment of algebraic objects, such as vectors, matrices, univariate and multivariate polynomials. Implementation details illustrate the performance of the approach.

1 Introduction

The need to combine symbolic and numeric computations is ubiquitous in many problems. Starting with an exact description of the equations, in most cases, we will eventually have to compute an approximation of the solutions. Even more, in many problems the coefficients of the equations may only be known with some inaccuracy (due, for instance, to measurement errors).

The aim of the SYNAPS[a] project is to provide a coherent platform for handling such problems, and integrating many of the nowadays freely available software in scientific computing. The kernel of this platform is based on the former library ALP[b] to which we incorporate new genericity and spe-

*PARTIALLY SUPPORTED BY THE PROJECT ECG IST-2000-26473
[a] `http://www-sop.inria.fr/galaad/logiciels/synaps/`
[b] `http://www-sop.inria.fr/galaad/logiciels/ALP/`

cialization mechanisms. This kernel provides data-structures and classes for the manipulation of basic objects, such as (dense, sparse, structured) vectors, matrices, univariate and multivariate polynomials. The aim of this paper is to describe the design of this kernel.

As SYNAPS is targeting different domains of application, it has to provide various internal representations for the abstract data-types required by algorithms specifications. Thus, this library makes it possible to define parameterized but efficient data structures for fundamental algebraic objects such as vectors, matrices, monomials and polynomials... which can be used easily in the construction of more elaborated algorithms. We pay a special attention to genericity, or more precisely to parameterized types (templates in C++ parlance) which allow us to tune easily the data structure to each specific problem at hand. The approach taken here is inspired by a recent paradigm of software developments called *active library*[5,6,7] and illustrated by the library STL[8]. The software components of such libraries take active parts of the configuration and generation of codes (at compile time), which combine parameterization and efficiency. Projects like BLITZ, SCITL, MTL, NTL successfully use those ideas.

We also carefully consider the problem of reusability of external or third-party libraries. Currently, we have connected LAPACK (Fortran library for numerical linear algebra), UMFPACK, SPARSELIB (respectively Fortran and C++ libraries for sparse matrices), SUPERLU (C library for solving sparse linear system), GMP[c](C library for extended arithmetics), GB[d], RS[e] ...Specialized implementations — for instance LAPACK routines — can coexist with generic ones, the choice being determined by the internal representation.

2 Tools for an Active Library Design

The fundamental characteristic of an active library is to provide general purpose abstractions along with domain-specific specializations, tuned to take advantage of configuration knowledge necessary to deliver optimal efficiency. The configuration knowledge of an active library must be bundled in a *uniform* and *extensible* framework.

2.1 Types Are Our Friends

SYNAPS extensively uses the *typefull programming* paradigm[9], whereby the motto is: *laws should be enforceable.* By that we mean, whenever some con-

[c] `http://www.swox.se/gmp/`

[d] `http://calfor.lip6.fr/~jcf/Software/Gb/index.html`

[e] `http://calfor.lip6.fr/~rouillie/Software/RS/index.html`

straints are known to hold, we use the type-system to express them[f] so that much errors can be detected very early in the development stage (*i.e.* compile-time). In doing so, the program translator (*i.e.* the compiler) can keep us from breaking hard-to-find constraints violations and save us considerable time. Another immediate benefit is that, when a program construct fails to type-check then the reasons of the failure are apparent. Last but not least, as compilers are improving, information gathered from static analysis of a program text may be profitably used by the compiler to emit efficient executable images (see e.g. [10]). Template expressions (see §2.4) illustrate typical uses of the type-system to implement (in library) domain-specific optimizations not available in common-place compilers.

One feature of symbolic and numeric computations is the ubiquity of parameterized data-types and algorithms. ¿From a symbolic perspective, the notion of polynomial may be defined for any coefficient data type that happens to meet the ring axioms. Therefore, symbolic computation naturally calls for polymorphism[11] as an effective programming tool. The C++ programming language[12] supports polymorphic data type and algorithms through class inheritance, virtual functions, template classes and template functions. Both programming paradigms are not exclusive even though the stress on parametric polymorphism is rather recent, compared to the conventional use of inclusion polymorphism, which through the use of virtual functions and *dynamic binding* yields a very flexible mechanism to plug new components in an existing environment. However, inclusion polymorphism may become cumbersome or even conflicts with situations where value semantics, efficiency and static-type constraints are premium.

The *template* mechanism is the way the C++ programming language implements parametric polymorphism. By definition, it is static-checking oriented; the program-translator uses static type information to generate program components which, in turn, are integrated into the program being translated. That means templates are the feature of choice where polymorphism, value semantics, execution efficiency and static constraints are discriminating criterions. Because templates operate on two levels, they embody a form of *meta-language* which has proven to be handy when it comes to use the type-system to express constraints, perform compile-time computations, or implement features (such as lazy evaluation) not directly supported by the language.

[f] Obviously, not all constraints are expressible in the type-system.

2.2 How Do We Find Our Algorithms?

A distinctive feature of the C++ programming language is the ability to name *scopes* which yields a powerful way of associating a declaration with the name of a function or variable. Understanding how this process, called *name lookup* works, what rules govern it, is crucial for large libraries and programs organization. There are three categories of name lookup rules: qualified name lookup, unqualified name lookup, and function argument dependent name lookup also known as Koenig lookup.

Qualified names are used when a particular scope (a class-scope or namespace-scope) is specified to designate where the program-translator should look for a name. When the qualified name has the form `N::m`, where `N` designates a namespace, then the set of declarations of `m` comprises all declarations for `m` in `N` and in the transitive closure of all namespaces nominated by using-directives in `N` and its used namespaces. Names brought from foreign namespaces through using-declarations are considered during qualified name lookup.

The rules for unqualified name lookup are rather simple. In general, a name shall have a declaration "in scope" before use (an exception to that general rule is that covered by the Koenig lookup to be discussed in the next paragraph). For that matter, declarations from namespaces nominated by using-directives are visible in the namespace enclosing the using-directives (e.g. `using DOMAIN`). For the name lookup machinery, names declared in namespaces nominated by using-directives are considered members of the enclosing namespace. That mechanism is known as *namespace composition*.

The third group of lookup rules, called Koenig lookup, works as follows. When looking up an unqualified name as a function-name in a function call, namespaces associated with the type of the arguments are also examined; in those namespaces the function is looked up as a qualified name where the qualifier is the associated namespace. Koenig lookup is applicable not only to operator names but also to any function-name used in an unqualified manner. Concretely, that means that given the following:

```
namespace Our {
template<typename T> class Matrix { /* ... */ };
template<typename T> T frobenius_norm(const Matrix<T>& m) { /* ... */ }
}
int main() {
   Our::Matrix<double> m; double n = frobenius_norm(m);  // ...
}
```

the function `Our::frobenius_norm` is found by Koenig lookup which considers the namespace `Our` associated with the type of the argument `m`. This may require the use of additional using-declarations such as `using std::function;` to make sure that the generic function (e.g. `frobenius_norm`) will also apply to built-in types.

2.3 Classes versus Namespaces

There are two kinds of "nameable" scope: class-scope and namespace-scope. These scopes, while sharing some properties, work in fundamentally different ways. Whereas namespaces are "open", classes are "closed": A namespace can be "re-open" for the purpose of adding new members but classes cannot, once the right-brace terminating their definitions have been seen. Another distinctive difference is the fact that namespaces cannot be used as template-arguments whereas classes can. Thus, the decision of using a class or a namespace to implement a particular instance of scope is usually guided by the intended use.

The first approach based on class-scope is used for instance in [13,14] [g]. The algorithms are enclosed in so-called kernels. A kernel is a class, containing other sub-classes representing data or function-objects or methods operating on these types. The extension of a given kernel is performed by the derivation of classes, but with some constraints.

In SYNAPS, the second approach based on namespace has been chosen, for its easy extension capability. Moreover, using Koenig lookup, this allows us to combine genericity and specialization, exploiting natural conventions of developments.

2.4 Template expression

Template expressions are type manipulations used to guide the compiler to produce optimized code. This technique is particularly interesting for some linear algebra operations, where intensive but independent instructions are performed. We illustrate it on the addition of vectors. Consider for instance the following instruction: `v = v1 + v2 + v3;` where `v, v1, v2, v3` are vectors (say of type `Vector`). A traditional implementation would define the operator `Vector operator+(const Vector & v1, Vector & v2);` so that the previous instruction will produce two temporary vectors which is time and memory consuming, especially if the vectors are of large size. This can be avoided by using template expression techniques, which we implement as:
`VAL<Op<'+',Vector,Vector> > operator+(const Vector & v1, Vector & v2);`
where `VAL<Op<'+',Vector,Vector> >` is the type of a data-structure which contains references to the two vectors `v1, v2`, but which does not

[g] `http://www-sop.inria.fr/prisme/CGAL/index.html`

compute the sum of these two vectors. The computation is effectively performed in the assignement operator, which does not involve temporary objects. A complete set of arithmetic unevaluated operations is available in the SYNAPS kernel, namely types of the form OP<c,T1,T2> with $c \in$ {'+','-','*','.','/','%','^'} can be used in the library. Such arithmetic operations will return an unevaluated arithmetic tree and the computation will be effectively performed only when the assignment operator will be used.

3 The Design

We have distinguished three levels of abstractions: **containers** attached to **domains**, **modules** and **views**, that we describe hereafter.

3.1 Containers and domains

A container specifies the internal representation of a SYNAPS object. It provides abstractions to access, scan, create, or transform that concrete representation and thus depends closely on the data-type being modeled. Containers are implemented as classes (often parameterized by coefficient types or index types), with few functions so that they can be easily rewritten or extended. These structures are adapted to the problem they aim to solve. The algorithm will be provided by other classes or namespaces.

Our approach is in the spirit of the STL library[8], which implements classical data structures. SYNAPS add its own containers like one-dimensional arrays **rep1d<C>** with generic coefficients, two-dimensional arrays **rep2d<C>** for dense matrices ... Specialized implementations are usually found in *domains* (concretely implemented by namespaces). For instance, the container **rep1d<C>** is in the namespace **linalg** and the usual way to call it, is **linalg::rep1d<C>**. Similarly, we have defined 2-dimensional containers **linalg::rep2d<C>** and **lapack::rep2d<C>**, for dense matrices. The last container is adapted to the internal representation of dense matrices in the LAPACK library. Adapted containers for the manipulations of structured matrices such as Toeplitz matrices (**linalg::toeplitz<C>**), Hankel matrices (**linalg::hankel<C>**) or sparse matrices (**linalg::sparse2d<C>**) are also available in the library. The job of iterating over data-structures is delegated to iterators.

Because namespaces are open, using this domain mechanism allows us to add easily specialized functions, and to transparently use them in elaborated algorithms. As we will see, using the Koenig lookup mechanism described in 2.2, these specialized functions are searched in the domain of the container, if they exist. Another interesting feature of this approach, is that functions and classes provided in a domain, can be compiled separately.

3.2 Modules

A module is a collection of generic implementations which apply to a family of objects sharing common properties (such as vectors, matrices, univariate polynomials, ...) implemented as *namespaces* and thus can be extended easily. For instance in the module (or *namespace*) `VECTOR`, we have gathered generic implementations for vectors, which are independent of the internal representation, such as `Print`, `Norm`, ... Another interesting feature of the approach, is that namespaces can be combined or extend naturally: `namespace UPOLY { using namespace VECTOR; ... }`. This yields a great flexibility in the construction of set of functions which applies for a families of objects. The main modules of the library are `VECTOR`, `MATRIX`, `UPOLY` (univariate polynomials), `MPOLY` (multivariate polynomials).

3.3 Views

The views specify how to interpret containers as mathematical objects and provide operations on the objects which are independent of the container. They are implemented as classes parameterized by the container type and sometimes by *trait* classes which precise the implementation. The only data available in such a class, via the member function `rep()`, is of container type.

The implementations of a view are searched in a module as it is illustrated below:

```
template <class C, class R>
UPolDse<C,R> operator*(const UPolDse<C,R>& v1, const UPolDse<C,R>& v2)
{
    UPolDse<C,R> w(Degree(v1)+Degree(v2)+1,AsSize());
    using namespace UPOLY; mul(w.rep(),v1.rep(),v2.rep());
    return w;
}
```

An eventual declaration of `mul` in `UPOLY` is added to the possible overload set constructed by Koenig lookup. `UPOLY::mul` is the naive implementation that works on all random-access containers. In the case the container `Domain::R` is specialized for an external library providing its notion of multiplication, the forwarding `mul` function can be declared as: `namespace Domain { void mul(R & s, const R & r1, const R & r2);}`. This is the scheme that has been used to connect the external libraries LAPACK, GMP, SUPERLU, ...

Here are the main examples of view classes currently implemented in the library, `R` stands for the container type: `VectDse<C,R>` (standard vectors), `MatrDse<C,R>` (dense matrices), `MatrStr<C,R>` (structured matrices),

MatSparse<R> (sparse matrices), **UPolDse<C,R>** (dense univariate polynomials), **UPolQuot<R>** (quotiented univariate polynomials), **Monom<C,E>** (monomials with coefficients in **C** and exponents in **E**), **MPol<C,O,R>** (multivariate polynomials where **O** defines the ordering on the monomials).

3.4 Large Objects Management: Copy On Write

Because objects management in the C++ programming language defaults to *value semantics*, we have to take explicit actions to avoid copying large objects — the time spent in copying large objects may be shown to be non negligible under usual circumstances. A popular way to solve that memory management problem is to use reference counting[15] coupled with a copy-on-write strategy. This scheme is implemented by the class **shared_object<R>**. Therefore, returning such objects (of type **VectDse**, **MatrDse**, **MatrStr**, **MatrSps**, **UPolDse** or **MPol**, ...) by value in a function does not induce a performance penalty. Indeed a copy and destruction of **W**, will in this case just increment or decrement a counter.

4 Available packages in the current distribution

We describe briefly the functions, algorithms, libraries which are available in the current distribution[h].

Input/output: Streams for printing objects in maple and latex format are available. An output stream for **vrml** format is under construction as well as a connection to **povray**.

Arithmetic: The extended arithmetic classes are based on the GMP and MPFR[i] libraries, and correspond to the predefined types **ZZ, QQ, RR, RRR, CC**. A simple implementation of modular numbers **Z<p>** is also available.

Linear algebra: Basic linear algebra operations such as LU, QR, Bareiss decomposition, determinant, svd, rank, eigenvalues and eigenvectors computations, FFT are available, either using the LAPACK routines for **double** number type or generic ones. In particular, the implementation of structured LU decomposition of sparse matrices[16] developed by J. Demmel, J.R. Gilbert and X.S. Li, in C for **float** and **double** number types has been extended to generic coefficients.

Resultants: A package devoted to resultant constructions including the Sylvester and Bezout matrices for univariate polynomials, the Macaulay, Bezoutian, Toric (integrating the C-library developed by I. Emiris) formulations for multivariate polynomials[2,1].

---

[h]accessible via `cvs -d :pserver:cvs@cvs-sop.inria.fr/CVS/galaad co synaps` or `wget http://www-sop.inria.fr/galaad/logiciels/synaps.tgz`

[i]`http://www.mpfr.org/`

Solvers: Regarding solvers for univariate polynomials, we implement subdivision methods based on Bézier representation and Descartes rule, or on exclusion functions, iterative methods such as Sebastio a Sylva method[3]. The solver `mpsolve` developed by D. Bini and G. Firorentino[17] is also connected to the environment. For multivariate polynomials, resultant based solvers, hiding a variable are also available. The algorithm[4,18] for solving zero-dimensional ideals, which computes first a normal form and then the eigenvalues or eigenvectors of the induced matrices of multiplications is implemented and illustrated in the next section.

Geometry: A package on algebraic curves and surfaces is also under constructions, including tools for intersecting them, computing their topology or drawing them.

5 Distributed computation

Our objective is not only to be able to integrate external libraries in a transparent way, but also to use SYNAPS from outside tools such as computer algebra systems like MAPLE, MATHEMATICA, ... Indeed, SYNAPS is an environment which allow to build efficient and dedicated tools, that can be used for distributed computation through the Inter Process Communication tool, called UDX.

It is a protocol for data exchange in binary mode based on a new patent[19]. It is fully customizable, since it can be used with so many supports as one wants: files, sockets, shared memory. Its implementation does not depend on the machine nor on the operating system considered (e.g. no information about the indianness is required). It performs an optimal number of operations (byte permutations), whatever the considered hardware (no operations are needed to exchange data between equivalent hardware), and without any additional information (such as byte ordering) to exchange basic scalar data (machine integers, IEEE floats) between heterogeneous machines. Coupled with a highly optimized buffering system, UDX turns out to be, not only portable, but also several dozen times faster than conventional strategies like XDR

The communication protocol of SYNAPS uses UDX as a basic tool for exchanging low level data structure and implements a full protocol that takes advantage of the general structure of the library for higher level objects. An UDX non atomic object is described by two components: the descriptor part (independent of the implementation) and the data part. For example, the descriptor for dense vectors which is a flag (`DENSE_VECTOR`) and the descriptor of the entries (recursive definition) are sent first, followed by the data part constituted by the number of entries, and the sequence of coefficients. The descriptor part, attached to each kind of objects, is used to facilitate the implementation of the abstract protocol for exchanging efficiently in a binary way, complex mathematical data structures.

6 Illustration

We have already shown in experimentations reported for instance in [4], that genericity is compatible with efficiency. We can push further this reasoning: genericity is in fact a matter of performance. Indeed as the code is not attached to any particular implementation of the containers, it is possible to change them to use the most efficient implementation available without any code rewriting. The environment has been used to implement the algorithms described in[4,18] which compute a normal form for zero-dimensional systems of polynomial equations in order to solve them by eigenvector computation. In this approach, we have two main needs in terms of linear algebra, which are fast sparse LU-decomposition, fast and robust eigenvector computations.

The SYNAPS framework allows us to use specialized implementations for these two points, namely SUPERLU[16] and LAPACK[20]. The results that we obtained when dealing with the Katsura equations (n is the number of variables) on an Pentium III 933MHz are shown below:

n	4	5	6	7	10	11
NF(mod)	0.01s	0.05s	0.17s	0.95s	256.81s	1412s
NF(double)	0.01s	0.02s	0.13s	0.54s		
Eg(double)	0.01s	0.02s	0.11s	2.33s		
$Max(\|f_i\|)$	10^{-13}	10^{-10}	10^{-8}	10^{-8}		

where NF(mod) (resp. NF(double)) is the time spent for the computation of normal form in modular (resp. double floating point) arithmetic, Eg(double) is the time spent for the computation of eigenvectors and $Max(|f_i|)$ is the maximum of the value of the normalized polynomials at the computed roots. During the normal form computations, the SUPERLU library (ported in C++ with template coefficients) is used and then LAPACK is called for the eigenvectors computation.

References

1. I.Z. Emiris and B. Mourrain. Matrices in Elimination Theory. *J. of Symb. Comp.*, 28(1&2):3–44, 1999.
2. I.Z. Emiris and J.F. Canny. Efficient incremental algorithms for the sparse resultant and the mixed volume. *J. Symb. Comp.*, 20(2):117–149, 1995.
3. J. P. Cardinal. On two iterative methods for approximating the roots of a polynomial. In J. Renegar, M. Shub, and S. Smale, editors, *Proc. AMS-SIAM Summer Seminar on Math. of Numerical Analysis, (Park City, Utah, 1995)*, volume 32 of *Lectures in Applied Math.*, pages 165–188. American Mathematical Society Press, Providence, 1996.

4. B. Mourrain and Ph. Trébuchet. Solving projective complete intersection faster. In C. Traverso, editor, *Proc. Intern. Symp. on Symbolic and Algebraic Computation*, pages 231–238. New-York, ACM Press., 2000.
5. K. Czarnecki, U. Eisenecker, R. Glück, D. Vandervoorde, and T. Veldhuizen. Generative Programming And Active Libraries. In *Dagstuhl Seminar on Generic Programming*, volume TBA of Lect. Notes in Comp. Science, 1998.
6. T. Veldhuizen and D. Cannon. Active Libaries: Rethinking the roles of compilers and libraries. In *SIAM Workshop on Object Oriented Methods for Inter-Operable Scientific and Engineering Computing*, 1998.
7. G. Dos Reis. Vers une nouvelle approche du calcul scientifique en C++. Rapport de Recherche 3362, INRIA, 1998.
8. D.R. Musser and A. Saini. *STL tutorial and reference guide : C++ programming with the standard template library.* Addison-Wesley, 1996.
9. L. Cardelli. Typeful programming. pages 471–522. Springer-Verlag, 1991.
10. Mark Mitchell. Type-based alias analysis. *Dr. Dobb's Journal*, October 2000.
11. L. Cardelli and P. Wegner. On understanding types, data abstraction, and polymorphism. *Computing Surveys*, 17(4):471–522, 1985.
12. Stroustrup B. *The C++ Programming Language, Special Edition.* Addison-Wesley, 2000.
13. J.-G. Dumas, W. Eberly, M. Giesbrecht, E. Kaltofen D. Saunders, and G. Villard. Linbox, a C++ library for symbolic linear algebra, particularly using black box matrix methods.
14. Susan Hert, Michael Hoffmann, Lutz Kettner, Sylvain Pion, and Michael Seel. An adaptable and extensible geometry kernel. In *Proc. Workshop on Algorithm Engineering*, volume 2141 of *Lecture Notes Comput. Sci.*, pages 79–90. Springer-Verlag, 2001.
15. R. Jones and R. Lins. *Garbage Collections.* John Wiley & Sons Ltd, 1996.
16. J.W. Demmel, J.R. Gilbert, and Xiaoye S. Li. *SuperLU Users' Guide*, 1997. `http://www.netlib.org/scalapack/prototype/`.
17. D. Bini. Numerical computation of polynomial zeros by means of aberth's method. *Numerical Algorithms*, 13, 1996.
18. B. Mourrain and Ph. Trébuchet. Algebraic methods for numerical solving. In *Proc. of the 3rd International Workshop on Symbolic and Numeric Algorithms for Scientific Computing'01 (Timisoara, Romania)*, 2002.
19. J.C. Faugère and F. Rouillier. Dispositif d'échanges de données entre matériels informatiques. Patent number 99 08172, 1999.
20. E. Anderson, Z. Bai, C. Bischof, J. Demmel, J. Dongarra, J. Du Croz, A. Greenbaum, S. Hammarling an A. McKenney, S. Ostrouchov, and D. Sorensen. *LAPACK Users' Guide.* SIAM, Philadelphia, 1992. `http://www.netlib.org/lapack/`.

COMPUTING ELLIPTIC CURVES USING KASH

MASANARI KIDA
Department of Mathematics
University of Electro-Communications
Chofu Tokyo 182-8585, Japan
E-mail: kida@sugaku.e-one.uec.ac.jp

The package TECC is a program library written in the KASH language aimed at doing basic arithmetic of elliptic curves defined over number fields. We shall outline the features of TECC.

1 Introduction

TECC (Tiny Elliptic Curve Calculator) is a small program package written in the KASH language for computing various arithmetical invariants of elliptic curves defined over algebraic number fields.

There are many good number theory packages we can use for computing elliptic curves, to name but a few, PARI-GP [1], SIMATH [2], Apecs [3] and MAGMA [4]. The main feature that makes our system TECC unique is that we do not make any assumptions on number fields over which elliptic curves are defined.

In writing a series of papers [5,6,7], I had to compute the reduction types of many elliptic curves defined over quadratic number fields. As far as I knew, this could be done only by Umegaki's program (see his paper [8]) at that time. Umegaki's program is written in C using his altered PARI library. Thus the program was not always easy to run on any computers and OS's. So I decided to write a portable program for computing the reduction types of elliptic curves over number fields. This is the prototype of TECC. The language I chose for the program is KASH (KAnt SHell), which is a user-friendly interface to the KANT library [9] developed under the project leadership of Professor Michael E. Pohst at Technische Universität Berlin. KASH is widely used in the number theory community in these days and is one of the most popular and the most powerful number theory program packages. I chose KASH, because it furnishes high-level functions for sophisticated computations in algebraic number fields, which we have to assume for computing elliptic curves over number fields. This strong property of KASH helps to make the size of our program small. It is also important that KASH is running on various OS's.

In this article, we give a brief introduction to the package TECC.

Acknowledgement. I would like to express my thanks, first of all, to the members of the KANT group, especially to Professor Michael E. Pohst for his leadership and Dr. Claus Fieker for his kind and valuable (and patient) discussion on KASH during the preparation of the program.

I owe the algorithms used in the program to many number theorists. Among others, I would like to thank Professor Joseph Silverman for his kind advice on computing height functions and to Professor Ian Connell for his nicely written program Apecs, whose source code is very helpful in writing TECC.

2 A short tour to TECC

In this section, we shall explain how to use TECC by some examples. We assume that the readers are familiar with KASH. Our explanation is based on TECC version 2.4 (September, 2000). For the complete description of the TECC functions, see TECC manual [10]. For the algorithms used in TECC, we refer to Cremona's book [11] and Cohen' book [12] and the references indicated in the manual [10].

2.1 Basic arithmetic

As an example, we consider the elliptic curve

$$E : y^2 = x^3 - 112x + 400$$

defined over $\mathbb{Q}$.

To use TECC functions, let KASH read them.

```
kash> Read("tecc.kash");
TECC Version 2.4
Copyright (C) 1998, 1999, 2000 Masanari Kida
TECC comes with NO WARRANTY.
```

Next we define the curve E by using the initialization function `ECInit`:

```
kash> E:=ECInit([0,0,0,-112,400],Z);
Elliptic Curve [a1,a2,a3,a4,a6]=[0,0,0,-112,400]
defined over Integer Ring
```

It is indispensable to give the base field (the order in the KANT language) explicitly. The `ECInit` function internally computes basic invariants such as b and c invariants, the discriminant and the j-invariant etc. These computed values are able to be accessed by using '.' operator:

```
kash> E.b4; # b4
-224
kash> Factor(E.disc);  # Factorization of the discriminant
[ [ 2, 12 ], [ 5077, 1 ] ]
kash> E.jinv; # j-invariant
37933056/5077
```

This tells us that E has potential multiplicative reduction at 5077. We check this by

```
kash> ECLocalReduction(E,5077);
[ 5077, 5, 1, 1, 1, 1, [ 1, 368, 0, 0 ] ]
```

The second element "5" in the above list means that the reduction of the curve at 5077 is of type I_1. By computing the reduction at 2, it turns out that 5077 is the only bad prime of E. Thus the conductor of E is 5077. It is known that E is the curve of smallest conductor with Mordell-Weil rank 3 over $\mathbb{Q}$ (see Cremona's elliptic curve data at **http://www.maths.nott.ac.uk/personal/jec/ftp/data/INDEX.html**). The generators of the Mordell-Weil group of E are given by

$$P_1 = (4, -4),\ P_2 = (0, -20),\ P_3 = (-4, -28).$$

Unfortunately, TECC does not have any function to find these generators. We have several other programs such as Cremona's mwrank, which is available from **http://www.maths.nott.ac.uk/personal/jec/packages.html** for doing this task.

Let k be the imaginary quadratic field $k = \mathbb{Q}(\sqrt{-5})$ and E_k the base change of E to the imaginary quadratic field k.

```
kash> k:=OrderMaximal(Z,2,-5); #  k = Q(√-5)
Generating polynomial: x^2 + 5
Discriminant: -20

kash> Ek:=ECMove(E,k);
Elliptic Curve [a1,a2,a3,a4,a6]=[0,0,0,-112,400]
defined over the order:
 Generating polynomial: x^2 + 5
Discriminant: -20
```

We search integral points on E_k in the set $\{a+b\sqrt{-5} \in \mathbb{Z}[\sqrt{-5}] \mid |a|, |b| < 10\}$:

```
kash> ECSearchIntegralPts(Ek,10,0);
[ [ 0, 20 ], [ 0, -20 ], [ 1, 17 ], [ 1, -17 ], [ 4, 4 ], [ 4, -4 ],
  [ -4, 28 ], [ -4, -28 ], [ -7, 29 ], [ -7, -29 ], [ 8, 4 ], [ 8, -4 ],
  [ -8, 28 ], [ -8, -28 ], [ 9, 11 ], [ 9, -11 ], [ [8, 2], [6, 10] ],
  [ [8, 2], [-6, -10] ], [ [8, -2], [6, -10] ], [ [8, -2], [-6, 10] ],
  [ [-8, 4], 52 ], [ [-8, 4], -52 ], [ [-8, -4], 52 ], [ [-8, -4], -52 ] ]
```

In TECC, a point on an elliptic curve is represented just by a list `[x,y]` of length 2. The point at infinity is exceptionally expressed by `[0]` of length 1. In the above, we have found several integral points not belonging to $\mathbb{Z}$.

In their paper [13], Smart and Stephens proved an algorithm to find all the integral points on elliptic curves defined over number fields. Applying their algorithm, they claimed in Example 1 of the paper that all the integral points in $E(k)$ are defined over $\mathbb{Z}$. Our computation shows that their assertion is apparently not true. Though their algorithm is correct, this mistake happens because their computation of the Mordell-Weil group $E(k)$ is not correct. In the remainder of this section, we determine all the integral points in $E(k)$.

We first compute the Mordell-Weil group $E(k)$. We need to know the Mordell-Weil group of the quadratic twist $E^{(-5)}$ of E corresponding to the quadratic extension $k/\mathbb{Q}$, which is given by

$$E^{(-5)}: \; Y^2 = X^3 - 2800X - 50000.$$

This model is computed by

```
kash> ET:=ECTwist(E,-5); # ET = E^(-5)
Elliptic Curve [a1,a2,a3,a4,a6]=[0,0,0,-2800,-50000]
defined over Integer Ring

kash> ECGlobalMinimalModel(ET);
[ [ Elliptic Curve [a1,a2,a3,a4,a6]=[0,0,0,-2800,-50000]
        defined over Integer Ring
    , [ 1, 0, 0, 0 ] ] ]
```

The second command ensures that the model of $E^{(-5)}$ is global minimal. The Mordell-Weil group of the curve $E^{(-5)}$ is of rank 1 and is generated by $Q_4 = (89, -637)$. We have an isomorphism ϑ defined over k from $E^{(-5)}$ to E sending (X, Y) to $\left(-\frac{X}{5}, -\frac{Y}{5\sqrt{-5}}\right)$. By this map, the point Q_4 is sent to

$$P_4 = \vartheta(Q_4) = \left(-\frac{89}{5}, -\frac{637\sqrt{-5}}{25}\right).$$

We check that P_4 is on the curve E_k.

```
kash> P4:=[-89/5,-637*Elt(k,[0,1])/25];
[ -89/5, [0, -637] / 25 ]
kash> ECPtIsOnCurve(Ek,P4);
true
```

The next task is putting together $E(k)$ from $E(\mathbb{Q})$ and $E^{(-5)}(\mathbb{Q})$. For this purpose, we consider the map

$$\begin{array}{rccc} \varphi\colon & E(\mathbb{Q})\langle 2\rangle + \vartheta(E^{(-5)}(\mathbb{Q}))\langle 2\rangle & \longrightarrow & E(k)\langle 2\rangle \\ & (P,Q) & \longmapsto & P+Q. \end{array}$$

Here and hereafter, for a finitely generated abelian group A written additively, let A_n be the kernel of the multiplication-by-n map and set $A\langle 2\rangle = A/A_{2^\infty}$.

The following properties of φ are shown in our previous paper [14]:

(i) φ is a group homomorphism.
(ii) $E(\mathbb{Q}) \cap \vartheta(E^{(-5)}(\mathbb{Q})) = E(\mathbb{Q})_2$. Hence the sum in the above definition is direct.
(iii) φ is injective.
(iv) $\mathrm{Coker}(\varphi)$ is killed by 2.

We also have

```
kash> ECTorsPts(Ek);
[ 1, [ 1 ], [ [ 0 ] ] ]
```

Namely the torsion part of $E(k)$ is trivial. Therefore, for the determination of $E(k)$, we have only to verify whether any possible sums of the points P_1, P_2, P_3, P_4 have a half point (i.e., a point Q such that $P = 2Q$ for a given point P) in $E(k)$ or not. The function `ECHalfPt` can be used for this computation.

```
kash> P1:=[4,-4];;P2:=[0,-20];;P3:=[-4,-28];;
kash> L:=[P1,P2,P3,P4];;
kash> while MultiCounterInc(M,1) do
>   P:=[0];
>    for j in [1..Length(L)] do
>      if M[j]=1 then
>        P:=ECAddPt(Ek,P,L[j]); #  ∑_{j=1}^4 M[j]P_j
>      fi;
>    od;
>    HP:=ECHalfPt(Ek,P);
>    if Length(HP)>0 then
>      Print(M," ",P," ",HP);
>    fi;
> od;
[ 1, 0, 1, 1 ] [ [39942, -7007] / 8978, [6522813, 4820179] / 1203052 ]
[ [ [8, 2], [6, 10] ] ]
```

This result means that the point $P_5 = (8 + 2\sqrt{-5}, 6 + 10\sqrt{-5})$ satisfies

$$2P_5 = P_1 + P_2 + P_4.$$

Applying the same procedure to the group generated by P_1, P_2, P_3 and P_5, we find that there is no further half point in the group. Thus we conclude

$$E(k) = \langle P_1\rangle \times \langle P_2\rangle \times \langle P_3\rangle \times \langle P_5\rangle$$

Thus all the integral points on $E(k)$ can be written in the form

$$n_1 P_1 + n_2 P_2 + n_3 P_3 + n_5 P_5 \text{ with } n_i \in \mathbb{Z}.$$

Smart-Stephens' algorithm gives the upper bound of n_i's. To apply their algorithm, we have to compute the elliptic logarithm of P_5 (other values are computed in their paper [13]). Let Ψ (resp. Ψ') be the elliptic logarithm on E (resp. $E^{(-5)}$). The value $z = \Psi(2P_5)$ can be computed as follows (see Section 2 in [14] for further details).

$$\begin{aligned} z &= \Psi(2P_5) \\ &= \Psi(P_1) + \Psi(P_3) + \Psi(P_4) \\ &= \Psi(P_1) + \Psi(P_3) + \sqrt{-5}\,\Psi'(Q_4). \end{aligned}$$

The actual value is given by

```
kash> Q4:=[89,-637];;
kash> Prec(10);;
kash> z:=ECElog(E,P1)+ECElog(E,P3)+Sqrt(-5)*ECElog(ECTwist(E,-5),Q4);
[ 1.43288874 + 0.989126120682*i ]
```

It is important to use a modest precision for computations involving floating point arithmetic. Otherwise the computation will be very slow. Now it is easy to see that the elliptic logarithm of P_5 agrees with one of

$$\frac{\Psi(P_5)}{2}, \frac{\Psi(P_5)}{2} + \frac{\omega_i}{2} \ (i = 1 \text{ or } 2) \text{ and } \frac{\Psi(P_5)}{2} + \frac{\omega_1 + \omega_2}{2},$$

where $\Lambda = \mathbb{Z}\omega_1 + \mathbb{Z}\omega_2$ is the period lattice for E. We can determine which is what we want by computing the inverse function of the elliptic logarithm. The following value is closest to P_5 among others:

```
kash> OM:=ECPeriods(E)[1];   # OM = [ω1, ω2]
[ 0.74027413412*i, 1.03792199 ]
kash> ECztoPoint(E,(z[1]+OM[1])/2); # The inverse function of ECElog
[ [ 8.00000091 + 4.47213653*i, 6.0000037 + 22.3606828*i ] ]
```

Thus we conclude

$$\Psi(P_5) = \frac{z + \omega_1}{2} \doteq 0.71644437 + 0.864700125\sqrt{-1}.$$

The canonical height of P_5 is

```
kash> P5:=[Elt(k,[8,2]),Elt(k,[6,10])];;
kash> ECCanonicalHeight(Ek,P5);
0.88209773
```

Note that TECC can compute the canonical height even if there is no global minimal model.

Now applying Smart-Stephens' algorithm, we show that the integral points in $\mathbb{Z}[\sqrt{-5}]$ are of the form

$$n_1P_1 + n_2P_2 + n_3P_3 + n_5P_5 \text{ with all } |n_i| \leq 10.$$

The following command finds all the integral points:

```
kash> ECSearchIntegralPtsFromList(Ek,[P1,P2,P3,P5],10);
 [ [ 4, -4 ], [ 56, -412 ], [ 0, -20 ], [ 12, -28 ], [ -7, -29 ], [ -4, -28 ],
  [ 9, -11 ], [ 8, 4 ], [ -8, 28 ], [ 1, 17 ], [ 44, 284 ], [ 3264, 186476 ],
  [ 1368, -50596 ], [ 208, 2996 ], [ 84, -764 ], [ [8, 2], [6, 10] ],
  [ [-64, 4], [-108, -224] ], [ 4, 4 ], [ 56, 412 ], [ 32, -172 ],
  [ 16, -52 ], [ -12, -4 ], [ 25, 115 ], [ 1624, 65444 ], [ [-8, 4], 52 ],
  [ [4, 12], [-84, 56] ], [ 0, 20 ], [ 32, 172 ], [ 148, -1796 ], [ 12, 28 ],
  [ -7, 29 ], [ 372, 7172 ], [ -4, 28 ], [ 16, 52 ], [ 148, 1796 ],
  [ 372, -7172 ], [ [-8, -4], -52 ], [ 9, 11 ], [ [8, -2], [-6, 10] ],
  [ 8, -4 ], [ -12, 4 ], [ 1, -17 ], [ 1368, 50596 ], [ 25, -115 ],
  [ 84, 764 ], [ 1624, -65444 ], [ [4, -12], [84, 56] ], [ -8, -28 ],
  [ 44, -284 ], [ 3264, -186476 ], [ 208, -2996 ], [ [-64, -4], [108, -224] ],
  [ [8, 2], [-6, -10] ], [ [-8, 4], -52 ], [ [-8, -4], 52 ],
  [ [8, -2], [6, -10] ], [ [4, -12], [-84, -56] ], [ [-64, -4], [-108, 224] ],
  [ [4, 12], [84, -56] ], [ [-64, 4], [108, 224] ] ]
kash>  Length(last);
60
```

Consequently, we have found 60 integral points in $\mathbb{Z}[\sqrt{-5}]$.

2.2 Complex multiplication

TECC has several functions related to complex multiplication.

First we create a field defined by the Hilbert class polynomial of discriminant -15, which is nothing but $\mathbb{Q}(\sqrt{5})$.

```
kash> f:=HilbertClassPoly(-15);
x^2 + 191025*x - 121287375
kash> O:=Order(f);
Generating polynomial: x^2 + 191025*x - 121287375

kash> t:=Elt(O,[0,1]);;
kash> k:=OrderMaximal(O);;j:=EltMove(t,k);;
```

The element j is a root of the Hilbert class polynomial f. Hence it is a j-invariant of the corresponding imaginary quadratic order.

```
kash> E:=ECGenericCurve(j);
Elliptic Curve [a1,a2,a3,a4,a6]=[0,0,0,[-30731133510, 49727382705],[3962170100
466270, -6410928347984160]]
defined over the order:
```

```
     F[1]
      |
    F[2]
    /
   /
Q
F   [ 1]      Given by transformation matrix
F   [ 2]      x^2 + 191025*x - 121287375
Discriminant: 5
kash> EltMinPoly(E.jinv);
x^2 + 191025*x - 121287375
```

As shown in the above, the function **ECGenericCurve** returns an elliptic curve with the given j-invariant. Therefore the above defined curve E has complex multiplication in the order of discriminant -15 as a matter of course. We can check this by

```
kash> ECHasCM(E);
-15
```

TECC keeps the data of singular j-invariants of degree up to four. By the theory of complex multiplication, the class number of the imaginary quadratic order of discriminant -15 is 2. Indeed, we have

```
kash> ImaginaryQuadraticOrderClassNumber(-15)[1];
2
```

as expected.

The root j of f generates the Hilbert class field H of $K = \mathbb{Q}(\sqrt{-15})$.

```
kash> K:=OrderMaximal(Z,2,-15);; # K = Q(√−15)
kash> H:=OrderMaximal(PolyMove(f,PolyAlg(K)));; # H = K(j)
kash> OrderDisc(H); # the relative discriminant
<1>
```

Since the coefficients of Hilbert class polynomials are very large, it is better to use Weber class polynomials, for example, to generate the Hilbert class fields of imaginary quadratic fields. See Section 7.6.3 in Cohen's textbook [12].

2.3 Reduction

Let E be the elliptic curve

$$y^2 + xy = x^3 - x^2 - 37x - 78$$

defined over $\mathbb{Q}$. The conductor of E is 49. Thus E has bad reduction only at 7 with Kodaira type $\mathrm{Type}(E, 7) = \mathrm{III}$. This can be verified by

```
kash> E:=ECInit([1,-1,0,-37,-78],Z);
Elliptic Curve [a1,a2,a3,a4,a6]=[1,-1,0,-37,-78]
defined over Integer Ring

kash> ECGlobalReduction(E);
[ [ 49, [ 1, 2, 0, 6 ] ], [ [ 7, 3, 2, 2, 2, 3, [ 1, 2, 0, 6 ] ] ] ]
kash> E.jinv;
16581375      # = (3·5·17)^3
```

Since the the j-invariant is integral, E has potential good reduction. (In fact, E has complex multiplication in the imaginary quadratic order of discriminant -28). In the following, we shall explicitly construct a finite extension where E has good reduction everywhere.

Let $K = \mathbb{Q}(\sqrt{-7})$ and $L = \mathbb{Q}(\sqrt[4]{-7})$. We have a tower of fields $\mathbb{Q} \subset K \subset L$. We compute the variation of the reduction type of E in this tower. This example is taken from Example 3 of the paper [15].

Let $\mathfrak{p}$ (resp. $\mathfrak{P}$) be the unique prime ideal of K (resp. L) lying above 7. We first go up to K.

```
kash> K:=OrderMaximal(Z,2,-7);; # Q(√-7)
kash> p:=Factor(7*K)[1][1];;    # p
kash> E_K:=ECMove(E,K);;        # E_K = E × K
kash> ECLocalReduction(E_K,p);
[ <[-1, 2]>, -1, 2, 2, 5, 6, [ 1, 2, 3, -1 ] ]
```

As before, the second element in the list stands for the reduction type. Here in this case, we have $\mathrm{Type}(E_K, \mathfrak{p}) = \mathrm{I}_0^*$.

Now we move to L.

```
kash> L:=OrderMaximal(Z,4,-7);; # Q(∜-7)
kash> P:=Factor(7*L)[1][1];;    # 𝔓
kash> E_L:=ECMove(E,L);;        # E_L = E × L
kash> ECLocalReduction(E_L,P);
[ <[0, -1, 0, 0]>, 1, 0, 1, 1, 0, [ [0, 1, 0, 0], 2, 3, -1 ] ]
kash> ECDisc(ECGlobalMinimalModel(E_L)[1][1]);
-1
```

Hence we have $\mathrm{Type}(E_L, \mathfrak{P}) = \mathrm{I}_0$. In other words, E has everywhere good reduction over L. Moreover the last command shows that it has a global minimal model with discriminant -1.

3 Pot-pourri

I would like to include several comments here.

- Besides functions dealing with elliptic curves over number fields, TECC also has functions concerning elliptic curves defined over finite fields. They have names starting with `ECFF`. These functions are designed mainly for the auxiliary use and are not efficiently implemented now.
- There is another elliptic package also based on KASH: Nils Bruin's Algae. Its main feature is computing Mordell-Weil groups of elliptic curves over number fields. Naturally Algae is larger than TECC. As he writes in his web page `http://www.cecm.sfu.ca/~bruin/`, Algae and TECC do not have common data structure for elliptic curves. Thus unfortunately these packages do not communicate with each other well.
- You can get more information on KASH from their web page:

  http://www.math.tu-berlin.de/algebra/.

  KASH is freely available from the above page.
- The newest version of TECC and information on TECC can be obtained from

  http://matha.e-one.uec.ac.jp/~kida/TECC.html.

References

1. C. Batut, K. Belabas, D. Bernardi, H. Cohen, and Olivier. *User's Guide to PARI-GP version 2.1.0.* Laboratoire A2X, U.M.R. 9936 du C.N.R.S., Université Bordeaux I, November 2000.
2. C. Hollinger and P. Serf. SIMATH—a computer algebra system. In *Computational number theory (Debrecen, 1989)*, pages 331–342. de Gruyter, Berlin, 1991.
3. Ian Connell. Apecs version 6, 2000. A Maple program.
4. Wieb Bosma, John Cannon, and Catherine Playoust. The Magma algebra system. I. The user language. *J. Symbolic Comput.*, 24(3-4):235–265, 1997. Computational algebra and number theory (London, 1993).
5. Masanari Kida and Takaaki Kagawa. Nonexistence of elliptic curves with good reduction everywhere over real quadratic fields. *J. Number Theory*, 66(2):201–210, 1997.
6. Masanari Kida. Reduction of elliptic curves over certain real quadratic fields. *Math. Comp.*, 68(228):1679–1685, 1999.
7. Masanari Kida. Computing elliptic curves having good reduction everywhere over quadratic fields. *Tokyo J. Math.*, 24:545–558, 2001.
8. Atsuki Umegaki. Computing elliptic curves over number fields. In *Proceeding of the 2nd Symposium on Algebra and Computation.* Tokyo Metropolitan University, 1997. (In Japanese).

9. M. Daberkow, C. Fieker, J. Klüners, M. Pohst, K. Roegner, M. Schörnig, and K. Wildanger. KANT V4. *J. Symbolic Comput.*, 24(3-4):267–283, 1997. Computational algebra and number theory (London, 1993).
10. Masanari Kida. *TECC manual version 2.4*. The University of Electro-Communications, September 2000.
11. J. E. Cremona. *Algorithms for modular elliptic curves*. Cambridge University Press, Cambridge, second edition, 1997.
12. Henri Cohen. *A course in computational algebraic number theory*. Springer-Verlag, Berlin, 1993.
13. N. P. Smart and N. M. Stephens. Integral points on elliptic curves over number fields. *Math. Proc. Cambridge Philos. Soc.*, 122(1):9–16, 1997.
14. Masanari Kida. Good reduction of elliptic curves over imaginary quadratic fields. *J. Théor. Nombres Bordeaux*, 13(1):201–209, 2001.
15. Masanari Kida. Potential good reduction of elliptic curves. To appear in J. Symbolic Comput.

COMPUTING THE ENDOMORPHISM TYPE OF ORDINARY ELLIPTIC CURVES OVER FINITE FIELDS WITH KANT V4

M. HENNINGSEN

Technische Universität Berlin, Institut für Mathematik,
Straße des 17. Juni 136, 10623 Berlin, Germany
E-mail: hennings@math.tu-berlin.de

Elliptic curve cryptography has received more and more attention from the security industry over the past years. Although some types of special curves have been successfully attacked, carefully chosen curves over prime fields, or over $\mathbb{F}_{2^m}$ with prime exponent m, seem to be intractable, and one cannot do much better than the Pollard-ρ algorithm for generic discrete logarithm problems. Other approaches are the Deuring lifting theorem [2] or the construction of isogenies between elliptic curves. Both attempts require explicit knowledge of the endomorphism ring of the original elliptic curve at some point. The basic ideas for computing the endomorphism type of ordinary elliptic curves over prime fields are outlined in this paper, and the computation of an example using the computer algebra system KANT V4 is presented.

1 Endomorphism Rings over Elliptic Curves

Since we consider prime fields with $p \geq 5$, we may assume that an elliptic curve E over $k = \mathbb{F}_p$ is given as a short Weierstrass model

$$E : y^2 = x^3 + Ax + B. \tag{1}$$

Let O denote the point at infinity and $E(k)$ the group of points (X, Y) satisfying (1).

An isogeny between two elliptic curves E_1 and E_2 is a rational map $\varphi : E_1 \to E_2$ over k with $\varphi(O) = O$. Two elliptic curves are called isogenous if there exists an isogeny between them. Isogenies are either constant or surjective [9]. Further every isogeny φ of elliptic curves respects the group structure of their points, thus φ is a homomorphism. The endomorphism ring $End(E)$ of an elliptic curve E is the set of all isogenies $\varphi : E \longrightarrow E$ which are defined over $\bar{k}$.

The j-invariant of an elliptic curve $E : Y^2 = X^3 + AX + B$ can be calculated easily via

$$j(E) = 1728 \frac{4A^3}{4A^3 + 27B^2}. \tag{2}$$

An important fact is that two elliptic curves defined over k are isomorphic if they have the same j-invariant.

We denote by π the Frobenius endomorphism

$$\pi : (x,y) \mapsto (x^p, y^p)$$

on an elliptic curve E/k. Being a zero of its characteristic polynomial, it satisfies the following relation

$$0 = X^2 - tX + p, \tag{3}$$

where t is the trace of π. If the t is divisible by p, $End(E)$ is a quaternion algebra and E is called super singular; otherwise E is called ordinary and $End(E)$ is isomorphic to $\mathbb{Z}$ or to an order in an imaginary quadratic number field [2]. In this paper we consider ordinary curves on prime fields.

Definition 1 *For $n \geq -1$ the nth division polynomial $\psi_n \in k[X,Y]$ of an elliptic curve $E : y^2 = x^3 + Ax + B$ over k of characteristic $\neq 2,3$ is defined as follows:*

- $\psi_{-1}(X,Y) = -1,$
- $\psi_0(X,Y) = 0,$
- $\psi_1(X,Y) = 1,$
- $\psi_2(X,Y) = 2Y,$
- $\psi_3(X,Y) = 3X^4 + 6AX^2 + 12BX - A^2,$
- $\psi_4(X,Y) = 4Y(X^6 + 5AX^4 + 20BX^3 - 5A^2X^2 - 4ABX - 8B^2 - A^3),$
- $\psi_{2n+1}(X,Y) = \psi_{n+2}(X,Y)\psi_n^3(X,Y) - \psi_{n+1}^3(X,Y)\psi_{n-1}(X,Y)$ *for* $n \geq 2,$
- $\psi_{2n}(X,Y) = \frac{\psi_n(X,Y)}{2Y}(\psi_{n+2}(X,Y)\psi_{n-1}^2(X,Y) - \psi_{n-2}(X,Y)\psi_{n+1}^2(X,Y))$ *for* $n \geq 3.$

Remark 2 *We observe that with the relation of $E : Y^2 = X^3 + AX + B$ each ψ_{2n+1} can be reduced to polynomials in X for $n \in \mathbb{N}$. Each ψ_{2n} can be reduced to a polynomial in X multiplied by Y.*

Another special type of endomorphism of E is the multiplication by n:

$$\begin{aligned} [n] :\ & E \longrightarrow E \\ & (x,y) \mapsto (x_n, y_n) = \left(x - \frac{(\psi_{n-1}\cdot\psi_{n+1})(x,y)}{\psi_n^2(x,y)}, \right. \\ & \left. \frac{(\psi_{n+2}\cdot\psi_{n-1}^2 - \psi_{n-2}\cdot\psi_{n+1}^2)(x,y)}{4y\cdot\psi_n^3(x,y)}\right) \end{aligned} \tag{4}$$

The map $[n]$ can be identified with $n \in \mathcal{O}_E$ where $\mathcal{O}_E \cong End(E)$ is either $\mathbb{Z}$ or an order in an imaginary quadratic field. The points of the subgroup

$$E[l] := \{P = (X, Y) \mid l \cdot P = O\} \tag{5}$$

in $E(k)$ are called the l-torsion points of E.

For a more detailed description of isogenies of elliptic curves see for example the books of Silverman [9] or Lang [5]. Here we just need the fact that any isogeny factors into a power of the Frobenius endomorphism and a separable isogeny. The degree of a separable isogeny $\varphi : E_1 \longrightarrow E_2$ is equal to the number of points in the subgroup $\ker(\varphi)$ [9]. A separable isogeny of degree d is called a d-isogeny.

2 The Endomorphism Type

The Frobenius endomorphism π is defined in any endomorphism ring of an elliptic curve, thus $\mathbb{Z}[\pi] \subseteq \mathcal{O}_E$. Let $\mathcal{O}_K$ be the maximal order in $K = \mathbb{Z}[\pi] \otimes \mathbb{Q}$ with discriminant D_K. Since

$$\mathbb{Z}[\pi] \subseteq \mathcal{O}_E \subseteq \mathcal{O}_K, \tag{6}$$

we know that $End(E) = \mathbb{Z}$ if $\pi \in \mathbb{Z}$, therefore we assume that $\pi \notin \mathbb{Z}$ in the following.

Any order $\mathcal{O} \subseteq \mathcal{O}_K$ has a positive integer c called the conductor of $\mathcal{O}$ in $\mathcal{O}_K$ characterized by

$$\mathcal{O} = \mathbb{Z} \oplus c \cdot \mathcal{O}_K. \tag{7}$$

Note that the conductor c is equal to the index $[\mathcal{O}_K : \mathcal{O}]$.

The order $\mathbb{Z}[\pi] \subseteq \mathcal{O}_K$ has discriminant $d = t^2 - 4p$ and conductor $m = \sqrt{d/D_K}$. There exists an $a \in \mathbb{Z}$ such that

$$\mathcal{O}_K = \mathbb{Z}\left[\frac{\pi - a}{m}\right]. \tag{8}$$

The value of a is determined by $a^2 - ta + p \equiv 0 \bmod m^2$ and $2a \equiv t \bmod m$. A simple calculation shows that we can choose $a = t/2$ if $D_K \equiv 0 \bmod 4$ and $a = (t+m)/2$ if $D_K \equiv 1 \bmod 4$.

2.1 $l \neq 2$

Let $\mathcal{O}_E$ be the order in $\mathcal{O}_K$ isomorphic to $End(E)$. For all primes $l \in \mathbb{P}$ dividing m, we need to find the biggest $n = l^r$ such that $(\pi - a)$ is still the zero map on $\mathcal{O}_E/n\mathcal{O}_E$. Note that l is coprime to p since $l \mid (t^2 - 4p)$ and

$0 < t < p$. For this to occur, there must be an $\alpha \in End(E)$ such that $(\pi - a) = n\alpha$, which is equivalent to $E[n] \subseteq \ker(\pi - a)$.

To do this, we search for the largest possible exponent r. Since ψ_n generates the ideal $E[n]-\{O\}$ we have $(\psi_n) = n\mathcal{O}_E$. Therefore we calculate $(\pi - a)$ modulo ψ_{l^r} starting with $r = 1$. The ring of polynomials

$$k[X,Y]/(\psi_n(X,Y), Y^2 - X^3 - AX - B) \tag{9}$$

reduces to a ring of polynomials in one variable X only.

With (4), we calculate the X- coordinate of $a \cdot P$ for a point $P = (X,Y)$ on E

$$[a](X) = X - \frac{(\psi_{a-1} \cdot \psi_{a+1})(X)}{\psi_a^2(X)}. \tag{10}$$

The Frobenius endomorphism for the X- coordinate is defined by $\pi_x(X) = X^p$. Therefore we calculate

$$\begin{aligned} h_r(X) &\equiv \psi_a^2(X)(\pi_x - a_x)(X) \bmod \psi_{l^r}(X) \\ &\equiv (X^p - X)\psi_a^2(X) + \psi_{a+1}\psi_{a-1}(X) \bmod \psi_{l^r}(X) \end{aligned} \tag{11}$$

Note that we can take $\tilde{a} \equiv a \bmod l^r$ instead of a in the calculation above.

If $h_r(X) = 0$, we increase the parameter r step by step until $h_r(X) \neq 0$ or until r exceeds the exponent of l in the conductor of $\mathbb{Z}[\pi]$. Then we have found the exponent r of the prime l in the index $[\mathcal{O}_E : \mathbb{Z}[\pi]]$.

Problems can arise if $n = l^r$ gets too large, since ψ_n is of degree $O(n^2)$. For small l but large exponent r the following method is very useful. If there are large prime divisors of the conductor m of $\mathbb{Z}[\pi]$ one should use another idea shown by Kohel [4] involving the class group of the endomorphism ring $\mathcal{O}_E$.

2.2 $l = 2$

For $l = 2$ we have the problem of dealing with $\psi_{2^r}(X,Y)$, which cannot be reduced to a polynomial in X (Remark 2). Therefore in this case we use a different method for calculating the index $[\mathcal{O}_E : \mathbb{Z}[\pi]]$ described by Kohel [4]. The algorithm outlined here is based on the following theorem and can thus be used for arbitrary primes $l \neq 2$ as well.

Given that $\mathbb{Z}[\pi] \subseteq \mathcal{O}_E \subseteq \mathcal{O}_K$, by an l-isogeny $\varphi : E \longrightarrow E'$ up we mean that for $End(E') \cong \mathcal{O}_{E'}$

$$[\mathcal{O}_K : \mathcal{O}_E] = l \cdot [\mathcal{O}_K : \mathcal{O}_{E'}]. \tag{12}$$

An l-isogeny down describes the opposite situation

$$[\mathcal{O}_K : \mathcal{O}_{E'}] = l \cdot [\mathcal{O}_K : \mathcal{O}_E]. \tag{13}$$

Theorem 3 (Number of Isogenies [3,4]) *Let E/k be an ordinary elliptic curve having endomorphism ring $\mathcal{O}$ where $\mathbb{Z}[\pi] \subseteq \mathcal{O} \subseteq \mathcal{O}_K$ and D its discriminant. Let l be a prime number and $(\frac{D}{l})$ the Legendre symbol.*

1. *If $l \nmid [\mathcal{O}_K : \mathcal{O}]$ then the number of l-isogenies of elliptic curves not isomorphic to E with endomorphism ring equal to $\mathcal{O}$ is $1 + (\frac{D}{l})$.*
2. *If $l \mid [\mathcal{O}_K : \mathcal{O}]$ then there is one l-isogeny up to an elliptic curve having a smaller conductor than E.*
3. *If $l \nmid\mid [\mathcal{O} : \mathbb{Z}[\pi]]$ then there are no l-isogenies down to elliptic curves with bigger conductor than E.*
4. *If $l \mid [\mathcal{O}_K : \mathcal{O}]$ and $l \mid [\mathcal{O} : \mathbb{Z}[\pi]]$ then the number of l-isogenies down is l.*
5. *If $l \mid [\mathcal{O}_K : \mathcal{O}]$ and $l \nmid [\mathcal{O} : \mathbb{Z}[\pi]]$ then the number of l-isogenies down is $l - (\frac{D}{l})$.*

Therefore we have to find sequences of l-isogenies to other elliptic curves. The j-invariants $j(E)$ of these elliptic curves can be computed via modular equations or polynomials $\Phi_l(X,Y) \in \mathbb{Z}[X,Y]$ [5,10,4,6]. The most important property for us is that the j-invariants of elliptic curves l-isogenous to an elliptic curve E are roots of

$$\Phi_l(X, j(E)) = 0. \tag{14}$$

The algorithm is described here for $l = 2$ only because the coefficients of the modular polynomial get very large, but the method can be used for other primes as well. We assume that $l = 2$ divides the conductor m of $\mathbb{Z}[\pi]$. First we compute the j-invariant of the elliptic curve E and the modular polynomial.

$$\begin{aligned}\Phi_2(X,Y) = {} & Y^3 + (-X^2 + 1488X - 162000)Y^2 \\ & +(1488X^2 + 40773375X + 8748000000)Y \\ & +X^3 - 162000X^2 + 8748000000X - 157464000000000\end{aligned} \tag{15}$$

In the next step we factor $\Phi_2(X, j(E))$, which is of degree $l + 1 = 3$. If there is only one root $j(E')$, the isogeny to the elliptic curve E' is going up and we know that $l = 2$ does not divide the index $[\mathcal{O}_E : \mathbb{Z}[\pi]]$. If $\Phi_2(X, j(E))$ factors completely, $l = 2$ divides $[\mathcal{O}_E : \mathbb{Z}[\pi]]$ at least once.

Setting $E_0 = E$ and $E_1 = E'$ we then construct sequences of $j(E_r)$ with $j(E_{r+1})$ a root of $\Phi_2(X, j(E_r))$ until we find an r such that $\Phi_2(X, j(E_r))$ has exactly one root or until r exceeds the exponent of $l = 2$ in m.

Since we might pick an isogeny that is going up (if we are in case 2. of Theorem 3) or staying on the same level (if we are in case 1.), we have to take two probes of sequences with two distinct $j(E_1)$ in the beginning.

If r exceeds the exponent of $l = 2$ in the conductor m of $\mathbb{Z}[\pi]$ in both cases, we conclude that l does not divide $[\mathcal{O}_K : \mathcal{O}_E]$. Otherwise the smaller r is the exponent of l in $[\mathcal{O}_E : \mathbb{Z}[\pi]]$.

Algorithm 1 (Calculation of the exponent of $l = 2$ in the index $[\mathcal{O}_E : \mathbb{Z}[\pi]]$ if $l = 2$ divides the conductor of $\mathbb{Z}[\pi]$)
Input: *E elliptic curve over the finite field $\mathbb{F}_p$,*
k exponent of $l = 2$ in the conductor m of $\mathbb{Z}[\pi]$
Output: *$r \in \mathbb{N}$ such that $2^r \mid [\mathcal{O}_E : \mathbb{Z}[\pi]]$ and $2^{r+1} \nmid [\mathcal{O}_E : \mathbb{Z}[\pi]]$*

1 *Set*

$$\Phi_2(X,Y) := Y^3 + (-X^2 + 1488X - 162000)Y^2 \\ +(1488X^2 + 40773375X + 8748000000)Y \\ +X^3 - 162000X^2 + 8748000000X - 157464000000000$$

For $i = 1, 2$ do steps 2 to 12.

2 *Set stop := false; $j(E_0) := j(E)$; and $r_i := 0$;*

3 *Factor $\phi(X) := \Phi_2(X, j(E_0))$.*

if $\phi(X)$ has exactly one root **then**

4 *return 0;*

else

5 *Set $j(E_{1_i}) :=$ (i^{th} root of $\phi(X)$);*
#This assures that $j(E_{1_1}) \neq j(E_{1_2})$

6 *Set prev := $j(E_0)$;*

7 *Set $r_i := r_i + 1$;* **fi;**

repeat

8 *Factor $\phi(X) := \Phi_2(X, j(E_{r_i}))$.*

if $\phi(X)$ has exactly one root **then**

9 *Set stop := true;*

else

10 *Set* $j(E_{r_i+1}) :=$ *(root of* $\phi(X)$ *which is* $\neq$ *prev);*

11 *Set* $prev := j(E_{r_i})$;

12 *Set* $r_i := r_i + 1$; **fi;**

until $r_i > k$ *or stop;*

if $r_1 > k$ and $r_2 > k$ **then**

13 *return* k; *#case 1. of Theorem 3*

else

14 *return* $Minimum(r_1, r_2)$; **fi;**

2.3 Complete algorithm

In the following we present the complete algorithm for determining the conductor c of the endomorphism ring of an elliptic curve $E : Y^2 = X^3 + AX + B$ over a finite field $\mathbb{F}_p$ of characteristic $\neq 2, 3$.

Algorithm 2 (Calculation of the conductor of the endomorphism ring)
Input: *E elliptic curve over finite field* $\mathbb{F}_p$
Output: $c \in \mathbb{N}$ *such that* $End(E) \cong \mathcal{O}_E = \mathbb{Z} + c \cdot \mathcal{O}_K$

1 *Calculate the trace t of the Frobenius endomorphism, for example with the SEA-algorithm* [7,8].

2 *Calculate the discriminant* D_K *of the maximal order* $\mathcal{O}_K \supseteq \mathcal{O}_E$.

3 *Set the discriminant of the order* $\mathbb{Z}[\pi]$ *to* $d := t^2 - 4p$ *via the Frobenius endomorphism* π *of E.*

4 *Set* $m := \sqrt{d/D_K}$; *and* $a := t/2$ *if* D_K *mod* $4 \equiv 0$ *or* $a := (t+m)/2$ *if* D_K *mod* $4 \equiv 1$;

5 *Set* $j := 1$; $c := 1$; *and factorize m.*

if $2|m$ **then**

6 *Compute the exponent r of* 2 *in* $[\mathcal{O}_E : \mathbb{Z}[\pi]]$ *with Algorithm 1.*

7 *Set* $c := c * 2^r$; **fi;**

if no odd primes divide m **then**

8 *return* c; **fi;**

repeat

9 *Set* $l :=$ *(*j^{th} *odd prime dividing* m*);* $\tilde{a} := a$ *mod* l;
$k :=$ *(exponent of* l *in* m*) and* $r := 0$;

repeat

10 $r := r + 1$;

11 *Compute:*

$$h(X) \equiv (X^p - X) \cdot \psi_{\tilde{a}}^2(X) + \psi_{\tilde{a}+1} \cdot \psi_{\tilde{a}-1}(X) \ \mathit{mod}\ \psi_{l^r}(X)$$

if $h(X) \neq 0$ **then**

12 *Set* $r := r - 1$; *and stop:=true;* **fi;**

until $r = k$ *or stop;*

13 $c := c * l^r$; # *c is the index* $[\mathcal{O}_E : \mathbb{Z}[\pi]]$.

14 $j := j + 1$;

until $l =$ *maximal odd prime dividing* m*;*

15 $c := m/c$; # *We are looking for* $[\mathcal{O}_K : \mathcal{O}_E]$

16 *return* c;

3 Computations with KANT V4

The following sample computations are done with the programming shell KASH of the computer algebra system KANT V4 [a].

Let $p = 277177626184296644424795796972184155491964623691$ and

$$E : Y^2 = X^3 + AX + B$$

over $\mathbb{F}_p$ with $A = 272541518603739663048376297783171386047488081741$ and $B = 155915327533246157338971654070259643256587149 9$. First we initialize the elliptic curve in KASH, and set the preliminaries of Algorithm 2.

[a] You can obtain the KANT V4 [1]- package at:
ftp://ftp.math.tu-berlin.de/pub/algebra/Kant/Kash/

```
kash> p:=277177626184296644424795796972184155491964623691;;
kash> A:=272541518603739663048376297783171386047488081741;;
kash> B:=155915327533246157338971654070259643256587 1499;;
kash> E:=EccInit(p,A,B);;
kash> t:=Trace(E);
-515828322256210498084 12
kash> d:=t^2-4*p;
-1106049716156770008232416433006518693385952533020
kash> D:=Disc(OrderMaximal(x^2-d));
-3413733691841882741458075410513946584524544855
kash> m:=IntRoot(d/D,2);
18
kash> a:=(t+m)/2;
-257914161128105249041 97
```

Next we factor m and since $m = (2 \cdot 3^2)$ we calculate the division polynomials up to the 9th polynomial.

```
kash> fac:=Factor(m);
[ [ 2, 1 ], [ 3, 2 ] ]
kash> DP:=EccDivisionPolys(E,9);;
```

We now do the calculation for $l = 2$.

```
kash> Zy:=PolyAlg(Zx);
Univariate Polynomial Ring in y over Univariate Polynomial Ring
in x over Integer Ring
kash>   PHI_2:=Poly(Zy,List([ [ 1 ], [ -1, 1488, -162000 ],
>   [ 1488, 40773375, 8748000000 ],
>   [ 1, -162000, 8748000000, -157464000000000 ] ] ,
>   x->Poly(Zx,x)));
y^3 + (-x^2 + 1488*x - 162000)*y^2 + (1488*x^2 + 40773375*x + 8\
748000000)*y + x^3 - 162000*x^2 + 8748000000*x - 157464000000000
kash> jinv:=EccJInv(E);
60368129946798606410901010093013026440262665943
kash> phi:=Eval(PHI_2,jinv);
x^3 + 246001087080537287287417208067725711778420336339*x^2 + 15\
560335799662549095819111662059674254829349791 4*x + 254370737584\
915898614371952293314118463679262623
kash> fac:=Factor(phi);
[ [ x + 135762359057552848562112296440686912015063305923, 1 ],
  [ x + 188365304178092657449871881545799381460179438904, 1 ],
  [ x + 199051050029188425700228827053423573795142215203, 1 ] ]
```

Since $\phi = \Phi_2(X, J(E))$ factors completely, we conclude that there is at least one 2-isogeny down to an elliptic curve E' thus

$$[\mathcal{O}_K : \mathcal{O}_{E'}] = 2 \cdot [\mathcal{O}_K : \mathcal{O}_E], \tag{16}$$

and therefore 2 does not divide the conductor c of $\mathcal{O}_E$.

Calculating $h_1(X)$ and $h_2(X)$ of Eq. (11), we find that $h_1(X) = 0$ and $h_2(X) \neq 0$ for $l = 3$ which means that $(\pi - a)$ is zero on $\mathcal{O}_E/3\mathcal{O}_E$ but not on $\mathcal{O}_E/9\mathcal{O}_E$ thus the conductor c of $\mathcal{O}_E$ is equal to 3.

References

1. M. Daberkow, C. Fieker, J. Klüners, M. Pohst, K. Roegner, M. Schörnig, K. Wildanger, KANT V4, J. Symb. Comp. Vol. 24 **3**, 267-283 (1997).
2. M. Deuring, *Die Typen der Multiplikatorenringe elliptischer Funktionenkörper*, Abh. Math. Sem. Hamburg **14**, 197-272 (1941).
3. S.D. Galbraith, *Constructing Isogenies between Elliptic Curves over Finite Fields*, LMS J. Comput. Math. **2**, 118-138 (1999).
4. D. Kohel, *Endomorphism rings of elliptic curves over finite fields*, Berkeley PhD thesis, 1996.
5. S. Lang, *Elliptic Functions*, Springer, New York (1973).
6. F. Morain, *Calcul du nombre de points sur une courbe elliptique dans un corps fini: aspects algorithmiques*, J. Théorie des Nombres de Bordeaux **7**, 219-254 (1995).
7. V. Müller, *Ein Algorithmus zur Bestimmung der Punktanzahl elliptischer Kurven über endlichen Körpern der Charakteristik größer drei*, Saarbrücken PhD thesis, 1995.
8. R. Schoof, *Counting points on elliptic curves over finite fields*, J. Théorie des Nombres de Bordeaux **7**, 219-254 (1995).
9. J.H. Silverman, *The Arithmetic of Elliptic Curves*, Springer, New York (1986).
10. J.H. Silverman, *Advanced Topics in the Arithmetic of Elliptic Curves*, Springer, New York (1994).

ON DISTRIBUTION OF THE GROUP OF RATIONAL POINTS OF REDUCTIONS OF AN ELLIPTIC CURVE

RYOUHEI TAKEUCHI

Department of Mathmatics, Tokyo Metropolitan University, Tokyo, Japan
E-mail: ryouhei@comp.metro-u.ac.jp

Let E be an elliptic curve over $\mathbb{Q}$ and $\bar{E}(\mathbb{F}_p)$ the group of rational points on the reduction of E modulo p for a prime p where E has good reduction. It is well-known that $\bar{E}(\mathbb{F}_p)$ is isomorphic to a direct product $\mathbb{Z}/n_1\mathbb{Z} \times \mathbb{Z}/n_2\mathbb{Z}$ of at most 2 cyclic groups where $n_1|n_2$. When p runs through good primes of a fixed elliptic curve E, the expectation of n_1 behaves differently according as E has complex multiplication or not. In the former case, it seems to diverge to infinity and in the latter, it seems to converge to a finite constant. This paper gives some theoretical reasoning and some supporting facts by computer experiments for the phenomena.

1 Introduction

We make a study of distribution of the group of rational points of reductions of an elliptic curve. Let E be an elliptic curve over $\mathbb{Q}$ and $\bar{E}(\mathbb{F}_p)$ the group of rational points on the reduction of E modulo p for a prime p where E has good reduction. It is well-known that $\bar{E}(\mathbb{F}_p)$ is isomorphic to a direct product $\mathbb{Z}/n_1\mathbb{Z} \times \mathbb{Z}/n_2\mathbb{Z}$ of at most 2 cyclic groups where $n_1|n_2$ in what follows we denote the n_1 by $n_1(E,p)$. In this paper, we concern a simple question: How large the expectation of n_1 should be if p runs through good primes of a fixed elliptic curve E. Namely, we investigate

$$\mathcal{E}(E) := \lim_{x\to\infty} \frac{\sum'_{p\leq x} n_1(E,p)}{\sum'_{p\leq x} 1} = \lim_{x\to\infty} \frac{\sum'_{p\leq x} n_1(E,p)}{\pi_E(x)},$$

where $\pi_E(x)$ is the number of good primes less than or equal to x and $\sum'$ means the summation over good primes of E. We conjecturally suppose that this limit should exist. We further propose three conjectures.

Conjecture 1. *If E has complex multiplication (CM) then $\mathcal{E}(E)$ diverges to infinity and if not it converges to a certain finite constant.*

We shall give some heuristics and discussions for the conjecture in the follow sections on the basis of Hooley's method [Ho67] applied to elliptic curves.

2 Heuristics for $\mathcal{E}(E)$

If $n_1(E,p)$ is equal to 1 then $\bar{E}(\mathbb{F}_p)$ is a cyclic group. Serre [Se77] and R. Murty [Mu83] studied how often $\bar{E}(\mathbb{F}_p)$ was cyclic for a given E and obtained the following result using Hooley's method.

Theorem (Serre, R. Murty). *Let μ be the Möbius function. Assuming the generalized Riemann hypothesis (GRH), one has*

$$\sharp\{\, p \le x \mid \bar{E}(\mathbb{F}_p) \textit{ is cyclic}\,\} \sim \delta_E \cdot \frac{x}{\log x},$$

where $\delta_E := \sum_{k=1}^{\infty}(\mu(k)/[\mathbb{Q}(E[k]) : \mathbb{Q}])$ and $E[k]$ is the set of k-th division points of E.

We shall consider a generalization of this statement. Let $\sharp N_E^{(t)}(x)$ be the number of those good primes p which are less than or equal to x and satisfy $n_1(E,p) = t$. Then it is natural to propose the following conjecture.

Conjecture 2. *One has*

$$\sharp N_E^{(t)}(x) \sim \delta_E^{(t)} \cdot \frac{x}{\log x}.$$

If $t = 1$ then Conjecture 2 is true under GRH by the theorem above. We shall show some supporting facts for Conjecture 2 by a computer experiment in Section 4. Since $\delta_E^{(t)}$ may be considered as the probability of $n_1(E,p) = t$ for randomly chosen p, we may give the expectation $\mathcal{E}(E)$ of n_1 by the following formula.

Conjecture 3. *Under Conjecture 2, one has*

$$\mathcal{E}(E) = \sum_{t=1}^{\infty} t \cdot \delta_E^{(t)}.$$

If E has CM then $\sum_{t=1}^{\infty} t \cdot \delta_E^{(t)}$ diverges to infinity and if not $\sum_{t=1}^{\infty} t \cdot \delta_E^{(t)}$ converges to the finite constant.

Our reasoning of the latter half of this conjecture will be explained in the next page. We shall show some supporting facts for Conjecture 3 by a computer experiment in Section 5. If the formula is true, we can obtain $\mathcal{E}(E)$ by computing $\delta_E^{(t)}$. Hence we investigate $\delta_E^{(t)}$ by using Hooley's method. Put

$$\sharp P_E^{(t)}(x) := \sharp\{\, p \le x \mid \bar{E}(\mathbb{F}_p) \text{ contains a } (t,t)\text{-type abelian group}\,\}.$$

Using an inclusion-exclusion argument [Mu83, Section 2], we see

$$\sharp N_E^{(t)}(x) = \sum_{k=1}^{\infty} \mu(k) \cdot \sharp P_E^{(kt)}(x).$$

We shall prepare the following lemma.

Lemma ([Mu83, Lemma 2]). *Suppose that E has good reduction at p. Then p splits completely in the extension $\mathbb{Q}(E[t])$ of $\mathbb{Q}$ if and only if $\bar{E}(\mathbb{F}_p)$ contains a (t,t)-type abelian group.*

It follows from this lemma that

$$\sharp P_E^{(t)}(x) = \sharp\{\, p \le x \mid p \text{ splits completely in } \mathbb{Q}(E[t])\,\},$$

and hence from Chebotarev's density theorem that

$$\sharp P_E^{(t)}(x) \sim [\mathbb{Q}(E[t]):\mathbb{Q}]^{-1}\cdot \frac{x}{\log x}.$$

Therefore we can refine Conjecture 2.

Conjecture 2′ (strict version). *One has*

$$\sharp N_E^{(t)}(x) \sim \delta_E^{(t)}\cdot\frac{x}{\log x}, \qquad \textit{where}\ \ \delta_E^{(t)} = \sum_{k=1}^{\infty}\frac{\mu(k)}{[\mathbb{Q}(E[kt]):\mathbb{Q}]}.$$

This is a natural generalization of Serre and Murty's theorem. If Conjecture 2′ and Conjecture 3 are true, we obtain

$$\mathcal{E}(E) = \sum_{t=1}^{\infty}\sum_{k=1}^{\infty}\frac{\mu(k)\cdot t}{[\mathbb{Q}(E[kt]):\mathbb{Q}]}.$$

We recall

$$[\mathbb{Q}(E[n]):\mathbb{Q}] \asymp \sharp(\mathcal{O}_K/n\mathcal{O}_K)^{\times},$$

if E has CM by an order $\mathcal{O}_K$ of an imaginary quadratic field K. If E does not have CM, Serre [Se72] gave

$$[\mathbb{Q}(E[n]):\mathbb{Q}] \asymp \sharp\mathrm{GL}_2(\mathbb{Z}/n\mathbb{Z}).$$

So each degree $[\mathbb{Q}(E[kt]):\mathbb{Q}]$ decomposes into the product of $[\mathbb{Q}(E[k]):\mathbb{Q}]$ and $[\mathbb{Q}(E[t]):\mathbb{Q}]$ asymptotically. Thus we have

$$\begin{aligned}
\mathcal{E}(E) &\asymp \sum_t\left(\frac{t}{[\mathbb{Q}(E[t]):\mathbb{Q}]}\sum_k\frac{\mu(k)}{[\mathbb{Q}(E[k]):\mathbb{Q}]}\right)\\
&\asymp C\times\sum_t\frac{t}{[\mathbb{Q}(E[t]):\mathbb{Q}]} \qquad (C:\text{constant})\\
&\asymp \begin{cases} C\times\displaystyle\sum_t\frac{t}{\sharp(\mathcal{O}_K/t\mathcal{O}_K)^{\times}} = \infty & (E:\text{CM}),\\[2ex] C\times\displaystyle\sum_t\frac{t}{\sharp\mathrm{GL}_2(\mathbb{Z}/t\mathbb{Z})} < \infty & (E:\text{non-CM}).\end{cases}
\end{aligned}$$

This is what we state in Conjecture 1.

3 Supporting facts for Conjecture 1

In this section, we describe the results of a computer experiment on Conjecture 1 for some elliptic curves. We employed PARI-GP [PARI] and implemented Miller's algorithm [Mi86] (the generalized algorithm for hyperelliptic curves is in [OS92]) for computing the group structure of $\bar{E}(\mathbb{F}_p)$.

3.1 CM case

The results of a computer experiment on Conjecture 1 for three CM curves E appear in Examples 1∼3. The three curves are $E : y^2 = x^3 + 17$ (Example 1) with CM by $\mathbb{Z}[\rho]$ where ρ is a primitive cubic root of unity, $E : y^2 = x^3 - 5x$ (Example 2) with CM by $\mathbb{Z}[i]$ where i is a primitive quartic root of unity and $E : y^2 = x^3 - 120x + 448$ (Example 3) with CM by $\mathbb{Z}[\sqrt{-2}]$. We observe from these examples that $\mathcal{E}(E)$ seems to diverge to infinity if E has CM.

3.2 non-CM case

The results of a computer experiment on Conjecture 1 for three non-CM curves E appear in Examples 4∼6. The three curves are $E : y^2 + y = x^3 - x^2$ (Example 4) with conductor 11, $E : y^2 + y = x^3 - x$ (Example 5) with conductor 37 and $E : y^2 + y = x^3 + x^2$ (Example 6) with conductor 43. We observe from these examples that $\mathcal{E}(E)$ seems to converge to a finite constant depending on E if E does not have CM.

4 Supporting facts for Conjecture 2

To check Conjecture 2′ by a computer experiment we need to have the degree $[\mathbb{Q}(E[n]) : \mathbb{Q}]$ explicitly. It is difficult to compute the degree $[\mathbb{Q}(E[n]) : \mathbb{Q}]$ explicitly in general. For some curves, however, we can compute the degrees explicitly.

4.1 CM case

In [Gu90], Gupta determined the degree $[\mathbb{Q}(E[n]) : \mathbb{Q}]$ for the elliptic curve $y^2 = x^3 - ax$ with CM by $\mathbb{Z}[i]$. By using Gupta's method, the author [T] determined the degree for Mordell's elliptic curve $y^2 = x^3 - a$ with CM by $\mathbb{Z}[\rho]$ where ρ is a primitive cubic root of unity.

Theorem 1. *Let E be the elliptic curve defined by $y^2 = x^3 - a$ with $a \in \mathbb{Z}_{\neq 0}$ and suppose $|a| = p_1 \cdots p_g$ (distinct primes) where $p_i \equiv 1, 5, 11 \pmod{12}$ for*

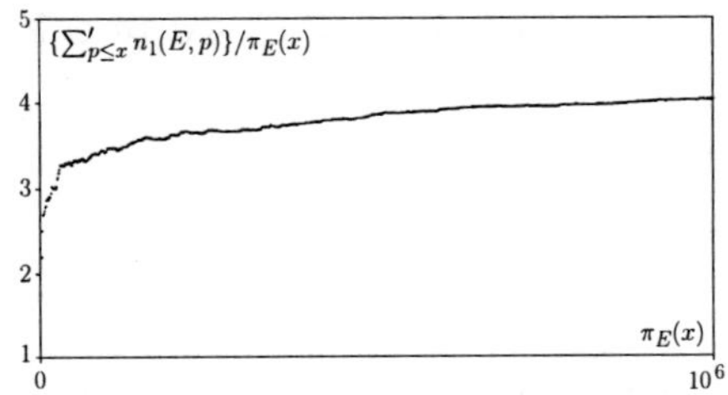

Example 1. $y^2 = x^3 + 17$ (CM)

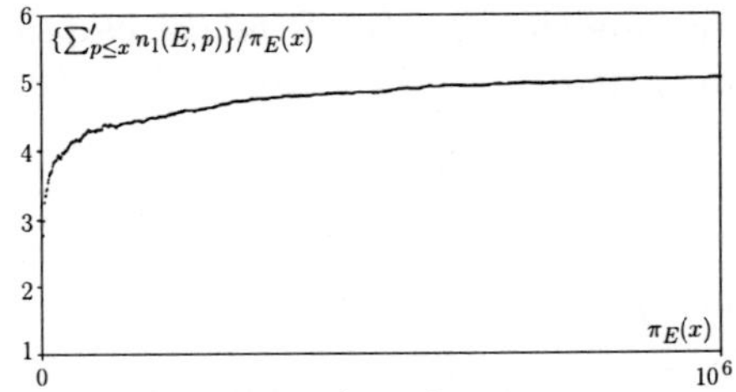

Example 2. $y^2 = x^3 - 5x$ (CM)

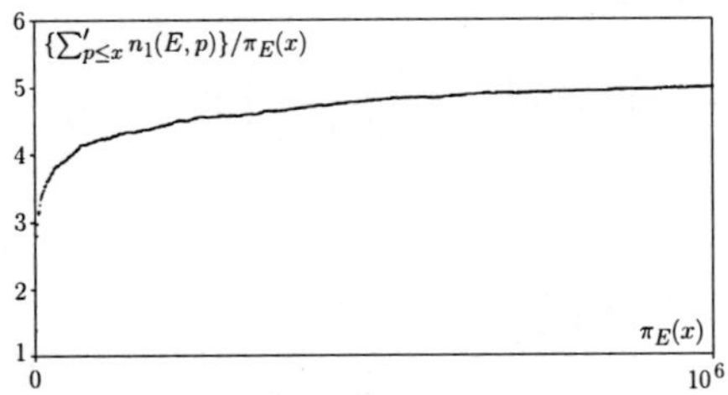

Example 3. $y^2 = x^3 - 120x + 448$ (CM)

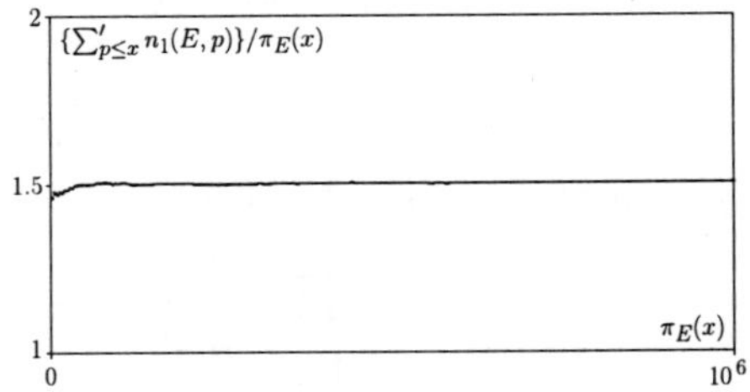

Example 4. $y^2 + y = x^3 - x^2$ (non-CM)

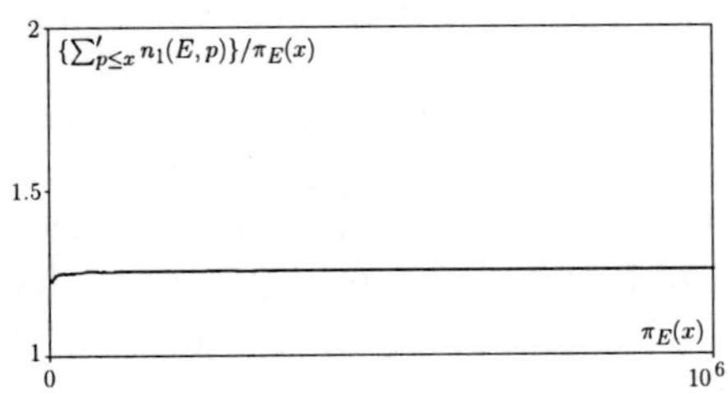

Example 5. $y^2 + y = x^3 - x$ (non-CM)

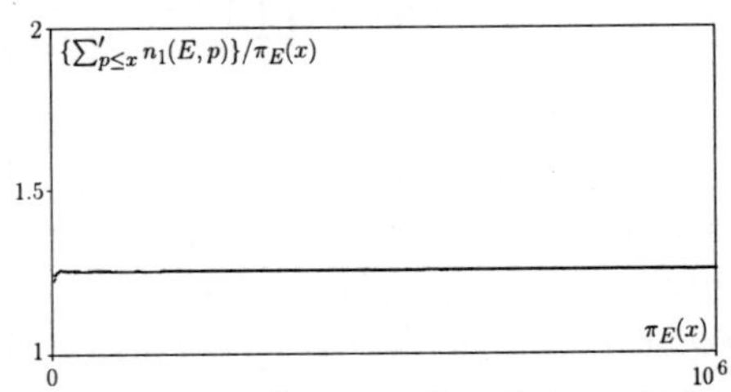

Example 6. $y^2 + y = x^3 + x^2$ (non-CM)

all i. For $n \geq 2$, we have

$$[\mathbb{Q}(E[n]) : \mathbb{Q}] = \begin{cases} 2\varphi_{\mathbb{Q}(\rho)}(n), & \text{if } 3f_2(a) \nmid n \text{ and } 2f_3(a) \nmid n, \\ \varphi_{\mathbb{Q}(\rho)}(n), & \text{if } 3f_2(a) \mid n \text{ and } 2f_3(a) \nmid n, \\ 2\varphi_{\mathbb{Q}(\rho)}(n)/3, & \text{if } 3f_2(a) \nmid n \text{ and } 2f_3(a) \mid n, \\ \varphi_{\mathbb{Q}(\rho)}(n)/3, & \text{if } 3f_2(a) \mid n \text{ and } 2f_3(a) \mid n, \end{cases}$$

where $\varphi_{\mathbb{Q}(\rho)}(n) := \sharp(\mathbb{Z}[\rho]/n\mathbb{Z}[\rho])^\times$ (Euler function of $\mathbb{Q}(\rho)$) and $f_2(a)$, $f_3(a)$ are the conductors of $\mathbb{Q}(\rho, \sqrt{3a})/\mathbb{Q}(\rho)$, $\mathbb{Q}(\rho, \sqrt[3]{a})/\mathbb{Q}(\rho)$ respectively.

This theorem enable us to compute $\sum_{k=1}^{\infty}\{\mu(k)/[\mathbb{Q}(E[kt]) : \mathbb{Q}]\}$ for some Mordell's curves E. For simplicity, we take a such that $|a|$ is prime and $a \equiv 19, 35 \pmod{36}$. Then the conductors $f_2(a)$ and $f_3(a)$ are equal to $|a|$. (The computation of $f_2(a)$ and $f_3(a)$ easy in general.) The infinite sum $\sum_{k=1}^{\infty}\{\mu(k)/[\mathbb{Q}(E[kt]) : \mathbb{Q}]\}$ is expressed by an Euler product.

Proposition 1. *Let E be the elliptic curve defined by $y^2 = x^3 - a$ ($a \in \mathbb{Z}$, $|a|$: prime, $a \equiv 19, 35 \pmod{36}$). Then*

$$\sum_{k=1}^{\infty} \frac{\mu(k)}{[\mathbb{Q}(E[k]) : \mathbb{Q}]} = \frac{1}{2} + \frac{1}{2} \cdot \{1 - v(|a|)\} \cdot C_{\mathbb{Q}(\rho)}$$

and, for $t \geq 2$,

$$\sum_{k=1}^{\infty} \frac{\mu(k)}{[\mathbb{Q}(E[kt]) : \mathbb{Q}]} = \frac{u(t)}{2t^2} \cdot \left\{1 + v\left(\frac{3|a|}{\gcd(3|a|, t)}\right)\right.$$
$$\left. +2v\left(\frac{2|a|}{\gcd(2|a|, t)}\right) + 2v\left(\frac{6|a|}{\gcd(6|a|, t)}\right)\right\} \cdot C_{\mathbb{Q}(\rho)},$$

where $v(1) := 1$ and, for $n \geq 2$,

$$u(n) := \prod_{p|n} \frac{p^2 - 1}{\varphi_{\mathbb{Q}(\rho)}(p) - 1}, \quad v(n) := \prod_{p|n} \frac{-1}{\varphi_{\mathbb{Q}(\rho)}(p) - 1},$$

$$C_{\mathbb{Q}(\rho)} := \prod_{p} \left(1 - \frac{1}{\varphi_{\mathbb{Q}(\rho)}(p)}\right) = 0.5011125958 \cdots .$$

From this proposition, we get an elementary formula of the natural density of $\sharp N_E^{(t)}(x)$ under Conjecture 2′. For example, we take $a = -17$, then a satisfies the condition in Proposition 1. We see from Table 1 that the conjectural density $\sum_{k=1}^{\infty}\{\mu(k)/[\mathbb{Q}(E[kt]) : \mathbb{Q}]\}$ of Proposition 1 is very close to the experimental density $\sharp N_E^{(t)}(x)/\pi_E(x)$ for $E : y^2 = x^3 + 17$ and $t \leq 10$.

4.2 non-CM case

In [Se72], Serre determined the degree $[\mathbb{Q}(E[n]) : \mathbb{Q}]$ for the non-CM curves $E_{37} : y^2 + y = x^3 - x$, $E_{43} : y^2 + y = x^3 + x^2$ with conductors 37, 43 respectively.
Theorem (Serre). *Let E_{37} and E_{43} be elliptic curves defined by $y^2 + y = x^3 - x$ and $y^2 + y = x^3 + x^2$ respectively. For $M = 37, 43$, we have*

$$[\mathbb{Q}(E_M[n]) : \mathbb{Q}] = \begin{cases} \sharp \mathrm{GL}_2(\mathbb{Z}/n\mathbb{Z}), & \text{if } 2M \nmid n, \\ \sharp \mathrm{GL}_2(\mathbb{Z}/n\mathbb{Z})/2, & \text{if } 2M \mid n. \end{cases}$$

This theorem enable us to compute $\sum_{k=1}^{\infty}\{\mu(k)/[\mathbb{Q}(E_M[kt]) : \mathbb{Q}]\}$ for $M = 37, 43$. The infinite sum is also expressed by an Euler product.
Proposition 2. *Let E_{37} and E_{43} be elliptic curves defined by $y^2 + y = x^3 - x$ and $y^2 + y = x^3 + x^2$ respectively. For $M = 37, 43$, we have*

$$\sum_{k=1}^{\infty} \frac{\mu(k)}{[\mathbb{Q}(E_M[kt]) : \mathbb{Q}]} = \frac{u(t)}{t^4} \cdot \left\{1 + v\left(\frac{2M}{\gcd(2M, t)}\right)\right\} \cdot C_{\text{non-CM}}$$

Table 1. Experimental data for $E : y^2 = x^3 + 17$ and its conjectural density

$\pi_E(x)$	$\frac{\sharp N_E^{(1)}(x)}{\pi_E(x)}$	$\frac{\sharp N_E^{(2)}(x)}{\pi_E(x)}$	$\frac{\sharp N_E^{(3)}(x)}{\pi_E(x)}$	$\frac{\sharp N_E^{(4)}(x)}{\pi_E(x)}$	$\frac{\sharp N_E^{(5)}(x)}{\pi_E(x)}$
10^3	0.7600000	0.1000000	0.0440000	0.0220000	0.0140000
10^4	0.7520000	0.0944000	0.0454000	0.0236000	0.0116000
10^5	0.7512000	0.0931700	0.0453500	0.0227900	0.0104400
10^6	0.7517980	0.0936330	0.0446220	0.0232880	0.0104470
10^7	0.7514785	0.0935087	0.0446906	0.0233559	0.0104908
Prop. 1	0.7514293	0.0935002	0.0446985	0.0233750	0.0104944

$\pi_E(x)$	$\frac{\sharp N_E^{(6)}(x)}{\pi_E(x)}$	$\frac{\sharp N_E^{(7)}(x)}{\pi_E(x)}$	$\frac{\sharp N_E^{(8)}(x)}{\pi_E(x)}$	$\frac{\sharp N_E^{(9)}(x)}{\pi_E(x)}$	$\frac{\sharp N_E^{(10)}(x)}{\pi_E(x)}$
10^3	0.0110000	0.0060000	0.0060000	0.0050000	0.0050000
10^4	0.0163000	0.0062000	0.0061000	0.0045000	0.0038000
10^5	0.0163800	0.0071000	0.0060400	0.0045200	0.0038300
10^6	0.0164530	0.0069950	0.0058140	0.0049780	0.0038370
10^7	0.0164071	0.0070517	0.0058688	0.0049678	0.0038965
Prop. 1	0.0164127	0.0070370	0.0058437	0.0049665	0.0039026

where $u(1) := 1, v(1) := 1$ *and, for* $n \geq 2$,

$$u(n) := \prod_{p|n} \frac{p^4-1}{(p^2-p)(p^2-1)-1}, \quad v(n) := \prod_{p|n} \frac{-1}{(p^2-p)(p^2-1)-1},$$

$$C_{\text{non-CM}} := \prod_p \left(1 - \frac{1}{(p^2-p)(p^2-1)}\right) = 0.8137519061\cdots.$$

From this proposition, we get an elementary formula of the natural density of $\sharp N_{E_M}^{(t)}(x)$ under Conjecture 2′ for $M = 37, 43$. We see from Tables 2 and 3 that the conjectural density $\sum_{k=1}^{\infty}\{\mu(k)/[\mathbb{Q}(E_M[kt]) : \mathbb{Q}]\}$ of Proposition 2 is very close to the experimental density $\sharp N_{E_M}^{(t)}(x)/\pi_{E_M}(x)$ for $M = 37$, 43 and $t \leq 5$.

5 Supporting facts for Conjecture 3

In this section, we check Conjecture 3 for the curves treated in Section 4 by comparing the experimental expectation $\{\sum_{p\leq x}' n_1(E,p)\}/\pi_E(x)$ with the conjectural expectation $\sum_{t=1}^{\infty}\sum_{k=1}^{\infty}\{\mu(k)\cdot t/[\mathbb{Q}(E[kt]) : \mathbb{Q}]\}$.

Table 2. Experimental data for $E_{37} : y^2 + y = x^3 - x$ and its conjectural density

$\pi_{E_{37}}(x)$	$\frac{\sharp N_{E_{37}}^{(1)}(x)}{\pi_{E_{37}}(x)}$	$\frac{\sharp N_{E_{37}}^{(2)}(x)}{\pi_{E_{37}}(x)}$	$\frac{\sharp N_{E_{37}}^{(3)}(x)}{\pi_{E_{37}}(x)}$	$\frac{\sharp N_{E_{37}}^{(4)}(x)}{\pi_{E_{37}}(x)}$	$\frac{\sharp N_{E_{37}}^{(5)}(x)}{\pi_{E_{37}}(x)}$
10^3	0.8270000	0.1460000	0.0110000	0.0030000	0.0030000
10^4	0.8175000	0.1525000	0.0154000	0.0081000	0.0015000
10^5	0.8142500	0.1525800	0.0168900	0.0093500	0.0017500
10^6	0.8140050	0.1524970	0.0170300	0.0095070	0.0016700
10^7	0.8138201	0.1525663	0.0171131	0.0095453	0.0016726
Prop. 2	0.8137519	0.1525783	0.0171001	0.0095361	0.0016961

Table 3. Experimental data for $E_{43} : y^2 + y = x^3 + x^2$ and its conjectural density

$\pi_{E_{43}}(x)$	$\frac{\sharp N_{E_{43}}^{(1)}(x)}{\pi_{E_{43}}(x)}$	$\frac{\sharp N_{E_{43}}^{(2)}(x)}{\pi_{E_{43}}(x)}$	$\frac{\sharp N_{E_{43}}^{(3)}(x)}{\pi_{E_{43}}(x)}$	$\frac{\sharp N_{E_{43}}^{(4)}(x)}{\pi_{E_{43}}(x)}$	$\frac{\sharp N_{E_{43}}^{(5)}(x)}{\pi_{E_{43}}(x)}$
10^3	0.8260000	0.1450000	0.0110000	0.0090000	0.0020000
10^4	0.8155000	0.1525000	0.0162000	0.0084000	0.0014000
10^5	0.8153900	0.1515700	0.0167800	0.0094900	0.0015000
10^6	0.8142310	0.1523450	0.0170700	0.0094250	0.0016880
10^7	0.8137796	0.1526011	0.0170759	0.0095215	0.0017170
Prop. 2	0.8137519	0.1525784	0.0171001	0.0095361	0.0016961

5.1 CM case

Let us see Mordell's curve $E : y^2 = x^3 + 17$ treated in Section 4.1. We see the experimental expectation $\{\sum'_{p\le x} n_1(E,p)\}/\pi_E(x)$ and the partial sum $\sum_{t=1}^{n}\sum_{k=1}^{\infty}\{\mu(k)\cdot t/[\mathbb{Q}(E[kt]) : \mathbb{Q}]\}$ in Table 4. The latter grows steadily as $\log n$. The former also grows proportionally as $\log \pi_E(x)$ for $\pi_E(x) \ge 10^3$.

5.2 non-CM case

Let us see the non-CM curves E_M for $M = 37$, 43 treated in Section 4.2. We see the experimental expectation $\{\sum'_{p\le x} n_1(E_M,p)\}/\pi_{E_M}(x)$ and the partial sum $\sum_{t=1}^{n}\sum_{k=1}^{\infty}\{\mu(k)\cdot t/[\mathbb{Q}(E_M[kt]) : \mathbb{Q}]\}$ in Tables 5 and 6. We see from these tables that each experimental and conjectural expectation seems to converge to the same finite constant.

Table 4. Experimental data for $E : y^2 = x^3 + 17$ and its conjectural expectation

$\pi_E(x)$	$\frac{\sum'_{p\le x} n_1(E,p)}{\pi_E(x)}$
10	1.200000000000
10^2	2.140000000000
10^3	2.213000000000
10^4	2.889200000000
10^5	3.466990000000
10^6	4.041835000000
10^7	4.602437200000

n	$\sum_{t=1}^{n}\sum_{k=1}^{\infty} \frac{\mu(k)\cdot t}{[\mathbb{Q}(E[kt]):\mathbb{Q}]}$
10	1.49670898595122260387263
10^2	2.40928798179478974466901
10^3	3.43728084256254497997534
10^4	4.45758426419407094124463
10^5	5.47744420715163944041556
10^6	6.49755807729855826445696
10^7	7.51769207296337122675677

Table 5. Experimental data for $E_{37} : y^2 + y = x^3 - x$ and its conjectural expectation

$\pi_{E_{37}}(x)$	$\frac{\sum'_{p\le x} n_1(E_{37},p)}{\pi_{E_{37}}(x)}$
10	1.000000000000
10^2	1.120000000000
10^3	1.225000000000
10^4	1.245600000000
10^5	1.255150000000
10^6	1.257112000000
10^7	1.257933800000

n	$\sum_{t=1}^{n}\sum_{k=1}^{\infty} \frac{\mu(k)\cdot t}{[\mathbb{Q}(E_{37}[kt]):\mathbb{Q}]}$
10	1.24874616383345875954423
10^2	1.25833967078234025179299
10^3	1.25845117524362018020124
10^4	1.25845231689297071290221
10^5	1.25845232831824828224338
10^6	1.25845232843254687022586
10^7	1.25845232843368988206028

Table 6. Experimental data for $E_{43} : y^2 + y = x^3 + x^2$ and its conjectural expectation

$\pi_{E_{43}}(x)$	$\frac{\sum'_{p\le x} n_1(E_{43},p)}{\pi_{E_{43}}(x)}$
10	1.000000000000
10^2	1.120000000000
10^3	1.228000000000
10^4	1.255900000000
10^5	1.254230000000
10^6	1.256924000000
10^7	1.258381500000

n	$\sum_{t=1}^{n}\sum_{k=1}^{\infty} \frac{\mu(k)\cdot t}{[\mathbb{Q}(E_{43}[kt]):\mathbb{Q}]}$
10	1.24874621238032271100809
10^2	1.25833866406877360568726
10^3	1.25844973378888600894793
10^4	1.25845087277020929899180
10^5	1.25845088416680799253277
10^6	1.25845088428079668832472
10^7	1.25845088428193658705863

6 Concluding remarks

We expect that it may be possible to prove Conjecture 2′ under GRH by using Murty's theorem in [Mu83]. If we want to prove Conjecture 3, however, we need some more detailed discussion (of course, for Conjecture 1 too).

It does not seem easy to study the distribution of $n_2(E,p) := n_2$ in $\bar{E}(\mathbb{F}_p) \simeq \mathbb{Z}/n_1\mathbb{Z} \times \mathbb{Z}/n_2\mathbb{Z}$ $(n_1|n_2)$. A reason is because this problem has the same difficulty as the Frobenius distribution problem for elliptic curves (that is, the Lang-Trotter conjecture [LT76]). Though we did not study n_2 in this paper, it may be one of the very interesting problems for elliptic curve cryptosystems.

References

[Gu90] R. Gupta, Division Fields of $Y^2 = X^3 - aX$. J. Number Theory **34**, 335-345 (1990)

[Ho67] C. Hooley, On Artin's conjecture. J. reine angew. Math. **225**, 209-220 (1967)

[LT76] S. Lang, H. Trotter, Frobenius distributions in GL_2-extensions. Lecture Notes in Math. **504** (1976)

[Mi86] V. Miller, Short programs for functions on curves. Unpublished manuscript (1986)

[Mu83] M.R. Murty, On Artin's conjecture. J. Number Theory **16**, 147-168 (1983)

[OS92] T. Okamoto, K. Sakurai, Efficient algorithms for the construction of hyperelliptic cryptosystems. Advances in cryptology—CRYPTO '91. Lecture Notes in Comput. Sci. **576**, 267–278, Springer, Berlin (1992)

[PARI] C. Batut, K. Belabas, D. Bernadi, H. Cohen, M. Olivier, PARI-GP 2.1.3. `ftp://megrez.math.u-bordeaux.fr/pub/pari/`

[Se72] J.-P. Serre, Propriétés galoisiennes des points d'ordre fini des courbes élliptiques. Invent. Math. **15**, 259-331 (1972)

[Se77] J.-P. Serre, Resumé de cours (1977). Oeuvres. Springer, Berlin-Heidelberg-New York (1986)

[T] R. Takeuchi, On cyclicity of Mordell's curve mod p. Preprint, Tokyo Metropolitan University Mathematics Preprint Series **7** (2002)

ALGORITHMIC CONSTRUCTIONS OF ELLIPTIC CURVES WITH COMPLEX MULTIPLICATION

ANDREAS ENGE

Laboratoire d'informatique, École polytechnique, 91128 Palaiseau Cedex, France
E-mail: enge@lix.polytechnique.fr
Web: http://www.lix.polytechnique.fr/Labo/Andreas.Enge

The theory of complex multiplication provides an attractive approach to the construction of elliptic curves with known cardinality over a finite field. It can be used to prove the primality of a number or to obtain a secure elliptic curve cryptosystem.

The cardinality of an elliptic curve is tied to its endomorphism ring. Hence, the basic idea of the complex multiplication approach is to select an elliptic curve over the complex numbers with a suitable endomorphism ring and to reduce it modulo a prime. By Deuring's theorem[1], the endomorphism ring will usually be preserved during this reduction. The method relies on the fact that non-trivial endomorphism rings are orders in imaginary-quadratic fields, and that the complex curves are actually defined over number fields, namely the corresponding Hilbert or ring class fields.

From an algorithmic point of view, the desired curve can be obtained by computing the *class polynomial*, that is the minimal polynomial, of a singular value of the modular invariant j. Such values generate the ring class field, and their minimal polynomials have integral rational coefficients. By reducing the polynomial modulo a rational prime and determining a root in the finite field, one obtains the j-invariant of the desired elliptic curve. From the j-invariant, it is then easy to deduce a curve equation.

Unfortunately, the j-function leads to minimal polynomials with huge coefficients, and it is advisable to replace it by other modular functions, so called *class invariants*, whose singular values also lie in the ring class field. This approach has first been taken by Atkin and Morain[2], who use functions examined by Weber[3]. These functions have been studied intensively in the literature.

Recently, Schertz, Morain and the speaker have proposed two parametric families of modular functions as class invariants, which can be used to construct elliptic curves with prescribed complex multiplication[4,5,6]. The associated modular curves are not necessarily of genus zero, as is the case for Weber's functions. Thus, it is not possible to retrieve the j-invariant as a rational function of these class invariants, but this difficulty can be overcome by computing the modular equations relating the functions to j[7]. The alternative class invariants yield polynomials with much smaller coefficients, whose size can be estimated quite accurately[8].

Class polynomials are usually computed from sufficiently accurate complex approximations to their roots. Explicit expressions for the roots as values of modular functions may be obtained from Shimura's reciprocity law[9]. This is worked out in much detail and generality by Schertz[10]. The speaker has implemented the algorithms with the help of a library for exact complex arithmetic developed in collaboration with Zimmermann[11].

The exposé briefly treats the theoretic foundations of complex multiplication of elliptic curves, presents the different families of class invariants and relates the speaker's expericences obtained from his implementation.

References

1. Max Deuring. Die Typen der Multiplikatorenringe elliptischer Funktionenkörper. *Abhandlungen aus dem mathematischen Seminar der hamburgischen Universität*, 14:197–272, 1941.
2. A. O. L. Atkin and F. Morain. Elliptic curves and primality proving. *Mathematics of Computation*, 61(203):29–68, July 1993.
3. Heinrich Weber. *Lehrbuch der Algebra*, volume 3. Chelsea Publishing Company, New York, 3rd edition, 1908.
4. Andreas Enge and Reinhard Schertz. Constructing elliptic curves from modular curves of positive genus. In preparation, 2001.
5. François Morain. Modular curves and class invariants. Preprint, June 2000.
6. Andreas Enge and François Morain. Further investigations of the generalised Weber functions. In preparation, 2001.
7. Andreas Enge and Reinhard Schertz. Modular curves of composite level. In preparation, 2001.
8. Andreas Enge and François Morain. Comparing invariants for class fields of imaginary quadratic fields. Accepted for ANTS V — Algorithmic Number Theory Symposium, Sydney, 2002.
9. Goro Shimura. *Introduction to the Arithmetic Theory of Automorphic Functions.* Iwanami Shoten and Princeton University Press, 1971.
10. Reinhard Schertz. Weber's class invariants revisited. To appear in *Journal de Théorie des Nombres de Bordeaux*, 2001.
11. Andreas Enge and Paul Zimmermann. `mpc` — a library for multiprecision complex arithmetic with exact rounding. Available from http://www.loria.fr/~zimmerma, 2002.

PARALLEL IMPLEMENTATION OF POLYHEDRAL CONTINUATION METHODS FOR SYSTEMS OF POLYNOMIAL EQUATIONS

MASAKAZU KOJIMA

Dept. of Math. and Comp. Sci, Tokyo Inst. of Tech., Tokyo 152-8552 Japan
E-mail: kojima@is.titech.ac.jp

YANG DAI

Dept. of Bioengineering, Univ. of Illinois at Chicago, Illinois 60607-7052
E-mail: yang.dai@uic.edu

KATSUKI FUJISAWA

Dept. of Architecture and Architectural Sys., Kyoto Univ., Kyoto 606-8501, Japan
E-mail: fujisawa@is-mj.archi.kyoto-u.ac.jp

SUNYOUNG KIM

Dept. of Math., Ewha Women's Univ., Seoul 120-750 Korea
E-mail: skim@ewha.ac.kr

AKIKO TAKEDA

Toshiba Corporation, Kawasaki, Kanagawa 212-8582, Japan
E-mail: akiko.takeda@toshiba.co.jp

Computing all isolated solutions of a system of polynomial equations with large dimensions requires a great deal of computing resources. When the cyclic-n polynomial system, a popular numerical test problem is solved for the dimension 10, 11, 12 and 13, the number of its isolated solutions amounts to $34,940$, $184,756$, $368,112$ and $2,704,156$, respectively [1,4]. Finding all isolated solutions of such systems successfully becomes increasingly difficult as the dimension grows. In this talk, we present a parallel implementation of the polyhedral homotopy continuation method [5] designed to compute all isolated solutions of a system of polynomial equations

$$\boldsymbol{f}(\boldsymbol{x}) = (f_1(x_1, \dots, x_n), \dots, f_n(x_1, \dots, x_n)) = \boldsymbol{0}$$

in complex variables $x_1, \dots, x_n \in \mathbb{C}$. The method can be divided into two phases. Implementing each phase in parallel is natural in the view of independent tasks in each stage.

The role of the first phase is to construct a class of homotopy functions $\boldsymbol{h}^1(\cdot,\cdot), \dots, \boldsymbol{h}^\ell(\cdot,\cdot)$ from $\mathbb{C}^n \times [0,1]$ into $\mathbb{C}^n$ satisfying the following properties:

(a) $\boldsymbol{h}^1(\boldsymbol{x}, 1) = \cdots = \boldsymbol{h}^\ell(\cdot, 1) = \boldsymbol{f}(\boldsymbol{x})$ for every $\boldsymbol{x} \in \mathbb{C}^n$.

(b) Each component $h_j^k(\boldsymbol{x}, t)$ of $\boldsymbol{h}^k(\boldsymbol{x}, t)$ $(j = 1, \ldots, n)$ is represented as the sum of terms of the form $cx_1^\alpha x_2{}^\beta \cdots x_n^\gamma t^\rho$ for some $c \in \mathbb{C}$, $\alpha \in \mathbb{Z}_+$ (the set of nonnegative integers), $\beta, \ldots, \gamma \in \mathbb{Z}_+$ and $\rho \geq 0$. Hence $h_j^k(\boldsymbol{x}, t)$ is nonlinear in the homotopy parameter $t \in [0, 1]$; in particular, when ρ is large, *e.g.* $\rho \geq 10,000$, the term $cx_1^\alpha x_2{}^\beta \cdots x_n^\gamma t^\rho$ (hence $h_j^k(\boldsymbol{x}, t)$) changes drastically as the value of t increases in an extremely small interval $(1.0 - \epsilon, 1.0]$ for some $\epsilon > 0$.

(c) Each component $h_j^k(\boldsymbol{x}, 0)$ of $\boldsymbol{h}^k(\boldsymbol{x}, 0)$ $(j = 1, \ldots, n)$ is a binomial, *i.e.* a polynomial consisting of two terms, in complex variables $x_1, \ldots, x_n \in \mathbb{C}$, so that we can easily compute all solutions, say $\boldsymbol{x}^{kp}$ $(p = 1, \ldots, m_k)$ of $\boldsymbol{h}^k(\boldsymbol{x}, 0)$.

(d) Each isolated solution $\boldsymbol{x}^*$ of $\boldsymbol{f}(\boldsymbol{x})$ is connected to a solution $\boldsymbol{x}^{pk}$ of the binomial system $\boldsymbol{h}^k(\boldsymbol{x}, 0) = \boldsymbol{0}$ for some $p = 1, \ldots, m_k$ and some $k = 1, \ldots, \ell$ through a solution curve of $\boldsymbol{h}^k(\boldsymbol{x}, t) = \boldsymbol{0}$, $(\boldsymbol{x}, t) \in \mathbb{C}^n \times [0, 1]$.

The construction of the class of homotopy functions $\boldsymbol{h}^1(\cdot, \cdot), \ldots, \boldsymbol{h}^\ell(\cdot, \cdot)$ is based on the polyhedral combinatorics, and it involves interesting combinatorial optimization problems [6].

In the second phase [1,2,3], each solution curve of $\boldsymbol{h}^k(\boldsymbol{x}, t) = \boldsymbol{0}$, $(\boldsymbol{x}, t) \in \mathbb{C}^n \times [0, 1]$ is traced numerically from a known $(\boldsymbol{x}^{kp}, 0)$ $(p = 1, \ldots, m_k, k = 1, \ldots, n)$ until t reaches 1. Because of the high nonlinearity of $\boldsymbol{h}^k(\boldsymbol{x}, t)$ with respect to the homotopy parameter $t \in [0, 1]$ mentioned above in (b), this phase also requires some new techniques in addition to the standard predictor-corrector path-following method.

References

1. Y. Dai, S. Kim and M. Kojima, B-373, Dept. of Math. and Comp. Sciences, Tokyo Inst. of Tech., Meguro, Tokyo 152-8552, November (2001).
2. S. Kim and M. Kojima, B-376, Dept. of Math. and Comp. Sciences, Tokyo Inst. of Tech., Meguro, Tokyo 152-8552, January (2002).
3. S. Kim and M. Kojima, B-378, Dept. of Math. and Comp. Sciences, Tokyo Inst. of Tech., Meguro, Tokyo 152-8552, March (2002).
4. M. Kojima, "http://www.is.titech.ac.jp/~kojima/polySys.html."
5. T.Y. Li, *Taiwan Journal of Mathematics* **3**, 251 (1999).
6. A. Takeda, M. Kojima, K. Fujisawa, *Journal of the Operations Research Society of Japan* **45**, 64 (2002).

CMPSM : A CONTINUATION METHOD FOR POLYNOMIAL SYSTEMS (MATLAB VERSION)

SUNYOUNG KIM

Dept. of Math., Ewha Women's University, Seoul 120-750 Korea
E-mail: skim@ewha.ac.kr

MASAKAZU KOJIMA

Dept. of Math. and Comp. Sci, Tokyo Inst. of Tech., Tokyo 152-8552 Japan
E-mail: kojima@is.titech.ac.jp

CMPSm is a MATLAB program using a predictor-corrector path-following method for tracing homotopy curves to approximate all isolated solutions of a system of polynomial equations. The program was originally developed for the cheater's homotopy, but it can be also used for linear and polyhedral homotopies. The software CMPSm, this manual and some numerical examples are available at http://www.is.titech.ac.jp/~kojima/CMPSm/

1 Introduction

We consider a system of polynomial equations $\boldsymbol{f}(\boldsymbol{x}) = \boldsymbol{0}$ from the n-dimensional complex space $\mathbb{C}^n$ into itself, where each $f_k(\boldsymbol{x}) \in \mathbb{C}$ of $\boldsymbol{f}(\boldsymbol{x}) = (f_1(\boldsymbol{x}), f_2(\boldsymbol{x}), \ldots, f_n(\boldsymbol{x})) \in \mathbb{C}^n$ denotes a polynomial in variables $x_1, x_2, \ldots, x_n \in \mathbb{C}$ and $\boldsymbol{x} = (x_1, x_2, \ldots, x_n)$ a variable vector in $\mathbb{C}^n$. As an example, we consider the cyclic-3 polynomial system [3] throughout the paper:

$$1 - x_1x_2x_3 = 0, \;\; x_1x_2 + x_2x_3 + x_3x_1 = 0, \;\; x_1 + x_2 + x_3 = 0 \tag{1}$$

Homotopy continuation methods [1,8,10,14,15,16,17] are known as powerful numerical methods for computing all isolated solutions of $\boldsymbol{f}(\boldsymbol{x}) = \boldsymbol{0}$. A common strategy behind the methods is to prepare a homotopy (polynomial) function $\boldsymbol{h} : \mathbb{C}^n \times [0,1] \to \mathbb{C}^n$ such that

(a) all solutions of *the starting polynomial system* $\boldsymbol{h}(\boldsymbol{x}, 0) = \boldsymbol{0}$ are easily attainable,

(b) *the target polynomial system* $\boldsymbol{h}(\boldsymbol{x}, 1) = \boldsymbol{0}$ coincides with $\boldsymbol{f}(\boldsymbol{x}) = \boldsymbol{0}$,

(c) for all t in $[0, 1)$, the system $\boldsymbol{h}(\boldsymbol{x}, t) = \boldsymbol{0}$ has only nonsingular solutions.

Then, starting from a known solution $\boldsymbol{x}^0$ of $\boldsymbol{h}(\boldsymbol{x}, 0) = \boldsymbol{0}$ with *the homotopy parameter* $t = 0$ and increasing the value of t, trace a solution curve of $\boldsymbol{h}(\boldsymbol{x}, t) = \boldsymbol{0}$ numerically in the space $\mathbb{C}^n \times [0, 1]$ to obtain a solution of the target system $\boldsymbol{h}(\boldsymbol{x}, 1) \equiv \boldsymbol{f}(\boldsymbol{x}) = \boldsymbol{0}$ at $t = 1$. Linear, polyhedral and the cheater's homotopies represent popular homotopies. A MATLAB program CMPSm is based on a predictor-corrector path-following method for tracing solution curves of such homotopy systems. Some technical details of the method were presented in the recent paper [4].

2 Homotopy systems

Let $\mathbb{R}$ and $\mathbb{Z}_+$ denote the set of real numbers and the set of nonnegative integers, respectively. For every $\boldsymbol{x} = (x_1, x_2, \ldots, x_n) \in \mathbb{C}^n$ and $\boldsymbol{a} = (a_1, a_2, \ldots, a_n) \in \mathbb{Z}^n_+$, we use the notation $\boldsymbol{x}^{\boldsymbol{a}}$ for the term $x_1^{a_1} x_2^{a_2} \cdots x_n^{a_n}$. Then we can write each $f_k(\boldsymbol{x})$ of $\boldsymbol{f}(\boldsymbol{x})$ as $f_k(\boldsymbol{x}) = \sum_{\boldsymbol{a} \in \mathcal{A}_k} c_k(\boldsymbol{a})\boldsymbol{x}^{a}$ for a finite subset $\mathcal{A}_k$ of $\mathbb{Z}^n_+$ and $c_k(\boldsymbol{a}) \in \mathbb{C}$ $(\boldsymbol{a} \in \mathcal{A}_k)$ $(k = 1, 2, \ldots, n)$. We call $\mathcal{A}_k$ the *support* of $f_k(\boldsymbol{x})$. For the cyclic-3 polynomial system (1), we take

$$\left.\begin{array}{l} \boldsymbol{a}_1^1 = (0,0,0),\ \boldsymbol{a}_1^2 = (1,1,1),\ \mathcal{A}_1 = \{\boldsymbol{a}_1^1, \boldsymbol{a}_1^2\}, \\ \boldsymbol{a}_2^1 = (1,1,0),\ \boldsymbol{a}_2^2 = (0,1,1),\ \boldsymbol{a}_2^3 = (1,0,1),\ \mathcal{A}_2 = \{\boldsymbol{a}_2^1, \boldsymbol{a}_2^2, \boldsymbol{a}_2^3\}, \\ \boldsymbol{a}_3^1 = (1,0,0),\ \boldsymbol{a}_3^2 = (0,1,0),\ \boldsymbol{a}_3^3 = (0,0,1),\ \mathcal{A}_3 = \{\boldsymbol{a}_3^1, \boldsymbol{a}_3^2, \boldsymbol{a}_3^3\}, \\ c_1(\boldsymbol{a}_1^1) = 1,\ c_1(\boldsymbol{a}_1^2) = -1,\ c_2(\boldsymbol{a}_2^1) = 1,\ c_2(\boldsymbol{a}_2^2) = 1,\ c_2(\boldsymbol{a}_2^3) = 1, \\ c_3(\boldsymbol{a}_3^1) = 1,\ c_3(\boldsymbol{a}_3^2) = 1,\ c_3(\boldsymbol{a}_3^3) = 1. \end{array}\right\} \quad (2)$$

Let $\widetilde{\mathcal{A}}_k \supseteq \mathcal{A}_k$ and $c_k(\boldsymbol{a}) = 0$ $(\boldsymbol{a} \in \widetilde{\mathcal{A}}_k \backslash \mathcal{A}_k,\ k = 1, 2, \ldots, n)$. Each component $h_k(\boldsymbol{x}, t)$ of the homotopy whose zeros (*i.e.* homotopy solution curve) can be traced by CMPSm is written as

$$h_k(\boldsymbol{x}, t) = \sum_{\boldsymbol{a} \in \widetilde{\mathcal{A}}_k} \left((1-t)\tilde{c}_k(\boldsymbol{a}) + t c_k(\boldsymbol{a})\right) \boldsymbol{x}^{\boldsymbol{a}} t^{\rho_k(\boldsymbol{a})}. \quad (3)$$

Here $\tilde{c}_k(\boldsymbol{a}) \in \mathbb{C}$ and $0 \leq \rho_k(\boldsymbol{a}) \in \mathbb{R}$ are given numbers $(\boldsymbol{a} \in \widetilde{\mathcal{A}}_k)$. To avoid possible degeneracy while tracing homotopy solution curves, a common practice is to assign randomly generated complex numbers to all (or some) of $\tilde{c}_k(\boldsymbol{a})$'s. Obviously $\boldsymbol{h}(\boldsymbol{x}, 1) = \boldsymbol{f}(\boldsymbol{x})$ for every $\boldsymbol{x} \in \mathbb{C}^n$; hence (b) holds. We call $\rho_k(\boldsymbol{a})$ $(\boldsymbol{a} \in \widetilde{\mathcal{A}}_k,\ k = 1, 2, \ldots, n)$ *powers* of t. Given solutions $\boldsymbol{x}^1, \boldsymbol{x}^2, \ldots, \boldsymbol{x}^\ell$ of $\boldsymbol{h}(\boldsymbol{x}, 0) = \boldsymbol{0}$, and the data $\widetilde{\mathcal{A}}_k$, $c_k(\boldsymbol{a})$ $(\boldsymbol{a} \in \widetilde{\mathcal{A}}_k)$, $\tilde{c}_k(\boldsymbol{a})$ $(\boldsymbol{a} \in \widetilde{\mathcal{A}}_k)$ and $\rho_k(\boldsymbol{a})$ $(\boldsymbol{a} \in \widetilde{\mathcal{A}}_k)$ $(k = 1, 2, \ldots, n)$, we can apply CMPSm to trace homotopy solution curves of $\boldsymbol{h}(\boldsymbol{x}, t) = \boldsymbol{0}$ numerically, starting from $\boldsymbol{x}^1, \boldsymbol{x}^2, \ldots, \boldsymbol{x}^\ell$ at $t = 0$. We describe three special homotopies, a linear homotopy, a polyhedral homotopy and cheater's homotopy.

2.1 Linear homotopies

Let $\rho_k(\boldsymbol{a}) = 0$ $(\boldsymbol{a} \in \widetilde{\mathcal{A}}_k,\ k = 1, 2, \ldots, n)$. Then

$$h_k(\boldsymbol{x}, t) = \sum_{\boldsymbol{a} \in \widetilde{\mathcal{A}}_k} \left((1-t)\tilde{c}_k(\boldsymbol{a})\boldsymbol{x}^{\boldsymbol{a}} + t c_k(\boldsymbol{a})\boldsymbol{x}^{\boldsymbol{a}}\right) = (1-t)h_k(\boldsymbol{x}, 0) + t f_k(\boldsymbol{x}).$$

Thus this leads to a linear homotopy $\boldsymbol{h}(\boldsymbol{x},t) = (1-t)\boldsymbol{h}(\boldsymbol{x},0) + t\boldsymbol{f}(\boldsymbol{x})$. See the literatures [1,8,17] for more details of linear homotopies. The usual form of a component of a linear homotopy function at $t = 0$ is $h_k(\boldsymbol{x},0) = x_k^{d_k} - r^{d_k}\exp(\alpha_k i)$ $(k = 1,2,\dots,n)$. Here $\mathbb{Z}_+ \ni d_k > 0$, $\mathbb{R} \ni r_k > 0$, $\alpha_k \in [0, 2\pi)$, and i denotes the imaginary unit. Then all the solutions of the starting system $\boldsymbol{h}(\boldsymbol{x},0) = \boldsymbol{0}$ are given by

$$\left(r_1 \exp\left(\frac{\alpha_1 + 2p_1\pi}{d_1} i\right), r_2 \exp\left(\frac{\alpha_2 + 2p_2\pi}{d_2} i\right), \dots, r_n \exp\left(\frac{\alpha_n + 2p_n\pi}{d_n} i\right)\right)$$

$(p_k = 0,1,\dots,d_k - 1,\ k = 1,2,\dots,n)$.

In addition to (2), let

$$\left.\begin{array}{l}
\boldsymbol{a}_1^3 = (4,0,0), c_1(\boldsymbol{a}_1^3) = 0, \boldsymbol{a}_2^4 = (0,3,0), c_2(\boldsymbol{a}_2^4) = 0, \boldsymbol{a}_2^5 = (0,0,0),\\
c_2(\boldsymbol{a}_2^5) = 0, \boldsymbol{a}_3^4 = (0,0,2), c_3(\boldsymbol{a}_3^4) = 0, \boldsymbol{a}_3^4 = (0,0,0), c_3(\boldsymbol{a}_3^5) = 0,\\
\mathcal{A}_1^L = \mathcal{A}_1 \cup \{\boldsymbol{a}_1^3\},\ \mathcal{A}_2^L = \mathcal{A}_2 \cup \{\boldsymbol{a}_2^4\},\ \mathcal{A}_3^L = \mathcal{A}_3 \cup \{\boldsymbol{a}_3^4\},\\
\tilde{c}_1^L(\boldsymbol{a}_1^1) = -r_1^4\exp(\alpha_1 i), \tilde{c}_1^L(\boldsymbol{a}_1^2) = 0, \tilde{c}_1^L(\boldsymbol{a}_1^3) = 1,\\
\tilde{c}_2^L(\boldsymbol{a}_2^1) = \tilde{c}_2^L(\boldsymbol{a}_2^2) = \tilde{c}_2^L(\boldsymbol{a}_2^3) = 0, \tilde{c}_2^L(\boldsymbol{a}_2^4) = 1, \tilde{c}_2^L(\boldsymbol{a}_2^5) = -r_2^3\exp(\alpha_2 i),\\
\tilde{c}_3^L(\boldsymbol{a}_3^1) = \tilde{c}_3^L(\boldsymbol{a}_3^2) = \tilde{c}_3^L(\boldsymbol{a}_3^3) = 0, \tilde{c}_3^L(\boldsymbol{a}_3^4) = 1, \tilde{c}_3^L(\boldsymbol{a}_3^5) = -r_3^2\exp(\alpha_3 i).
\end{array}\right\} \quad (4)$$

Then we have a linear homotopy system

$$h_k^L(\boldsymbol{x},t) = \sum_{\boldsymbol{a}\in\mathcal{A}_k^L} \left((1-t)\tilde{c}_k^L(\boldsymbol{a}) + tc_k(\boldsymbol{a})\right)\boldsymbol{x}^{\boldsymbol{a}}\ (k = 1,2,3) \quad (5)$$

for the cyclic-3 polynomial system (1). All data necessary to execute CMPSm for the linear homotopy (5) is in the directory 3cycL. Specifically, $\mathcal{A}_k^L$, $c_k(\boldsymbol{a})$ and $\tilde{c}_k^L(\boldsymbol{a})$ $(\boldsymbol{a} \in \mathcal{A}_k^L,\ k = 1,2,3)$ are described in the *coefficient file* 3cycL/3cycL.coef. This file contains the six keywords "n = ", "m = ", "a$k.\ell$ = ", "coef$k.\ell$r = ", "coef$k.\ell$i = ", "Rcoef$k.\ell$r = ", and "Rcoef$k.\ell$i = ", where k and ℓ stand for positive integers. A single space is necessary before and after = in each key word. A line should start with one of the keywords to be taken as valid data by CMPSm. Following two lines of comments in the beginning, the dimension 3 of the problem is specified by "n = 3", and then the cardinalities of $\mathcal{A}_1^L$, $\mathcal{A}_2^L$, $\mathcal{A}_3^L$ by "m = 3 5 5". It is followed by the description of the supports $\mathcal{A}_1^L$, $\mathcal{A}_2^L$, $\mathcal{A}_3^L$ of the linear homotopy system $\boldsymbol{h}^L(\boldsymbol{x},t)$, where each line denotes a$k.\ell$ $= \boldsymbol{a}_k^\ell \in \mathcal{A}_k^L$ $(k = 1,2,3)$. Then each of the next several lines denotes

coef$k.\ell$r $=$ the real part of $c_k(\boldsymbol{a}_k^\ell)$ or coef$k.\ell$i $=$ the imag. part of $c_k(\boldsymbol{a}_k^\ell)$,

and each line of the last part means

Rcoef$k.\ell$r $=$ the real part of $\tilde{c}_k(\boldsymbol{a}_k^\ell)$ or Rcoef$k.\ell$i $=$ the imag. part of $\tilde{c}_k(\boldsymbol{a}_k^\ell)$, respectively.

The solutions of the starting polynomial system $\boldsymbol{h}^L(\boldsymbol{x}, 0) = \mathbf{0}$ are described in a *cell file of initial points* 3cycL/3cycL. This file includes the three key words "NumOfSols = ", "sol$k$ = ", and "= ", where k stands for positive integers. Any line starting without one of the key words is neglected. As specified by NumOfSols = 24, this file contains 24 solutions, sol1, sol2, ... sol24 of the starting polynomial system $\boldsymbol{h}^L(\boldsymbol{x}, 0) = \mathbf{0}$. The kth solution, sol$k$ = (x_1, x_2, x_3) is described such that

sol$k$ = the real part of x_1

the imaginary part of x_1

...

the imaginary part of x_3

Each line of the last part of the file provides the powers $\rho_k(\boldsymbol{a})$ $(\boldsymbol{a} \in \mathcal{A}_k^L)$ of the homotopy parameter t $(k = 1, 2, 3)$ that are zeros in the linear homotopy.

2.2 Polyhedral homotopies

Let $\widetilde{\mathcal{A}}_k = \mathcal{A}_k$ $(k = 1, 2, \ldots, n)$, where $\mathcal{A}_k$ denotes the support of $f_k(\boldsymbol{x})$. To compute all isolated solutions of a polynomial system $\boldsymbol{f}(\boldsymbol{x}) = \mathbf{0}$, we produce multiple polyhedral homotopy systems for an auxiliary polynomial system $\tilde{f}_k(\boldsymbol{x}) = \sum_{\boldsymbol{a} \in \mathcal{A}_k} \tilde{c}_k(\boldsymbol{a}) \boldsymbol{x}^a = 0$ $(k = 1, 2, \ldots, n)$ with randomly generated coefficients $\tilde{c}_k(\boldsymbol{a})$ $(\boldsymbol{a} \in \mathcal{A}_k,\ k = 1, 2, \ldots, n)$ based on the polyhedral homotopy theory [2,6,10,14,15,16], and then we construct a linear homotopy system between the auxiliary polynomial system with its solutions and the target polynomial system $\boldsymbol{f}(\boldsymbol{x}) = \mathbf{0}$. See [10,11,13,16] for numerical methods for constructing polyhedral homotopy systems. Therefore, for the case of the polyhedral homotopy, CMPSm is used twice to find the solutions of the target polynomial systems. First, we apply CMPSm to each polyhedral homotopy system $h_k(\boldsymbol{x}, t) = \sum_{\boldsymbol{a} \in \mathcal{A}_k} \tilde{c}_k(\boldsymbol{a}) \boldsymbol{x}^a t^{\rho_k(\boldsymbol{a})} = 0$ $(k = 1, 2, \ldots, n)$, where the set of powers $\rho_k(\boldsymbol{a})$ $(\boldsymbol{a} \in \mathcal{A}_k,\ k = 1, 2, \ldots, n)$ differs from one polyhedral homotopy system to another but $\tilde{c}_k(\boldsymbol{a})$ $(\boldsymbol{a} \in \mathcal{A}_k,\ k = 1, 2, \ldots, n)$ remain the same for all polyhedral homotopy systems. Then, we apply CMPSm again to a linear homotopy system $\sum_{\boldsymbol{a} \in \mathcal{A}_k} \left((1-t)\tilde{c}_k(\boldsymbol{a}) + t c_k(\boldsymbol{a})\right) \boldsymbol{x}^a = 0$ $(k = 1, 2, \ldots, n)$ with the solutions of the auxiliary polynomial system.

The distinctive feature of a polyhedral homotopy system $\boldsymbol{h}(\boldsymbol{x}, t)$ is that exactly two of $\rho_k(\boldsymbol{a})$ $(\boldsymbol{a} \in \mathcal{A}_k)$ are zeros and all the others are positive for each $k = 1, 2, \ldots, n$. As a result, the starting polynomial system $\boldsymbol{h}(\boldsymbol{x}, 0) = \mathbf{0}$ is a binomial system, which we can easily solve. When applying CMPSm, we need to scale the homotopy parameter t so that all positive $\rho_k(\boldsymbol{a})$'s are not less than 1. Such a scaling is always possible. In fact, replace the homotopy parameter t by $t = s/\rho^*$ where ρ^* denotes the minimum of all positive $\rho_k(\boldsymbol{a})$'s ($\boldsymbol{a} \in \mathcal{A}_k$, $k = 1, 2, \ldots, n$). Then the modified homotopy $\hat{\boldsymbol{h}}(\boldsymbol{x}, s) = \boldsymbol{h}(\boldsymbol{x}, s/\rho^*)$ satisfies the requirement.

In the case of the cyclic-3 polynomial system (1), we constructed two polyhedral homotopies based on the polyhedral homotopy theory; the one has the set of powers

$$\left.\begin{array}{l}\rho_1(\boldsymbol{a}_1^1) = \rho_1(\boldsymbol{a}_1^2) = 0,\ \rho_2(\boldsymbol{a}_2^1) = \rho_2(\boldsymbol{a}_2^2) = 0,\ \rho_2(\boldsymbol{a}_2^3) = 4.75143, \\ \rho_3(\boldsymbol{a}_3^1) = \rho_3(\boldsymbol{a}_3^2) = 0,\ \rho_3(\boldsymbol{a}_3^3) = 1.9309\end{array}\right\} \quad (6)$$

stored with 3 starting solutions in a cell file of initial points 3cycP/3cycP1, and the other has the set of powers

$$\left.\begin{array}{l}\rho_1(\boldsymbol{a}_1^1) = \rho_1(\boldsymbol{a}_1^2) = 0,\ \rho_2(\boldsymbol{a}_2^1) = 1.9309,\ \rho_2(\boldsymbol{a}_2^2) = \rho_2(\boldsymbol{a}_2^3) = 0, \\ \rho_3(\boldsymbol{a}_3^1) = 0,\ \rho_3(\boldsymbol{a}_3^2) = 4.75143,\ \rho_3(\boldsymbol{a}_3^3) = 0\end{array}\right\} \quad (7)$$

together with 3 starting solutions in another cell file of initial points 3cycP/3cycP2. All other data $\mathcal{A}_k$ ($k = 1, 2, 3$), $c_k(\boldsymbol{a}) = \tilde{c}_k(\boldsymbol{a})$ ($\boldsymbol{a} \in \mathcal{A}_k$, $k = 1, 2, 3$) and $\tilde{c}_k(\boldsymbol{a})$ ($\boldsymbol{a} \in \mathcal{A}_k$, $k = 1, 2, 3$) are located in the coefficient file 3cycP/3cycP.coef. The structures of 3cycP/3cycP.coef and 3cycP/3cycP1 (and 3cycP/3cycP2) are the same as 3cycL/3cycL.coef and 3cycL/3cycL. Only small differences lie in the last parts of 3cycP/3cycP1 and 3cycP/3cycP2 where powers of the homotopy parameter t are described.

2.3 Cheater's homotopies

We take $\widetilde{\mathcal{A}}_k = \mathcal{A}_k$ ($k = 1, 2, \ldots, n$), where $\mathcal{A}_k$ denotes the support of $f_k(\boldsymbol{x})$. Then, a homotopy which we use in CMPSm may be regarded as a combination of a linear and a polyhedral homotopy with randomly generated values $\tilde{c}_k(\boldsymbol{a})$ and $c_k(\boldsymbol{a})$ ($\boldsymbol{a} \in \mathcal{A}_k$, $k = 1, 2, \ldots, n$). The use of this type of homotopy was proposed by Li [10] in the framework of the polyhedral homotopy continuation method. He called this homotopy cheater's homotopy. See [9] for the original definition of the cheater's homotopy, and also [12] for coefficient-parameter polynomial continuation methods that include the cheater's homotopy as a special case. Exactly two of $\rho_k(\boldsymbol{a})$ ($\boldsymbol{a} \in \mathcal{A}_k$) of the cheater's homotopy are zeros and all the others are positive for each $k = 1, 2, \ldots, n$. This enables CMPSm to start from solutions of a binomial system with the coefficient $\tilde{c}_k(\boldsymbol{a})$'s at $t = 0$ and attain solutions of the target polynomial system at $t = 1$ by tracing homotopy solution curves. All data to implement the cheater's homotopies for the cyclic-3 polynomial system (1) are stored in the directory 3cycC. Specifically, the coefficient file 3cycC/3cyc.coef includes the data $c_k(\boldsymbol{a})$ and $\tilde{c}_k(\boldsymbol{a})$ ($\boldsymbol{a} \in \mathcal{A}_k$, $k = 1, 2, 3$), while the cell files of initial points 3cycC/3cycC1 and 3cycC/3cycC2 contain the same powers of t and the same initial solutions as the files 3cycP/3cycP1 and 3cycP/3cycP2. As in the polyhedral homotopy, we need to scale the homotopy parameter t so that all positive $\rho_k(\boldsymbol{a})$'s are not less than 1.

3 Parameter file

A user can specify the parameters accINfVal, accInNewtonDir, beta, divMagOFx, dTauMax, minEigForNonsing, NewtonDirMax and predItMax which control the behavior of CMPSm. The below is the parameter file, para_3cycL.dat used when CMPSm is applied to the cyclic-3 polynomial system (1) by a linear homotopy with the files 3cycL.coef and 3cycL mentioned in Section 2.1.

```
# para3cycL.dat
accINfVal= 1.e-10
accInNewtonDir= 1.e-8
# take beta = 1 for linear homotopy case
beta=1;
divMagOFx= 1.0e+3
dTauMax= 0.1
minEigForNonsing= 1.e-12
NewtonDirMax= 0.1
predItMax= 2000
```

Here "accINfVal=", "accInNewtonDir=", "beta=", "divMagOFx=", "dTauMax=", "minEigForNonsing=", "NewtonDirMax=", "predItMax=" are keywords at the first column of a line, and lines begin with '♯' are comments which CMPSm neglects. "accINfVal", "accInNewtonDir" and "minEigForNonsing" are used as stopping criteria; when an approximate solution $\boldsymbol{x}$ satisfying either

$$\begin{cases} \max\{|f_k(\boldsymbol{x})| : k = 1, 2, \dots\} \leq \text{accINfVal}, \\ \|\boldsymbol{Df}(\boldsymbol{x})^{-1}\boldsymbol{f}(\boldsymbol{x})\| \leq \text{accInNewtonDir, and} \\ \text{the minimum eigenvalue of } \boldsymbol{Df}(\boldsymbol{x})^*\boldsymbol{Df}(\boldsymbol{x}) \geq \text{minEigForNonsing} \end{cases}$$

or

$$\begin{cases} \max\{|f_k(\boldsymbol{x})| : k = 1, 2, \dots, n\} \leq \text{accINfVal, and} \\ \text{the minimum eigenvalue of } \boldsymbol{Df}(\boldsymbol{x})^*\boldsymbol{Df}(\boldsymbol{x}) < \text{minEigForNonsing.} \end{cases}$$

CMPSm decides that the homotopy curve has converged to a nonsingular solution in the former case, and a singular solution in the latter case. When the 2-norm of an iterate $\boldsymbol{x}$ becomes larger than divMagOFx or when the predictor iteration exceeds predItMax, CMPSm stops. The parameter "dTauMax" provides an upper bound for increases in $t^{\rho_k(\boldsymbol{a})}$ $(\boldsymbol{a} \in \widetilde{\mathcal{A}}_k, k = 1, 2, \ldots, n)$ in the predictor procedure. If the 2-norm $||d\boldsymbol{x}||$ of the Newton direction is greater than the value NewtonDirMax$\times||\boldsymbol{x}||$, the corrector iteration is discarded and then a predictor iteration with a smaller step is retried. The smaller the values of these two parameters are, the more accurately CMPSm traces a homotopy curve, to avoid an illegal jump from the curve into another curve.

The parameter beta ≥ 1 serves as a scaling for the homotopy parameter t. However, it is not relevant in the linear homotopy since all powers $\rho_k(\boldsymbol{a})$ $(\boldsymbol{a} \in \widetilde{\mathcal{A}}_k,\ k = 1, 2, \ldots, n)$ are zero. For the polyhedral or the cheater's homotopy, we suggest to take

$$1 \leq \text{ beta } \approx \gamma \times \text{ "the maximum of powers, } \rho_k(\boldsymbol{a})\text{'s over all homotopies"},$$
$$0.01 \leq \gamma \leq 0.1$$

for computational efficiency and numerical stability. When large dimensional problems are solved by the polyhedral or the cheater's homotopy, the maximum of $\rho_k(\boldsymbol{a})$'s over all homotopies can be quite large, for example, it can exceed 200,000. Taking an appropriate beta improves the performance of CMPSm significantly for such cases. When the maximum is small, for example, less than 100, choose beta $= 1$. A more advanced method to deal with high powers of the continuation parameter t can be found in [5].

Remark. Three additional parameters "informationLevel", "coeffSW" and "modSW" with the usages are embedded in the main program CMPSm.m.

4 Execution of CMPSm

We briefly explain the procedure to solve the cyclic-3 polynomial system using the linear homotopy (5) with (2) and (4):

(i) Place the following files in a directory:

- CMPSm.m, traceOneCurve.m — MATLAB program files,
- para3cycL.dat — a parameter file,
- 3cycL.coef, 3cycL — coefficient and cell files of initial points.

(ii) Run MATLAB.

(iii) Specify three input arguments for CMPSm in the MATLAB environment as

```
>>parameterFile='para3cycL.dat';
>>coefficientFile='3cycL.coef';
>>cellFileInitPt='3cycL';
```

(iv) Execute CMPSm as

```
>>CMPSm(parameterFile,coefficientFile,cellFileInitPt);
```

Alternatively, combining (iii) and (iv) above, we can execute
`>>CMPSm('para3cycL.dat','3cycL.coef','3cycL');`

When we use the polyhedral or the cheater's homotopy, we need to specify several "cellFileInitPt"s for input arguments. CMPSm can accept two more input arguments, "startCellNo" and "endCellNo" to handle those cases. For solutions of the cyclic-3 problem by the cheater's homotopy with the data files 3cycC.coef, 3cycC1, 3cycC2 and the parameter file para3cycC.dat, issue a command as
`>>CMPSm('para3cycC.dat','3cycC.coef','3cycC',1,2);`
Here we take startCellNo= 1 and endCellNo= 2.

5 Output files

CMPSm generates two output files, *.stat file and *.sol file. Here * stands for the file name "cellFileInitPt" specified as the third input argument. The following is an example of the *.stat file from the cyclic-9 polynomial system stored in the directory 9cycC.

```
# 9cycC1-978.stat
#  cell   init statusP  pIT   TcIT    cpu hValError      normOFx         minEig
     ...
     49      1  +3        65    135    4.4  4.18e-15  6.8541e+00   +6.9948e-01
     ...
     49     10  +4        69    224    4.6  2.13e-12  2.6180e+00   +3.6930e-13
     ...
     50      1  -2       217    993   19.4  1.02e-01  7.3932e+01   +2.8480e-16
```

Here "cell" denotes a positive integer attached to a cell file of initial points, which takes a value from "startCell" to "endCell" (the last two input arguments). "init" means the initial solution number. "statusP" indicates whether the homotopy path converged to a nonsingular solution (statusP = 3), converged to a singular solution (statusP = 4) or diverges (statusP = -2). Also statusP can have other values in different circumstances. The meaning of the values is described at the top of the *.stat file. "pIT", "TcIT" and "cpu " stand for the total number of predictor iterations, the total number of corrector iterations and the cpu time in seconds spent to trace the homotopy path, respectively. "hValError", "normOFx" and "minEig" denote the 1-norm of errors in function values of an approximate solution $\boldsymbol{x}$ computed, the 2-norm of $\boldsymbol{x}$ and the minimum eigenvalue of $\boldsymbol{Df}(\boldsymbol{x})^*\boldsymbol{Df}(\boldsymbol{x})$, respectively. These values are meaningful only when statusP is either +3 or +4.

The other output file is solution file. For the cyclic-9 polynomial system, we have

Table 1. Numerical results from economic n problems by the cheater's homotopy

n	6	7	8	9	10
No.paths	16	32	64	128	256
Av.pred.it.	107.1	94.9	98.0	108.8	134.1
Av.corr.it.	208.6	190.7	201.8	228.3	285.2
Av.CPU	1.6	1.8	2.2	3.0	4.3
No.sol.	16	32	64	128	256

```
# 9cycC1-978.sol
+7.6604444311e-01 . . . +9.3781448372e-02  +4.1789657759e-15  49  1
    ...
+1.7364863502e-01 . . . -3.7616321836e-01  +2.1321577253e-12  49 10
```

Each line without comment mark # shows an approximate solution $\boldsymbol{x}$, the 1-norm of errors in function values, the cell number of initial points and the initial solution number, where the last three numbers correspond "hValError', "cell" and "init" in the *.stat file, such that

$$\text{real}(x_1), \text{imag}(x_1), \ldots, \text{real}(x_n), \text{imag}(x_n), \text{hValError}, \text{cell}, \text{init}.$$

6 Numerical results

Table 1 shows numerical results on the economic polynomial with dimensions 6, 7, ..., 10 solved by the cheater's homotopy with the data stored in the directories 6ecoC, 7ecoC, ..., 10ecoC. The computation was done using MATLAB version 6.2 and Pentium III 1.8GHz CPU. Here the following notation is used:

n : the number of variables.
No.paths : the number of paths traced
Av.pred.it. : the average number of predictor iterations per path.
Av.corr.it. : the average number of corrector iterations per path.
Av.CPU : the average CPU time in seconds per path.
No.sol. : the number of (nonsingular) solutions computed.

7 Concluding remarks

CMPSm may not be robust to find all solutions of a polynomial system for some cases, although the rate of missing a solution is very low, less than 0.1% from our experience. Two effective techniques exist to recover missing solutions. The one is to recompute the homotopy curves that have converged to a common solution with more conservative parameters ; take dTauMax= 0.01 and NewtonDirMax= 0.01 instead of their default values 0.1 and 0.1. The other is to apply CMPSm to more than a set of homotopy functions for a polynomial system to be solved, and merge multiple sets of the solutions obtained into a set of the solutions; choosing a different $\beta \geq 1$ in the parameter file and/or reseting coeffSW to -1 in CMPSm would yield different homotopies. See [4] for more details.

An important advantage of homotopy continuation methods lies in parallel computation. Indeed, we can trace all homotopy paths simultaneously in parallel if the input data are split in a consistent manner. CMPSm has been designed to benefit from this feature. For example, if three MATLAB environments are set up in three different cpus, we can execute each of the following commands

```
>>CMPSm('para9cycC.dat','9cycC.coef','9cycC',1,300);
>>CMPSm('para9cycC.dat','9cycC.coef','9cycC',301,600);
>>CMPSm('para9cycC.dat','9cycC.coef','9cycC',601,978);
```

in one of three MATLAB environments, instead of issuing a command

```
>>CMPSm('para9cycC.dat','9cycC.coef','9cycC',1,978);
```

in a single MATLAB environment to solve the economic problem of the dimension 9 whose data is stored in 9cycC.

CMPSm has been developed as a part of a joint project [4,13] of (parallel) implementation of polyhedral homotopy continuation methods with Yang Dai, Katsuki Fujisawa, Sunyoung Kim, Masakazu Kojima and Akiko Takeda. A partial outcome of this project was reported in [4] with some numerical results on the cyclic polynomial system of the dimensions $8, 9, \ldots, 12$, which was obtained by a C++ code implementing the cheater's homotopy continuation method. They also succeeded in approximating all isolated solutions of the cyclic-13 polynomial system by a parallel implementation of the C++ code. It resulted in 2,704,156 isolated solutions (counting multiplicity). See the webpage [7]. CMPSm itself is not as fast as the C++ code for larger dimensional polynomial systems. Recently the authors have released CMPSc, a C++ version of CMPSm, which is available at the webpage [7].

Acknowledgments

The authors would like to thank Prof. Dai and Dr. Takeda for making most of input data available with their programs. They also would like to express sincere thanks to an anonymous referee for valuable comments and suggestions. Sunyoung Kim's research was supported by Kosef R004-2001-00200 and BK 21.

References

1. E. Allgower and K. Georg, *Numerical continuation methods,* Springer-Verlag, 1990.
2. D. N. Bernshtein, *Functional Analysis and Appl.* **9**(3), 183 (1975).
3. G. Björck and R. Fröberg, *Journal Symbolic Computation* **12**, 329 (1991).
4. Y. Dai, S. Kim and M. Kojima, B-373, Dept. of Math. and Comp. Sciences, Tokyo Inst. of Tech., September (2001).
5. T. Gao, T. Y. Li, J. Verschelde and M. Wu, *Applied Mathematics and Computation* **114**, 233 (2000).
6. B. Huber and B. Sturmfels, *Mathematics of Computation* **64**, 1541 (1995).
7. M. Kojima, "http://www.is.titech.ac.jp/~kojima/polySys.html."
8. T. Y. Li, *The mathematical intelligencer*, **9**(3), 33 (1987).
9. T. Y. Li, T. Sauer and J. A. York, *SIAM J. Numer. Anal.*, **26**, 1241 (1989).
10. T. Y. Li, *Taiwan Journal of Mathematics* **3**, 251 (1999).
11. T. Y. Li and X. Li, *Foundation of Computational Mathematics* **1**, 161 (2001).
12. A. P. Morgan and A. J. Sommese, *Appl. Math. Comput.* **29**(2), 123 (1989).
13. A. Takeda, M. Kojima, and K. Fujisawa, *J. of Operations Society of Japan* **45**(1), 64 (2002).
14. J. Verschelde, The database of polynomial systems is in his web site: "http://www.math.uic.edu/~jan/."
15. J. Verschelde, P. Verlinden and R. Cools, *SIAM J. Numerical Analysis,* **31**, 915 (1994).
16. J. Verschelde, *ACM Trans. Math. Softw.* **25**, 251 (1999).
17. L. T. Watson, M. Sosonkina, R. C. Melville, A. P. Morgan, and H. F. Walker, *ACM Trans. Math. Softw.* **23**(4), 514 (1997).

GENERATING FUNCTIONS THAT COUNT T-DESIGNS WITH GIVEN AUTOMORPHISM GROUP: ALGORITHMS AND STRUCTURE *

JESÚS A. DE LOERA

Dept of Mathematics, Univ. of California, Davis
email: deloera@math.ucdavis.edu

Let $t \leq k \leq v$ be fixed positive integers. We outline algorithms for calculating generating functions, parametrized by λ, that count $t-(v,k,\lambda)$-designs with given automorphism group. Along the way we generalize a theorem of R. Wilson [25] by showing that there exists an integer λ_0 such that for any $\lambda > \lambda_0$ and any permutation subgroup G of v elements there is a $t-(v,k,\lambda)$-design with automorphism group G.

1 Introduction

Let $v \geq k \geq t$ be fixed positive integers. Given the set $[v] = \{1, \ldots v\}$. A $t-(v,k,\lambda)$ *design* D is a finite collection of possibly repeated k-subsets of $[v]$, the *blocks* of D, with the property that any t-subset of $[v]$ is contained in exactly λ blocks. The constant λ is called its *incidence number.* A t-design is *simple* if and only if the blocks are not repeated. A permutation π of the elements of $[v]$ is an *automorphism* of the t-design D if π maps blocks of D into blocks of D. Note that any finite permutation group G of $[v]$ also induces an action on the set of t-subsets of $[v]$ and the k-subsets of $[v]$. In Figure 1 we show an example, where the points along each segment and the circle denote the blocks of size 3. We refer the reader to [4] for all details on combinatorial design theory.

Let $d_{t,v,k}(G,\lambda)$ denote the number of $t-(v,k,\lambda)$ designs with automorphism group G. In particular, if $G = \{1\}$, $d_{t,v,k}(1,\lambda)$ is the number of $t-(v,k,\lambda)$ designs. We are interested in finding a close formula for the generating function, $F_{t,v,k}(G,x) = \sum d_{t,v,k}(G,\lambda)x^{\lambda}$. In this note we demonstrate that it is possible to compute $F_{t,v,k}(G,x)$. We look at the problem of counting and constructing t-designs as a problem of studying lattice points on certain polyhedral cones and polytopes. This paper was inspired by recent efforts to use polyhedral geometry in the theory of designs (see [17,18]). We take the point of view of linear homogeneous Diophantine equations presented in Chapter 4 of [22]. This is already useful to prove some theoretical results about

*RESEARCH SUPPORTED BY NSF GRANT DMS-0073815.

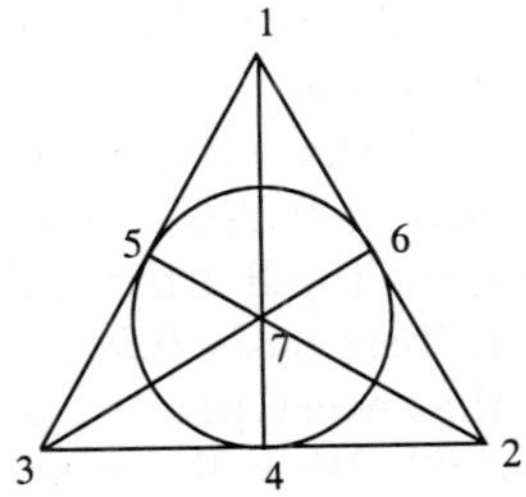

Figure 1. The Fano plane: A $2-(7,3,\lambda)$ design with D_3 action. $\lambda = 1$ in this case.

t-designs. For instance, we can easily deduce a strengthening of a result of Wilson [25]:

Theorem 1.1 *For any non-negative integers t, v, k and any subgroup G of the symmetric group S_v there exist an integer λ_0 such that there is always a $t-(v,k,\lambda)$ design with automorphism group G for any $\lambda > \lambda_0$.*

Section 2 contains the background material and ends with a proof of Theorem 1.1. Section 3 describes in detail one algorithm for computing $F_{t,v,k}(G,x)$ and outlines three others. All the methods we present use polyhedral combinatorics [22] and computational algebraic geometry [10], to compute the generating formulas or their coefficient for individual terms.

2 Background

In what follows we will assume basic knowledge of polyhedral combinatorics as in Chapters 8 and 16 of [21]. Let A be an integer $d \times n$ matrix, we study pointed cones of the form $C = \{x | Ax = 0, x \geq 0\}$. A cone is *pointed*, if it does not contain any linear subspace besides the origin. It is well-known that pointed cones also admit a representation as the set of all possible non-negative *real* linear combinations of finitely many vectors, the so called *extreme rays* of the cone. For a cone C we are interested in $S_C = C \cap \mathbb{Z}^n$, the *semigroup of the cone* C. For any rational polyhedron R, either cone or polytope, in $\mathbb{R}^n$ we define following rational generating function in n variables:

$$f(R,z) \quad = \quad \sum_{u \in R \cap \mathbb{Z}^n} z^u$$

where z^u denotes, from now on, $z_1^{u_1} z_2^{u_2} \ldots z_n^{u_n}$. For example, if R is a polytope,the $f(R,z)$ is actually a multivariate polynomial.

An element v of S_C is called *irreducible* if a decomposition $v = v_1 + v_2$ for $v_1, v_2 \in S_C$ implies that $v_1 = 0$ or $v_2 = 0$. A *Hilbert basis* for C is a finite set of vectors $HB(C)$ in S_C such that every other element of S_C is a positive integer combination of elements in $HB(C)$. A *minimal Hilbert basis* $HB(C)$ is inclusion minimal with respect to all other Hilbert bases of C. As a result all elements of the minimal Hilbert basis $HB(C)$ are irreducible and $HB(C)$ is unique. It is well-known that every pointed polyhedral cone has indeed a Hilbert basis (see page 232 of [21]). We will now specify the cones for t-designs:

Denote by $[v]$ the set $\{1, \ldots, v\}$. We can think of $t-(v, k, \lambda)$ designs as the solutions of a system of linear Diophantine equations $M_{t,v,k}x = \lambda(1, \ldots, 1)^T$ where the matrix $M_{t,v,k} = (m_{i,j})$ is an $\binom{v}{t} \times \binom{v}{k}$ matrix where $m_{i,j} = 1$ exactly when the i-th t-subset is contained in the j-th k-subset and otherwise it is zero. This is certainly a rather large matrix already for very small values of t, v, k. Fortunately, we can reduce the size of the matrix by using a group action. If we have a finite permutation group G acting on $[v]$, G induces also an action on the set of t-subsets of $[v]$ and the k-subsets of $[v]$, let us denote by $Orb_j(t), Orb_j(k)$ the j-th orbit of t-subsets (respectively k-subsets) and $\#Orb(t), \#Orb(k)$ the total number of orbits. Define now a matrix $M_{t,v,k}(G) = (n_{i,j})$ where $n_{i,j}$ denotes the number of elements in $Orb_j(k)$ which contain a representative of the orbit $Orb_i(t)$. This matrix has $\#Orb(t)$ rows and $\#Orb(k)$ columns. A useful lemma for us is a slight strengthening of a theorem of Kramer and Mesner [15]. The proof is rather simple and it is left for the reader.

Lemma 2.1 *There is $t-(v, k, \lambda)$ design with group of automorphisms G if and only if there is a non-negative integer solution x of the Diophantine system $M_{t,v,k}(G)x = \lambda(1, 1, \ldots, 1)^T$ $x \geq 0$. The integer solutions of the system are in bijection with the $t-(v, k, \lambda)$ designs.*

Similarly, construct the matrix $A_{t,v,k}(G)$ from the matrix $M_{t,v,k}(G)$ by equating the first row linear equation of $M_{t,v,k}(G)$ to the second row equation, setting second row equation equal to the third row equation and so on. We are interested on the convex pointed polyhedral cone $C_{t,v,k}(G) = \{y | A_{t,v,k}(G)y = 0,\ y \geq 0\}$. This cone is the union of the polytopes $P_{t,v,k,\lambda}(G) = \{M_{t,v,k}(G)y = \lambda(1, 1, \ldots, 1)^T\ y \geq 0\}$ as λ takes non-negative real values. When $\lambda = 1$ the integer hull of these polytopes has been studied under the name of *design polytopes* by Moura [17,18].

Example: We consider the example of $2-(7, 3, \lambda)$ designs. From Figure 1 it is clear that the dihedral group D_3, whose generators are the permutations $\pi = (1, 2, 3)(4, 5, 6)$ and $\sigma = (2, 3)(5, 6)$, acts on at least one element of the family of t-designs. The resulting matrix $M_{2,7,3}(D_3)$, up to a permutation of the columns and rows, is given by

$$M_{2,7,3}(D_3) = \begin{bmatrix} 1 & 2 & 1 & 1 & 0 & 0 & 0 & 0 & 0 & 0 \\ 0 & 2 & 0 & 0 & 1 & 2 & 0 & 0 & 0 & 0 \\ 0 & 1 & 1 & 0 & 0 & 1 & 1 & 1 & 0 & 0 \\ 0 & 0 & 0 & 2 & 1 & 0 & 0 & 2 & 0 & 0 \\ 0 & 0 & 0 & 0 & 1 & 0 & 0 & 2 & 0 & 2 \\ 0 & 0 & 0 & 0 & 0 & 2 & 1 & 0 & 1 & 1 \end{bmatrix}$$

where the orbits for 3-subsets label the columns of the matrix, and the orbits of 2-subsets label the rows. From the matrix $M_{2,7,3}(D_3)$ we can construct the matrix $A_{2,7,3}(D_3)$ that defines the cone. We remark that the way we number the orbits is not going to be relevant to the computations below, which give a permutation of the columns of the matrix $A_{2,7,3}(D_3)$ and thus an isomorphic cone.

$$A_{2,7,3}(D_3) = \begin{bmatrix} 0 & 0 & 0 & 2 & 1 & -2 & -1 & 2 & -1 & -1 \\ 0 & -2 & 0 & 0 & 0 & -2 & 0 & 2 & 0 & 2 \\ 1 & 2 & 1 & 1 & -1 & 0 & 0 & -2 & 0 & -2 \\ 0 & 1 & -1 & 0 & 1 & 1 & -1 & -1 & 0 & 0 \\ 0 & 1 & 1 & -2 & -1 & 1 & 1 & -1 & 0 & 0 \end{bmatrix}$$

For the purpose of this note, the following is an important lemma about the generating function $F_{t,v,k}(G,x) = \sum_{\lambda=1}^{\infty} d_{t,v,k}(G,\lambda)x^{\lambda}$:

Lemma 2.2 *Let $v \geq k \geq t$ be fixed positive integers. Let G be a finite group of permutations of $[v] = \{1,\ldots,v\}$. Then*

The generating function $F_{t,v,k}(G,x) = \sum d_{t,v,k}(G,\lambda)x^{\lambda}$ is a rational generating function in the variable x. Moreover the function $d_{t,v,k}(G,\lambda)$ is a quasi-polynomial on λ. This means that there is an integer $N > 0$ and polynomials $f_0, f_1, \ldots, f_{N-1}$ such that $d_{t,v,k}(G,\lambda) = f_i(\lambda)$ for all those λ congruent with i modulo N.

Proof: From the earlier discussion, the $t-(v,k,\lambda)$ designs with t,v,k fixed and λ changing, have the structure of the lattice points of a pointed polyhedral cone $C_{t,v,k}(G) = \{A_{t,v,k}(G)y = 0, y \geq 0\}$, with $A_{t,v,k}(G)$ a $d \times n$ matrix and Hilbert basis $HB(C_{t,v,k}(G))$. Thus, from Lemma 4.6.11 in [22], we know that the multivariate generating function $f(C_{t,v,k}(G),z)$, can be written as

$$f(C_{t,k,v}(G),z) = P(z)/\Pi_{\alpha \in HB(C)}(1 - z^{\alpha})$$

There is a simple substitution that yields the the univariate series $F_{t,v,k}(G,x)$: choose a row of the matrix $M_{t,v,k}(G)$, it has n entries $(r_1,\ldots,r_n)$, substitute z_i by x^{r_i}. This way, a multivariate monomial x^a coming from a $t-(v,k,\lambda)$-design D gets assigned a univariate monomial x^λ, where λ is, by construction, the incidence number of D. We conclude that $F_{t,v,k}(G,x)$ is a rational function in the variable x. Note that by construction every root of its denominator is a root of unity because the denominator factors as the product of terms $(1-x^r)$. Therefore, by Proposition 4.4.1 of [22], the function $d_{t,v,k}(G,\lambda)$, which is the coefficient of x^λ in $F_{t,v,k}(G,x)$, is then a quasi-polynomial. □

Proof of Theorem 1.1: Wilson's theorem [25] is an easy application of the second part of the above Lemma: For large positive values of λ a polynomial $p(\lambda)$ cannot have any real roots. Thus for large enough λ none of the quasi-polynomials will vanish. This ensures a t-design for large enough values of λ. □

From the theory in Chapter 4 of [22] we know that the degree of the quasipolynomial $d_{t,k,v}(G,\lambda)$ is equal to the dimension of the cone $C_{t,v,k}(G)$ minus one.

3 Algorithms

Now we present algorithms to deduce the generating function $F_{t,v,k}(G,x)$ from the input data of G,t,v,k. For any cone $C_{t,v,k}(G)$ of t-designs the existence of a Hilbert basis means:

Corollary 3.1 *For any choice of $t \le k \le v$ and subgroup G of S_v there is a finite set $H_{t,v,k}(G)$ of $t-(v,k,\lambda)$-designs such that any other can be expressed as a finite multiset union of t-designs in $H_{t,v,k}(G)$, i.e possibly some of the t-designs appear with repeated copies. When the t-designs are written as integer vectors then the vectors associated to $H_{t,v,k}(G)$ generate all other t-designs are non-negative integer combinations.*

It is interesting to remark that all t-designs that are Steiner systems (i.e. $\lambda = 1$) are forced to belong to the Hilbert basis as they are necessarily irreducible. However, we see in the example below that they are not sufficient to generate **every** other t-design in the family and that the Hilbert basis will usually contain t-designs with higher incidence number and with repeated blocks. A natural question is, *how can we compute the minimal Hilbert basis of a cone C?* There is a wide variety of algorithms [7,9,13,20,23]. In our calculations of minimal Hilbert bases we used the project-and-lift algorithm presented in [14] and implemented in `MLP` by R. Hemmecke (see these proceedings).

Example: We continue with the example of the cone of $2-(7,3,\lambda)$ designs. Using `MLP` we find its unique, minimal Hilbert basis. From now on the degree of an integral vector in the cone C of t-designs is simply its incidence number

when we think of it as a design, e.g. if the degree is one then the t-design is a Steiner system. In our example there are exactly two degree one elements. They correspond to the two ways of labeling the Fano plane such that the group D_3 is an automorphism group.
$[0,0,1,0,1,0,0,0,1,0]$ and $[1,0,0,0,1,0,1,0,0,0]$, both of degree 1, $[2,0,0,0,0,1,0,1,0,0]$ and $[0,1,0,0,0,0,0,1,2,0]$ of degree 2, $[0,0,2,1,1,1,0,0,0,1]$ and $[0,1,0,1,1,0,2,0,0,1]$ of degree 3, $[0,1,1,1,0,1,0,1,1,1]$ and $[1,1,0,1,0,1,1,1,0,1]$ of degree 4, and finally $[0,1,2,2,0,2,0,1,0,2]$ and $[0,2,0,2,0,1,2,1,0,2]$ of degree 6.

With any d-dimensional rational pointed polyhedral cone $C = \{Ax = 0, x \geq 0\}$ and a field k we associate a *semigroup ring*, $R_C = k[y^a : a \in S_C]$, where there is one monomial $y_1^{a_1} y_2^{a_2} \ldots y_d^{a_d}$ in the ring for each element $a = (a_1, \ldots, a_d)$ of the semigroup S_C. By the definition of a Hilbert basis we know that every element of S_C can be written as a finite linear combination $\sum \mu_i h_i$ where the μ_i are non-negative integers. Thus R_C is in fact a finitely generated k-algebra, with one generator per element of a Hilbert bases. Therefore R_C can be written as the quotient $k[x_1, x_2, \ldots, x_N]/I_C$ where N is the number of Hilbert basis elements, and I_C is an ideal generated by binomials obtained as the kernel of the map $\phi : k[x_1, x_2, \ldots, x_N] \longrightarrow k[t_1, t_2, \ldots t_d]$, where $\phi(x_i) = t^{h_i}$ and for $h_i = (a_1, a_2, \ldots, a_d)$. Remember that $t^{h_i} = t_1^{a_1} t_2^{a_2} \ldots t_d^{a_d}$. There are standard techniques for computing the kernel of a polynomial map (see [23] and references).

For our cones of t-designs we can give a natural grading to R_C. A t-design array can be thought of as a monomial on the ring R_C and its degree will be its incidence number. For example, all Steiner systems have degree one. Once we have a graded k-algebra we can talk about its decomposition into the direct sum of its graded components $R_C = \bigoplus R_C(i)$, where each $R_C(i)$ collects all elements of degree i and it is a k-vector space (where $R_C(0) = k$). The function $H(R_C, i) = dim_k(R_C(i))$ is the *Hilbert function* of R_C. Similarly one can construct the *Hilbert-Poincaré series* of R_C, this is $H_{R_C}(t) = \sum_{i=0}^{\infty} H(R_C, i) t^i$.

Lemma 3.2 *Let R_C be the semigroup ring obtained from the minimal Hilbert basis of the cone $C = C_{t,v,k}(G)$ of t-designs with automorphism group G. The number of $t-(v,k,\lambda)$ designs with automorphism group G, $d_{t,v,k}(G,\lambda)$, equals the value of the Hilbert function $H(R_C, \lambda)$.*

Proof: By the definition of a Hilbert basis we have that every $t-(v,k,\lambda)$ design in the cone C can be written as a linear integer combination of the elements of the Hilbert basis. The elements of $HB(C) = \{h_1, h_2, \ldots, h_N\}$ are not affinely independent therefore there are different combinations that produce the same magic array. We have some dependencies of the form $\sum a_i h_i = \sum a_j h_j$ where the sums run over some subsets of $\{1,\ldots,N\}$. We consider such identities as giving a single t-design. The dependencies are precisely the elements of the toric ideal I_C, that give $R_C = k[x_1, x_2, \ldots, x_N]/I_C$. Every such dependence is a linear combination of binomial generators of any Gröbner basis of the ideal I_C [23]. Thus, if we encode a t-design X as a monomial in variables $x_1, \ldots, x_N$ whose exponents are the coefficients of the corresponding Hilbert basis elements that add to X, we are counting the equivalence classes modulo I_C. These are called *standard monomials.* Finally, it is known that the number of standard monomials of graded degree i equals the dimension of $R_C(i)$ as a k-vector space (see Chapter 9 of [10]). □

To compute the Hilbert-Poincaré series we relied on the computer algebra package CoCoA [8] that has implementations for different algorithms of Hilbert series computations [5]. The basic idea, common to all algorithms, comes from the theory of Gröbner bases (see Chapter 9 of [10]). It is known that the initial ideal of I_C with respect to any monomial order gives a monomial ideal J and the Hilbert functions of $k[x_1, x_2, \ldots, x_N]/I_C$ and $k[x_1, x_2, \ldots, x_N]/J$ are equal. Thus the computation reduces to an inclusion-exclusion process.

Algorithm 1: *Hilbert Bases Algorithm* to obtain the generating function $F_{t,v,k}(G,x)$ generating formula proceeds as follows: **(1)** Identify the matrix $A_{v,k,t}(G)$ defining the cone $C = \{x : A_{v,k,t}(G)x = 0, x \geq 0\}$ of $t-(v,k,\lambda)$ designs that have a group G as an automorphism group.**(2)** Compute the Hilbert basis H of the cone C. **(3)** Compute the toric ideal I_C induced by the Hilbert basis H. **(4)** Compute the Hilbert-Poincaré series of the graded ring $k[x_1, x_2, \ldots, x_N]/I_C$.

Example: Using what we know so far for $2-(7,3,\lambda)$ designs with automorphism group D_3 we have that the ring R_C is the quotient ring of $k[x_1, \ldots, x_n]/I_C$. The ideal I_C can be computed using CoCoA from the Hilbert basis' elements we saw earlier. They are given to CoCoA as the columns of a matrix:

```
Use R::=Q[x[1..10]];
J:=Toric(Mat([[0,1,0,2,0,0,0,1,0,0], [0,0,0,0,1,1,1,1,1,2],
              [1,0,2,0,0,0,1,0,2,0], [0,0,1,0,1,0,1,1,2,2],
              [1,1,1,0,1,0,0,0,0,0], [0,0,1,1,0,0,1,1,2,1],
              [0,1,0,0,2,0,0,1,0,2], [0,0,0,1,0,1,1,1,1,1],
              [1,0,0,0,0,2,1,0,0,0], [0,0,1,0,1,0,1,1,2,2]]));
Print J;
Ideal(x[2]x[7] - x[1]x[8], x[4]x[5] - x[2]x[8], -x[3]x[6] + x[1]x[7],
-x[2]x[3]x[6] + x[1]^2x[8], -x[3]x[7] + x[1]x[9], -x[3]x[8] + x[2]x[9],
-x[5]x[8] + x[2]x[10], -x[5]x[7] + x[1]x[10], x[7]^2 - x[6]x[9],
-x[8]^2 + x[4]x[10], x[5]x[9] - x[3]x[10])
```

There are ten variables because there are ten Hilbert basis elements. Using `CALI` and the corresponding incidence numbers as degrees for the variables we get a rational function expression for $F_{2,7,3}(D_3, x)$.

$$F_{2,7,3}(D_3, x) = \frac{(t^14+2t^13+3t^12+5t^11+9t^10+10t^9+12t^8+12t^7+12t^6+10t^5+9t^4+5t^3+3t^2+2t+1)}{(t^2-t+1)^2(t^2+t+1)^3(t+1)^4(t-1)^5}$$

which expands to $1 + 2t + 5t^2 + 10t^3 + 20t^4 + 31t^5 + 55t^6 + 79t^7 + \ldots$. This means, for example, that there are fifty five $2-(7,3,6)$ designs whose group of automorphism is D_3.

There are at least three other algorithms that we could use instead of the one presented here. All of them have in common that they use algebraic and geometric interpretations, just as Algorithm 1. Due to lack of space we devoted only a short description to each of them. We are implementing all these algorithms in the forthcoming software system `LATTE`, for lattice point enumeration in rational convex polyhedra, written by the author and students. In our previous discussion we worked with the matrix $A_{t,v,k}(G)$ which was derived from the matrix $M_{t,v,k}(G)$. Obviously one can make sense of most of our calculations for arbitrary rational polyhedral cones and polytopes as long as the entries of defining matrices are integer. As input data we get $d \times n$ matrices M, A, that define a cone $C = \{\alpha : A\alpha = 0, \alpha \geq 0,\}$ or a convex polytope $P(M,b) = \{y : \ My = (b_1, b_2, \ldots, b_d)^T \ y \geq 0\}$ and we wish to output a short formula for the multivariate generating functions $f(C, z)$ or $f(P(M,b), z)$. From this generating function several pieces of enumerative information can be deduced, for example, in Algorithm 1 we obtained essentially $f(C, z)$, but after a subtitution of the variables z_i with powers of a single parameter x, thus it a becomes univariate function.

Algorithm 2: Elliot-MacMahon Reduction

This method [16] obtains from matrix A a short rational function expression for $F_A(z)$. From the columns of the matrix A, $A_1, \ldots A_n$, construct an auxiliary generating function $G(y, z) = \Pi_{j=1}^{n} \frac{1}{(1-y^{A_j} z_j)}$. Note that $F_A(z) = G(y, z)|_{y=0}$ because z^b will be in the new expression if and only if $Ab = 0$, thus it is in the y-free part of $G(y, z)$. To extract the y-free part of $G(y, z)$ one uses the identity

$$\frac{1}{(1-p)(1-q)} = \frac{1}{(1-pq)}\left(\frac{1}{1-p} + \frac{1}{1-q} - 1\right).$$

The reduction process does not appear to be very efficient in practice due to the explotion on the number of new terms.

Algorithm 3: Barvinok-Brion-Lawrence method

Brion proved that for any rational convex polytope P in $\mathbb{R}^n$,

$$f(P,z) \quad = \sum_{v \in \text{vertices}(P)} f(\text{cone}(P,v), z),$$

where $\text{cone}(P,v) = v + \{u \in \mathbb{R}^n : v + \delta u \in P$ for all sufficiently small $\delta > 0\}$. This says that the polynomial $f(P,z)$ can be written as a sum of rational functions. Note that evaluating the polynomial $f(P,z)$ at $z = (1,1,1,\ldots,1)$ gives the number of integer points in P.

Therefore we need to compute and add together the series , $f(\text{cone}(P,v),z)$. Barvinok's key idea [1] was to decompose the cones at vertices into unimodular cones. A *unimodular* cone is a pointed simplicial cone with generators $\{u_1,\ldots,u_k\}$ that form a basis for the lattice $\mathbb{R}\{u_1,\ldots,u_k\} \cap \mathbb{Z}^n$. For unimodular cones, the rational generating function $f(\text{cone}(P,v),z)$ takes a simple form: $f(\text{cone}(P,v),z) \quad = \quad \prod_{i=1}^{k} \frac{z^v}{1-z^{u_i}}$. Thus directly from Barvinok's decomposition we obtain a close form for $f(P,z)$. See [11] for applications, examples, and explicit computations using this method.

Algorithm 4: Residue methods: Beck,Pemantle,Vergne

Let M_i denote the columns of the matrix M. We can interpret $P(M,b) \cap \mathbb{Z}^n$ as the Taylor coefficient of z^b for the function $\Pi_{j=1}^{n} \frac{1}{(1-z^{M_j})}$. One approach to obtain the particular coefficient is to use the residue theorems. For example, it was seen in [2] that if M_i denotes the i-th column of the defining matrix M. Then

$$P(M,b) \cap \mathbb{Z}^n = \frac{1}{(2\pi i)^m} \int_{|z_1|=\epsilon_1} \cdots \int_{|z_m|=\epsilon_m} \frac{z_1^{-b_1-1} \ldots z_m^{-b_m-1}}{(1-z^{M_1})\cdots(1-z^{M_d})} \, dz \ .$$

Here $0 < \epsilon_1, \ldots, \epsilon_m < 1$ are different numbers such that we can expand all the $\dfrac{1}{1-z^{M_k}}$ into power series about 0. It is possible to do a partial fraction decomposition of the integrand into a sum of simple fractions. This was done very successfully to carry on very hard computations in [3]. Vergne and have recently developed a powerful general theory about the multivariate rational functions $\Pi_{j=1}^{n} \frac{1}{(1-z^{M_j})}$ [24]. In practice they used this to speed up things further. Pemantle [19] has pursued an even more general computational theory of rational generating functions where the denominators are not necessarily products of linear forms. So far this method has the top performance in computation.

References

1. Barvinok, A.I. and Pommersheim, J. *An algorithmic theory of lattice points in polyhedra*, in: *New Perspectives in Algebraic Combinatorics* (Berkeley, CA, 1996-1997), 91-147, Math. Sci. Res. Inst. Publ. 38, Cambridge Univ. Press, Cambridge, 1999.
2. Beck M.*Counting lattice points by means of the residue theorem* Ramanujan Journal, 4, no. 3 (2000), 299-310
3. Beck M. and Pixton, D.*The Ehrhart polynomial of the Birkhoff polytope*, manuscript 2002.
4. Beth. T., Jungnickel, D., and Lenz, H. *Design theory* volumes I and II, second edition, Encyclopedia of mathematics and its Applications, Cambridge Univ. Press, 1999.
5. Bigatti, A.M *Computation of Hilbert-Poincaré Series*, J. Pure Appl. Algebra, 119/3, (1997), 237–253.
6. Brion, M. *Points entiers dans les polyèdres convexes*, *Ann. Sci. École Norm. Sup.* **21** (1988) 653-663.
7. Bruns, W. and Koch, R. `NORMALIZ`, *computing normalizations of affine semigroups*, Available via anonymous ftp from ftp//ftp.mathematik.uni-onabrueck.de/pub/osm/kommalg/software/
8. Capani, A., Niesi, G., and Robbiano, L., `CoCoA`, *a system for doing Computations in Commutative Algebra*, Available via anonymous ftp from `cocoa.dima.unige.it`, (2000).
9. Contejean, E. and Devie, H. *Resolution de systemes lineaires d'equations diophantienes* C. R. Acad. Sci. Paris, 313: 115-120, 1991. Serie I
10. Cox, D., Little, J., and O'Shea, D. *Ideals, varieties, and Algorithms*, Springer Verlag, Undergraduate Text, 2nd Edition, 1997.
11. De Loera J. and Sturmfels, B *Algebraic unimodular counting*, to appear Math. Programming Ser. B. Preprint available at (`arXiv:math.CO/0104286`), 2001.
12. Gräbe, H. G. `CALI` *a* `REDUCE` *package for commutative algebra version 2.2.1* http://www.uni-koeln.de/REDUCE/3.6/doc/cali/
13. Henk, M. and Weismantel, R. *On Hilbert bases of polyhedral cones*, Results in Mathematics, 32, (1997), 298-303.
14. Hemmecke, R. *On the Computation of Hilbert Bases of Cones*, These proceedings. Software implementation `MLP` available from http://www.testsets.de.
15. Kramer E.S. and Mesner D.M *t-Designs on hypergraphs*, Discrete Math. 15, (1976), 263–296.

16. MacMahon, P.A. *Combinatorial Analysis*, Vol I and II, Chelsea, 1960, reprint of 1915 edition.
17. Moura, L. *Polyhedral aspects of combinatorial designs* Ph.D Thesis, Dept. of Computer Science, Univ. of Toronto, 1999.
18. Moura, L. *Polyhedral methods in design theory*, 227–254 in *Computational and Constructive Design Theory* W.D. Wallis Editor, Kluwer Academic Press, 1996.
19. Pemantle, R. and Wilson, M. *Asymptotics of multivariate sequences, part I: smooth points of the singular variety.* J. Comb. Th. Ser. A. , to appear.
20. Pottier, L. *Minimal solutions of linear Diophantine systems: bounds and algorithms*, In Rewriting techniques and applications (Como, 1991), 162–173, Lecture Notes in Comput. Sci., 488, Springer, Berlin, 1991.
21. Schrijver, A. *Theory of Linear and Integer Programming.* Wiley-Interscience, 1986.
22. Stanley, R.P. *Enumerative Combinatorics*, Volume I, Cambridge, 1997.
23. Sturmfels, B. *Gröbner bases and convex polytopes*, university lecture series, vol. 8, AMS, Providence RI, (1996).
24. Vergne M. and Baldoni-Silva W. *Residues formulae for volumes and Ehrhart polynomials of convex polytopes.* manuscript 81 pages. available at math.ArXiv, CO/0103097.
25. Wilson R.M. *The necessary conditions for t-designs are sufficient for something* Util. Math. 4, (1973), 207–215.

ON THE COMPUTATION OF HILBERT BASES OF CONES

RAYMOND HEMMECKE
Dept. of Mathematics, Univ. of California, Davis
E-mail: raymond@hemmecke.de

In this paper we present a novel project-and-lift approach to compute the set of minimal generators of the semigroup $(\Lambda \cap \mathbb{R}^n_+, +)$ for lattices $\Lambda \subseteq \mathbb{Z}^n$. This problem class includes the computation of Hilbert bases of cones $\{z : Az = 0, z \in \mathbb{R}^n_+\}$ for integer matrices A. A similar approach could be used to compute the extreme rays of such cones. Finally, some combinatorial applications and computational experience are presented.

1 Introduction

Lattice points in polyhedral cones arise as interesting objects in many branches of mathematics as, for example, combinatorics, integer programming, computational algebra, or topology. Often one is interested in a finite subset of lattice points in the cone that generate all other lattice points in the cone as non-negative integer linear combinations.

Definition 1.1 *Let $\mathcal{C} \subseteq \mathbb{R}^n$ be a polyhedral cone with rational generators and let $\Lambda \subseteq \mathbb{Z}^n$ be a lattice. Then we call a finite set $H = \{h_1, \ldots, h_t\} \subseteq \Lambda \cap \mathcal{C}$ a generating set of the monoid $(\Lambda \cap \mathcal{C}, +)$ if for every $z \in \Lambda \cap \mathcal{C}$ there are non-negative integral multipliers $\lambda_1, \ldots, \lambda_t$ such that $z = \sum_{i=1}^t \lambda_i h_i$.*

Note that for every pointed rational polyhedral cone $\mathcal{C} \subseteq \mathbb{R}^n$ and every lattice $\Lambda \subseteq \mathbb{Z}^n$ there exists a unique generating set of the monoid $(\Lambda \cap \mathcal{C}, +)$ that is minimal with respect to inclusion ([13], Chapter 7). For $\Lambda = \mathbb{Z}^n$ this generating set is also called a Hilbert basis [14]. It is in general a hard problem to compute Hilbert bases. In this paper we will give a novel project-and-lift algorithm to compute minimal generating sets for $\mathcal{C} = \mathbb{R}^n_+$ and arbitrary lattices $\Lambda \subseteq \mathbb{Z}^n$. One prominent special case, for which many algorithms have been proposed (e. g. [2,3,5,9,10,11,12,13,15]), is the case of $\Lambda = \ker_{\mathbb{Z}}(A)$ for matrices $A \in \mathbb{Z}^{d \times n}$, where $\ker_{\mathbb{Z}}(A) = \{z : Az = 0, z \in \mathbb{Z}^n\}$ denotes the integral kernel of A. In these algorithms, the Hilbert basis is extracted from a (usually much) bigger superset, in some cases even additional variables have to be introduced. The advantages of the algorithm that we propose here are that we do not need additional variables (we actually compute with fewer variables most of the time), that we may throw away unnecessary vectors in early and intermediate stages of the algorithm, and that we therefore arrive at a smaller superset from which the set of minimal generators of the monoid can be extracted. Similar to most of the above mentioned algorithms we also use a completion procedure ([4]) at the core of our algorithm. However, in Section 3 we present a selection strategy for the completion procedure that allows us to get rid of the time-consuming reduction process involved, and to replace it by a much faster reducibility test.

Note that a similar project-and-lift approach can be used to compute the extreme rays of cones given as $\{z : Az = 0, z \in \mathbb{R}^n_+\}$ and for the computation of truncated Graver bases. The latter improves the algorithm given in [8]. We omit the details here for lack of space.

Some implementations (e. g. NORMALIZ [3]) of algorithms to compute Hilbert bases of cones are freely available over the internet. The latter code, however, needs a generating set of the cone $\{z : Az = 0, z \in \mathbb{R}^n_+\}$ as input. Thus, no direct computational comparison between NORMALIZ and the algorithm presented below is possible.

The main objective of this paper is an algorithmic solution to the following problem.

Problem 1.2 *Given $\{p_1, \ldots, p_s\} \subseteq \mathbb{Z}^n$ that generate a lattice $\Lambda \subseteq \mathbb{Z}^n$ over $\mathbb{Z}$, compute the unique minimal generating set $H(\Lambda)$ of the monoid $(\Lambda \cap \mathbb{R}^n_+, +)$.*

In Section 3 we present a selection strategy for the S-vector that has to be considered next in the Completion Algorithm 2.3 to compute the minimal generators of the monoid. This strategy leads to a tremendous speed-up of our project-and-lift algorithm. In Section 4 we collect some applications of Hilbert bases and extreme ray computations. These include, for example, the computation of dual cones. The extreme rays of the dual cone of a cone $\mathcal{C}$, however, are exactly the normal vectors of the facets of $\mathcal{C}$. Finally, in Section 5, we report on an implementation, MLP, and some computational experience while working on the problem of counting magic arrays [1].

2 Computation of Minimal Generators

In order to state the algorithm below we introduce some useful notation.

Definition 2.1 *For any $m \geq j$ let $\pi^m_j : \mathbb{R}^m \to \mathbb{R}^j$ be the projection onto the first j coordinates. Moreover, let $K_j := \{\pi^n_j(v) : v \in \Lambda\}$, $K^+_j := K_j \cap (\mathbb{R}^{j-1}_+ \times \mathbb{R}_+)$, $K^-_j := K_j \cap (\mathbb{R}^{j-1}_+ \times \mathbb{R}_-)$.*

Note that K_j is a sublattice of $\mathbb{Z}^j$ and that K^+_j and K^-_j are intersections of a lattice, K_j, with pointed rational cones, that is, $(K^+_j, +)$ and $(K^-_j, +)$ are monoids. Let H^+_j and H^-_j denote their unique inclusion minimal generating sets. Clearly, $H(\Lambda) = H^+_n$ since $K^+_n = \Lambda \cap \mathbb{R}^n_+$.

We may assume that the generators $\{p_1, \ldots, p_s\} \subseteq \mathbb{Z}^n$ of Λ have the following structure, which is clearly achievable by suitable elementary integral row operations and permutations of columns:

$$\begin{array}{rl} p_1 = & (p_{1,1}, p_{1,2}, \ldots, \ldots, p_{1,s}, \ldots, p_{1,n}), \\ p_2 = & (0, p_{2,2}, \ldots, \ldots, p_{2,s}, \ldots, p_{2,n}), \\ p_3 = & (0, 0, p_{3,3} \ldots, p_{3,s}, \ldots, p_{3,n}), \\ \vdots & (\vdots, \vdots, \ldots, \ddots, \vdots, \ldots, \vdots), \\ p_s = & (0, 0, \ldots, 0, p_{s,s}, \ldots, p_{s,n}), \end{array}$$

with $p_{i,i} > 0$, $i = 1, \ldots, s$.

In what follows, we will start with $H_1^+ = \{(p_{1,1})\}$, which is easily computed, and compute $H_2^+, H_3^+, \ldots, H_n^+$ inductively. For each step $H_j^+ \to H_{j+1}^+$ we will employ a completion procedure [4]. The input sets for $j \geq s$ and $j < s$ differ slightly due to the fact that for $j < s$ the map π_j^n is not injective. In particular, there are non-zero vectors $v \in \Lambda$ with $\pi_j^n(v) = 0$ if $j < s$.

As $H_j^+ \subseteq K_j$, we know that for every $h \in H_j^+$ there is at least one vector $v \in \Lambda$ with $h = \pi_j^n(v)$. Thus, for every $h \in H_j^+$ there is some $h' \in \mathbb{Z}$ such that $(h, h') \in K_{j+1}$. If there are several choices for h', which can only happen if $j < s$, we will choose the smallest non-negative integer h' such that $(h, h') \in K_{j+1}$ to form the input set F below. For $j < s$ we have to add the vectors $\pi_{j+1}^n(p_{j+1})$ and $-\pi_{j+1}^n(p_{j+1})$ to the input set F.

Lemma 2.2 *The set*

$$F = \begin{cases} \bigcup_{h \in H_j^+} \{(h, h') : (h, h') \in K_{j+1}\} & \text{if } j \geq s, \\ \bigcup_{h \in H_j^+} \{(h, h') : (h, h') \in K_{j+1}\} \cup \{\pi_{j+1}^n(p_{j+1}), -\pi_{j+1}^n(p_{j+1})\} & \text{if } j < s, \end{cases}$$

generates $K_{j+1}^+ \cup K_{j+1}^-$ over $\mathbb{Z}_+$, that is, every $z \in K_{j+1}^+ \cup K_{j+1}^-$ can be written as a non-negative integer linear combination of elements of F.

Proof. Let $\pi_{j+1}^n(v) \in K_{j+1}^+ \cup K_{j+1}^-$, where $v \in \Lambda$. Since H_j^+ is a generating set for the monoid $(K_j^+, +)$, we can write $\pi_j^n(v) = \sum \alpha_i \pi_j^n(v_i)$ for some $\pi_j^n(v_i) \in H_j^+$ and $\alpha_i > 0$. Therefore, $\pi_{j+1}^n(v) - \sum \alpha_i \pi_{j+1}^n(v_i)$ has zeros in the first j components and must be the zero vector if $j \geq s$, or an integral multiple of $\pi_{j+1}^n(p_{j+1})$ if $j < s$. This concludes the proof. □

In what follows, dashed variables, like h', at the end of a vector, like (h, h'), always represent integer numbers.

Algorithm 2.3 *(Algorithm to Compute $H^+_{j+1} \cup H^-_{j+1}$)*

Input: F *as in Lemma 2.2*

Output: a set G *which contains* $H^+_{j+1} \cup H^-_{j+1}$

$G := F$

$C := \bigcup_{f,g\in G}$ S-vectors(f,g)

while $C \neq \emptyset$ *do*

  $s :=$ *an element in* C

  $C := C \setminus \{s\}$

  $f :=$ normalForm(s,G)

  if $f \neq 0$ *then*

    $C := C \cup \bigcup_{g\in G}$ S-vectors(f,g)

    $G := G \cup \{f\}$

return G.

It remains to define the set S-vectors(f,g) and the function normalForm(s,G).

$$\text{S-vectors}((v,v'),(w,w')) := \begin{cases} \{(v+w, v'+w')\} & \text{if } v'w' < 0, \\ \emptyset & \text{otherwise.} \end{cases}$$

In case that one is interested only in those elements of $H(\Lambda)$ whose components lie below certain bounds $u \in (\mathbb{Z}_+ \cup \{\infty\})^n$, that is $H(\Lambda) \cap \{z : z \leq u, z \in \mathbb{Z}^n_+\}$, these upper bounds u can easily be used to speed-up the computation by considering only the following set of S-vectors:

$$\text{S-vectors}((v,v'),(w,w')) := \begin{cases} \{(v+w, v'+w')\} & \text{if } v'w' < 0 \text{ and } v+w \leq u, \\ \emptyset & \text{otherwise.} \end{cases}$$

This choice will be justified below after the proof of Lemma 2.5.

Behind the function normalForm(s,G) there is the following algorithm, wherein $u \sqsubseteq_{j+1} v$ if and only if $u^{(i)} \leq v^{(i)}$, $i = 1,\ldots,j$, $|u^{(j+1)}| \leq |v^{(j+1)}|$, and $u^{(j+1)}v^{(j+1)} \geq 0$. Note that in case of $u \sqsubseteq_{j+1} v$, u and v lie in the same orthant of $\mathbb{R}^{j+1}$.

Algorithm 2.4 *(Normal Form Algorithm)*

Input: a vector s*, a set* G *of vectors*

Output: a normal form of s *with respect to* G

while there is some $g \in G$ *such that* $g \sqsubseteq_{j+1} s$ *do*

  $\alpha := \min\{\lfloor s^{(i)}/g^{(i)} \rfloor : i = 1,\ldots,j+1, g^{(i)} \neq 0\}$

  $s := s - \alpha g$

return s

Note that the first j components of s always remain non-negative during the normal form algorithm. Moreover, the $(j+1)$st component of s either remains non-negative or non-positive. Since $\|s - \alpha g\|_1 < \|s\|_1$, the normal form algorithm always terminates.

Lemma 2.5 *Algorithm 2.3 terminates and returns a set containing* $H_{j+1}^+ \cup H_{j+1}^-$.

Proof. First, let us show termination of the algorithm. For this define for any $r \in \mathbb{R}$ the symbols $r^+ := \max\{r, 0\}$ and $r^- := \max\{-r, 0\}$. Next, consider the sequence $G \setminus F = \{(g_1, g_1'), \ldots\}$ as it is generated by Algorithm 2.3. This sequence fulfills $(g_i, g_i') \not\sqsubseteq_{j+1} (g_k, g_k')$ whenever $i < k$. Therefore, $(g_i, (g_i')^+, (g_i')^-) \not\leq (g_k, (g_k')^+, (g_k')^-)$ whenever $i < k$. Applying the Gordan-Dickson Lemma (see for example [6]) to the sequence $\{(g_1, (g_1')^+, (g_1')^-), \ldots\} \subseteq \mathbb{Z}_+^{j+2}$, we conclude that this sequence must be finite and thus Algorithm 2.3 terminates. It remains to show correctness of the algorithm.

By G denote the set that is returned by Algorithm 2.3 and let $(z, z') \in H_{j+1}^+ \cup H_{j+1}^-$. By Lemma 2.2 and $F \subseteq G$, we can write (z, z') as a finite positive integer linear combination $(z, z') = \sum \alpha_i (v_i, v_i')$ for some $\alpha_i \in \mathbb{Z}_{>0}$ and vectors $(v_i, v_i') \in G$ with $0 \leq v_i \leq z$ for all i. From the set of all such positive integer linear combinations $\sum \alpha_i (v_i, v_i')$ choose one such that $\sum \alpha_i |v_i'|$ is minimal. Note that, by the triangle inequality, $\sum \alpha_i |v_i'| \geq |z'|$ with equality if and only if $v_i' \sqsubseteq_1 z'$ for all i, that is, if and only if z' and all v_i' are either both non-negative or both non-positive.

If $\sum \alpha_i |v_i'| = |z'|$, then we get $(v_i, v_i') \sqsubseteq_{j+1} (z, z')$ for all i. Since $(z, z') \in H_{j+1}^+ \cup H_{j+1}^-$, the representation $(z, z') = \sum \alpha_i (v_i, v_i')$, $\alpha_i \in \mathbb{Z}_{>0}$ must be trivial, that is, $(z, z') = 1 \cdot (z, z') \in G$, and consequently, $(z, z') \in G$ and nothing is left to prove.

Hence, we will assume on the contrary that $\sum \alpha_i |v_i'| > |z'|$ holds. Therefore, there must exist $(v_{i_1}, v_{i_1}'), (v_{i_2}, v_{i_2}')$ such that $v_{i_1}' v_{i_2}' < 0$. The sum $(v_{i_1}, v_{i_1}') + (v_{i_2}, v_{i_2}')$ was reduced to 0 during the run of Algorithm 2.3 which gives a integer linear combination $(v_{i_1}, v_{i_1}') + (v_{i_2}, v_{i_2}') = \sum \beta_k (w_k, w_k')$ for some positive integers β_k and some $(w_k, w_k') \in G$.

Moreover, $\beta_k (w_k, w_k') \sqsubseteq_{j+1} (v_{i_1}, v_{i_1}') + (v_{i_2}, v_{i_2}')$ for all k, implying that $0 \leq w_k \leq z$ and that

$$\sum \beta_k |w_k'| = |\sum \beta_k w_k'| = |v_{i_1}' + v_{i_2}'| < |v_{i_1}'| + |v_{i_2}'|$$

holds. But then $\sum \beta_k (w_k, w_k') + (\alpha_{i_1} - 1)(v_{i_1}, v_{i_1}') + (\alpha_{i_2} - 1)(v_{i_2}, v_{i_2}') + \sum_{i \neq i_1, i_2} \alpha_i (v_i, v_i')$ contradicts the minimality of $\sum \alpha_i |v_i'|$. We conclude $\sum \alpha_i |v_i'| = |z'|$ and the claim follows. □

Note that if there are upper bounds u on z, the relations $v_{i_1} + v_{i_2} \leq z \leq u$ always holds in the above proof. This justifies the above mentioned restriction of necessary S-vectors that need to be considered if upper bounds u on the variables are given.

3 Improving the Project-and-Lift Algorithm

In this section we improve Algorithm 2.3 such that we need not reduce the S-vectors from C with respect to G. Instead, it suffices to check each S-vector only for reducibility with respect to G. Moreover, the set C of S-vectors that still have to be considered for reducibility need not be stored in memory explicitly. Both properties speed-up the computation, save computer memory and thus, allow us to solve bigger problems as by application of Algorithm 2.3 alone.

The major idea of the improved algorithm is to consider the elements (v, v') in C by increasing norm $\|v\|_1 = \sum_{i=1}^{j} v^{(i)}$. To this end, let $\mathcal{G}_i := \{(v, v') \in H_{j+1}^+ \cup H_{j+1}^- : \|v\|_1 = i\}$. Suppose we have already computed $\mathcal{G}_0, \mathcal{G}_1, \ldots, \mathcal{G}_k$, for some $k \geq 1$. Then the following observation allows us to avoid all the reduction steps in order to compute $\mathcal{G}_{k+1}$.

Lemma 3.1 *Every vector $(z, z') \in K_{j+1}^+ \cup K_{j+1}^-$ with $\|z\|_1 \leq k$ can be written as a finite positive integer linear combination $\sum \alpha_i(g_i, g_i')$ with $\alpha_i \in \mathbb{Z}_{>0}$, $(g_i, g_i') \in \mathcal{G}_{\leq k} := \mathcal{G}_0 \cup \mathcal{G}_1 \cup \ldots \cup \mathcal{G}_k$, and $(g_i, g_i') \sqsubseteq_{j+1} (z, z')$ for all i.*

Proof. Let $(z, z') \in K_{j+1}^+ \cup K_{j+1}^-$ with $\|z\|_1 \leq k$. Clearly, by definition of $H_{j+1}^+ \cup H_{j+1}^-$, (z, z') can be written as a finite positive integer linear combination $\sum \alpha_i(g_i, g_i')$ with $\alpha_i \in \mathbb{Z}_{>0}$, $(g_i, g_i') \in H_{j+1}^+ \cup H_{j+1}^-$, and $(g_i, g_i') \sqsubseteq_{j+1} (z, z')$ for all i. But $(g_i, g_i') \sqsubseteq_{j+1} (z, z')$ implies $\|g_i\|_1 \leq \|z\|_1 \leq k$, which, together with $(g_i, g_i') \in H_{j+1}^+ \cup H_{j+1}^-$, gives $(g_i, g_i') \in \mathcal{G}_{\leq k}$ and the proof is complete. □

Note that it is crucial for the proof of the following corollary that, in case of $j < s$, we always choose the smallest non-negative integer h' such that $(h, h') \in K_{j+1}$ when we form the input set F to Algorithm 2.3. Moreover, note that $\mathcal{G}_0 = \emptyset$ if $j \geq s$ and that $\mathcal{G}_0 = \{\pi_{j+1}^n(p_{j+1}), -\pi_{j+1}^n(p_{j+1})\}$ if $j < s$.

Corollary 3.2 *Let $(z, z') \in K_{j+1}^+ \cup K_{j+1}^-$, $\|z\|_1 = k + 1$, and $|z'| < p_{j+1,j+1}$. Then there are precisely two possible situations:*

1. *There exists a vector $(v, v') \in \mathcal{G}_{\leq k}$ with $(v, v') \sqsubseteq (z, z')$. Then we already know in advance that* normalForm$((z, z'), \mathcal{G}_{\leq k}) = 0$.
2. *There does not exist a vector $(v, v') \in \mathcal{G}_{\leq k}$ with $(v, v') \sqsubseteq (z, z')$. Then $(z, z') \in \mathcal{G}_{k+1}$.*

Proof. First, suppose that there exists $(v, v') \in \mathcal{G}_{\leq k}$ with $(v, v') \sqsubseteq (z, z')$. Clearly, if $j < s$, $(v, v') \sqsubseteq_{j+1} (z, z')$ and $|z'| < p_{j+1,j+1}$ imply $(v, v') \notin \mathcal{G}_0 = \{\pi^n_{j+1}(p_{j+1}), -\pi^n_{j+1}(p_{j+1})\}$. Thus we have $v \neq 0$ and consequently $\|z - v\|_1 \leq k$. We conclude by Lemma 3.1 that there are finitely many (not necessarily different) elements $(g_i, g'_i) \in \mathcal{G}_{\leq k}$ such that $(z - v, z' - v') = \sum(g_i, g'_i)$ and $(g_i, g'_i) \sqsubseteq_{j+1} (z - v, z' - v') \sqsubseteq_{j+1} (z, z')$ for all i. Choosing (v, v') together with these (g_i, g'_i) in the algorithm normalForm we obtain normalForm$((z, z'), \mathcal{G}_{\leq k}) = 0$.

On the other hand, if there does not exist a vector $(v, v') \in \mathcal{G}_{\leq k}$ with $(v, v') \sqsubseteq_{j+1} (z, z')$, then $(z, z') \in \mathcal{G}_{k+1}$, that is, the vector (z, z') cannot be written as a sum $(v_1, v'_1) + (v_2, v'_2)$ with $(v_i, v'_i) \in (K^+_{j+1} \cup K^-_{j+1}) \setminus \{0\}$ and $(v_i, v'_i) \sqsubseteq_{j+1} (z, z')$, $i = 1, 2$. To see this, assume on the contrary that such vectors (v_1, v'_1) and (v_2, v'_2) do exist. Note that both vectors lie in the same orthant of $\mathbb{R}^{j+1}$ as (z, z'). From $(v_i, v'_i) \sqsubseteq_{j+1} (z, z')$, $i = 1, 2$, and $|z'| < p_{j+1,j+1}$ we conclude that $v_i \neq 0$, $i = 1, 2$. Therefore, $\|v_i\|_1 \leq k$, $i = 1, 2$ which implies by Lemma 3.1 that (v_1, v'_1) and (v_2, v'_2) can both be written as finite positive integer linear combinations of elements from $\mathcal{G}_{\leq k}$ that all lie in the same orthant as (v_1, v'_1) and (v_2, v'_2). Put together, both combinations allow a finite positive integer linear representation of (z, z') by elements from $\mathcal{G}_{\leq k}$ that all lie in the same orthant as (z, z'). But all the summands $(v, v') \in \mathcal{G}_{\leq k}$ in this representation of (z, z') fulfill $(v, v') \sqsubseteq_{j+1} (z, z')$ in contradiction to our initial assumption that no such non-zero vector (v, v') exists, and thus, $(z, z') \in \mathcal{G}_{k+1}$. □

In view of Corollary 3.2, considering the elements (v, v') in C by increasing norm $\|v\|_1$ leads to an algorithm to compute $H^+_{j+1} \cup H^-_{j+1}$ that has the following advantages:

- The set C of S-vectors need not be stored, as for $k = 2, 3, \ldots$, the elements in $\mathcal{G}_{k+1}$ are generated and immediately checked for reducibility within a simple loop. No superfluous vectors are stored.
- Thus, precisely $H^+_{j+1} \cup H^-_{j+1}$ is computed.
- There is no reduction of vectors, only much cheaper reducibility tests are done.
- Upper bounds on variables can be easily be used directly to truncate the output set and to speed-up the computation.

It remains to give a criterion of when $H^+_{j+1} \cup H^-_{j+1} = \mathcal{G}_{\leq k}$ that allows us to stop the inductive construction of the $\mathcal{G}_k$.

Lemma 3.3 *Let $k \in \mathbb{Z}_{>0}$ satisfy $\mathcal{G}_{k+1} = \ldots = \mathcal{G}_{2k} = \emptyset$. Then $H_{j+1}^+ \cup H_{j+1}^- = \mathcal{G}_{\leq k}$.*

Proof. Note that the input set F defined in Section 2 satisfies $F \subseteq \mathcal{G}_{\leq k}$. Therefore, by Lemma 2.2, also $\mathcal{G}_{\leq k}$ generates $K_{j+1}^+ \cup K_{j+1}^-$ over $\mathbb{Z}_+$. Thus, by Lemma 2.5, Algorithm 2.3 would return a set contain $H_{j+1}^+ \cup H_{j+1}^-$ if it were started with $\mathcal{G}_{\leq k}$ as input set. However, by Lemma 3.1 and Corollary 3.2, all S-vectors reduce to 0 with respect to $\mathcal{G}_{\leq k}$ and Algorithm 2.3 returns exactly $\mathcal{G}_{\leq k}$. As all elements in $\mathcal{G}_{\leq k}$ are minimal, we must have $\mathcal{G}_{\leq k} = H_{j+1}^+ \cup H_{j+1}^-$. □

For $j \geq s$, one question remains open from a practical point of view: As the remaining variables (with index bigger than j) can be freely permuted and then re-permuted at the end of the computation, what is a good selection strategy to choose the $(j+1)^{\text{st}}$ variable to be appended next? First computational experiments show that huge differences in sizes of intermediate sets H_j^+ and in running times do indeed occur.

4 Applications

4.1 Decomposition of Vectors

Problem. Given $A \in \mathbb{Z}^{d \times n}$ and $u \in \mathbb{Z}_+^n$, find a representation $u = \sum_{i=1}^t \lambda_i h_i$ with $h_i \in H(\ker_{\mathbb{Z}}(A))$, $i = 1, \ldots, t$, and positive integral multipliers $\lambda_1, \ldots, \lambda_t$.

Solution. Compute a truncated Hilbert basis $\bar{H}(\ker_{\mathbb{Z}}(A))$ with upper bounds u on the variables. Once $\bar{H}(\ker_{\mathbb{Z}}(A))$ has been computed, u can be easily decomposed as a non-negative integer linear combination of elements of $\bar{H}(\ker_{\mathbb{Z}}(A))$ as follows: Find $v \in \bar{H}(\ker_{\mathbb{Z}}(A))$ such that $v \leq u$ and replace u by $u - v$. Note that $u - v$ can also be written as a non-negative integer linear combination of elements of $\bar{H}(\ker_{\mathbb{Z}}(A))$ as $u - v \leq u$.

This solution approach can be sped-up once an element of some H_j^+ is found that extends to a vector $v \in \ker_{\mathbb{Z}}(A) \cap \mathbb{R}_+^n$ with $v \leq u$. Note that this vector v must belong to H_n^+ as already its projection onto the first j components is not decomposable. In case such a vector v has been found, we may replace u by $u - v$ and use the stronger upper bounds $u - v$ instead of u. Note that the computation of the truncated Hilbert basis need not be started again. It suffices to throw away unnecessary vectors from H_j^+ (that do not lie below the new upper bounds $u - v$) and to continue with the computation of H_{j+1}^+. If H_n^+ is reached, we can completely decompose the current u or prove that it is not further decomposable and therefore $u \in \bar{H}(\ker_{\mathbb{Z}}(A)) \subseteq H(\ker_{\mathbb{Z}}(A))$.

4.2 Counting Magic Arrays and t-designs

Problem. Given a d dimensional $n \times n \times \cdots \times n$ array with n^d non-negative integer entries, wherein the sum of the entries in any coordinate (and in the main diagonal) direction equals a magic constant s. Find a formula for the number of magic arrays with given magic constant s.

Start of Solution. The method presented in [1] computes in a first step all non-decomposable magic arrays. This step is equivalent to the computation of the Hilbert basis of the cone of all magic arrays with non-negative real entries. It turned out that the computation of Hilbert bases is currently not the computational bottleneck in the method described in [1].

Hilbert basis computations are also employed in counting t-designs [7].

4.3 Integer Programming

Problem. Given $A \in \mathbb{Z}^{d\times n}$, $b \in \mathbb{Z}^d$, $c \in \mathbb{R}^n$, consider the integer program

$$\min_z \{c^\intercal z : Az = b, z \in \{0,1\}^n\}.$$

Given a 0-1 solution z_0 to $Az = b$, find a cheaper solution or prove optimality of z_0.

Solution. Transform the problem via $z^{(i)} \to 1-(z')^{(i)}$ whenever $z_0^{(i)} = 1$, $z^{(i)} \to (z')^{(i)}$ otherwise, into a 0-1 problem with problem matrix A', cost vector c', and current feasible solution $z'_0 = 0$. The above problem reduces to the computation of an element $v \leq 1$ in the truncated Hilbert basis of $\ker_{\mathbb{Z}}(A') \cap \mathbb{R}^n_+$ (with upper bounds 1) that has a strictly negative objective value $(c')^\intercal v$.

4.4 Dual cones

Problem. Given a cone $\mathcal{C}$ generated by integral vectors $p_1, \ldots, p_s$ in $\mathbb{R}^n$. Compute the generators $R(\mathcal{C}^D)$ and the minimal Hilbert basis $H(\mathcal{C}^D)$ of the dual cone

$$\mathcal{C}^D := \{v \in \mathbb{R}^n : z^\intercal v \geq 0, \forall z \in \mathcal{C}\} = \{v \in \mathbb{R}^n : p_i^\intercal v \geq 0, i = 1, \ldots, s\}.$$

Note the $R(\mathcal{C}^D)$ gives exactly the set of normal vectors of the facets of $\mathcal{C}$.

Solution. Let I_s denote the $s \times s$ identity matrix and let P be the $s \times n$ matrix that has $p_1^\intercal, \ldots, p_s^\intercal$ as rows. Moreover, let $L := \{(v,u) \in \mathbb{R}^{n+s} : Pv - I_s u = 0\}$ be generated over $\mathbb{R}$ by the vectors $(e_1, Pe_1), \ldots, (e_n, Pe_n)$. Note that these vectors also generate $\ker_{\mathbb{Z}}(P|-I_s)$ over $\mathbb{Z}$. Therefore, we have $\mathcal{C}^D = \pi_n^{n+s}(L \cap (\mathbb{R}^n \times \mathbb{R}^s_+))$ and thus

$$R(\mathcal{C}^D) = \{v \in \mathbb{R}^n : u = Pv \in R(P^\intercal)\}, H(\mathcal{C}^D) = \{v \in \mathbb{Z}^n : u = Pv \in H(P^\intercal)\},$$

where $R(P^\intercal)$ and $H(P^\intercal)$ denote the set of extreme rays and the Hilbert basis of the cone which is the intersection of $\mathbb{R}^s_+$ with the subspace of $\mathbb{R}^s$ spanned over $\mathbb{R}$ by the columns of P (rows of $P^\intercal$).

4.5 Hilbert Bases of Cones given by their Generators

Problem. Given a cone $\mathcal{C}$ generated by integral vectors $p_1, \ldots, p_s$ in $\mathbb{R}^n$, compute the minimal Hilbert basis $H(\mathcal{C})$ of $\mathcal{C}$.

Solution. Since $\mathcal{C} = (\mathcal{C}^D)^D$, this problem can be solved by first computing the generators $R(\mathcal{C}^D)$ of the dual cone and by then computing the Hilbert basis of the dual of the dual cone $\mathcal{C}^D$, which is the Hilbert basis of $\mathcal{C}$.

5 Computational Experience

The algorithm presented in Sections 2 and 3 were implemented into a computer program that we called MLP. It is written in C and downloadable from `http://www.testsets.de`. Besides (truncated) Graver test sets [8], MLP allows the computation of a (truncated) set of minimal generators of the monoid $\Lambda \cap \mathbb{R}^n_+$ for any lattice $\Lambda \subseteq \mathbb{Z}^n$, in particular for $\Lambda = \ker_{\mathbb{Z}}(A)$.

In [1], MLP was used to solve some non-trivial instances. For example, it took about 10 days on a 1GHz PC with 4GB RAM running linux to compute the set of all minimal 6×6 magic squares, yielding a Hilbert basis containing 522,347 elements. These examples can be found on the benchmark section of `http://www.testsets.de`.

Acknowledgments. The author wants to thank Jesús De Loera, UC Davis, for many helpful discussions and remarks. Moreover, the author gratefully acknowledges the support by AFOSR Grant F49620-01-0327 and by NSF Grant DMS-0073815.

References

1. Ahmed, M.; De Loera, J.; Hemmecke, R.: Polyhedral cones of magic cubes and squares, to appear in "New Directions in Combinatorial Geometry", Springer Verlag.
2. Ajili, F.; Contejean, E.: Avoiding slack variables in the solving of linear Diophantine equations and inequations, Theoretical Computer Science, 173, (1997), 183-208.
3. Bruns, W.; Koch, R.: `NORMALIZ`, computing normalizations of affine semigroups, Available from `ftp://ftp.mathematik.uni-onabrueck.de/pub/osm/kommalg/software/`.

4. Buchberger, B.: History and basic features of the critical-pair/completion procedure, Journal of Symbolic Computation, 2 (1987), 3-38.
5. Contejean, E.; Devie, Hervé, An efficient incremental algorithm for solving systems of linear Diophantine equations, Information and Computation, 113, (1994), 143-172.
6. Cox, D.; Little, J.; O'Shea, D.: Ideals, Varieties, Algorithms, Springer-Verlag, 1992.
7. De Loera, J.: Generating Functions that Count t-Designs with Given Automorphism Group: Algorithms and Structure, These proceedings.
8. Hemmecke, R.: On the Positive Sum Property of Graver Test Sets, Preprint SM-DU-468, University of Duisburg, Germany, 2000, Downloadable from `http://www.hemmecke.de/raymond`.
9. Henk, M.; Weismantel, R.: On Hilbert bases of polyhedral cones, Results in Mathematics, 32, (1997), 298-303.
10. Pasechnik, D. V.: Computing the Hilbert base via the Elliott-MacMahon algorithm, Theoretical Computer Science, 263, (2001), 37-46.
11. Pisón, P.; Vigneron, A.: First Syzygies of Toric Varieties and Diophantine Equations in Congruence, Comunications in Algebra, Vol. 29 (4), April 2001.
12. Pottier, L.: Euclide's algorithm in dimension n, Research report, ISSAC 96, ACM Press, 1996.
13. Rosales, J. C.; García-Sánchez, P. A.: Finitely Generated Commutative Monoids. Nova Science, Inc., New York, 1999.
14. Schrijver, A.: Theory of Linear and Integer Programming, Wiley, Chichester, 1986.
15. Sturmfels, B.: Gröbner Bases and Convex Polytopes, American Mathematics Society, Providence, Rhode Island, 1995.

ENCAPSULATE!

DORON ZEILBERGER*
DEPARTMENT OF MATHEMATICS, RUTGERS UNIVERSITY (NEW BRUNSWICK), HILL CENTER-BUSCH CAMPUS, 110 FRELINGHUYSEN RD., PISCATAWAY, NJ 08854-8019, USA.
ZEILBERG@MATH.RUTGERS.EDU,
HTTP://WWW.MATH.RUTGERS.EDU/~ZEILBERG/.

Abstract: The "pre-history" of mathematics, that is about to end, may also be called "naked-brain" mathematics, in analogy to 'naked-eye" astronomy. But the significace of the telescope to the development of astronomy eclipses in comparison to the future impact of the computer on mathematical research. To take full advantage of this mega-revolution, we have to get rid of many working habits and ways of thinking that we were forced to adopt in our trivial, pre-computer, search for mathematical truths. In particular, we should start to think in a computerized way, and most importantly, learn to Encapsulate objects and concepts, so that the computer can take it from there.

*SUPPORTED IN PART BY THE NSF.

COMPUTER ALGEBRA LIBRARY FOR THE CONSTRUCTION OF THE MINIMAL TELESCOPERS

S. A. ABRAMOV

Russian Academy of Science,
Dorodnicyn Computing Centre,
Vavilova st. 40, 119991, Moscow, GSP-1, Russia
E-mail: abramov@ccas.ru

K. O. GEDDES

Symbolic Computation Group
University of Waterloo
Waterloo, N2L 3G1, Canada
E-mail: kogeddes@scg.math.uwaterloo.ca

H. Q. LE

Symbolic Computation Group
University of Waterloo
Waterloo, N2L 3G1, Canada
E-mail: hqle@scg.math.uwaterloo.ca

This paper is an exposition of various recently-developed results related to the construction of the minimal telescopers for hypergeometric terms. A Maple library which includes an implementation of these results is described. A comparison between this implementation and other well-known implementations is also given.

1 Introduction

The sequences that are named *hypergeometric terms* (or simply *terms*) are very often involved in various combinatorial sums. The characteristic property of a term $T(k)$ is that the ratio $T(k+1)/T(k)$ is a rational function in k. This rational function, denoted by $\mathcal{C}_k(T)$, is the *certificate* of $T(k)$. A term $T(n,k)$ in two variables n and k has two certificates $\mathcal{C}_n(T) = T(n+1,k)/T(n,k)$ and $\mathcal{C}_k(T) = T(n,k+1)/T(n,k)$. They are named the n-certificate and the k-certificate, respectively. These certificates are rational functions in n and k. By using the notations E_n, E_k for the shift operators w.r.t. n and k, we obtain $\mathcal{C}_n(T) = E_nT/T$ and $\mathcal{C}_k(T) = E_kT/T$.

Given a term $T(n,k)$ as input, Zeilberger's algorithm [11,13], also known as the method of creative telescoping, tries to construct for $T(n,k)$ a Z-pair (L, G) which consists of a linear recurrence operator with coefficients which are polynomials in n over $\mathbb{C}$

$$L = a_\rho(n)E_n^\rho + \cdots + a_1(n)E_n^1 + a_0(n)E_n^0, \quad (1)$$

i.e., $L \in \mathbb{C}[n, E_n]$, and a term $G(n,k)$ such that

$$LT(n,k) = (E_k - 1)G(n,k). \quad (2)$$

It is proven in [13] that if there exist Z-pairs for $T(n,k)$ then Zeilberger's algorithm terminates with one of the Z-pairs and the telescoper L in the returned Z-pair is of minimal order. The minimal-order telescoper is unique up to a left-hand factor $P(n) \in \mathbb{C}[n]$, and is called the minimal telescoper. The Z-pair (L, G) where L is the minimal telescoper is called the minimal Z-pair.

Zeilberger's algorithm, named hereafter as $\mathcal{Z}$, has a wide range of applications which include finding closed forms of definite sums of hypergeometric terms, verification of combinatorial identities, and asymptotic estimate [11,13,10].

For a given term $T(n,k)$, it was for a quite long period of time that the question whether there exists a Z-pair for T (or whether $\mathcal{Z}$ terminates in finite time given T as input) was not conclusively answered, although a sufficient condition was known via the "fundamental theorem" [11,12]. It states that a Z-pair for T exists if T is a proper term (see Section 3 for a definition).

$\mathcal{Z}$ uses an item-by-item examination on the order ρ of the telescopers L. It starts with the value of 0 for ρ and increases ρ until it is successful in finding a Z-pair (L, G) for T, provided that such a pair exists. It is easy to observe that this strategy has two deficiencies. First, $\mathcal{Z}$ tries to compute a Z-pair for T when such a pair might not exist (Examples 2, 5). It is well-known that the "fundamental theorem" does not provide a necessary condition for the existence of a Z-pair. Secondly, let ρ be the order of the minimal telescoper for T, then $\mathcal{Z}$ simply wastes resources trying to compute a Z-pair where the guessed orders of the telescopers are less than ρ (Examples 4, 7).

The recent results [1,5,9] supply a theoretical foundation and algorithms to overcome the aforementioned problems. The main focus of this paper is a complete Maple implementation of these algorithms, in particular of a function which constructs the minimal Z-pairs by combining all these algorithms.

We first summarize the results that help guarantee the existence and expedite the construction of the minimal telescopers.

2 Additive Decomposition of Terms

We describe in this section the result on additive decomposition of terms in one variable [2]. This decomposition is used as a basis for the follow-up algorithms.

We consider an additive decomposition of a term $T(k)$ (see Theorem 1 below) assuming that the certificate of $T(k)$ is a rational function in k over an arbitrary field K of characteristic 0.

Definition 1 [3] *Let $R \in K(k)$ be a nonzero rational function. If there exist nonzero polynomials $f_1, f_2, v_1, v_2 \in K[k]$ such that*

(i) $R = F \cdot \frac{E_k V}{V}$ *where* $F = \frac{f_1}{f_2}$, $V = \frac{v_1}{v_2}$, *and* $\gcd(v_1, v_2) = 1$,

(ii) $\gcd(f_1, E_k^h f_2) = 1$ *for all* $h \in \mathbb{Z}$,

then $F \cdot \frac{E_k V}{V}$ *is a* rational normal form (RNF) *of* R.

Note that every rational function has an RNF [3] which in general is not unique. See [3] for a description of a construction of such a form.

As presented in [2], the algorithm to solve the additive decomposition problem for a term $T(k)$ constructs two terms $T_1(k)$, $T_2(k)$ such that

$$T(k) = (E_k - 1)\,T_1(k) + T_2(k), \tag{3}$$

and either T_2 vanishes or $\mathcal{C}_k(T_2)$ has an RNF

$$\frac{f_1}{f_2}\,\frac{E_k(v_1/v_2)}{(v_1/v_2)} \tag{4}$$

with v_2 of minimal degree. Any RNF of $\mathcal{C}_k(T_2)$ of the form (4) has $v_2 \in K[k]$ of the same minimal degree.

Theorem 1 [2] *Let $T(k)$ be a term and equality (3) be valid for some terms $T_1(k), T_2(k)$. Suppose that $T_2(k) \neq 0$. Let (4) be an RNF of $\mathcal{C}_k(T_2)$. Then (3) is an additive decomposition of $T(k)$ iff for each irreducible p from $K[k]$ such that $p \mid v_2$, the following three properties hold:*

$$\mathbf{Pa} : E_k^h p \mid v_2 \Rightarrow h = 0, \;\; \mathbf{Pb} : E_k^h p \mid f_1 \Rightarrow h < 0, \;\; \mathbf{Pc} : E_k^h p \mid f_2 \Rightarrow h > 0. \tag{5}$$

When working with terms in two variables n and k over $\mathbb{C}$, we can consider n as a parameter, and hence can construct an additive decomposition w.r.t. k:

$$T(n, k) = (E_k - 1)\,T_1(n, k) + T_2(n, k). \tag{6}$$

If (4) is an RNF w.r.t k of $\mathcal{C}_k(T_2)$ with $f_1, f_2, v_1, v_2 \in \mathbb{C}[n, k]$, then for each irreducible $p \in \mathbb{C}[n, k]$ such that $p \mid v_2$, properties (5) hold. Here K is $\mathbb{C}(n)$, and $K(k)$ is $\mathbb{C}(n, k)$.

Example 1 Consider the term

$$T(n, k) = -\frac{nk}{nk + 1}\binom{n+1}{k} + \binom{n+1}{k+1}.$$

An additive decomposition (T_1, T_2) of $T(n,k)$ w.r.t. k that satisfies (6) is

$$\left(\frac{2nk+2n-1}{2n} \prod_{w=0}^{k-1} \frac{n-w+1}{w+2}, \frac{n^2k+5nk-2k+5n-1}{4n(nk+1)} \prod_{w=0}^{k-1} \frac{n-w+1}{w+3} \right).$$

It follows from (6) that if the term T_2 vanishes, then $(1, T_1)$ is the minimal Z-pair for T. For the remainder of this paper, we assume that T_2 is not identically zero. Also note that a telescoper for T exists iff a telescoper for T_2 exists, and the minimal telescopers for these two terms are equal to each other.

3 Applicability of Zeilberger's Algorithm

This section provides a description of the result related to the applicability of $\mathcal{Z}$ to terms [1].

Definition 2 *A polynomial* $\alpha(n,k) \in \mathbb{C}[n,k]$ *is* integer-linear *if it has the form* $a\,n + b\,k + c$ *where* $a,\ b \in \mathbb{Z}$ *and* $c \in \mathbb{C}$.

Definition 3 [11,12] *A term* $T(n,k)$ *is* proper *if it can be written in the form*

$$P(n,k) \frac{\prod_{i=1}^{l} \Gamma(\alpha_i(n,k))}{\prod_{i=1}^{m} \Gamma(\beta_i(n,k))} u^n v^k \quad , \tag{7}$$

where $\alpha_i(n,k), \beta_i(n,k)$ *are integer-linear polynomials,* $l, m \in \mathbb{N}$, $u, v \in \mathbb{C}$.

The following theorem provides a necessary and sufficient condition for the termination of $\mathcal{Z}$.

Theorem 2 [1] *Let* $T(n,k)$ *be a term in* n *and* k*, and (6) be an additive decomposition of* T*. Let (4) be an RNF w.r.t.* k *of* $\mathcal{C}_k(T_2)$ *with* $v_2 \in K[n,k]$*. Then a telescoper for* $T(n,k)$ *exists iff each factor of* $v_2(n,k)$ *irreducible in* $\mathbb{C}[n,k]$ *is an integer-linear polynomial, i.e., iff* $T_2(n,k)$ *is a proper term.*

For a given polynomial $f(n,k) \in \mathbb{C}[n,k]$, a decision procedure for the factorability of f into integer-linear polynomials is described in [6]. This procedure does not require a complete factorization of f into irreducible factors.

Example 2 Consider the term $T(n,k)$ in Example 1. It follows from the computed additive decomposition of T that an RNF of $\mathcal{C}_k(T_2)$ of the form (4) has $v_2 = (n^2 + 5\,n - 2)(nk+1)$ which cannot be written as a product of integer-linear polynomials. This is an example where the input term T is not a proper term, and $\mathcal{Z}$ is not applicable to T either.

Example 3 Consider the term

$$T(n,k) = \frac{1}{nk+n+1} \binom{n}{k+1} + \frac{nk-2k-n+2}{n^2k+2nk^2-nk+2k+n-1} \binom{n}{k}.$$

An additive decomposition of $T(n,k)$ has $v_2 = (2k+n-3)$ which is integer-linear. Hence, even though $T(n,k)$ itself is not proper [4], $\mathcal{Z}$ is applicable to T.

4 Efficient Algorithms to Compute the Minimal Telescopers

Let $T(n,k)$ be a term. In this section we assume that $\mathcal{Z}$ is proven applicable to T. For the case where T is also a rational function in n and k (the class of rational functions is a proper subset of the class of terms), there exists a direct algorithm [9] which constructs the minimal telescoper for T efficiently without using item-by-item examination. For the case where T is a non-rational term, there exists an algorithm [5] which computes a lower bound μ for the order of the telescopers for T. This helps save the time to compute a telescoper of order less than μ.

4.1 Rational Function Case: a Direct Algorithm

Let $T(n,k) \in \mathbb{C}(n,k)$. Consider an additive decomposition of T of the form (6). First one constructs a special form of representation for T_2 as stated in the following theorem.

Theorem 3 [9] *Set*

$$T_2 = \sum_{i=1}^{t}\sum_{j=1}^{m_i} \frac{r_{ij}(n)}{(a_i n + b_i k + c_i)^j}, \quad a_i, b_i \in \mathbb{Z}, b_i > 0, c_i \in \mathbb{C}, \gcd(a_i, b_i) = 1, \tag{8}$$

$r_{ij}(n) \in \mathbb{C}(n)$. *Then* $T_2(n,k)$ *can be represented in the form*

$$M_1 F_1 + \cdots + M_s F_s, \tag{9}$$

where each $M_i \in \mathbb{C}(n)[E_n, E_k, E_k^{-1}]$, *each* $F_i = 1/(a_i n + b_i k + c_i)^{m_i}$ *is such that*

$$a_i, b_i \in \mathbb{Z},\ b_i > 0,\ c_i \in \mathbb{C},\ \gcd(a_i, b_i) = 1,\ m_i \in \mathbb{N} \setminus \{0\}, \tag{10}$$

and for all $i \neq j$, *at least one of the following four relations is not satisfied:*

$$m_i = m_j, a_i = a_j, b_i = b_j, c_i - c_j \in \mathbb{Z} \setminus \{0\}. \tag{11}$$

T_2 can be written in the form (8) since $\mathcal{Z}$ is assumed to be applicable to T. Once the representation (9) is constructed, one can compute the minimal telescopers for each member $M_i F_i \in \mathbb{C}(n,k)$ directly and efficiently [9]. The minimal Z-pair for $T_2(n,k)$, and subsequently for $T(n,k)$, can then be constructed using Least Common Left Multiple (LCLM) computation. This direct algorithm is in general more efficient than the original $\mathcal{Z}$. See Example 6 for a result of our experimentation.

4.2 Hypergeometric Case: a Lower Bound

Let $T(n,k)$ be a non-rational term. Consider an additive decomposition of T of the form (6). Since the minimal telescopers for T and T_2 are the same, the focus can be shifted to computing a lower bound for the order of the telescopers for the term T_2. Let an RNF w.r.t. k of $\mathcal{C}_k(T_2)$ be of the form (4). For each irreducible p such that $p \mid v_2$, the three properties **Pa, Pb, Pc** in (5) hold.

Definition 4 [5] *Let $M \in \mathbb{C}[n, E_n]$ be such that $MT_2 \neq 0$, and there exists an RNF $F' \frac{E_k V'}{V'}$, $V' = \frac{v_1'}{v_2'}$ of $\mathcal{C}_k(MT_2)$ such that each of the irreducible factors of v_2' does not have at least one of the three properties* **Pa, Pb, Pc**. *Then M is a* crushing operator *for T_2. The minimal crushing operator is a crushing operator of minimal order.*

It is simple to show that if L is a telescoper for T_2, then L is also a crushing operator for T_2. Hence, the problem of computing a lower bound for the order of the telescopers for T_2 is reduced to the problem of computing a lower bound for the order of the minimal crushing operator for T_2.

Theorem 4 [5] *Let $\mathcal{C}_k(T_2)$ have an RNF w.r.t. k $F(E_k V)/V$ of the form (4), $f_1, f_2, v_1, v_2 \in \mathbb{C}[n,k]$, and $D = d_1(n,k)/d_2(n,k)$ be such that $\mathcal{C}_n(T_2) = D(E_n V)/V$. Let there exist a crushing operator for T_2 of order ρ. Then for each integer-linear factor p of v_2, $\deg_k p = 1$, there exists an integer h such that*

$$E_k^h\, p \mid E_n v_2 \cdot E_n^2 v_2 \cdots E_n^{\rho} v_2 \cdot d_2 \cdot E_n d_2 \cdots E_n^{\rho-1} d_2. \tag{12}$$

As a consequence, if ρ_p is the minimal positive value of ρ such that there exists an h satisfying (12), then the order of any crushing operator for T_2 is not less than $\mu = \max_{p \mid v_2} \rho_p$.

Since $\mathcal{Z}$ is assumed to be applicable to the input $T(n,k)$, it follows from Theorem 2 that the polynomial $v_2 \in K[n,k]$ factors into integer-linear polynomials. By [4], the polynomial $d_2 \in K[n,k]$ in Theorem 4 also factors into integer-linear polynomials. An algorithm, called *LowerBound*, which realizes Theorem 4 is described in [5]. Once each of the two polynomials v_2, d_2 is written as a product of integer-linear polynomials (this does require a complete factorization of monic univariate polynomials into irreducible factors, see [9]), the algorithm is reduced to solving bivariate linear diophantine equations, a very inexpensive operation.

Example 4 Consider the term

$$T(n,k) = \frac{(n+k+2)!}{(n^2+k+2)(k+3)!} - \frac{(n+k+1)!}{(n^2+k+1)(2+k)!} + \frac{(n+k)!}{(n+7\,k-2)\,k!}.$$

The computed lower bound for T as the result of applying *LowerBound* to $T(n,k)$ is 7 which is also equal to the order of the minimal telescoper for T.

5 Implementation

We have implemented the algorithms described in Sections 2, 3, 4 in the computer algebra system Maple 7. They are grouped together into a module named `Telescopers`.

```
> eval(Telescopers);
```

module()
export AdditiveDecomposition, IsZApplicable, ZpairDirect,
        LowerBound, Zeilberger, MinimalZpair;
option package;
description "Algorithms to compute the minimal telescopers for
        hypergeometric terms";
end module

5.1 Functionalities

The exported local variables indicate the functions that are accessible to users. Due to space limitations, we only briefly describe the functions in the module. See Section 6 for information on how to obtain the specifications for these functions.

- AdditiveDecomposition(T, k) computes an additive decomposition of the term $T(k)$. The output is a list of two elements $[T_1, T_2]$ representing the two terms T_1, T_2 in the decomposition (6);
- IsZApplicable(T, n, k) returns *true* if $\mathcal{Z}$ is applicable to the term $T(n, k)$, *false* otherwise;
- ZpairDirect(T, n, k, E_n) computes the minimal Z-pair for the rational function $T(n, k)$. The output is a list of two elements $[L, G]$ representing the minimal Z-pair (L, G) for T, or an error message if it is proven that $\mathcal{Z}$ is not applicable to T;
- LowerBound(T, n, k) returns $\mu \in \mathbb{N}$ which is the computed lower bound for the order of the telescopers for the term $T(n, k)$, or an error message if it is proven that $\mathcal{Z}$ is not applicable to T;
- Zeilberger(T, n, k, E_n) returns a list of two elements $[L, G]$ representing the minimal Z-pair (L, G) for the input term $T(n, k)$. Note that an upper bound ρ for the order of the telescopers for $T(n, k)$ needs to be specified in advance (the default value is 6). The function returns an error message if no telescoper of order less than or equal to ρ exists.

The main function of the module is MinimalZpair. It has the calling sequence "MinimalZpair(T, n, k, E_n)" where T is a term in n and k, and E_n denotes the shift operator which acts on n. This function combines the functionalities of all functions in the above list. For an input term $T(n,k)$, the execution steps can be described as follows.

1. determine the applicability of $\mathcal{Z}$ to T;
2. if it is proven in step 1 that a Z-pair for T does not exist, return the conclusive error message "There does not exist a Z-pair for T"; Otherwise,
   a. if T is a rational function in n and k, apply the direct algorithm to compute the minimal Z-pair;
   b. T is a non-rational term. First compute a lower bound μ for the order of the telescopers for T. Then compute the minimal Z-pair using $\mathcal{Z}$ with μ as the starting value for the guessed orders.

For case 2b, since the term T_2 in the additive decomposition (6) is "simpler" than T in some sense, we first apply $\mathcal{Z}$ to T_2 to obtain the minimal Z-pair (L, G) for T_2. It is easy to show that $(L, LT_1 + G)$ is the minimal Z-pair for the input term T (Example 7).

Example 5 Consider the term

$$T(n,k) = \frac{1}{nk+1}\binom{2\,n}{2\,k}.$$

Apply MinimalZpair and Zeilberger to $T(n,k)$, and record the time required [a].

MinimalZpair first checks for the applicability of $\mathcal{Z}$ to T (step 1). It recognizes that $\mathcal{Z}$ is not applicable to T and returns the conclusive answer "Error, (in MinimalZpair) There does not exist a Zpair for T" in 0.56 seconds. Zeilberger, on the other hand, does not know if a Z-pair (L, G) for T exists. It tries to compute one and returns the inconclusive answer "Error, (in Zeilberger) No telescoper of order 6 was found" in 30.55 seconds. Since there does not exist a Z-pair for T, the higher the value of the upper bound for the order of L is set, the more time and memory are wasted.

Example 6 (rational function case) This example is a comparison between the original Zeilberger's algorithm and the direct algorithm (case 2a of MinimalZpair). The test samples are the same as those used in Example 5 in [9]. Three set of tests each of which consists of 20 rational functions in n

[a] All the reported timings were obtained on a 400Mhz SUN SPARC SOLARIS with 1Gb RAM.

and k were randomly generated. Each rational function where the numerator and the denominator are in expanded forms is of the form (9). We ran MinimalZpair ($\mathcal{M}$), Zeilberger ($\mathcal{Z}$) on these tests, and collected resource requirements. We also enforced a limit of 2,000 seconds on each input rational function in the tests. Note that we only recorded the time and space requirements for the tests that ran under this time limit.

Table 1 shows the time and space requirements to run the three sets of tests S_1, S_2 and S_3.

Table 1. Time and space requirements for $\mathcal{M}$ and $\mathcal{Z}$ (Example 6).

	Completed		Timing (seconds)		Memory (kilobytes)	
	$\mathcal{M}$	$\mathcal{Z}$	$\mathcal{M}$	$\mathcal{Z}$	$\mathcal{M}$	$\mathcal{Z}$
S_1	20	15	12.15	3127.84	54,159	8,095,930
S_2	20	18	12.43	2635.94	54,653	7,873,146
S_3	20	0	959.07	–	3,864,026	–

Example 7 (hypergeometric case) For $b \in \mathbb{N} \setminus \{0\}$, $j \in \{1, 3\}$, let

$$T_1 = \frac{1}{(nk-1)(n-bk-2)^j(2n+k+3)!}, \; T_2 = \frac{1}{(n-bk-2)(2n+k+3)!}.$$

Consider the term $T(n,k) = (E_k - 1)\, T_1(n,k) + T_2(n,k)$. This example is a comparison between the original Zeilberger's algorithm and case 2b of MinimalZpair. The computed lower bound for the order of the telescopers is b, while the order of the minimal telescoper is $b+1$. Let $\mu \in \mathbb{N}$ be the starting value for the guessed order of the telescopers. Recall that the function Zeilberger applies $\mathcal{Z}$ to the input term T with $\mu = 0$, while MinimalZpair applies $\mathcal{Z}$ to the term T_2 in the decomposition (6) with $\mu = b$. Table 2 shows the time and space requirements. As one can easily notice, as b and/or j increase, the relative performance of Zeilberger (compared to MinimalZpair) quickly worsens.

5.2 A Comparison

There exist different Maple implementations of $\mathcal{Z}$ such as `Zeil` in the EKHAD package [11], `sumrecursion` in the sumtools package [8], `SummandToRec` in the HYPERG package [7]. A Mathematica implementation (the function `Zb`) is described in [10]. Due to the lack of a criterion for the applicability of $\mathcal{Z}$ at the time these programs were implemented, the item-by-item examination strategy is employed (these programs are in principal equivalent to the program

Table 2. Time and space requirements for $\mathcal{M}$ and $\mathcal{Z}$ (Example 7).

		Timing (seconds)		Memory (kilobytes)	
j	b	$\mathcal{M}$	$\mathcal{Z}$	$\mathcal{M}$	$\mathcal{Z}$
1	1	6.49	5.35	27,838	24,702
	2	8.34	34.64	33,066	142,889
	3	11.13	124.53	44,233	535,736
	4	14.46	570.02	56,410	1,882,730
	5	25.79	2999.22	97,506	6,536,309
3	1	14.64	16.40	62,566	73,830
	2	17.24	228.59	68,304	770,529
	3	20.15	1286.51	78,701	3,074,051
	4	24.08	8771.08	91,844	10,766,646
	5	38.60	77663.68	139,823	33,423,168

`Zeilberger` in our package). This strategy leads to the two deficiencies which are discussed in Section 1, and which are illustrated by the examples in this paper.

For the case where the input is a rational function, a program such as `Zb` "accepts an input if the irreducible factors of the denominator are integer-linear" [10]. This is equivalent to the condition that the input be a proper term. By Theorem 2, such a program prevents the computation of a Z-pair when such a pair exists. Note that we also implemented in the program `MinimalZpair` a direct and efficient algorithm to compute the minimal Z-pairs.

For the case where the input $T(n, k)$ is a non-rational term, all the aforementioned programs apply $\mathcal{Z}$ directly to T. On the other hand, MinimalZpair first computes a lower bound μ for the order of the telescopers (a fairly low-cost operation), and then applies $\mathcal{Z}$ to the term T_2 in the additive decomposition (6) using μ as the starting value for the guessed orders of the telescopers (note that the existence of a Z-pair is guaranteed). The minimal Z-pair for T can then be easily obtained. Experimentation shows that this proposed approach helps expedite the construction of the minimal Z-pairs.

6 Availability

Information on the availability of the library archive, functional specifications, test samples used in this paper can be found at the URL
`www.scg.math.uwaterloo.ca/~hqle/code/Telescopers/Telescopers.html`

Acknowledgements

This work is partially supported by the French-Russian Lyapunov Institute under grant 98-03, by Natural Sciences and Engineering Research Council of Canada Grant No. RGPIN8967-01, and No. CRD215442-98. The authors wish to express their thanks to R.F. Burger for his careful reading of a preliminary version of this paper.

References

1. S.A. Abramov, Applicability of Zeilberger's Algorithm to Hypergeometric Terms, to appear in *Proc. ISSAC'2002.*
2. S.A. Abramov, M. Petkovšek, Minimal Decomposition of Indefinite Hypergeometric Sums, *Proc. ISSAC'2001*, 2001, 7–14.
3. S.A. Abramov, M. Petkovšek, Canonical Representations of Hypergeometric Terms. *Proc. FPSAC'2001*, 2001, 1–10.
4. S.A. Abramov, M. Petkovšek, Proof of a Conjecture of Wilf and Zeilberger. Preprint Series of the Institute of Mathematics, Physics and Mechanics **39**, 2001, no. 748, Ljubljana, March 9, 2001.
5. S.A. Abramov, H.Q. Le, A Lower Bound for the Order of Telescopers for a Hypergeometric Term, to appear in *Proc. FPSAC'2002.*
6. S.A. Abramov, H.Q. Le, Applicability of Zeilberger's Algorithm to Rational Functions, *Proc. FPSAC'2000*, Springer–Verlag LNCS, 2000, 91–102.
7. B. Gauthier, HYPERG, Maple Package, User's Reference Manual. Version 1.0, `http://www-igm.univ-mlv.fr/~gauthier/HYPERG.html`.
8. W. Koepf, *Hypergeometric Summation: An Algorithmic Approach to Summation and Special Function Identities*, Vieweg, 1998.
9. H.Q. Le, A Direct Algorithm to Construct Zeilberger's Recurrences for Rational Functions, *Proc. FPSAC'2001*, 2001, 303–312.
10. P. Paule, M. Schorn, A Mathematica Version of Zeilberger's Algorithm for Proving Binomial Coefficient Identities. *J. Symb. Comput.* **20**, 1995, 673–698.
11. M. Petkovšek, H. Wilf, D. Zeilberger, *A=B*, A.K. Peters, Wellesley, Massachusetts, 1996.
12. H. Wilf, D. Zeilberger, An Algorithmic Proof Theory for Hypergeometric (ordinary and "q") Multisum/Integral Identities, *Inventiones Mathematicae,* 108, 1992, 575–633.
13. D. Zeilberger, The Method of Creative Telescoping, *J. Symb. Comput.* **11**, 1991, 195–204.

TOPCOM: TRIANGULATIONS OF POINT CONFIGURATIONS AND ORIENTED MATROIDS

JÖRG RAMBAU

Zuse-Institut Berlin, Takustr. 7, 14195 Berlin, Germany

TOPCOM is a package for computing triangulations of point configurations and oriented matroids. For example, for a point configuration one can compute the chirotope, components of the flip graph of triangulations, enumerate all triangulations. The core algorithms implemented in TOPCOM are described, and implentation issues are discussed.

1 Introduction

TOPCOM[1] uses a combinatorial model of point configurations in order to compute their triangulations in arbitrary dimensions: the oriented matroid.[2,3,4] This allows to avoid round-off errors in all calculations once the oriented matroid of the point configuration has been computed. If the point configuration is has rational coordinates then its oriented matroid can be computed with exact arithmetics. The number of operations in exact arithmetics necessary to compute the relevant oriented matroid data for a particular problem is minimal in the sense that any coordinate-based method achieving the same goal needs at least as many operations in exact arithmetics. Naïve coordinate-based approaches like checking intersections of simplices by linear programming need usually a lot more.

2 Mathematical Background

The basic idea utilized in TOPCOM is the following combinatorial characterization of triangulations:

Theorem 2.1 (Rambau[5]). *Let $\mathcal{A}$ be a full-dimensional point configuration in $\mathbb{R}^d$. A set T of full-dimensional $(d+1)$-subsets (simplices) of $\mathcal{A}$ is a triangulation of $\mathcal{A}$ iff*

(IP) For every pair S, S' of simplices in T, there exists no circuit (Z^+, Z^-) in $\mathcal{A}$ with $Z^+ \subseteq S$ and $Z^- \subseteq S'$ (Intersection Property).

(UP) For each facet of a simplex S in T there is another simplex $S' \neq S$ having F as a facet, or F is contained in a facet of $\mathcal{A}$ (Union Property).

With the following characterization, we only need the chirotope representation of the oriented matroid of $\mathcal{A}$ in order to check a triangulation:

Theorem 2.2 (de Loeara et al.[6]). *Let $\mathcal{A}$ be a full-dimensional point configuration in $\mathbb{R}^d$. A set T of full-dimensional $(d+1)$-subsets (simplices) of $\mathcal{A}$ is a triangulation fo $\mathcal{A}$ iff*

(CP) For every interior facet F, there are exactly two simplices $F \cup s$ and $F \cup s'$ in T such that s and s' have different signs in the cocircuit defined by F (Cocircuit-Property).

(EP) $\mathcal{A}$ contains a simplicial facet or there is an extension in the relative interior of $\mathcal{A}$ in general position that is contained in the interior of exactly one simplex (Extension Property).

3 Main Algorithms

We present the main algorithms implemented in TOPCOM together with rough estimates for the asymptotic run-time complexity.

We denote by n the number of points in $\mathcal{A}$, by r the rank of $\mathcal{A}$, which equals the dimension (denoted by d) plus one, and by $c := n - r$ the corank of $\mathcal{A}$. In all practically solvable cases you may think of n and r as small numbers compared to the total number of triangulations of $\mathcal{A}$.

3.1 Compute the Chirotope

First, we describe how we compute the interface from geometric information in the coordinates of the points to purely combinatorial data. The chirotope χ is a function assigning a sign in $\{0, +, -\}$ to each r-subset of $\mathcal{A}$; this describes the orientation of the corresponding sequence of points.

$$\chi : \begin{cases} \binom{\mathcal{A}}{d+1} \to \{+, -, 0\} \\ (i_1, i_2, \ldots, i_{d+1}) \mapsto \operatorname{sign}\big(\det(a_{i_1}, a_{i_2}, \ldots, a_{i_{d+1}})\big) \end{cases} \tag{1}$$

Naïvely, we have to compute $\binom{n}{r}$ determinants of size r, each requiring $O(r^3)$ steps if we proceed by an elimination algorithm (unit cost per arithmetic operation assumed).

If we organize the computation in a certain tree we can reuse some elimination operations. The nodes at level k of the computation tree are matrices M_k in column normal form corresponding to a subset A_k of k columns of $\mathcal{A}$. A child A_{k+1} of a node is produced by adding a column to the right of A_k to M_k and performing elimination steps until we arrive at a matrix M_{k+1} in column normal form. The leaves of the tree are all possible $(r \times r)$-submatrices of $\mathcal{A}$ in equivalent triangular form, and their determinant can be computed easily.

Counting nodes and operations in the tree, we get that the number of operations per determinant is essentially $O(\frac{r+c}{1+c}r^2)$. Why? We first count the nodes in the levels of the computation tree as follows: Let $g(k,j) = \binom{j-1}{k-1}$ be the number of submatrices of A with k columns and rightmost column at index j. Then the number of nodes in level k of the tree is equal to $\sum_{j=k}^{n-r+k} g(k,j)$. We estimate the number of operations in each node of level k by rk because of the following considerations. We have to eliminate $k-1$ values in the kth column. We start the elimination in row 1 and proceed from top to bottom. If the mth row, $1 \leq m < k$, of the mth column is non-zero we can eliminate (i.e., set to zero) the non-zero in the mth row of the newly added kth column by computing $r-m \leq r$ new numbers. If m is the smallest index such that the mth row of column m is zero, we swap column k and column m. The first $k-1$ columns of the resulting matrix are again in column normal form, and row m of column k is zero now. Therefore, and we can proceed with the next m. The total number of operations is therefore at most rk. The total number of nodes in the tree is given by $\sum_{k=1}^{r}\sum_{j=k}^{n-r+k} g(k,j)$. Hence, the number of operations is at most $G(n,r) := \sum_{k=1}^{r}\sum_{j=k}^{n-r+k} g(k,j)kr$. The number of operations per determinant is then $G(n,r)/\binom{n}{r}$. The analysis of this expression can be done most easily in a computer algebra program to get the result $G(n,r) = \frac{(c+r+1)r^2}{c+2}$, which asymptotically has the claimed growth.

This proves that the computation tree yields savings of order r per determinant whenever the codimension is not much smaller than the dimension.

The chirotope values are stored in a hash table so that they can be retrieved in amortized constant time. Alternatively, one could compute the lexicographic index of each simplex and store its value in an array, yielding access to each chirotope value constant times computing the lexicographic index.

3.2 Check a Potential Triangulation

From now on, we assume that the chirotope of $\mathcal{A}$ has been preprocessed. Given the combinatorial characterizations of triangulations, checking the correctness of a triangulation is, in principle, straight-forward. The only issue is efficiency. TOPCOM uses characterization 2.2 for checking the correctness of triangulations:

- (CP) requires checking the link of all d-facets in T. The computation of all links can be accomplished in $O(r|T|)$ operations by scanning all the simplices in T and collecting their contributions to the links of their facets in a hash map (supporting unique insert in amortized constant time). If the link of some facet F contains only one point, we need to check whether F is in the boundary of $\mathcal{A}$. This can be done by retrieving at most c chirotope values. In case the link has more than two elements, we output 'invalid' immediately. In the case that there are two elements in the link of a facet, the orientations of the two points in the corresponding cocircuit can be checked by retrieving two chirotope values. Summarizing, the number of operations needed to check (CP) is $O(cr|T|)$.

- The existence of a simplicial facet is not checked. Instead, (EP) is always checked by utilizing a lexicographic extension p in the interior of a some simplex in T. The formula for the chirotope values involving p requires the retrieval of at most r other chirotope values (only one if $\mathcal{A}$ is in general position).[4] For all simplices in T we need to check whether p is in its interior. This can be done by retrieving r chirotope values involving p. Thus, for the whole procedure, we need at most $O(r^2|T|)$ steps.

Thus, in at most $O(\max\{c, r\}r|T|)$ we can check the validity of a triangulation.

3.3 Construct a Placing Triangulation

The construction of a placing triangulation[7] is an incremental process. During the process, we maintain in each step k

- the set of points $\mathcal{A}_k$ that is currently triangulated
- a placing triangulation T_k of $\mathcal{A}_k$
- a set $\mathcal{F}_k$ of all boundary facets of T_k that are interior in $\mathcal{A}$, i.e., those facets
  - that are facets of exactly one simplex in T_k
  - that are not in the boundary of $\mathcal{A}$, i.e., the corresponding cocircuits are neither positive nor negative in $\mathcal{A}$.

The first r points to be added have to form a valid (i.e., full-dimensional) simplex. Finding such a simplex can be integrated in the computation of the chirotope at the cost of storing a simplex S with non-zero chirotope. Consequently, $\mathcal{A}_r$ equals S, T_r consists of the simplex S, and $\mathcal{F}_r$ contains those facets of S that are not in the boundary of $\mathcal{A}$.

Adding a new point a_{k+1} to $\mathcal{A}_k$, yielding $\mathcal{A}_{k+1}$, for some $k \geq r$ will add all simplices $F \cup a_{k+1}$ to $\mathcal{T}_k$ for which $F \in \mathcal{F}_k$ is visible from a_{k+1}, i.e., the sign of a_{k+1} in the cocircuit defined by F is opposite to those signs of the points in $\mathcal{A}_k$ that are non-zero: this yields T_{k+1}. (There are points in $\mathcal{A}$ with non-zero sign since $\mathcal{A}_k$ is full-dimensional for all $k \geq r$.) The visible facets in $\mathcal{F}_k$ are removed and the new non-boundary facets of the new simplices are added to $\mathcal{F}_k$: this yields $\mathcal{F}_{k+1}$.

The process stops as soon as $\mathcal{F}_k$ is empty for some $k \geq r$. This happens at the latest when $k = n$, but maybe earlier, in which case not all points are used for the placing triangulation. If one wants a triangulation that uses all the points the missing points are added one by one by performing stellar subdivisions inside existing simplices: the points are "flipped-in".

The number of operations on the sets $\mathcal{F}_k$ is bounded by the number of all $(d-1)$-faces of the resulting placing triagulation T, which is at most $O(r|T|)$. The use of universal hashing facilitates unique insert and deletion in amortized constant time. For one visibilty check we need to retrieve at most c chirotope values (if $\mathcal{A}$ is in general position two values suffice). Each facet occuring in some $\mathcal{F}_k$ may have to be checked for visibility in each but one step of the construction.

Thus, in total we need at most $O(rc^2|T|)$ operations (resp. $O(rc|T|)$ operations if $\mathcal{A}$ is in general position) for all visibility checks. This dominates the computation of a placing triangulation.

3.4 Explore a Flip-Graph Component

Flips are stored in TOPCOM as pairs of outgoing and incoming simplices (flip-out sets and flip-in sets). TOPCOM uses a standard Breadth-First-Search (BFS) procedure to explore the flip graph.[8] Since the graph is not given explicitely, we need to construct edges dynamically during the exploration. To this end, we maintain for each triangulation a node containing its set of flips, i.e., a set of edges leading out of this node. Given a flip in a triangulation we can compute the resulting triangulation in amortized time $O(r)$ by using a special data structure for triangulations that allows for unique insert and deletion in amortized constant time (e.g., hash tables with universal hashing or something derived thereof).

New flips are found using the chirotope as follows: Assume, $\mathcal{S}$ is the set of flips in T, and we have used one of those flips $f \in \mathcal{S}$ to discover a new triangulation T'. Then the set $\mathcal{S}'$ of flips of T' inherits all those flips in $\mathcal{S}$ whose flip-out sets are disjoint from f's flip-out set. Finding these flips takes time $O(r|\mathcal{S}|)$.

Moreover, there are some new flips. Their flip-out sets have always non-empty intersection with the flip-in set of f. Thus, for finding new flips we can restrict ourselves to potential flips containing simplices of f's flip-in set. To this end, we first build all potential supports (sets of involved vertices) of flips by collecting all $(r+1)$-supersets of simplices in the flip-in set that might support a flip. This can be done in $O(r|T'|)$ by finding adjacent simplices in T'; alternatively one can do this in $O(rn)$ by adding all possible points to an in-flip simplex, which results, however, in more non-flippable supports. Since checking whether or not a support is actually flippable is the expensive operation, TOPCOM uses the first method. The maximal number of possible supports returned by this procedure is bounded from above by the number of boundary facets of the flip-in set of f which is $O(r)$.

To check the flippability of an $(r+1)$-set, we first compute its circuit signature

$Z = (Z^+, Z^-)$ by retrieving $r + 1$ chirotope values. The two possible triangulations $T^+(Z)$ and $T^-(Z)$ can be computed from this data in time $O(r)$. Then we need to check whether one, say $T^+(Z)$, of the two possible circuit triangulations is a subcomplex of T'. If Z is in general position (no chirotope value is zero) then we need to check containment in T' for all the at most $O(r)$ simplices in $T^+(Z)$. This can be done in amortized time $O(r)$ by using a special data structure for triangulations that allows for membership test in amortized constant time. If Z is not in general position we need to compute the links of all (maximal) simplices in T^+ in T'. This takes at most time $O(r|T'|)$. If all simplices in T^+ have the same non-empty link then Z is indeed flippable, and the flip-out and flips-in sets are the unions of the simplices in $T^+(Z)$ resp. $T^-(Z)$ with the common link; they can therefore be deduced in time $O(r^2)$ (rank of $\mathcal{A}$ times corank of Z).

Summarizing, we need time $O(r^2|T'|)$ for finding new flips in T'; this reduces to $O(r|T'| + r^2)$ when $\mathcal{A}$ is in general position. Thus, generating the new flip set $\mathcal{S}'$ takes time $O(r^2|T'| + r|\mathcal{S}|)$, which is in $O(r^2(|T'| + |T|))$.

Note that computing flips from scratch for each node would yield a time complexity of $\Omega(r|T'|^3)$—assuming the same algorithm is used for checking flippability of a potential support of a flip.

Using a data structure for triangulations that allows to retrieve an adjacent simplex in constant time (the adjacency graph[9]) can reduce the time complexity in the non-general-position case. This, however, would result in a higher complexity when computing an adjacent node, a higher complexity for equality checks for triangulations, and—most important—in a substantially higher memory consumption. Moreover, the sizes of problems for which an asymptotically superior data structure returns the result faster is out of reach memory-wise anyway because for small problems the TOPCOM data structures facilitate very small constants in the computing time bounds. This, however, has to be carefully monitored in the future as the computers are getting more and more powerful.

3.5 Explore the Partial Order of Partial Triangulations

Here, TOPCOM employs a standard Depth-First-Search (DFS) in the Hasse-diagram of all partial triangulations, partially ordered by inclusion. Partial triangulations are sets of simplices satisfying (IP). Every node in the DFS-tree contains a partial triangulation T, a set of simplices $A(T)$ (admissibles) of simplices that can be added to T without violating (IP), and a the set $F(T)$ of facets of simplices in the boundary of T that are in the interior of $\mathcal{A}$. The leaves in the DFS-tree are the non-extendable partial triangulations, and some of them are in fact triangulations, namely those T with $F(T) = \emptyset$. This is essentially the method of enumerating maximal independet sets[9] in the intersection graph of all simplices, except that, first, we don't proceed via adjacent simplices only and, second, we use the chirotope for checking proper intersections of simplices. The basic method goes back to 1980.[10]

Edges (i.e., new simplices) are explored in lexicographic order. Consider a node $(T, A(T))$ and add a simplex $S \in A(T)$ to T in order to discover node $(T', A(T')$. Then $A(T)$ is updated to $A(T) \setminus S$, so that the none of the other branches above $(T, A(T)$ can ever reach a partial triangulation with subcomplex $T' = T \cup S$.

This way, we only need the memory to store one complete branch in the DFS-tree. If M is the maximal number of simplices in a triangulation of $\mathcal{A}$, then storing $O(M)$ simplices on a stack for the partial triangulations plus storing $O(M\binom{n}{r})$ simplices in the admissibles fields suffice. Important is the fact that the memory requirements do not depend on the number of triangulations of $\mathcal{A}$.

TOPCOM uses a preprocessed hash map with all possible $\binom{n}{r}$ simplices as keys. The value $A(S)$ of a simplex S is the set of all simplices S' such that S and S' form an admissible pair, i.e., $\{S, S'\}$ satisfies (IP). This preprocessing takes time $O(\binom{n}{r}^2)$. The table allows for the following fast update algorithm: if T' is discovered by adding simplex S to T then $A(T') = A(T) \cap A(S)$. This step can be accomplished in time $O(|A(T)|)$, which is—crudely estimated—in $O(\binom{n}{r})$.

The set of all interior facets of $\mathcal{A}$ is also preprocessed by checking for each $(r-1)$-subset the facet property. This needs time $O(r\binom{n}{r-1})$. The set $F(T)$ can be updated by adding the interior boundary facets of S to $F(T)$ modulo 2. These are can be done in time $O(r|F(T)|)$, which is in $O(r^2M)$.

In total, we need time $O(\binom{n}{r}^2)$ preprocessing time and time $O(r\binom{n}{r})$ per ouput node. The time needed per triangulation is way higher because there are a lot more proper partial triangulations than triangulations.

3.6 Check Regularity (Requires External LP Solver with Exact Arithmetics)

Checking regularity of a triangulation T requires checking the feasibility of a linear program with a constraint for each interior facet F in T. The constraint induced by F adjacent to the simplices S and S' in T expresses that S and S' can be "folded" at F such that the resulting signed volume spanned by S and S' is positive.

This is a condition that can be expressed as a condition on the determinant of the homogeneous coordinate vectors of the points in $S \cup S'$. This determinant develops into a linear constraint in the height variables for each point. The coefficients turn out to be the same determinants that specify the signs for the chirotope values on all subsets of $S \cup S'$. If we therefore save the determinants rather than the signs in the chirotope then we can form each constraint in time $O(r)$, resulting in a time complexity of $O(r^2|T|)$ for setting up the complete linear program.

This effort is currently dominated by actually solving the linear program in exact arithmetics. TOPCOM employs for this task the cdd package.[11]

3.7 Symmetry-Handling

TOPCOM can exploit symmetries that are given by the user in form of generating permutations. The main savings using symmetries can be implemented in the flip graph exploration. In the BFS procedure there are always three kinds of nodes:[8] unknown nodes (white nodes), nodes that have been discovered but may have unknown neighbors (gray nodes), and discovered nodes all of whose neighbors have also been discovered (black nodes). Black nodes can be removed from memory.

The BFS procedure behaves friendly w.r.t. symmetries. That means it is possible (by an equivalent-edge marking procedure) to store only one representative per symmetry class and to remove the black nodes from memory. This works without ever discovering a white node equivalent to a black one.[3] Using the symmetries that leave a triangulation unchanged one can reduce the effort of finding flips.

Symmtries are also observed in checking (CP) in a potential triangulation. In contrast to this, (EP) has to be checked for all simplices—not only for all symmetry classes of simplices.

4 Data Structures and Implementation Details

In this section, we informally explain a selection of low level details in the current implementation of TOPCOM.

4.1 Simplicial Complexes

Simplicial complexes are regarded as the sets of their maximal simplices. TOPCOM aims at using set data structures that perform fast on the following tasks: equality check, membership test, unique insert, deletion, intersection, union.

Simplices are implemented as dynamic bitsets. The largest number of points ever handled in TOPCOM was 324 in dimension six (the Santos triangulation). This still requires only 41 Bytes per simplex, as opposed to $7 \cdot 4 = 28$ Bytes for an array representation. For all potentially accessible enumeration problems the number of points must be much smaller: in most applications 32 bits suffice, in which case also the memory consumption is superior.

Equality check needs $\lceil n/32 \rceil$ integer equality checks. Membership test, unique insert, and deletion just need a single (very fast) bit operation each. Intersections and unions are linear in n; however, on a normal PC, 32 bits can be processed at a time, thus this is practically very fast for reasonable values of n.

Simplicial complexes are based on dynamic bitsets where each bit represents a simplex. The assignment of bit positions to simpleces remains fixed during one complete run of the program. When a simplex occurs for the first time it gets the next free bit assigned; this assignment is stored in a dictionary. Retrieving the simplex corresponding to a bit is implemented via an array indexed by integers; getting the bit position of a simplex is accomplished by a hash table look-up. Both can be done (theoretically and practically) in amortized constant time.

This way, operations on simplicial complexes can work fast via bitset operations. Another advantage is the excellent behaviour in terms of memory cache misses (a main bottleneck on nowaday's computers): retrieving even a long bitstring from memory is very fast whenever the bits are stored consecutively. If we enumerate a flip graph component then for each possible simplex there is at least one triangulation (usually many) that contains that simplex. Thus, we need to store each possible simplex at least once, whence the use of the TOPCOM data structure does not harm memory-wise.

The main draw-back of this structure is that equality checks take time $O(\binom{n}{r})$, with small constants, though. The plausible observation is that for dense simplicial complexes (number of simplices is of order $\binom{n}{r}$), the structure is performing very well; for very sparse complexes, one should use an ordinary representation.

When the sizes of accessible problems grow, TOPCOM will probably switch to an asymptotically faster date structure. For (most) instances solvable on nowaday's computers, the bitstring technique works fastest.

4.2 Chirotopes

The chirotope values for simplices are stored in a hash map, allowing for amortized constant time retrieval. If the problem is very large, the table can be assigned a fixed size and works as a cache with random evict.

4.3 Node in the Flip-Graph

TOPCOM stores the triangulation T as a simplicial complex and the flips of T as a hashmap assigning marked or unmarked to the representation of the flip.

A flip's internal representation is simply a dependent set with $r+1$ points, which is stored as a simplex. By deleting $r+1$ bits and retrieving $r+1$ chirotope values, one can obtain from this a flip represented by a pair of simplicial complexes: the flip-in and the flip-out simplices. Thus, deleting the flip-out set from and adding the flip-in set to a triangulation works on bitset level. Marking of a particular flip would work in amortized constant time in hash maps. Because the set of flips assigned to one triangulation is not that large, TOPCOM now rather uses sorted maps for flips (logarithmic time for marking, but less overhead).

4.4 Node in the Tree of Partial Triangulations

The triangulation T, the set A(T) of admissible simplices, and the set of uncovered interior facets F(T) are all stored as simplicial complexes. This way the computing the intersection of admissible sets is a bitset operation. Moreover, adding interior facets modulo two means an xor-operation on bitset level, which is also very fast in practice.

4.5 Symmetries

Symmetries are stored as arrays σ of integers, where $\sigma[i]$ is the image of vertex i under the symmetry σ. The images of simplices and simplicial complexes are computed vertex by vertex. Here is certainly potential for improvement by using more sophisticated structures.

5 Computational Experiments

We present some results of experiments with the data structures used in TOPCOM-0.10.1 in Table 1. In Table 2, we reproduce a table from[3] containing some large enumeration problems solved by TOPCOM for the first time (to the best of our knowledge).

	relative CPU-time/s for different data structures representing a simplicial complex		
operation	STL set(vec(int))	STL set(bitstr.)	TOPCOM
insert 99 simplices	100%	58%	50%
assignment	100%	49%	0.3%
delete 89 simplices	100%	57%	40%
intersect (~ 50 simplices)	100%	56%	4.7%
lex. compare (~ 50 simplices)	100%	53%	64%
equality false (~ 50 simplices)	100%	100%	600%
equality true (w.r.t. line above)	72100%	39250%	750%

Table 1. Operations in simplicial complexes (1.8GHz Pentium IV, 1GByteX RAM, SuseLinux 7.3, simplices chosen randomly) using three different data structures for simplicial complexes: STL (=C++ Standard Template Library) set(vec(int))=STL set, based on red-black trees, of arrays of integers; STL set(bitstr.)=STL set of TOPCOM's dynamic bitstrings; TOPCOM= TOPCOM's bitstrings specifying simplex indices w.r.t. a table of essentially all possible simplices; parameters comparable to a triangulations of the 4-cube; note that equality checks in STL sets are fast when the cardinalities of the operands differ, as was the case in the equality-false test instances above; the genuin equality checks for TOPCOM's simplicial complexes are much faster but cannot take advantage of the cardinality function

what	*configuration*	*description*	*result*	*CPU-time/s*
# triangulations	C(12,5)	cyclic polytope	5,049,932	2,104
# triangulations	C(13,7)	cyclic polytope	6,429,428	4,055
# fine triangulations	(4 × 5)-lattice	two-dim. lattice, 20 points	2,822,648	501
# flip graph component	$\Delta_3 \times \Delta_3$	product of tetrahedra	4,533,408	499
# flip graph component	C^4	four-dimensional cube	92,487,256	15,624
check (IP') & (UP')	Santos triang.	six-dim. construction	okay	70
# flips	Santos triangulation	six-dim. construction	0	30

Table 2. Some figures computed by TOPCOM for the first time (1.8GHz Pentium IV, 1GByte RAM, SuseLinux 7.3); cyclic polytopes have connected flip graphs;[5] so have two-dimensional point sets; fine triangulations use all the vertices

References

1. J. Rambau. TOPCOM—Triangulations of Point Configurations and Oriented Matroids. `http://www.zib.de/rambau/TOPCOM.html`, 1999–2002.
2. J. A. de Loera. *Triangulations of Polytopes and Computational Algebra*. PhD thesis, Cornell University, 1995.
3. J. Pfeifle and J. Rambau. Computing triangulations using oriented matroids. ZIB-Report 02-02, Zuse-Institute Berlin, 2002.
4. A. Björner, M. Las Vergnas, B. Sturmfels, N. White, and G. M. Ziegler. *Oriented matroids*. Cambridge University Press, Cambridge, 1993.
5. J. Rambau. Triangulations of cyclic polytopes and higher Bruhat orders. *Mathematika*, 44:162–194, 1997.
6. J. A. de Loera, S. Hoşten, F. Santos, and B. Sturmfels. The polytope of all triangulations of a point configuration. *Documenta Mathematika*, 1:103–119, 1996.
7. C. Lee. Triangulations of polytopes. In J. E. Goodman and J. O'Rourke, editors, *CRC Handbook of Discrete and Computational Geometry*, chapter 14, pages 271–190. CRC Press LLC, Boca Raton, 1997.
8. T. H. Cormen, C. E. Leiserson, and R. L. Rivest. *Introduction to Algorithms*. MIT Press, 1990.
9. F. Takeuchi and H. Imai. Enumerating triangulations for products of two simplices and for arbitrary configurations of points. In *Computing and Combinatorics*, pages 470–481, 1997.
10. E. Lawler, J. Lenstra, and A. H. G. Rinnooy Kan. Generating all maximal independent sets: NP-hardness and polynomial-time algorithms. *SIAM Journal on Computing*, 9:558–565, 1980.
11. K. Fukuda. cdd—double description method and LP solver. `http://www.ifor.math.ethz.ch/~fukuda/cdd_home/cdd.html`.

PARALLEL BACKTRACK COMPUTING OF ASSOCIATION SCHEMES USING CLASSROOM PC'S

IZUMI MIYAMOTO

Yamanashi University 4-3-11 Takeda Kofu 400-8511 JAPAN
E-mail:izumi@esi.yamanashi.ac.jp

A parallel backtrack search visits its searching tree in a breadth-first fashion. The nodes at each level are visited parallel by the processors and again their children are redistributed among the processors. A cluster of classroom PC's is not suitable for parallel computing, because its communication capacity is not so rich or reliable. In this paper we introduce a method to search the subtree rooted by the node distributed to each processor in a depth-first fashion with a certain time limit. If there remain some subtrees not visited thoroughly, then we collect the nodes which are the roots of the subtrees and redistribute the children of these nodes in the next step. This reduces the times of sending messages among the processors. We apply this method to computing association schemes. We also show what data is stored in order to restart the parallel computation when it is interrupted.

1 Introduction

In parallel computing we usually expect that the network connection is reliable and not interrupted during the computation. However most clusters of workstations are used for the education of students. Among such a cluster of workstations the network connection is sometimes interrupted temporally. So a parallel computation using such computers may be interrupted during the computation and not finish successfully. Although such computing systems may not be used for theoretical parallel computing, we want to enjoy the computing power of such systems, because we sometimes have computing problems of large size. Association schemes are algebraically combinatorial objects and backtrack methods are popular in computing combinatorial objects. The aim of the present paper is to study how to compute a backtrack algorithm parallel to construct association schemes using such a cluster of workstations practically. We think that the searching trees of the backtrack algorithm in our problem are common. So we expect that our method is available to other problems. We note that our original programs are developed for ordinary sequential computing, as we are not studying theoretical parallel computing. Our method proposed in the present paper requires only a slight modification in order to use them for parallel computing.

In Miyamoto [10] an integer factorization by elliptic curve method is computed using classroom PC's parallel. In elliptic curve method many curves are computed to find a curve that gives a factorization. So we only send a messages to each processor to generate a curve randomly and receive the result whether it gives a factorization. We continued this process until we received a successful message. Hence in this problem the situation is quite easy. Even if the computation is interrupted, we only start to compute again and wait for a successful curve before the next interruption. Using ParGAP [3] and GMP-ECM [12], a new record in the factorizations by elliptic curve method is obtained in Miyamoto [10] (cf. Web page of R. Brent [2]).

A. Hanaki and the author [6,7] have been classifying the isomorphism classes of association schemes of small order using computers. The isomorphism classes of association schemes of order up to 30 are obtained. Especially A. Hanaki improved the program and classified the cases of order 29 and 30 recently. We constructed association schemes that covered the isomorphism classes by a backtrack algorithm. The association schemes are generated from various input patterns. It happened that it took a very long computing time for a small number of input patterns. We would like to show how to compute the backtrack algorithm of constructing association schemes from an input pattern parallel. At the present time no new results are obtained by parallel computing. However our program could not compute one input pattern of order 29 in the classification of association schemes. The classification of this case depended on the classification of strongly regular graphs by E. Spence [11]. By parallel computing we finished the computation of this case.

GAP [4] is a software package for algebraic and symbolic computation. ParGAP(Parallel GAP/MPI) [3] is a share package of GAP, which run GAP parallel. ParGAP is designed so that the messages are sent and received between one master process and slave processes. The last program constructing association schemes was written in C-language by A. Hanaki. The author modified Hanaki's program for the present experiment. We use ParGAP as an interface running the modified Hanaki's program parallel.

General situation of the backtrack search to compute association schemes satisfies the following conditions. An input pattern determines the (maximal) height of the searching tree. The tree is of bounded degree. So the size of the tree is bounded. Initially only the root of the tree is known and the children of a node are made only after the node itself is visited. If the tree attains the maximal height, then the leaves at the deepest level give the association schemes associated with the input pattern.

A parallel backtrack search is rather straitforward if it visits the searching tree in a breadth-first fashion. Then we can visit the nodes at a certain level parallel by the processors and again redistribute their children among the processors. Theoretical parallel computing of a backtrack method based on this strategy is seen in Herley *et al* [8] using prefix sums in JáJá [9], Goldberg and Zwick [5] for load balancing. In our problem there are too many nodes at some levels. This yields to increase the memory use and message communications. Furthermore our computations were interrupted by some troubles in message communications and it is not clear in the above algorithm what data should be saved in order to restart the computation. If the number of the nodes at some level is not too large but fairly large comparing to that of the processors, we may expect that if we distribute these nodes to the processors to visit the subtrees rooted by these nodes, the load among the processors may be balanced, as some processor visits many subtrees of small size and some processor visits a subtree of large size. In our problem this is not well attained.

In our method we set a certain time limit. As stated in the previous paragraph, if we obtain a sufficient number of nodes at a certain level, then we visit the subtrees rooted by the nodes in a depth-first fashion up to the time limit and redistribute only the children of the nodes of which subtrees were not visited completely within the time limit. We repeat this process until all the subtrees are visited. This process yields that if a node of which subtree is large, then we compute the children of the node and distribute them to the processors in the next step to visit their subtrees parallel. In our case the number of the nodes treated at a certain level is large enough comparing that of the processors and one step stated above is suitable for parallel computing. Actually our experiments of this parallel computing were interrupted during the computation and we had to restart the processes nearly from the interrupted point of the computation. So we also designed our parallel algorithm so that we could restart the process in such a way easily.

In our department we have about 75 PC's for the education of our students. So our computing system consists of (about) 75 processors with distributed memories and a shared file system. We use these machines parallel for the present experiment. The machines operate under Linux with Pentium III 800MHz and 256MB memory. The network is 100 base LAN.

2 Computing association schemes

For the definition of association schemes readers may refer to Bannai and Ito [1]. We would like to explain what is needed here. Let $(X, \{R_i\}_{0 \leq i \leq d})$ be an association scheme on a set X and let A_0, $A_1, \cdots, A_d$ be its adjacency matrices with $A_0 =$ *identity matrix*. So each entry of A_i is 0 or 1. Its relation matrix A is defined by $A = 0A_0 + 1A_1 + 2A_2 + \cdots + dA_d$. Each A_i has a constant column and row sum, say, v_i. As $A_0 + A_1 + A_2 + A_d = J$ (all

1 matrix), each row and column of the relation matrix A consists of v_i i's. Let $n = |X|$, which is the size of the relation matrix A. For each adjacency matrix A_i there exists i^* such that $A_{i^*} = {}^tA_i$.

Our input pattern to construct association schemes is a couple of lists $[v_1, v_2, \cdots, v_d]$ and $[1^*, 2^*, \cdots, d^*]$. So the first list is a partition of the integer $n-1$. The lower triangular part of A is determined by its upper triangular part and the list $[1^*, 2^*, \cdots, d^*]$ in a input. We fix a lexicographic order on relation matrices. Then the list $[v_1, v_2, \cdots, v_d]$ determines the first row of the relation matrix A. Our backtrack method proceed in a depth-first fashion to search the possibilities of the entries (i, j) of the relation matrix A from $(2, 3)$, $(2, 4)$, $\cdots$, to $(2, n)$, and then from $(3, 4)$, $(3, 5)$, $\cdots$, to $(n-1, n)$. Hence the node of the searching tree of our backtrack methods is denoted by a relation matrix A of which entries are determined partially, for instance, the entries $(2, 3)$, $(2, 4)$, $\cdots$ and (i, j) are determined. We note that Hanaki's program saves the relation matrix periodically during the computation in order to restart computing from this matrix in case of interruption of computing.

3 Parallel Computing

For parallel computing we modify the program in the following way. If the program visits a node at the deepest level which corresponds to the $(n-1, n)$ entry of some relation matrix A successfully, then we have an association scheme. If we limit the searching to a certain level, say $(5, n)$ entry of A, then we have all the relation matrices constructed up to the 5-th row. If our backtrack method followed a breadth-first fashion, we could have such sets of relation matrices at each level. So the above modification is considered to convert a depth-first search to a breadth-first search partially. Next we start the program from one of the relation matrix obtained above. The program finishes if it returned to the root, $(1, n)$ entry of A. So we modify the program so that it finishes if it returned to the certain level, $(5, n)$ entry of A. Then we can visit the subtree rooted by a node at level $(5, n)$ entry of A. We also modify the program so that it checks a time limit whenever it visits the next node and that it finishes computing if the time is over.

In the first step of our algorithm, we distribute the nodes at a certain level, for instance, level $(5, n)$ to the processors and visit the subtrees rooted by these nodes in a depth-first fashion up to a certain time limit. If there remain some subtrees not visited thoroughly, then we collect the nodes which are the roots of such subtrees. These nodes will be distributed in the next step. This is the first step of our algorithm.

In the second step we distribute these nodes to the processors, visit these nodes and redistribute their children (or descendents at another certain level, say level $(6, n)$). Then we visit the subtrees rooted by the redistributed children up to a certain time limit and if there are some subtrees not visited thoroughly, then we collect the children which are the roots of such subtrees. These children are the nodes distributed in the next step.

In the following steps we repeat the process in the second step by setting appropriate levels, until all the subtrees are visited within the time limit. The levels $(5, n)$ and $(6, n)$ stated above are chosen by experiments. Since in every step we collect the children of which subtrees are not visited thoroughly within the time limit and pass them to the next step, the number of the nodes in each step does not become too large. So we save these nodes in every step. In the first part of our computation it may take a while to obtain the nodes at level $(5, n)$. So we modified the program to compute a certain number of nodes, say 20 nodes, and distribute to the processors these 20 nodes and the computation of the next 20 nodes. We call the set of such 20 nodes a lot.

In the first step we save the lot-number when all the subtrees rooted by the nodes in a lot are visited up to the time limit. We also save the last node of every lot. We compute the next lot from the last node of the present lot. If the computation is interrupted in the first step, we find the smallest lot-number not saved. Then using the data of the last nodes, we restart the computation. In the second step and after we save the node-number when all the subtrees rooted by the children of the node are visited up to the time limit. Then if the computation is interrupted, we restart it to distribute the nodes of which numbers are not saved. We note that, by setting a time limit, we are free from troubles during a long time search of a large subtree.

4 Experiments

Case of input pattern [11, 11], [2, 1] of order 23: The size of relation matrix is 23. In the tables the times are real times in seconds. It took about 35 minutes by sequential computing. It took 7 seconds to compute all the 164 relation matrices up to the 5-th row (ie. the nodes at level $(5, 23)$). The number 164 may be not large enough to distribute the matrices to 75 processors but we distributed them and searched the subtrees rooted by them parallel without a time limit as the first experiment. In Table 1 the first column shows the time ranges for searching subtrees, the second column shows the numbers of the nodes of which subtrees were searched within the time ranges and the third column shows the total times for the parallel searches achieved within the time range.

In Table 2 we set the time limit 30 seconds. As above we had the 164

Table 1. order 23 without a time limit

time	number of nodes	total time
0	71	0
1∼9	30	97
10∼19	20	300
20∼49	29	930
50∼120	12	851
155	1	155
215	1	215
parallel computing		255

Table 2. order 23 with a time limit

level	nodes	time	remained nodes	all nodes
$(5,23)$	164	42	38	164
$(6,23)$	190	44	10	601
$(7,23)$	69	19	0	3769
total time		105		

nodes at level $(5,23)$ and distributed them to the processors to search their subtrees up to the time limit of 30 seconds. We collected the nodes of which subtrees remained not to be searched thoroughly. Their number is shown in the fourth column. In the next step we computed their children (descendents at level $(6,23)$) parallel and redistributed the children to search their subtrees up to the time limit of 30 seconds, and again collected the children of which subtrees remained not to be searched thoroughly. We repeated this process at level $(7,23)$ and finished the computation. The numbers of the children in the second step and later are denoted in the second column of row 2 and below. The number of all the nodes at each level is denoted in column 5, which was computed sequentially. If we compute the backtrack search parallel in a simple breadth-first fashion, then the numbers in column 5 of nodes are distributed to the processors, while in our method the numbers in columns 2 and 4 of nodes are distributed. We obtained 104 relation matrices in this case.

Case of input pattern $[13,13]$,$[2,1]$ of order 27: In this case our experiments were similar as above. First, we computed the 1067 nodes (relation matrices) at level $(5,27)$ and distributed them to 75 processors without a time limit. This is shown in Table 3. It took about 48 hours for the parallel computing and it took about 40 hours for the longest subtree search. About

Table 3. order 27 without a time limit

time	number of nodes	total time
0	395	0
1~9	185	222
10~99	18	879
100~999	74	37043
1000~9999	248	1079405
10000~99999	144	3851808
100000~142807	3	372690
	parallel computing	170775

Table 4. order 27 with a time limit

level	nodes	time	remained nodes	all nodes
(5, 27)	1067	15312	359	1067
(6, 27)-1	*3616	39320	*164	15794
(6, 27)-2	6664	47616	381	–
(7, 27)	9751	5961	0	428728
	total time	108209		

3 minutes after the start of the process all the processors became busy and almost all processors were busy for about next 16 hours. After that no tasks were distributed to the processors and we waited for the longest task of 142807 seconds.

In the next experiment we set the time limit 25 minutes. In Table 4 we computed the 1067 nodes at level $(5, 27)$ and distributed them to the processors to search their subtrees up to the time limit of 25 minutes. We collected the 359 nodes of which subtrees were not searched thoroughly and distributed them to compute their children at level $(6, 27)$. Then we redistributed the children to the processors to search the subtrees rooted by them up to the time limit of 25 minutes again. This search was interrupted during the computation and we had to restart it, which is shown in rows 2 and 3 of Table 4. As in Table 2, the fourth column in Table 4 shows the numbers of the remained nodes and the second column shows the numbers of the children of these nodes. Before the interruption 3616 children of 217 nodes among the 359 nodes at level $(5, 27)$ were generated and redistributed, but 195 node-numbers were saved. So we restarted the computation of this step with the remaining 164 nodes at level $(5, 27)$. There were 10053 nodes at level $(6, 27)$ which were

Table 5. order 29

level	nodes	time		remained nodes	all nodes
(5, 29)	1449	8239		249	1449
(6, 29)-1	7794	15941	137	94	
(6, 29)-2	1183	19104	107	283	
(6, 29)-3	572	17342	91	204	
(6, 29)-4	5064	129461	*3053	2472	44988
(7, 29)-1	1242	12413	3003	10	
(7, 29)-2	21725	175506	1932	192	
(7, 29)-3	45959	179691	1049	139	
(7, 29)-4	7399	18477	906	0	
(7, 29)-5	6078	14260	807	0	
(7, 29)-6	36086	102037	462	0	
(7, 29)-7	34334	96891	*341	0	778291
(8, 29)	4969	5725		0	-
total time		794331			

the children of the 359 remained nodes at level $(5, 27)$. The number of the obtained relation matrices was 4508.

Case of input pattern $[14, 14], [1, 2]$ of order 29: The isomorphism classes of association schemes given by this input pattern are obtained from the classification of strongly regular graphs by E. Spence [11], and our program can not compute this pattern without parallel computing. The data is shown in Table 5. In this case the computation was interrupted so many times. So we inserted a new column in the fourth. The numbers with an asterisk in column 4 show the total numbers of the remained nodes at levels $(6, 29)$ and $(7, 29)$. The other numbers in column 4 show that these numbers of nodes are remained after the interruption of the computing. The numbers of the remained nodes to be distributed in the next step are denoted in column 5. The second column shows the numbers of their children, each of which subtrees are visited in a depth-first fashion. So these numbers are considered to indicate how often troubles occur during the computation. We obtained 194 relation matrices in total. It took about 13 calendar days for the whole experiment, while the total time of the computation was a little shorter than 10 days. It took about 30 minutes to compute all the 44988 nodes at level $(6, 29)$ and about 15 hours to compute all the 778291 nodes at level $(7, 29)$ So we did not compute sequentially all the nodes at level $(8, 29)$.

We note that the numbers of the relation matrices are not the number of isomorphism classes of association schemes with the given input patterns, since our program to construct relation matrices do not check isomorphisms completely.

Acknowledgements

The author would like to express his thanks to A. Hanaki for using his program in the present experiments.

References

1. E. Bannai and T. Ito, Algebraic Combinatorics I: Association Schemes, Menlo Park, CA: Benjamin/Cummings, 1984.
2. R. Brent, http://www.comlab.ox.ac.uk/oucl/work/richard.brent/factors.html, ftp://ftp.comlab.ox.ac.uk/pub/Documents/techpapers/Richard.Brent/champs.txt
3. G. Cooperman, http://www.ccs.neu.edu/home/gene/pargap.html
4. The GAP Group, *GAP – Groups, Algorithms, and Programming, Version4*, Lehrstuhl D f Mathematik, Rheinisch Westfälische Technische Hochschule, Aachen, Germany and School of Mathematical and Computational Sciences, Univ. St. Andrews, Scotland, 1997.
5. T. Goldberg, U. Zwick, Optimal deterministic approximate parallel prefix sums and their applications, Proc. of 3th Israel Symposium on Theory of Computing and Systems, (1995), 220-228.
6. A. Hanaki. Data of association schemes, published at WWW (1999~): http://kissme.shinshu-u.ac.jp/as/.
7. A. Hanaki and I. Miyamoto, *Classification of association schemes with* 16 *and* 17 *vertices*, Kyushu J. Math. **52** (1998) 383-395. *Classification of association schemes with* 18 *and* 19 *vertices*, Korean J. Comp. App. Math. **5** (1998) 543-551. *Classification of association schemes of order up to* 22, Kyushu J. Math. **54** (2000) 81-86.
8. K. Herley, A. Pietracaprina, and G. Pucci, *Deterministic parallel backtrack search*, Theoretical Computer Science, 2001. To appear.
9. J. JáJá, Introduction to Parallel Algorithms, Addison Wesley, Reading MA, 1992.
10. I. Miyamoto, Experiments on ParallelGAP, (in Japanese) in Computer Algebra - Algorithms, Implementations and Applications, RIMS kokyuroku to appear.
11. E. Spence, http://gauss.maths.gla.ac.uk/~ted/srgraphs.html
12. P. Zimmermann, http://www.loria.fr/ zimmerma/records/ecmnet.html

ON CANONICAL REPRESENTATIONS OF CONVEX POLYHEDRA

DAVID AVIS
School Of Computer Science, MGill University, Canada
E-mail: avis@cs.mcgill.ca

KOMEI FUKUDA
School Of Computer Science, MGill University, Canada
E-mail: fukuda@cs.mcgill.ca

STEFANO PICOZZI
Institute of Mathematics, EPFL, Switzerland
E-mail: stefano.picozzi@epfl.ch

Every convex polyhedron in the Euclidean space $\mathbb{R}^d$ admits both H-representation and V-representation. When working with convex polyhedra, in particular large-scale ones in high dimensions, it is useful to have a canonical representation that is minimal and unique up to some elementary operations. Such a representation allows one to compare two H-polyhedra or two V-polyhedra efficiently. In this paper, we define such representations that are simple and can be computed in polynomial time. The key ingredients are redundancy removal for linear inequality systems and affine transformations of polyhedra.

1 Introduction

A *convex polyhedron* or simply *polyhedron* in $\mathbb{R}^d$ is the set of solutions to a finite system of inequalities with real coefficients in d real variables. For a matrix $A \in \mathbb{R}^{m\times d}$ and a vector $b \in \mathbb{R}^m$, a pair (b, A) is said to be an *H-representation* of a convex polyhedron P if $P = \{x \in \mathbb{R}^d \mid b + Ax \geq 0\}$. Motzkin's decomposition theorem (see, e.g. [4,6]) states that every polyhedron has another representation called a V-representation. For matrices $V \in \mathbb{R}^{p\times d}$ and $R \in \mathbb{R}^{q\times d}$, a pair (V, R) is said to be a *V-representation* of a polyhedron P if $P = conv(V) + cone(R)$, where $conv(M)$ ($cone(M)$, respectively) denotes the convex hull (the nonnegative hull) of the row vectors of the matrix M. In each of representations, there are trivial transformations that preserve the represented polyhedron. For an H-representation, each inequality can be multiplied with any positive numbers. For a V-representation, each row of R can be scaled by any positive number. Also, any permutation of inequalities and the row vectors within V and R does not modify the polyhedron. Two representations are considered *equal* if they are the same up to these transformations.

One can easily see that neither V- nor H-representations are unique for a given polyhedron. First of all both V- and H-representations can have an unlimited number of redundant vectors or inequalities. Furthermore, even without redundancy, representations are not unique. Namely there are infinite many nonredundant V-representations of polyhedra without extreme points, and infinite many nonredundant H-representation of polyhedra that are not full-dimensional, see Figure 1 . The main objective of the present paper is to

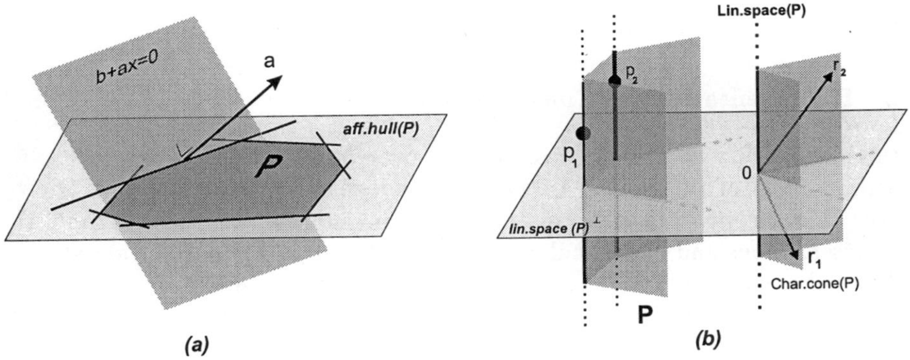

Figure 1. (a) A 2-dimensional H-polyhedron P in $\mathbb{R}^3$, and its affine hull $\mathit{aff}(P)$. One has infinite many possible choices for vector a which preserve the polyhedron P. Requiring that a belongs to the linear subspace parallel $\mathit{aff}(P)$ leads to the representation discussed in the book of Schrijver. (b) A 3-dimensional non pointed polyhedron $P = conv(p_1, p_2) + cone(r_1, r_2) + lin.space(P)$, and its characteristic cone $char.cone(P) = cone(r_1, r_2) + lin.space(P)$. The V-representation of P is clearly not unique. As discussed by Schrijver, requiring that p_1, p_2, r_1 and r_2 belong to $lin.space(P)^{\perp}$ ensures the uniqueness of the representation.

define a unique V-representation (H-representation, respectively) of a polyhedron that can be computed in polynomial time from any V-representation (H-representation). We shall review the unique polyhedral representations discussed in the book of Schrijver [4], and propose a variation that is computationally simpler and guaranteed to have certain sparsity property. As a consequence, it is possible to check in polynomial time whether two H- (or V-) representations define the same polyhedron or not. We also show that redundancy removal is polynomially equivalent to the linear programming. This justifies the use of linear programming algorithms for redundancy removal.

Our main motivation arises from the continuing effort by the first two authors to develop polyhedral computation codes, such as cddlib [2] and lrslib

[1], that perform basic transformations of polyhedral representations. While the main transformation of these codes is the conversion between H- and V-representations, it is increasing important to make an efficient implementation of algorithms to compute canonical representations. This would allow users to compare two H-polyhedra (or two V-polyhedra) quickly without applying the (often too) expensive conversion. In addition, the computation of a canonical representation can be considered as a preprocessing step for the main transformation, which in some cases results in a drastic speed up.

Due to space limitations, many of the proofs are omitted, and will be included in the full version of the paper.

2 Representations of Convex Polyhedra

In this section, we present basic results on representations of convex polyhedra. Most of the results are classical and not very hard to prove. However there is no thorough treatment of this subject for both V- and H-representations and in particular no study with computational considerations. Throughout this section, P is a non-empty polyhedron in $\mathbb{R}^d$.

The notation for H-representation and V-representation given in the last section, although standard, hides the underlying duality of polyhedral cones. A natural way to unify the two representations is through homogenization. We encode a V-representation (V, R) as a pair of a vector and a matrix $\left(\begin{bmatrix} e \\ 0 \end{bmatrix}, \begin{bmatrix} V \\ R \end{bmatrix}\right)$, where e is the column vector of all 1's. In this representation written as (b, A), each row index i with $b_i = 1$ indicates that the corresponding i^{th} row vector A_i of A is a convex hull generator, and each $b_i = 0$ indicates A_i is a cone generator. This way, both representations are of the same form. Actually we must go even further to define a unique representation of any given polyhedron, in both V and H formats.

Let A be an $m \times d$ matrix, b an m-dimensional vector and $\{I, L\}$ be a partition of $[m] := \{1, \ldots, m\}$. We define a quadruple (b, A, I, L) to be an H-representation of the polyhedron

$$P_H = \{x \in \mathbb{R}^d \mid b_I + A_I x \geq \mathbf{0},\, b_L + A_L x = \mathbf{0}\}.$$

We define a quadruple (b, A, I, L) to be a V-representation of the polyhedron

$$P_V = \{x \in \mathbb{R}^d \mid x = A^T y,\, y_I \geq \mathbf{0},\, b^T y = 1\}.$$

The set L is called the *linearity* of the representation. Two representations (b, A, I, L) and (b', A', I', L') are said to be *equivalent* if the represented polyhedra are equal. Each row index $i \in [m]$ is called *redundant in a representation*

(b, A, I, L) if the represented polyhedron stays unchanged after deleting the ith data, namely removing b_i, A_i, i from I or L whichever contains i, and shifting the rest of the data accordingly. A representation is *nonredundant* if it has no redundant row index. For technical reasons, if the zero vector appears in a V-representation it will be considered non redundant.

A V-representation (b, A, I, L) is called *standard* if b is binary (i.e. $b_i \in \{0, 1\}$) and $b_L = \mathbf{0}$. A V-representation can be transformed to an equivalent standard V-representation in quadratic time. For this reason, we assume for the rest of the paper that any V-representation is standard. Each vector A_i in a representation is called a *generator*. It is a *point generator* if $b_i = 1$, a *ray generator* if $b_i = 0$ and $i \notin L$, and a *line generator* if $b_i = 0$ and $i \in L$.

We say two representations (b, A, I, L) and (b', A', I', L') *equal* if $b_L + A_L x = 0 \Leftrightarrow b_{L'} + A_{L'} x = 0$ and if there is a permutation π of I such that $\pi(I) = I'$ and each (b'_i, A'_i) is a positive multiple of $(b_{\pi(i)}, A_{\pi(i)})$ for any $i \in I$. Clearly, if two representations of the same type are equal, they are equivalent and thus define the same polyhedron. In particular for V-representation, $b_L = 0$ and then the first equivalence coincides with the statement $\{x \in \mathbb{R}^d \mid x = (A_L)^T y\} = \{x \in \mathbb{R}^d \mid x = (A_{L'})^T y\}$.

While the equivalence of two representations is nontrivial to check, the equality is easy to check using Gaussian elimination. Our main goal is to discuss an efficient procedure to reduce any representation to a uniquely defined equivalent representation. In particular, we want to extend the definition of the representation discussed in the book of Schrijver [4] in order to get a canonical representation guaranteeing a certain sparsity property. Extending this definition uses the notion of compatible and complementary linear subspaces of $\mathbb{R}^d$.

2.1 Compatible and Complementary linear subspaces

We say two linear subspaces S_1 and S_2 of $\mathbb{R}^d$ having same dimension are *compatible* if the orthogonal projection of S_1 onto S_2 is S_2 itself, or, equivalently, if the orthogonal projection of S_2 onto S_1 is S_1 itself. We say two linear subspaces S_1 and S_2 of $\mathbb{R}^d$ are *complementary* if any basis of S_1 and any basis of S_2 form, together, a basis of $\mathbb{R}^d$. It follows from elementary linear algebra that S_1 and S_2 are complementary if and only if $\dim(S_1) + \dim(S_2) = d$ and $S_1 \cap S_2 = \{\mathbf{0}\}$. Similarly any $r \in \mathbb{R}^d$ can be written uniquely as $r = r_1 + r_2$, where $r_1 \in S_1$ and $r_2 \in S_2$. For any two complementary linear subspaces S_1 and S_2 of $\mathbb{R}^d$, we let $\mathcal{P}_{S_1 S_2}$ denote the *projector onto* S_1 *along* S_2 defined as

$$\begin{aligned} \mathcal{P}_{S_1 S_2} : \mathbb{R}^d &\longrightarrow S_1 \\ r &\longmapsto \mathcal{P}_{S_1 S_2}(r) = r_1. \end{aligned}$$

It is easy to see that $\mathcal{P}_{S_1S_2}$ is a projector in the sense of the linear algebra. By linearity, this definition can be extended to any subset of $\mathbb{R}^d$.

2.2 Canonical V-Representation

To define a unique V-representation of a general (non-pointed) polyhedron we need a few more definitions. The *linearity space* $lin.space(P)$ of P is defined as

$$lin.space(P) = \{z \in \mathbb{R}^d \mid x + \alpha z \in P \text{ for all } x \in P \text{ and } \alpha \in \mathbb{R}\}.$$

The *characteristic cone* $char.cone(P)$ of P is

$$char.cone(P) = \{z \in \mathbb{R}^d \mid x + \alpha z \in P \text{ for all } x \in P \text{ and } \alpha \geq 0\}.$$

The following defines the V-representation discussed in the book of Schrijver:
Lemma 2.1. *A nonredundant V-representation* (b, A, I, L) *of polyhedron* P *satisfying the following exists and is unique:*

(1) $lin.space(P) = \{z \in \mathbb{R}^d \mid z = (A_L)^T y_L\}$, *and*

(2) all row vectors A_i $(i \in I)$ *are orthogonal to* $lin.space(P)$.

This lemma gives a theoretically satisfactory definition of the canonical representation. Its weakness is that it tends to make the resulting representation dense. The next theorem gives more general criteria which allows one to look for unique V-representations satisfying certain sparsity properties.
Theorem 2.2. *A nonredundant V-representation* (b, A, I, L) *of polyhedron* P *satisfying the following exists and is unique:*

(1) $lin.space(P) = \{z \in \mathbb{R}^d \mid z = (A_L)^T y_L\}$, *and*

(2) all row vectors A_i $(i \in I)$ *are orthogonal to* S,

where S *is any fixed subspace of* $\mathbb{R}^d$ *compatible with* $lin.space(P)$.

The V-representation of Lemma 2.1 is the special case of Theorem 2.2 when S is chosen to be $lin.space(P)$ itself.

2.3 Canonical H-Representation

In this section, we use the notion of compatible spaces to find a general canonical H-representation. We denote by $\mathit{aff}(P)$ the *affine hull* of P. The H-representation discussed in the book of Schrijver is:
Lemma 2.3. *A nonredundant H-representation* (b, A, I, L) *of a nonempty polyhedron satisfying the following exists and is unique:*

(1) $aff(P) = \{x \in \mathbb{R}^d \mid b_L + A_L x = 0\}$, *and*

(2) each row of A_I *is the linear space parallel to* $aff(P)$.

Again this representation might be computationally inconvenient. The next theorem gives more flexible notion of canonical H-representations which allows one to look for those satisfying additional favourable properties such as sparsity.

Theorem 2.4. *A nonredundant H-representation* (b, A, I, L) *of polyhedron* P *satisfying the following exists and is unique:*

(1) $aff(P) = \{x \in \mathbb{R}^d \mid b_L + A_L x = 0\}$, *and*

(2) each row of A_I *is in* S,

where S *is any linear subspace compatible with the linear space parallel to* $aff(P)$.

3 Computing Canonical Representations

Given any V- (resp. H-) representation (b, A, I, L) of a polyhedron P in $\mathbb{R}^d$, our goal is to compute efficiently the unique equivalent representation (b', A', I', L') satisfying the following :

(1) it is nonredundant;

(2) $lin.space(P) = \{x \in \mathbb{R}^d \mid x = (A'_{L'})^T y\}$ (resp. $aff(P) = \{x \in \mathbb{R}^d \mid b'_{L'} + A'_{L'} x = 0\}$);

(3) the rows of $A'_{I'}$ are orthogonal to a subspace S compatible with $lin.space(P)$ (resp. are contained in a subspace S compatible with $\mathcal{P}_0$).

Let $\bar{P}_0$ denote $lin.space(P)$ in V-format, and $aff(P)$ in H-format, and let $\bar{S}$ denote $S^{\perp}$ in V-format, and S in H-format. Then, (3) can be replaced by :

(3') the rows of $A'_{I'}$ are the projections onto $\bar{S}$ along $\bar{P}_0$ of any equivalent set of generators of P.

We now discuss two special choices for the linear subspace S.

Any linear subspace is compatible with itself. Choosing $\bar{S}$ as $\bar{P}_0$ itself leads to the *orthogonal representation* discussed in the book of Schrijver [4]. In this case, the projector $\mathcal{P}_{\bar{S}\bar{P}_0}$ onto $\bar{S}$ along $\bar{P}_0$ coincides with the orthogonal projector $\mathcal{P}_{\bar{S}}$ onto $\bar{S}$. This representation is perhaps mathematically the most natural but it tends to lead to a very dense representation, (see e.g. Section 6).

Let us consider another natural choice for S. A *coordinate subspace* is any vector subspace of $\mathbb{R}^d$ generated by some unit vectors e^j $(j = 1, \ldots, d)$. Selecting $\bar{S}$ as a coordinate subspace guarantees a certain sparsity of the projected vectors. Moreover, selecting $\bar{S}$ as the lexicographically smallest subspace complementary with $(\bar{P_0})$ guarantees that the nonzero coordinates are indexed by indices as small as possible. The resulting representation will be called the *lexicographically smallest* or *lexico-smallest* representation.

Computationally, the orthogonal representation requires the computation of orthogonal basis of a linear subspace, which amounts to doing the Gram-Schmidt orthogonalization procedure. The lexico-smallest representation needs only to check whether a unit e^j is linearly independent of a given set of chosen vectors, which amounts to applying Gaussian eliminations.

4 Redundancy Removal

4.1 Redundancy Removal in V-representation

The techniques used for solving redundancy removal are very similar in both V and H formats. Therefore, the discusssion in H-representation will be omitted. A complete description of the redundancy removal techniques in both H- and V-respresentation can be found in the Polyhedral Computation FAQ [3] of K. Fukuda.

For any given V-representation (b, A, I, L) of a polyhedron P, we let $I_0 = \{i \in I \mid b_i = 0\}$ and $I_1 = \{i \in I \mid b_i = 1\}$. It follows from the definitions, that $P = conv(A_{I_1}) + cone(R)$, where $R := \begin{bmatrix} A_{I_0} \\ A_L^{\pm} \end{bmatrix}$ and $A_L^{\pm} := \begin{bmatrix} +A_L \\ -A_L \end{bmatrix}$. Let (b, A, I, L) be any V representation of polyhedron P. The definitions of point, ray and line generators of P imply that,

(1) $k \in I_0$ is redundant if and only if A_k is redundant in $cone(R)$;
(2) $k \in L$ is redundant if and only if A_k and $-A_k$ are redundant in $cone(R)$.
(3) $k \in I_1$ is redundant if and only if A_k is redundant in $conv(V) + cone(R)$.

For any row index $k \in I \cup L$, we define the linear program R.LPV(b_k, A_k) as

$$\begin{array}{lll} \max 0 & & \\ \text{s.t.} & (A_k)^T = b_k \sum_{i \neq k} (A_i)^T x_i + \sum_{l \neq k} (R_l)^T y_l, & i \in I_1,\ l \in I_R \\ & b_k = b_k \sum_{i \neq k} x_i, & i \in I_1 \\ & x_i \geq 0, \quad y_l \geq 0, & \forall i \in I,\ \forall l \in I_R \end{array}$$

where I_R is the index set of the rows of matrix R.

Theorem 4.1. *Let (b, A, I, L) be any V-representation P. Then,*

(1) $k \in I$ is redundant if and only if R.LPV(b_k, A_k) has optimal value zero,

(2) $k \in L$ is redundant if and only if both R.LPV$(0, A_k)$ and R.LPV$(0, -A_k)$ have optimal value zero.

Removing sequentially the redundancies from any representation (b, A, I, L) of a polyhedron P using this theorem leads to a nonredundant representation (b', A', I', L') in $\mathcal{O}(m \times \mathrm{LP}(m, d))$ time, where $m = |I \cup L|$. The size of the LPs can be reduced to that of the resulting nonredundant system by using the techniques given by Ottmann et al. [5]

4.2 Computing lin.space(P) and aff(P)

The canonical V- (resp. H-) representation of polyhedron P is defined so that the rows of the matrix A_L form a basis of $S(A_L)$. In this section, we present how one can decide efficiently whether any given row index $k \in I \cup L$ from any nonredundant representation (b, A, I, L) of P belongs to the linearity of the canonical representation of P. Solving this decision problem for every $k \in I \cup L$ leads to a set of indices, say L'', such as $(A_{L''})^T y_L \in lin.space(P)$ (resp. such as $b_{L''} + A_{L''}x = \mathbf{0}$). The linearity L' of the canonical representation of P then arises from L'' by removing the redundancies.

This decision problem can be easily reduced to linear programming. But it may also be (equivalently) solved using the redundancy recognition. This latter method has a nice and simple geometric interpretation in both H and V formats.

We first consider V-representations. Clearly, we only have to discuss the case of the ray generators of P. Any ray generator A_k of P belongs to $lin.space(P)$ if and only if both A_k and $-A_k$ belong to $lin.space(P)$. In other words,

Theorem 4.2. *Let (b, A, I, L) be any V-representation of polyhedron $P = conv(A_{I_1}) + cone(R)$, where I_1 and R are defined as in Section 4.1. Then, any row A_k of matrix A belongs to $lin.space(P)$ if and only if $k \in L \cup L_I$, where $L_I = \{i \in I_0 \mid A_k$ is redundant in $cone(-R)\}$.*

Now, consider H-representations. One has to decide whether $\{x \in \mathbb{R}^d \mid b_k + A_k x = 0\} \supseteq aff(P)$. It is clearly the case if and only if $P \subseteq \{x \in \mathbb{R}^d \mid b_k + A_k x = 0\}$. In other words,

Theorem 4.3. *Let (b, A, I, L) be any H-representation of polyhedron P. Then, any row A_k of matrix A is such as $\{x \in \mathbb{R}^d \mid b_k + A_k x = 0\} \supseteq aff(P)$ if and only if $k \in L \cup L_I$, where $L_I = \{i \in I \mid -b_i - A_i x \geq 0$ is redundant with $\{b_I + A_I x \geq 0\} \cup \{b_L + A_L x = 0\}\}$.*

5 Computational Equivalence of Problems

The results of the last section show how a canonical representation may be computed efficiently. The following theorem, which directly follows from the preceding results, characterizes the complexity of this computation.

Theorem 5.1. *Redundancy removal and the computation of a canonical representation are polynomially equivalent problems.*

As we saw in Section 4, redundancy removal has complexity $\mathcal{O}(m \times \mathrm{LP}(m,d))$, where $m = |I \cup L|$. Solving these linear programs is expensive in time. The natural question that arises concerns the necessity of solving linear programs to remove redundancies. It is well known that checking feasibility of a linear system is polynomially equivalent to linear programming. The following lemma then states the polynomial equivalence of redundancy removal and the linear programming.

Lemma 5.2. *Let $Ax + b \geq \mathbf{0}$, where $A \in \mathbb{R}^{m \times d}$ and $b \in \mathbb{R}^d$, be any linear system of inequalities. Then, $Ax+b \geq \mathbf{0}$ is feasible if and only if the inequality $\{x_0 \leq 1\}$ is not redundant with $Ax + bx_0 \geq \mathbf{0}$.*

Proof. Consider the homogenized system $Ax + bx_0 \geq \mathbf{0}$. Clearly, $Ax + b \geq \mathbf{0}$ is feasible if and only if it exist a vector $\bar{x} \in \mathbb{R}^d$ such that the pair $(x_0 = 1; x = \bar{x})$ is a feasible solution of $bx_0 + Ax \geq \mathbf{0}$. This is the case if and only if $(x = \alpha\bar{x}; x_0 = \alpha)$ is a feasible solution of $bx_0 + Ax \geq \mathbf{0}$ for any $\alpha \geq 0$. In particula, $b + Ax \geq \mathbf{0}$ is feasible if and only if it exist an $x \in \mathbb{R}^d$ such that $(x; x_0 > 1)$ is a feasible solution of $bx_0 + Ax \geq \mathbf{0}$. On the other hand, the system $bx_0 + Ax \geq \mathbf{0}$, $x_0 \leq 1$ is feasible. Then, one may check whether the inequality $\{x_0 \leq 1\}$ is redundant with $bx_0 + Ax \geq \mathbf{0}$ or not. In particular, it is not hard to see that $bx_0 + Ax \geq 0$, $x_0 > 1$ is feasible if and only if $\{x_0 \leq 1\}$ is redundant with $bx_0 + Ax \geq \mathbf{0}$. □

It follows from the above results that we may check the feasibility of any linear system $Ax \leq b$ in polynomial time by performing redundancy removal. As a direct consequence one has the argued result that,

Theorem 5.3. *Redundancy removal, the computation of a canonical representation and linear programming are polynomially equivalent problems.*

6 Some Examples

We start by illustrating the difference between the two canonical representations described in Section 6. Consider the following (b, A, I, L) V-representation:

$$b = \begin{bmatrix} 1 \\ 1 \\ 1 \end{bmatrix}, A = \begin{bmatrix} 0&1&1&1&1&0 \\ 1&0&1&1&0&1 \\ 1&1&0&0&1&1 \end{bmatrix}, I = \{1,2,3\}, L = \emptyset.$$

(This corresponds to the three hamiltonian circuits for the complete graph on four vertices.) An orthogonal H-representation in this case is given by:

$$b = \begin{bmatrix} -2 \\ -2 \\ -2 \\ -2 \\ 1/3 \\ 1/3 \\ 1/3 \end{bmatrix}, A = \begin{bmatrix} 1 & 1 & 1 & 0 & 0 & 0 \\ 1 & 0 & 0 & 1 & 1 & 0 \\ 0 & 1 & 0 & 1 & 0 & 1 \\ 0 & 0 & 1 & 0 & 1 & 1 \\ -1/3 & 1/6 & 1/6 & 1/6 & 1/6 & -1/3 \\ 1/6 & -1/3 & 1/6 & 1/6 & -1/3 & 1/6 \\ 1/6 & 1/6 & -1/3 & -1/3 & 1/6 & 1/6 \end{bmatrix}, I = \{5,6,7\}, L = \{1,2,3,4\}.$$

Note that apart from the linearity space, the matrix is completely dense. The lexico-smallest H-representation is guaranteed to contains a square 3 by 4 submatrix of zeroes in the rows indexed by I. It is given by:

$$b = \begin{bmatrix} -2 \\ -2 \\ -2 \\ -2 \\ 1 \\ 1 \\ -1 \end{bmatrix}, A = \begin{bmatrix} 1 & 1 & 1&0&0&0 \\ 1 & 0 & 0&1&1&0 \\ 0 & 1 & 0&1&0&1 \\ 0 & 0 & 1&0&1&1 \\ -1 & 0 & 0&0&0&0 \\ 0 & -1 & 0&0&0&0 \\ 1 & 1 & 0&0&0&0 \end{bmatrix}, I = \{5,6,7\}, L = \{1,2,3,4\}.$$

Redundancy removal can have a significant effect in speeding up the running time of vertex enumeration algorithms. This is particularly pronounced on algorithms which enumerate bases, such as pivot based algorithms, and other algorithms that use symbolic perturbation to resolve degeneracy. In fact, redundancy causes a problem when redundant hyperplanes contain vertices of the polyhedron. For such vertices, even after perturbation, redundancy causes the number of bases to increase, often exponentially. This is because the redundant hyperplanes can be used to make up the d tight inequalities that define vertex. Since the complexity of the algorithms in the class described depends at least linearly on the number of bases, removal of the redundant inequalities is highly beneficial. Improvement is also noticeable for double description algorithms. Here we give a few examples, using lrs [1] as an example of a pivot based method, and cdd [2] as an example of a double description based method.

Consider the metric cone which for any integer $n \geq 3$ is a polyhedron in $R^{n(n-1)/2}$ defined by the triangle inequalities $x_{ij} - x_{ik} - x_{jk} \geq 0$ for all distinct $1 \leq i, j, k \leq n$, where $x = (x_{ij})$ $1 \leq i < j \leq n$. The metric cone is extremely degenerate. Vectors x satisfying these inequalities are known as semimetrics: they are nonnegative and satisfy all triangle inequalities. Often the nonnegativity condition $x_{ij} \geq 0$ is explicitly specified although these inequalities are in fact redundant. The program lrs has a nonnegative option for specifying these additional inequalities, but using it here causes a big increase in the running time. For example, with $n = 6$ and without the redundant constraints, lrs generates 203,956 bases to find the 296 extreme rays. With the nonnegative option, it generates 1,960,411 bases and so the running time is nearly 10 times longer. The effect on cdd is not so pronounced but still noticeable: including the redundant inequalities nearly doubles the computation time.

An other way redundancy occurs is when the programs are used to generate rays lying on lower dimensional faces. This is easily performed in both lrs and cdd by including a linearity option which specifies that certain inequalities should be treated as equations. This normally results in some of the original inequalities becoming redundant. Taking one such linearity causes all rays on a facet to be generated. For the metric cone with $n = 6$, a facet contains 113 extreme rays. Without removing redundant inequalities, lrs generates 121,215 bases. There are 14 redundant inequalities, and after there removal lrs generates 38,119 bases, a speed up of a factor of about 4. Choosing two linearities causes ray enumeration on a ridge. Here the effect of redundancy removal is much more pronounced, as 30 of the original constraints are redundant. lrs runs more than 50 times faster after redundancy removal. The effect of redundancy removal on cdd is also noticeable, with speed ups of the order of about 3 1/2 and 6 times respectively.

References

1. D. Avis. *lrs Homepage*, 2001. School of Computer Science, McGill University, Canada. http://cgm.cs.mcgill.ca/~avis/C/lrs.html.
2. K. Fukuda. *cddlib reference manual, cddlib Version 092a.* Swiss Federal Institute of Technology, Switzerland, 2001. http://www.ifor.math.ethz.ch/~fukuda/cdd_home/cdd.html.
3. K. Fukuda. *Frequently Asked Questions in Polyhedral Computation* Swiss Federal Institute of Technology, Switzerland, 2000 http://www.ifor.math.ethz.ch/~fukuda/polyfaq/polyfaq.html
4. A. Schrijver. *Theory of Linear and Integer Programming.* John Wiley & Sons, New York, 1986.
5. Th. Ottmann and S. Schuierer and S. Soundaralakshmi. *Enumerating extreme points in higher dimensions.* STACS 95: 12th Annual Symposium on Theoretical Aspects of Computer Science. Springer-Verlag,1995.
6. G.M. Ziegler. *Lectures on polytopes.* Graduate Texts in Mathematics 152. Springer-Verlag, 1994.

DECOMPOSING POLYNOMIAL SYSTEMS INTO STRONG REGULAR SETS

YONG-BIN LI

[1]*Institute of Mathematics, Shantou University,*
Shantou, Guangdong, 515063, P. R. China
[2]*Mathematical College, Sichuan University*
E-mail:yuengpen@netease.com

JING-ZHONG ZHANG AND LU YANG

Chengdu Institute of Computer Applications,
Chinese Academy of Sciences, Chengdu 610041, P. R. China
E-mail:zjz20001@sina.com; luyang@guangztc.edu.cn

A polynomial system is a pair of polynomial sets in $\mathbf{K}[\mathbf{x}]$. This paper presents a method of zero decomposition for any polynomial system into strong regular sets in $\mathbf{K}[\mathbf{x}, t]$, where t is a new variable.

1 Introduction

Let $\mathbf{K}$ be a computable field of characteristic 0 and $\mathbf{K}[x_1, \cdots, x_n]$ (or $\mathbf{K}[\mathbf{x}]$ for short) the ring of polynomials in the variables $(x_1, \cdots, x_n)$ with coefficients in $\mathbf{K}$. A *polynomial set* is a finite set $\mathbf{P}$ of nonzero polynomials in $\mathbf{K}[\mathbf{x}]$. For any polynomial $P \notin \mathbf{K}$, the biggest index p such that $\deg(P, x_p) > 0$ is called the *class*, denoted by $\text{cls}(P)$, and x_p is called the *leading variable*, denoted by $\text{lv}(P)$. A finite nonempty ordered set $\mathbf{T} = \{f_1, \cdots, f_s\}$ of polynomials in $\mathbf{K}[\mathbf{x}]\backslash\mathbf{K}$ is called a *triangular set* if $\text{cls}(f_1) \prec \cdots \prec \text{cls}(f_s)$. Let a triangular set $\mathbf{T}$ be written as the following form

$$\mathbf{T} = \{f_1(u_1, u_2, \cdots, u_r, y_1), \cdots, f_s(u_1, u_2, \cdots, u_r, y_1, y_2, \cdots, y_s)\},$$

where $(u_1, u_2, \cdots, u_r, y_1, y_2, \cdots, y_s)$ is a permutation of $(x_1, x_2, \cdots, x_n)$. $I_j = \text{ini}(f_j)$ denotes the leading coefficient of f_j in y_j for each j, and $\text{ini}(\mathbf{T})$ denotes the set of all I_j. Let $\bar{\mathbf{K}}$ be an arbitrary extension field of $\mathbf{K}$. $\bar{\mathbf{K}}$ is the field $\mathbf{C}$ of complex numbers in this paper. While speaking about a *polynomial systems*[10], we refer to a pair $[\mathbf{P}, \mathbf{Q}]$ of polynomial sets. For any nonempty polynomial set $\mathbf{P}$, a famous zero decomposition was developed by Wu[1]:

Theorem 1.1. Let $\mathbf{P}$ be a polynomial set. There is an algorithm, according to which one may arrive after a finite number of steps at a zero decomposition of the form $\text{Zero}(\mathbf{P}) = \bigcup_i \text{Zero}(\mathbf{T}_i/\text{ini}(\mathbf{T}_i))$, in which $\mathbf{T}_i$ is an ascending set.

Let $\mathbf{x}_i$ stand for $(x_1, \cdots, x_i)$ or $x_1, \cdots, x_i$, and similarly for $\bar{\mathbf{x}}_i$. A polynomial system $[\mathbf{T}, \mathbf{U}]$ is called a *triangular system* if $\mathbf{T}$ is a triangular set and $I(\bar{\mathbf{x}}_i) \neq 0$ for any $I \in \mathrm{ini}(\mathbf{T})$ of class i and $\bar{\mathbf{x}}_i \in \mathrm{Zero}(\mathbf{T}^{(i)}/\mathbf{U})$ where $\mathbf{T}^{(i)} = \mathbf{T} \cap \mathbf{K}[x_1, \cdots, x_i]$. A triangular system $[\mathbf{T}, \mathbf{U}]$ is said to be *fine* if $\mathrm{prem}(\mathbf{U}, \mathbf{T}) \neq \{0\}$. The following theorem was presented by Wang[10,11,12].

Theorem 1.2. Let $[\mathbf{P}, \mathbf{Q}]$ be a polynomial system. There are two algorithms, by which one can decompose $[\mathbf{P}, \mathbf{Q}]$ into e fine triangular systems $[\mathbf{T}_i, \mathbf{U}_i]$ such that $\mathrm{Zero}(\mathbf{P}/\mathbf{Q}) = \bigcup_{i=1}^{e} \mathrm{Zero}(\mathbf{T}_i/\mathbf{U}_i)$.

Definition 1.1. Let $\{f_1, \cdots, f_s\}$ be a triangular set as above and g a polynomial in $\mathbf{K}[\mathbf{x}]$. The polynomial $\mathrm{resultant}(\cdots \mathrm{resultant}(g, f_s, y_s), \cdots, f_1, y_1)$ is called the *resultant* of triangular set $\{f_1, \cdots, f_s\}$ with respect to g, denoted by $\mathrm{res}(g, f_s, \cdots, f_1)$ or $\mathrm{res}(g, \{f_1, \cdots, f_s\})$.

The concept of *regular sets* was introduced by Yang and Zhang[2] under the name of *proper ascending chains* in 1991. Independently, it was also introduced by Kalkbrener[5] under the name of *regular chains* in 1993.

Definition 1.2.[2,7] A triangular set $\mathbf{T} = \{f_1, \cdots, f_s\}$ is called a regular set, if $I_1 \neq 0$ and $\mathrm{res}(I_j, \mathbf{T}) \neq 0$, for $j = 2, 3, \cdots, s$.

Definition 1.3.[15] Let $\mathbf{T}$ be a triangular set and g a polynomial. $\mathbf{T}$ is said to be *strongly independent* of g if $\mathrm{Zero}(\{g\}) \cap \mathrm{Zero}(\mathbf{T}) = \emptyset$.

It is obvious that $\mathbf{T}$ is strongly independent of g if $\mathrm{res}(g, \mathbf{T}) \in \mathbf{K}\backslash\{0\}$.

Kalkbrener[5] and Yang, Zhang and Hou[7] respectively presented two different algorithms for decomposing $\mathbf{P}$ into a finite number of regular sets $\mathbf{T}_1, \cdots, \mathbf{T}_e$ such that $\mathrm{Zero}(\mathbf{P}) = \bigcup_{i=1}^{e} \mathrm{Zero}(\mathrm{sat}(\mathbf{T}_i))$, where $\mathrm{sat}(\mathbf{T}_i)$ is the saturation of $\mathbf{T}_i$. The set $\{\mathbf{T}_1, \cdots, \mathbf{T}_e\}$ is also called a *regular series* of $\mathbf{P}$.

One of the algorithms mentioned in Theorem 1.2 is RegSer[12], by which one can decompose any polynomial system into finitely many *regular systems.*

Definition 1.4.[12] A triangular system $[\mathbf{T}, \mathbf{U}]$ is called a regular system if $\mathrm{cls}(T) \neq \mathrm{cls}(U)$ for any $T \in \mathbf{T}$ and $U \in \mathbf{U}$, and $I(\bar{\mathbf{x}}_i) \neq 0$ for any $I \in \mathrm{ini}(\mathbf{U})$ of class i and $\bar{\mathbf{x}}_i \in \mathrm{Zero}(\mathbf{T}^{(i)}/\mathbf{U}^{(i)})$ where $\mathbf{U}^{(i)} = \mathbf{U} \cap \mathbf{K}[x_1, \cdots, x_i]$. A triangular set $\mathbf{T}$ is said to be *regular* if there exists a polynomial set $\mathbf{U}$ such that $[\mathbf{T}, \mathbf{U}]$ is a regular system. A finite set Ψ of regular systems is called a *regular series* of a polynomial system $[\mathbf{P}, \mathbf{Q}]$ if

$$\mathrm{Zero}(\mathbf{P}/\mathbf{Q}) = \bigcup_{[\mathbf{T},\mathbf{U}] \in \Psi} \mathrm{Zero}(\mathbf{T}/\mathbf{U}).$$

Theorem 1.3.[12,13] Let $[\mathbf{T}_1, \mathbf{U}_1], \cdots, [\mathbf{T}_e, \mathbf{U}_e]$ be a a regular series of polynomial system $[\mathbf{P}, \mathbf{Q}]$ computed by $\mathrm{RegSer}(\mathbf{P}, \mathbf{Q})$. Then

1) there exists an integer $d > 0$ such that $\mathrm{prem}(P^d, \mathbf{T}_i) = 0$ for all $P \in \mathbf{P}$ and $1 \leq i \leq e$;

2) for any integers $m > 0$, $1 \leq i \leq e$ and polynomial $Q \in \mathbf{Q}$, $\mathrm{prem}(Q^m, \mathbf{T}_i) \neq 0$;

3) $$\mathrm{Zero}(\mathbf{P}/\mathbf{Q}) = \bigcup_{i=1}^{e} \mathrm{Zero}(\mathbf{T}_i/\mathrm{ini}(\mathbf{T}_i) \cup \mathbf{Q}).$$

Remark 1.1. The condition of Definition 1.4 is slightly strong in verifying $I \in \mathrm{ini}(\mathbf{U})$ of each class i. It maybe causes some trivial results such as

$$\begin{aligned}\mathrm{RegSer}(\{x_1\}, \{x_2x_3x_5 + x_4\}) = \{\ & [\{x_1, x_2x_3\}, \{x_2, x_4\}], [\{x_1, x_2\}, \{x_4\}], \\ & [\{x_1\}, \{x_2x_3x_5 + x_4, x_2x_3, x_2\}]\ \}.\end{aligned}$$

Moreover, it is likely to be true that $\mathrm{Zero}(\mathbf{T}_i/\mathbf{U}_i) \cap \mathrm{Zero}(\{Q\}) \neq \emptyset$ for some $Q \in \mathbf{Q}$ and $1 \leq i \leq e$ in Theorem 1.3. RegSer is very efficient in computing regular sets of nonempty polynomial set $\mathbf{P}$ by $\mathrm{RegSer}(\mathbf{P}, \emptyset)$. In this paper, $\mathrm{Reg}(\mathbf{P})$ denotes a regular series of a nonempty polynomial set $\mathbf{P}$ computed by one of the above-mentioned algorithms. In order to decompose any polynomial system $[\mathbf{P}, \mathbf{Q}]$, we present an approach of splitting every regular set in $\mathrm{Reg}(\mathbf{P})$ with respect to every polynomial in $\mathbf{Q}$. Because a good helpful theory in the next section is used, the approach has a few advantages.

Remark 1.2. If one computes $\mathrm{Reg}(\mathbf{P})$ by $\mathrm{RegSer}(\mathbf{P}, \emptyset)$, then all $\mathbf{U}_i$ are abandoned. It seems to cause a waste, but the $\mathbf{U}_i$ have less importance in our approach as $\mathrm{Zero}(\mathbf{T}_i/\mathbf{U}_i) \subset \mathrm{Zero}(\mathbf{T}_i/\mathrm{ini}(\mathbf{T}_i))$. In fact we may obtain something similar to the $\mathbf{U}_i$ which may be simpler in most cases.

2 Strong regular sets

It is well known that there is a problem of excluding some redundant zeros in most zero decompositions. The theory of the *weakly nondegenerate condition*[3,4] may deduce a useful result to help forward solving the problem.

Definition 2.1.[3,4] Let $\mathbf{T} = \{f_1, \cdots, f_s\}$ be a regular set, where $f_j = c_{j0}y_j^{n_j} + c_{j1}y_j^{n_j-1} + \cdots + c_{jn_j}$, for $j = 1, \cdots, s$. A zero $\mathbf{z}_0 \in \mathrm{Zero}(\mathbf{T})$ is called a *normal zero* if $c_{j0}(\mathbf{z}_0) \neq 0$ for $j = 1, \cdots, s$, also said to be satisfying the *nondegenerate condition*; the zero is called a *quasi-normal zero* if there exists some $k_0 \in \{0, 1, \cdots, n_j\}$ such that $c_{jk_0}(\mathbf{z}_0) \neq 0$ for $j = 1, \cdots, s$, also said to be satisfying the weakly nondegenerate condition.

Theorem 2.1. Let $\mathbf{T} = \{f_1, \cdots, f_s\}$ be a regular set and g a polynomial. If there exists an integer $d > 0$ such that $\mathrm{prem}(g^d, \mathbf{T}) = 0$, then every quasi-normal zero of $\mathbf{T}$ is also a zero of g.

Definition 2.2. Let $\mathbf{T}$ be a regular set. $\mathbf{T}$ is called a *strong regular set* if every zero of $\mathbf{T}$ is also a quasi-normal zero.

Remark 2.1. There exists a great difference between a strong regular set and a regular system. One is a polynomial set, while the other is a polynomial system. A regular system $[\mathbf{T}, \emptyset]$ implies a strong regular set $\mathbf{T}$. But a strong regular set $\mathbf{T}$ does not imply that $[\mathbf{T}, \emptyset]$ is a regular system, even a triangular system. For example, $\{x_2^2 - x_1, x_2x_3x_4^2 + x_4 - x_1\}$ is a strong regular set, but $[\{x_2^2 - x_1, x_2x_3x_4^2 + x_4 - x_1\}, \emptyset]$ is not a triangular system at all. One must construct a triangular set $\mathbf{U} = \{x_2, x_2x_3\}$ such that $[\{x_2^2 - x_1, x_2x_3x_4^2 + x_4 - x_1\}, \mathbf{U}]$ is a regular system.

Proposition 2.1.[15] Let $\mathbf{T} = \{f_1, \cdots, f_s\}$ be a regular set with the above notation. If $\mathrm{Zero}(\{c_{10}, \cdots, c_{1n_1}\}) = \emptyset$ and $\mathrm{Zero}(\{c_{j0}, \cdots, c_{jn_j}\}) = \emptyset$ or $\mathrm{Zero}(\{RI_{j0}, \cdots, RI_{jn_j}\}) = \emptyset$ for $j = 2, \cdots, s$, where $RI_{jk} = \mathrm{res}(c_{jk}, \mathbf{T})$ for $k = 0, \cdots, n_j$, then $\mathbf{T}$ is a strong regular set.

Theorem 2.2.[15] Let $\mathbf{T}$ be a strong regular set and g a polynomial. If there exists an integer $d > 0$ such that $\mathrm{prem}(g^d, \mathbf{T}) = 0$, then every zero of $\mathbf{T}$ is also a zero of g. Moreover, $\mathrm{Zero}(\mathrm{sat}(\mathbf{T})) = \mathrm{Zero}(\mathbf{T})$.

Example 2.1. Let $\mathbf{P} = \{P_1, P_2, P_3\}$ with $P_1 = z(x^2 + y^2 - c) + 1$, $P_2 = y(x^2 + z^2 - c) + 1$, $P_3 = x(y^2 + z^2 - c) + 1$. This set of polynomials has been considered by Gao and Chou[9] and Wang[11,13]. Under the variable ordering $c \prec z \prec y \prec x$, $\mathbf{P}$ can be decomposed into 7 fine triangular systems $[\mathbf{T}_1, \mathbf{U}_1], \cdots, [\mathbf{T}_7, \mathbf{U}_7]$ such that $\mathrm{Zero}(\mathbf{P}) = \bigcup_{i=1}^{7} \mathrm{Zero}(\mathbf{T}_i/\mathbf{U}_i)$, where

$$
\begin{aligned}
&\mathbf{T}_1 = \{2cz^4 - 2z^3 - c^2z^2 - 2cz - 1, (cz + 1)y + cz^2 - z, 2z^2x + cz + 1\},\\
&\mathbf{T}_2 = \{2z^4 - 3cz^2 + z + c^2, zy - z^2 + c, x - z\},\\
&\mathbf{T}_3 = \{z^3 - cz - 1, (z^2 - c)y^2 + y - cz^2 + z + c^2, yx - z^2 + c\},\\
&\mathbf{T}_4 = \{2z^4 - 3cz^2 + z + c^2, (2z^3 - 2cz + 2)y - cz^2 - z + c^2, P_3\},\\
&\mathbf{T}_5 = \{2z^3 - cz + 1, y - z, 2z^2x - cx + 1\},\\
&\mathbf{T}_6 = \{c, 2z^3 + 1, y - z, 2z^2x + 1\},\\
&\mathbf{T}_7 = \{4c^3 - 27, 9z + 2c^2, 6cy^2 - 9y - 4c^2, 3yx + 2c\};\\
&\mathbf{U}_1 = \{c, z, cz + 1\},\;\; \mathbf{U}_2 = \{z, z^2 - c, 2z^2 - c\},\;\; \mathbf{U}_3 = \{z^2 - c, y\},\\
&\mathbf{U}_4 = \{z^2 - c, z^3 - cz + 1, z^3 - cz - 1\},\\
&\mathbf{U}_5 = \{z, 2z^2 - c\},\;\; \mathbf{U}_6 = \{z\},\;\; \mathbf{U}_7 = \{c, y\}.
\end{aligned}
$$

It is easy to verify that $\mathbf{T}_i$ is a strong regular set for each $i \neq 2$. Thus, we have

$$\text{Zero}(\mathbf{P}) = \bigcup_{i=1, i\neq 2}^{7} \text{Zero}(\mathbf{T}_i) \cup \text{Zero}(\mathbf{T}_2/\mathbf{U}_2),$$

by Theorem 2.2. In fact $\mathbf{T}_7$ can be removed according to the result used by Chou and Gao[8].

3 Decompose polynomial systems into strong regular sets

In order to decompose any polynomial system into strong regular sets, we describe first the following algorithm RegToStr by which one can decompose one regular set into finitely many strong regular sets without excluding any zero.

3.1 Algorithm RegToStr

Definition 3.1.1. For any polynomial set $\mathbf{P}$ in $\mathbf{K}[x_1, \cdots, x_n, t]$ (or $\mathbf{K}[\mathbf{x}, t]$ for short) with an ordering for variables $x_1 \prec \cdots \prec x_n \prec t$, the *projection* of $\text{Zero}(\mathbf{P})$ onto $\mathbf{x} = (x_1, x_2, \cdots, x_n)$ is defined to be

$$\begin{aligned}\text{Proj}_{\mathbf{x}}\text{Zero}(\mathbf{P}) = \{(\bar{x}_1, \bar{x}_2, \cdots, \bar{x}_n) \in \bar{\mathbf{K}}^n : \exists\, \bar{t} \in \bar{\mathbf{K}} \text{ such that} \\ (\bar{x}_1, \bar{x}_2, \cdots, \bar{x}_n, \bar{t}) \in \text{Zero}(\mathbf{P})\}.\end{aligned}$$

Hereinafter, $\mathbf{T}$ is called a triangular set in $\mathbf{K}[\mathbf{x}, t]$ if $\mathbf{T} = \{f_1, \cdots, f_s\}$ or $\mathbf{T} = \{f_1, \cdots, f_s, \mu_0 t - 1\}$, where each f_i and $\mu_0 \in \mathbf{K}[\mathbf{x}]$.

Theorem 3.1.1. Let $\mathbf{T}$ be a regular set, but not a strong regular set in $\mathbf{K}[\mathbf{x}, t]$ with the notation above. There exist a strong regular set $\mathbf{T}^*$ and a finite nonzero set $\{\mu_1, \cdots, \mu_{m_\mathbf{T}}\} \subset \mathbf{K}[u_1, \cdots, u_r]$ such that

$$\text{Zero}(\mathbf{T}) = \bigcup_{i=1}^{m_\mathbf{T}} \text{Zero}(\mathbf{T} \cup \{\mu_i\}) \cup \text{Zero}(\mathbf{T}^*).$$

Proof: Consider the case in which $\mathbf{T} = \{f_1, \cdots, f_s\}$. As $\mathbf{T}$ is not a strong regular set, there exists a nonempty index set η such that $\text{Zero}(\{c_{10}, \cdots, c_{1n_1}\}) \neq \emptyset$ if $1 \in \eta$ or $\text{Zero}(\{c_{j0}, \cdots, c_{jn_j}, f_1, \cdots, f_{j-1}\}) \neq \emptyset$ if $1 < j \in \eta$. We pick a nonzero element from $\{c_{j0}, \cdots, c_{jn_j}\} \cap \mathbf{K}[u_1, \cdots, u_r]$ or $\{RI_{j0}, \cdots, RI_{jn_j}\}$, and write it as μ_j^* for any $j \in \eta$. Collect all μ_j^* and denote them by $\{\mu_1, \cdots, \mu_{m_\mathbf{T}}\}$. Then, put $\mathbf{T}^* = \{f_1, \cdots, f_s, (\prod_{i=1}^{m_\mathbf{T}} \mu_i)t - 1\}$, where t is a new variable. $\mathbf{T}^*$ is a strong regular set in $\mathbf{K}[\mathbf{x}, t]$.

The case in which $\mathbf{T} = \{f_1, \cdots, f_s, \mu_0 t - 1\}$ is proved analogously, just replacing $\mathbf{T}^*$ by $\{f_1, \cdots, f_s, (\prod_{i=0}^{m_\mathbf{T}} \mu_i)t - 1\}$ in the above case.

Remark 3.1. Continued from Remark 1.2, let $[\mathbf{T}, \mathbf{U}]$ be a regular system computed by RegSer$(\mathbf{P}, \emptyset)$; Theorem 3.1.1 holds also true if one picks μ_j^* from $\mathbf{U}$. Since some elements of $\mathbf{U}$ are likely to be superabundant and not simple, we have no choice but to abandon $\mathbf{U}$. Refer to the example in Remak 2.1, the decomposition $\text{Zero}(\{x_2^2 - x_1, x_2x_3x_4^2 + x_4 - x_1\}) = \text{Zero}(\{x_2^2 - x_1, x_2x_3x_4^2 + x_4 - x_1, x_2^2x_3t - 1\}) \cup \text{Zero}(\{x_2^2 - x_1, x_2x_3x_4^2 + x_4 - x_1, x_2\}) \cup \text{Zero}(\{x_2^2 - x_1, x_2x_3x_4^2 + x_4 - x_1, x_2x_3\})$ is not necessary if one does not abandon $\mathbf{U}$.

Algorithm RegToStr: $\Psi \leftarrow \text{RegToStr}(\mathbf{T})$. Given a regular set $\mathbf{T}$, but not a strong regular set in $\mathbf{K}[\mathbf{x}, t]$, this algorithm computes a finite set Ψ of strong regular sets in $\mathbf{K}[\mathbf{x}, t]$ such that

$$\text{Zero}(\mathbf{T}) \subseteq \bigcup_{\mathbf{T}^* \in \Psi} \text{Zero}(\mathbf{T}^*).$$

RTS1. Set $\Psi \leftarrow \emptyset$, and $\Phi \leftarrow \{\mathbf{T}\}$.

RTS2. While $\Phi \neq \emptyset$, do:

RTS2.1. Let $\mathbf{T}^\star$ be an element of Φ, set $\Phi \leftarrow \Phi \backslash \{\mathbf{T}^\star\}$. If $\mathbf{T}^\star$ is a strong regular set, then $\Psi \leftarrow \Psi \cup \{\mathbf{T}^\star\}$ and go to RTS2.

RTS2.2. If $\mathbf{T}^\star$ is not strong regular set, then one can compute a strong regular set $\{\mathbf{T}^{\star\star}\}$ and a polynomial set $\{\mu_1^*, \cdots, \mu_{m_{\mathbf{T}^\star}}^*\}$ by Theorem 3.1.1. Set $\Psi \leftarrow \Psi \cup \{\mathbf{T}^{\star\star}\}$ and

$$\Phi \leftarrow \Phi \cup \bigcup_{i=1}^{m_{\mathbf{T}^\star}} \text{Reg}(\mathbf{T}^\star \cup \{\mu_i^*\}).$$

Example 3.1.1 Continued from Example 2.1. $\mathbf{T}_2$ is not a strong regular set in $\mathbf{Q}[c, z, y, x, t]$. According to Theorem 3.1.1, we can get $\eta = \{2\}$, and then pick $\mu_2^* = RI_{20} = c^2$, thereby obtain a strong regular set $\mathbf{T}_{2,0} = \{2z^4 - 3cz^2 + z + c^2, zy - z^2 + c, x - z, c^2t - 1\}$ and a polynomial set $\{c^2\}$ in $\mathbf{Q}[c, z, y, x, t]$. Compute $\text{Reg}(\mathbf{T}_2 \cup \{c^2\}) = \{\mathbf{T}_{2,1}, \mathbf{T}_{2,2}\}$, where $\mathbf{T}_{2,1} = \{c, z, x\}$ and $\mathbf{T}_{2,2} = \{c, 2z^3 + 1, zy - z^2, x - z\}$. Thus $\text{RegToStr}(\mathbf{T}_2) = \{\mathbf{T}_{2,0}, \mathbf{T}_{2,1}, \mathbf{T}_{2,2}\}$ and $\text{Zero}(\mathbf{T}_2) \subseteq \bigcup_{i=0}^{2} \text{Zero}(\mathbf{T}_{2,i})$ in $\mathbf{Q}[c, z, y, x, t]$.

3.2 Algorithm RSplit

We are given an arbitrary strong regular set $\mathbf{T}$ in $\mathbf{K}[\mathbf{x}, t]$ and a polynomial g in $\mathbf{K}[\mathbf{x}]$. By the following algorithm RSplit, one can decompose $\mathbf{T}$ into two sets Φ_1 and Φ_2 of strong regular sets in $\mathbf{K}[\mathbf{x}, t]$ with respect to g.

Definition 3.2.1. Let $\mathbf{T}$ be a triangular set in $\mathbf{K}[\mathbf{x}, t]$ and g a polynomial in $\mathbf{K}[\mathbf{x}]$. One can form a sequence of polynomials $g_{-1}, g_0, g_1, \cdots, g_{m-1}, g_m,$

where $g_{-1} = g$ and $g_0 = \text{prem}(g_{-1}, \mathbf{T})$ by the rule: for any $i \geq 0$, if $g_i = 0$ and $g_{i-1} \neq 0$ then put $g_m := g_{i-1}$; else if $g_i \in \mathbf{K}[u_1, \cdots, u_r] \backslash \{0\}$ then put $g_m := g_i$; otherwise put $g_i := \text{prem}(\text{prem}(f_{\text{cls}(g_{i-1})}, g_{i-1}, \text{lv}(g_{i-1})), \mathbf{T})$. The g_m is called the *expanding pseudo-remainder*[14] of g with respect to $\mathbf{T}$, denoted simply by $\text{Eprem}(g, \mathbf{T})$.

Proposition 3.2.1. Let $\mathbf{T}$ be a triangular set in $\mathbf{K}[\mathbf{x}, t]$ and g a polynomial in $\mathbf{K}[\mathbf{x}]$. If $\text{Eprem}(g, \mathbf{T}) \in \mathbf{K} \backslash \{0\}$, then $\text{Proj}_{\mathbf{x}}\text{Zero}(\mathbf{T}) \cap \text{Zero}(\{g\}) = \emptyset$.

Theorem 3.2.1. Let $\mathbf{T}$ be a strong regular set in $\mathbf{K}[\mathbf{x}, t]$ and g a polynomial in $\mathbf{K}[\mathbf{x}]$ such that $\text{prem}(g, \mathbf{T}) \neq 0$. Assume $\text{lv}(g^*) = y_k$ $(1 \leq k \leq s)$, where $g^* = \text{Eprem}(g, \mathbf{T})$. If $\text{res}(\text{ini}(g^*), \mathbf{T}) \in \mathbf{K} \backslash \{0\}$, then $\mathbf{T}$ can be split up into two strong regular sets $\mathbf{T}_1$ and $\mathbf{T}_2$ in $\mathbf{K}[\mathbf{x}, t]$ with respect to g such that $\text{Zero}(\mathbf{T}) = \text{Zero}(\mathbf{T}_1) \cup \text{Zero}(\mathbf{T}_2)$.

Proof: Consider the case in which $\mathbf{T} = \{f_1, \cdots, f_s\}$. From the definition of the expanding pseudo-remainder, the form $g^* = \text{Eprem}(g, \mathbf{T})$ implies that $\text{prem}(\text{prem}(f_k, g^*, y_k), \mathbf{T}) = 0$. It follows that

$$I_0^{c_0} f_k = f_{k,1} f_{k,2} + R, \ \text{prem}(R, \{f_1, \cdots, f_{k-1}\}) = 0,$$

where $f_{k,1} = g^*$, $I_0 = \text{ini}(f_{k,1})$, $f_{k,2} \in \mathbf{K}[\mathbf{x}]$ and c_0 is some nonnegative integer. Obtain two triangular sets $\mathbf{T}_1$ and $\mathbf{T}_2$ by substituting f_k in $\mathbf{T}$ for $f_{k,1}$ and $f_{k,2}$ respectively.

Next we prove that $\text{Zero}(\mathbf{T}) = \text{Zero}(\mathbf{T}_1) \cup \text{Zero}(\mathbf{T}_2)$. For any $\mathbf{z_0} \in \text{Zero}(\mathbf{T})$. It follows from Theorem 2.2 that $f_{k,1}(\mathbf{z_0}) f_{k,2}(\mathbf{z_0}) = 0$. Thus we have $\text{Zero}(\mathbf{T}) \subset \text{Zero}(\mathbf{T}_1) \cup \text{Zero}(\mathbf{T}_2)$. In the other direction, consider any $\mathbf{z_0} \in \text{Zero}(\mathbf{T}_1) \cup \text{Zero}(\mathbf{T}_2)$. The fact $\text{res}(I_0, \mathbf{T}) \in \mathbf{K} \backslash \{0\}$ implies $f_k(\mathbf{z_0}) = 0$, thus $\text{Zero}(\mathbf{T}) \supset \text{Zero}(\mathbf{T}_1) \cup \text{Zero}(\mathbf{T}_2)$. Therefore, $\text{Zero}(\mathbf{T}) = \text{Zero}(\mathbf{T}_1) \cup \text{Zero}(\mathbf{T}_2)$. The proof of the assertion that $\mathbf{T}_1$ and $\mathbf{T}_2$ are strong regular sets is somewhat simple and trivial, so we omit the details.

The case in which $\mathbf{T} = \{f_1, \cdots, f_s, \mu_0 t - 1\}$ is proved similarly.

Theorem 3.2.2. Let $\mathbf{T}$ be a strong regular set in $\mathbf{K}[\mathbf{x}, t]$ and g a polynomial in $\mathbf{K}[\mathbf{x}]$ such that $\text{prem}(g, \mathbf{T}) \neq 0$. If g^* or $\text{res}(\text{ini}(g^*), \mathbf{T}) \in \mathbf{K}[u_1, \cdots, u_r] \backslash \mathbf{K}$ with $g^* = \text{Eprem}(g, \mathbf{T})$, then there exist a strong regular set $\mathbf{T}_1$ and a triangular set $\mathbf{T}_2$ with respect to g such that $\text{Zero}(\mathbf{T}) = \text{Zero}(\mathbf{T}_1) \cup \text{Zero}(\mathbf{T}_2)$.

Proof: Consider the case in which $\mathbf{T} = \{f_1, \cdots, f_s\}$. Set $c_0 = g^*$ if $g^* \in \mathbf{K}[u_1, \cdots, u_r] \backslash \mathbf{K}$; $c_0 = \text{res}(\text{ini}(g^*), \mathbf{T})$ otherwise. Put $\mathbf{T}_1 = \{f_1, \cdots, f_s, c_0 t - 1\}$ and $\mathbf{T}_2 = \mathbf{T} \cup \{c_0\}$. $\mathbf{T}_1$ is a strong regular set in $\mathbf{K}[\mathbf{x}, t]$.

The case in which $\mathbf{T} = \{f_1, \cdots, f_s, \mu_0 t - 1\}$ is proved analogously, just replacing $\mathbf{T}_1$ by $\{f_1, \cdots, f_s, c_0\mu_0 t - 1\}$ in the above case.

Theorem 3.2.3. Let $\mathbf{T}$ be a strong regular set in $\mathbf{K}[\mathbf{x}, t]$ and g a polynomial in $\mathbf{K}[\mathbf{x}]$ such that $\mathrm{prem}(g, \mathbf{T}) \neq 0$. If $\mathrm{lv}(g^*) = y_k$, $1 \leq k \leq s$ and $\mathrm{res}(\mathrm{ini}(g^*), \mathbf{T}) = 0$ with $g^* = \mathrm{Eprem}(g, \mathbf{T})$, then there exist two polynomial sets $\mathbf{T}_1$ and $\mathbf{T}_2$ with respect to g such that $\mathrm{Zero}(\mathbf{T}) = \mathrm{Zero}(\mathbf{T}_1) \cup \mathrm{Zero}(\mathbf{T}_2)$.

Proof: Consider the case in which $\mathbf{T} = \{f_1, \cdots, f_s\}$. Put $\mathbf{T}_1 = \{f_1, \cdots, f_s, c_0 t - 1\}$ and $\mathbf{T}_2 = \mathbf{T} \cup \{c_0\}$, where $c_0 = \mathrm{ini}(g^*)$.

The case in which $\mathbf{T} = \{f_1, \cdots, f_s, \mu_0 t - 1\}$ is proved analogously, just replacing $\mathbf{T}_1$ by $\{f_1, \cdots, f_s, c_0\mu_0 t - 1\}$ in the above case.

Algorithm RSplit: $[\Phi_1, \Phi_2] \leftarrow \mathrm{RSplit}(\mathbf{T}_0, P)$. Given a strong regular set $\mathbf{T}_0$ in $\mathbf{K}[\mathbf{x}, t]$ and a polynomial P in $\mathbf{K}[\mathbf{x}]$, this algorithm computes two sets Φ_1 and Φ_2 of strong regular sets in $\mathbf{K}[\mathbf{x}, t]$ such that

$$\mathrm{Zero}(\mathbf{T}_0) \subseteq \bigcup_{\mathbf{T}^* \in \Phi_1 \cup \Phi_2} \mathrm{Zero}(\mathbf{T}^*),$$

and there exists an integer $d > 0$ such that $\mathrm{prem}(P^d, \mathbf{T}^*) = 0$ for any $\mathbf{T}^* \in \Phi_1$; $\mathrm{Proj}_{\mathbf{x}}\mathrm{Zero}(\mathbf{T}^*) \cap \mathrm{Zero}(\{P\}) = \emptyset$ for any $\mathbf{T}^* \in \Phi_2$.

SP1. Set $\Phi_1 \leftarrow \emptyset$, $\Phi_2 \leftarrow \emptyset$ and $\Psi \leftarrow \{\mathbf{T}_0\}$.

SP2. While $\Psi \neq \emptyset$, do:

SP2.1. Let $\mathbf{T}$ be an element of Ψ, set $\Psi \leftarrow \Psi \backslash \{\mathbf{T}\}$. If there exists an integer $d_0 > 0$ such that $\mathrm{prem}(P^{d_0}, \mathbf{T}) = 0$, then set $\Phi_1 \leftarrow \Phi_1 \cup \{\mathbf{T}\}$, and go to SP2.

SP2.2. If $\mathrm{Eprem}(P, \mathbf{T}) \in \mathbf{K} \backslash \{0\}$, then set $\Phi_2 \leftarrow \Phi_2 \cup \{\mathbf{T}\}$ and go to SP2.

SP2.3. Now the condition of Theorem 3.2.1 holds; then one can obtain two strong regular sets $\mathbf{T}_1$ and $\mathbf{T}_2$ by Theorem 3.2.1. Set $\Psi \leftarrow \Psi \cup \{\mathbf{T}_1, \mathbf{T}_2\}$ and go to SP2.

SP2.4. Now the condition of Theorem 3.2.2 holds; then one can obtain a strong regular sets $\mathbf{T}_1$ and a triangular set $\mathbf{T}_2$ by Theorem 3.2.2. Compute $\mathrm{Reg}(\mathbf{T}_2) = \Phi^*$, and then set

$$\Psi \leftarrow \Psi \cup \bigcup_{\mathbf{T}^* \in \Phi^*} \mathrm{RegToStr}(\mathbf{T}^*) \cup \{\mathbf{T}_1\}.$$

SP2.5. Now the condition of Theorem 3.2.3 holds; then one can obtain two polynomial sets $\mathbf{T}_1$ and $\mathbf{T}_2$ by Theorem 3.2.3. Compute $\mathrm{Reg}(\mathbf{T}_1) = \Phi_1^*$ and $\mathrm{Reg}(\mathbf{T}_2) = \Phi_2^*$, and then set

$$\Psi \leftarrow \Psi \cup \bigcup_{\mathbf{T}^* \in \Phi_1^* \cup \Phi_2^*} \mathrm{RegToStr}(\mathbf{T}^*).$$

Example 3.2.1. Continued from Example 3.1.1, $\text{RSplit}(\mathbf{T}_{2,0}, P) = [\{\mathbf{T}_{2,0}\}, \emptyset]$; $\text{RSplit}(\mathbf{T}_{2,1}, P) = [\emptyset, \{\mathbf{T}_{2,1}\}]$; $\text{RSplit}(\mathbf{T}_{2,2}, P) = [\{\mathbf{T}_{2,2}\}, \emptyset]$, for any $P \in \mathbf{P}$.

3.3 Algorithm Dec

Algorithm Dec: $\Phi \leftarrow \text{Dec}(\mathbf{P}, \mathbf{Q})$. Given a polynomial system $[\mathbf{P}, \mathbf{Q}]$ in $\mathbf{K}[\mathbf{x}]$, this algorithm computes a set Φ of strong regular sets in $\mathbf{K}[\mathbf{x}, t]$ such that

$$\text{Zero}(\mathbf{P}/\mathbf{Q}) = \bigcup_{\mathbf{T} \in \Phi} \text{Proj}_{\mathbf{x}}\text{Zero}(\mathbf{T}),$$

and there exists an integer $d > 0$ such that $\text{prem}(P^d, \mathbf{T}) = 0$ and $\text{Proj}_{\mathbf{x}}\text{Zero}(\mathbf{T}) \cap \text{Zero}(\{Q\}) = \emptyset$ for any $\mathbf{T} \in \Phi$, $P \in \mathbf{P}$ and $Q \in \mathbf{Q}$.

D1. Compute $\text{Reg}(\mathbf{P}) = \Phi_0$ and set

$$\Phi \leftarrow \emptyset, \quad \Psi \leftarrow \bigcup_{\mathbf{T} \in \Phi_0} \text{RegToStr}(\mathbf{T}).$$

D2. While $\Psi \neq \emptyset$, do:
- D2.1. Let $\mathbf{T}$ be an element of Ψ, set $\Psi \leftarrow \Psi \backslash \{\mathbf{T}\}$.
- D2.2. For any $P \in \mathbf{P}$ and $Q \in \mathbf{Q}$ do:
  - D2.1.1. If $\text{Eprem}(P, \mathbf{T}) \in \mathbf{K} \backslash \{0\}$, then go to D2.
  - D2.1.2. If there exists an integer $k > 0$ such that $\text{prem}(Q^k, \mathbf{T}) = 0$, then go to D2.
  - D2.1.3. If $\text{prem}(P, \mathbf{T}) \neq 0$, then compute $\text{RSplit}(\mathbf{T}, P) = [\Phi_1, \Phi_2]$. Set $\Psi \leftarrow \Psi \cup \Phi_1$, and then go to D2.
  - D2.1.4. If $\text{prem}(Q, \mathbf{T}) \neq 0$ and $\text{Eprem}(Q, \mathbf{T}) \notin \mathbf{K}$, then compute $\text{RSplit}(\mathbf{T}, Q) = [\Phi_1, \Phi_2]$. Set $\Psi \leftarrow \Psi \cup \Phi_2$, and then go to D2.
- D2.3. Set $\Phi \leftarrow \Phi \cup \{\mathbf{T}\}$.

Example 3.3.1. Continued from Example 3.2.1, we have

$$\text{Zero}(\mathbf{P}) = \bigcup_{i=1, i \neq 2}^{6} \text{Proj}_{\mathbf{x}}\text{Zero}(\mathbf{T}_i) \cup \text{Proj}_{\mathbf{x}}\text{Zero}(\mathbf{T}_{2,0}) \cup \text{Proj}_{\mathbf{x}}\text{Zero}(\mathbf{T}_{2,2}),$$

where $\mathbf{x} = (c, z, y, x)$.

The algorithm is different from RegSer which uses subresultant subchain and allows simultaneously splitting one polynomial system into several subsystems. $\text{Dec}(\mathbf{P}, \mathbf{Q})$ performs expanding pseudo-division and splitting step-by-step each regular set in Reg($\mathbf{P}$) if there exists any common zero, without computing GCDs, into finitely many strong regular sets in $\mathbf{K}[\mathbf{x}, t]$.

Remark 3.2. If a regular series $[\mathbf{T}_1, \mathbf{U}_1], \cdots, [\mathbf{T}_e, \mathbf{U}_e]$ is computed by RegSer from a polynomial system $[\mathbf{P}, \mathbf{Q}]$ in $\mathbf{K}[\mathbf{x}]$, in this case let v_i be the product of all the polynomials in $\mathbf{U}_i$ and $\mathbf{T}_i^* = \mathbf{T}_i \cup \{v_i t - 1\}$ where t is a new variable, then similarly we have

$$\mathrm{Zero}(\mathbf{P}/\mathbf{Q}) = \bigcup_{i=1}^{e} \mathrm{Zero}(\mathbf{T}_i/\mathbf{U}_i) = \bigcup_{i=1}^{e} \mathrm{Proj}_{\mathbf{x}}\mathrm{Zero}(\mathbf{T}_i^*).$$

But the result of Dec($\mathbf{P}, \mathbf{Q}$) is better than that of RegSer($\mathbf{P}, \mathbf{Q}$), because $\mathrm{Proj}_{\mathbf{x}}\mathrm{Zero}(\mathbf{T}) \cap \mathrm{Zero}(\{Q\}) = \emptyset$ for any $\mathbf{T} \in \mathrm{Dec}(\mathbf{P}, \mathbf{Q})$ and $Q \in \mathbf{Q}$. On the other hand, Dec($\mathbf{P}, \mathbf{Q}$) may be simpler in some cases: for example

$$\begin{aligned}\mathrm{Zero}(\{x_1\}/\{x_2x_3x_5 + x_4\}) = \ & \mathrm{Proj}_{\mathbf{x}}\mathrm{Zero}(\{x_1, x_2x_3, x_2x_4t - 1\}) \\ & \cup \mathrm{Proj}_{\mathbf{x}}\mathrm{Zero}(\{x_1, (x_2x_3x_5 + x_4)x_2^2x_3t - 1\}) \\ & \cup \mathrm{Proj}_{\mathbf{x}}\mathrm{Zero}(\{x_1, x_2, x_4t - 1\})\end{aligned}$$

by RegSer($\{x_1\}, \{x_2x_3x_5 + x_4\}$), but

$$\mathrm{Zero}(\{x_1\}/\{x_2x_3x_5 + x_4\}) = \mathrm{Proj}_{\mathbf{x}}\mathrm{Zero}(\{x_1, (x_2x_3x_5 + x_4)t - 1\})$$

by Dec($\{x_1\}, \{x_2x_3x_5 + x_4\}$).

Remark 3.3. The algorithm Dec can be implemented by using existing mathematical software; its efficiency heavily depends upon the algorithm for Reg. We recommend one to use Wang's algorithm RegSer. All of the above examples in this paper have been computed by Maple system in an interactive way. We will further experiment and compare our algorithm with some of the other efficient algorithms implemented with Maple system when the package of RegSer is available.

Acknowledgments

The authors thank Dr. D. Wang specially for his helpful comments and suggestions. The first author[a] is grateful to Prof. J.-H. Mai for his help.

References

[1] Wu W-T. On the decision problem and the mechanization of theorem-proving in elementary geometry. *Scientia Sinica*, 1978, 21: 159-172

[a]Supported by Sichuan Communications Polytechnic.

[2] Yang L, Zhang J-Z. Search dependency between algebraic equations: An algorithm applied to automated reasoning. Technical Report IC/91/6, International Atomic Energy, Miramare, Trieste, 1991

[3] Zhang J-Z, Yang L, Hou X-R. A note on Wu Wen-Tsün's nondegenerate condition. Technical Report IC/91/160, International Atomic Energy, Miramare, Trieste, 1991

[4] Zhang J-Z, Yang L, Hou X-R. A note on Wu Wen-Tsün's nondegenerate condition. *Chinese Science Bulletin,* 1993, 38:1, 86-87

[5] Kalkbrener M. A generalized Euclidean algorithm for computing triangular representations of algebraic varieties. *J. Symb. Comput.,* 1993, 15: 143-167

[6] Yang L, Zhang J-Z, Hou X-R. An efficient decomposition algorithm for geometry theorem proving without factorization. MM Res. Preprints, 1993, 9: 115-131

[7] Yang L, Zhang J-Z, Hou X-R. Non-Linear equation systems and automated theorem proving. Shanghai: Shanghai Sci. Tech. Education Publ, 1996 [in Chinese]

[8] Chou S-C, Gao X-S. Ritt-Wu's decomposition algorithm and geometry theorem proving. In: Stickel, M. E. (ed.): 10th International Conference on Automated Deduction. Springer, Berlin Heidelberg New York Tokyo, 1990. 207-220 (Lecture notes in computer science, vol. 449)

[9] Gao X-S, Chou S-C. Solving parametric algebraic systems. In: Proceedings ISSAC '92, Berkeley, July 27-29, 1992. Association for Computing Machinery, New York, 1992. 335-341

[10] Wang D. An elimination method for polynomial systems. *J. Symb. Comput.,* 1993, 16: 83-114

[11] Wang D. Decomposing triangular systems into simple systems. *J. Symb. Comput.,* 1998, 25: 295-314

[12] Wang D. Computing triangular systems and regular systems. *J. Symb. Comput.,* 2000, 30: 221-236

[13] Wang D. Elimination methods. Wien/New York: Springer, 2001

[14] Li Y-B. The expanding WE algorithm of mehchanical geometry therorem proving. *Journal of Sichuan University(Natural Science Edition)*, 2000, (37)3: 331-335 [in Chinese]

[15] Li Y-B. The expanding WE algorithm and the approach of strong regular decomposition on nonlinear algebraic equation system. Ph.D. thesis, Mathematical College, Sichuan University, Chengdu, China, 2001

WU WEN-TSÜN FORMULAE FOR THE BLENDING OF QUADRATIC SURFACES

NA LEI TIE-RU WU GUO-CHEN FENG
(INSTITUTE OF MATHEMATICS, JILIN UNIVERSITY, CHANGCHUN, 130012, P.R.CHINA)

Abstract. We generalize the Wu Wen-tsün Formulae in [1] to the general case of the two-way blending of the implicit defined quadratic surfaces. The formulae obtained disclose the relationship between the existence condition of the two-way blending and the intrinsic geometric parameters of the primary quadratic surfaces and the clipping planes.

Keywords. Wu Wen-tsün Formulae, blending, quadratic surfaces.

1 Introduction

One of the central questions in CAGD is blending of implicit algebraic surfaces. Wu Wen-tsün [1] studied the problem by using the characteristic set method and derived a sufficient and necessary condition for the existence of a GC^1 blending cubic surface of two cylinders whose axes meet at a right angle and the intersections are perpendicular to the axes,

$$\begin{cases} g_1 = y^2 + z^2 - r_1^2, \\ h_1 = x - d_1, \end{cases} \qquad \begin{cases} g_2 = x^2 + z^2 - r_2^2, \\ h_2 = y - d_2, \end{cases}$$

and proved that there exists a unique cubic blending surface if and only if

$$r_1^2 + d_1^2 = r_2^2 + d_2^2, \tag{1}$$

where r_i are radii of the pipes, d_i are the distances between the clipping planes and the intersection point of the axes of the two cylinders.

C. Hoffmann and J. Hopcroft[2] derived a constructive method for the blending of two quadrics by using ideal theory. J. Warren[3] introduced a new notion of geometric continuity that he called *rescaling* continuity, and he discussed the algebraic structure of the blending surfaces. Wu and Zhou[4] discussed n-way blending surfaces of a symmetric form. Wu Tie-ru et al[5] derived a general algorithm on blending of two surfaces.

In this paper we further develop the theoretical results in [5], derive the sufficient and necessary condition for the existence of the GC^0 quadratic blending surfaces, and study the sufficient and necessary condition for the existence of the GC^1 blending cubic surfaces provided that there exist the GC^0 quadrics. In particular, we generalize Wu Wen-tsün's Formula (1) to the case that the axes of two quadratic surfaces may meet at any angle and the clipping planes may be chosen arbitrarily.

Let g_1 and g_2 be irreducible quadratic polynomials, which determine two distinct surfaces $V(g_1)$ and $V(g_2)$. Choose two linear functions h_1 and h_2 which determine two different planes $V(h_1)$ and $V(h_2)$ such that $V(h_i)$ and $V(g_i)$ intersect transversally with an irreducible quadratic intersection $V(g_i, h_i)(i = 1, 2)$. The problem is to seek the condition for h_1 and h_2 such that there exists a cubic polynomial f such that the surface $V(f)$ meets $V(g_i)$ with GC^1 continuity along the curves $V(g_i, h_i)(i = 1, 2)$.

2 Canonical Expansion

As shown in [5], f can be written in the form

$$f = u_1 g_1 + a_1 h_1^2 = u_2 g_2 + a_2 h_2^2, \tag{2}$$

and $\deg(u_i) = 1$, $\deg(a_i) = 1$. (2) is a non-linear system of 20 equations and 28 variables on the coefficients of unknown u_i, a_i, g_i and $h_i(i = 1, 2)$. Our purpose is to seek the condition for the existence of solutions of the system (2) and to obtain some kind of solutions. Because the coefficients are not real numbers but abstract symbols, we have no available way to solve the system. In the present paper, by virtue of the special structure of the system, a new method is used to obtain the sufficient and necessary condition for the existence of solutions of the equation system (2) in some important cases.

When clipping planes $V(h_1)$ and $V(h_2)$ are not parallel, we can expand g_1 and g_2 to the following form

$$g_1 = s_{00}^{(1)} + s_{10}^{(1)} h_1 + s_{01}^{(1)} h_2 + s_{20}^{(1)} h_1^2 + s_{11}^{(1)} h_1 h_2 + s_{02}^{(1)} h_2^2,$$

$$g_2 = s_{00}^{(2)} + s_{10}^{(2)} h_1 + s_{01}^{(2)} h_2 + s_{20}^{(2)} h_1^2 + s_{11}^{(2)} h_1 h_2 + s_{02}^{(2)} h_2^2,$$

where $\deg(s_{ij}^{(l)}) \leq 2 - i - j$ and $s_{ij}^{(l)} \in Span_R(N(I))(l = 1, 2)$. Here $Span_R(N(I))$ is the R-linear combination of the elements which compose a basis of quotient ring $R[x, y, z]/I$ of the ideal I. Denote $s_{ij}^{\lambda} = s_{ij}^{(1)} - \lambda s_{ij}^{(2)}$. Then obtain

$$g_1 - \lambda g_2 = s_{00}^{\lambda} + s_{10}^{\lambda} h_1 + s_{01}^{\lambda} h_2 + s_{20}^{\lambda} h_1^2 + s_{11}^{\lambda} h_1 h_2 + s_{02}^{\lambda} h_2^2.$$

Theorem 1. There exists an irreducible quadratic ruled surface $S(g)$ (i.e. a surface which contains curves $S(g_1, h_1), S(g_2, h_2)$) if and only if there exist a real number $\lambda \neq 0$ such that

$$s_{00}^{\lambda} = 0. \tag{3}$$

Proof. If $S(g)$ contains curves $S(g_1, h_1)$ and $S(g_2, h_2)$, then g may be expressed as $g = \lambda_1 g_1 + c_1 h_1 = \lambda_2 g_2 + c_2 h_2$, where λ_i are non-zero real numbers, c_i are linear functions. Substitute the canonical expansion g_i into the expression of g and check the constant term with respect to h_1 and h_2, we have $\lambda_1 s_{00}^{(1)} = \lambda_2 s_{00}^{(2)}$, namely (3), where $\lambda = \frac{\lambda_2}{\lambda_1}$.

Conversely, if (3) holds, then set $c_1 = -s_{10}^{\lambda} - s_{20}^{\lambda} h_1$, $c_2 = -s_{01}^{\lambda} + s_{11}^{\lambda} h_1 - s_{20}^{\lambda} h_2$ and the corresponding $g = \lambda_1 g_1 + c_1 h_1 = \lambda_2 g_2 + c_2 h_2$ is a ruled surface.

Theorem 2. If $V(h_1)$ intersects with $V(h_2)$, then (2) has solutions satisfying the condition

$$u_i = u_{i1} h_1 + u_{i2} h_2, \quad (i = 1, 2) \tag{4}$$

if and only if there exist real numbers $\lambda \neq 0$, $\mu \neq 0$ such that

$$s_{00}^{\lambda} = 0, \quad s_{10}^{\lambda} = \mu s_{01}^{\lambda}. \tag{5}$$

Proof. By virtue of the property of the prime ideal $\langle h_i \rangle$, we proved that if (2) has solutions satisfying (4), then there exists a quadratic ruled surface $S(g)$. Now compute the canonical expansion of $u_1 g_1 - u_2 g_2$. Let the coefficients of h_1, h_2 and $h_1 h_2$ be zero and use Theorem 1, we have (5).

Conversely, if (5) holds, set $u_1 = \mu h_1 - h_2$, $c_1 = -s_{10}^{\lambda}\mu - s_{20}^{\lambda}\mu h_1 + (s_{20}^{\lambda} - s_{11}^{\lambda}\mu) h_2$, $u_2 = \lambda u_1$, $c_2 = -s_{01}^{\lambda} + (s_{02}^{\lambda}\mu - s_{11}^{\lambda}) - s_{02}^{\lambda} h_2$. then we construct a solution satisfying (2) and (4).

3 Blending of Quadrics with Circular Transverse Cross Sections

Now we apply Theorem 2 to the problem on the blending of two quadrics with coplanar axes and the transverse cross section of the surfaces are circular. Two quadrics whose axes meet at an angle θ can be stated as the affine coordinate expression:

$$g_1 = \begin{cases} (y \sin\theta)^2 + z^2 - r_1^2, & (\textit{cylinder}) \\ (y \sin\theta)^2 + z^2 - 2p_1(x + y\cos\theta - x_0), & (\textit{circular paraboloid}) \\ y^2 + z^2 + (x - x_0)^2 - r_1^2, & (\textit{sphere}) \\ (y \sin\theta)^2 + z^2 - (x + y\cos\theta - x_0)^2 c_1^2, & (\textit{circular cone}) \end{cases} \tag{6}$$

$$g_2 = \begin{cases} (x \sin\theta)^2 + z^2 - r_2^2, & (\textit{cylinders}) \\ (x \sin\theta)^2 + z^2 - 2p_2(x\cos\theta + y - y_0), & (\textit{circular paraboloid}) \\ x^2 + z^2 + (y - y_0)^2 - r_2^2, & (\textit{sphere}) \\ (x \sin\theta)^2 + z^2 - (x\cos\theta + y - y_0)^2 c_2^2, & (\textit{circular cone}) \end{cases} \tag{7}$$

where x-axis and y-axis denote the axes of g_1 and g_2 respectively and $z = \bar{z}$.

If $V(h_i)$ and $V(g_i)$ intersect transversally with an irreducible quadratic intersection $(i = 1, 2)$, then the clipping planes can be expressed by

$$\begin{cases} h_1 = x + h_{12}y + h_{13}z + h_{10}, \\ h_2 = h_{21}x + y + h_{23}z + h_{20}. \end{cases} \tag{8}$$

In order to describe the geometric relation between the clipping planes h_1 and h_2, we introduce the notation $\Delta_{ij} = \det \begin{pmatrix} h_{1i} & h_{1j} \\ h_{2i} & h_{2j} \end{pmatrix}$ $(0 \le i \ne j \le 3)$, where $h_{11} = h_{22} = 1$. Obviously they can determine h_1, h_2, and they can be regarded as basic parameters. From Theorem 2 we prove that, if $V(h_1)$ intersects with $V(h_2)$, $\Delta_{12} \ne 0$, $h_{13} = h_{23} = 0$, then the sufficient and necessary conditions for the existence of the GC^0 quadratic blending surfaces(ruled surfaces) is the same as the sufficient and necessary conditions for the existence of the GC^1 blending cubic surfaces satisfying (4). The condition for the existence are the generalization of Wu Wen-tsün Formula (1). We also call them Wu Wen-tsün Formulae.

Theorem 3.(Wu Wen-tsün Formulae) If $\Delta_{12} \ne 0$, $h_{13} = h_{23} = 0$, then the sufficient and necessary conditions for the existence of the GC^0 quadratic blending surfaces(ruled surfaces) are the same as the sufficient and necessary conditions for the existence of the GC^1 blending cubic surfaces satisfying (4), which are

$$(1.1)\quad r_1^2\Delta_{12}^2 + \Delta_{20}^2 \sin^2\theta = r_2^2\Delta_{12}^2 + \Delta_{10}^2 \sin^2\theta,$$

$$(1.2)\quad \Delta_{10}^2 \sin^2\theta + 2p_1x_0\Delta_{12}^2 - p_1\Delta_{20}\Delta_{12} + p_1\Delta_{10}\Delta_{12}\cos\theta = \Delta_{20}^2\sin^2\theta - r_2^2\Delta_{12}^2,$$

$$(1.3)\quad \Delta_{10}^2 + x_0^2\Delta_{12}^2 - r_1^2\Delta_{12} - 2x_0\Delta_{20}\Delta_{12} + r_2^2\Delta_{12}^2 = 0, \quad (\text{take } \theta = \tfrac{\pi}{2} \text{ in } g_1)$$

$$(1.4)\quad \begin{aligned} &\Delta_{20}^2\sin^2\theta - r_2^2\Delta_{12}^2 = \Delta_{10}^2(\sin^2\theta - \cos^2\theta c_1^2) - \Delta_{20}^2c_1^2 - x_0^2c_1^2\Delta_{12}^2 \\ &+2x_0\Delta_{20}\Delta_{12}c_1^2 - 2x_0\Delta_{10}\Delta_{12}\cos\theta c_1^2 + 2\Delta_{10}\Delta_{20}\cos\theta c_1^2, \end{aligned}$$

$$(2.2)\quad \begin{aligned} &\Delta_{10}^2\sin^2\theta + 2p_1x_0\Delta_{12}^2 - 2p_1\Delta_{20}\Delta_{12} + 2p_1\Delta_{10}\Delta_{12}\cos\theta \\ &= \Delta_{20}^2\sin^2\theta + 2p_2y_0\Delta_{12}^2 + 2p_2\Delta_{10}\Delta_{12} - 2p_2\Delta_{20}\Delta_{12}\cos\theta, \end{aligned}$$

$$(2.3)\quad \begin{aligned} &\Delta_{10}^2 + \Delta_{20}^2 + x_0^2\Delta_{12}^2 - r_1^2\Delta_{12} - 2x_0\Delta_{20}\Delta_{12} \\ &= \Delta_{20}^2 + 2p_2y_0\Delta_{12}^2 + p_2\Delta_{10}\Delta_{12}, \quad (\text{take } \theta = \tfrac{\pi}{2} \text{ in } g_1) \end{aligned}$$

$$(2.4)\quad \begin{aligned} &\Delta_{10}^2(\sin^2\theta - \cos^2\theta c_1^2) - \Delta_{20}^2c_1^2 - x_0^2c_1^2\Delta_{12}^2 + 2x_0\Delta_{20}\Delta_{12}c_1^2 - 2x_0\Delta_{10}\Delta_{12}\cos\theta c_1^2 \\ &+2\Delta_{10}\Delta_{20}\cos\theta c_1^2 = \Delta_{20}^2\sin^2\theta + 2p_2y_0\Delta_{12}^2 + p_2\Delta_{10}\Delta_{12} - p_2\Delta_{20}\Delta_{12}\cos\theta, \end{aligned}$$

$$(3.3) \quad x_0^2\Delta_{12} - r_1^2 - 2x_0\Delta_{20} = y_0^2\Delta_{12} - r_2^2 + 2y_0\Delta_{10},$$

$$(3.4) \quad -\Delta_{20}^2c_1^2 - x_0^2c_1^2\Delta_{12}^2 + 2x_0\Delta_{20}\Delta_{12}c_1^2 = \Delta_{20}^2 + y_0^2\Delta_{12}^2 - r_2^2\Delta_{12} + 2y_0\Delta_{10}\Delta_{12},$$

$$\begin{aligned}(4.4) \quad &\Delta_{10}^2(\sin^2\theta - \cos^2\theta c_1^2) - \Delta_{20}^2c_1^2 - x_0^2c_1^2\Delta_{12}^2 + 2x_0\Delta_{20}\Delta_{12}c_1^2 - 2x_0\Delta_{10}\Delta_{12}\cos\theta c_1^2 \\ &+2\Delta_{10}\Delta_{20}\cos\theta c_1^2 = \Delta_{20}^2(\sin^2\theta - \cos^2\theta c_2^2) - \Delta_{10}^2c_2^2 - y_0^2c_2^2\Delta_{12}^2 - 2y_0\Delta_{10}\Delta_{12}c_2^2 \\ &+2x_0\Delta_{20}\Delta_{12}\cos\theta c_2^2 + 2\Delta_{10}\Delta_{20}\cos\theta c_2^2,\end{aligned}$$

where we denote $(m.n)$ the blending problem of two quadratic surfaces in which the first quadric is defined by the m-th equation of (6) and the second quadric is defined by the n-th equation of (7).

Corollary 1.(Wu Wen-tsün Formulae) Under the assumption and denotation of Theorem 3, we suppose that the two quadrics whose axes meet at a right angle and the intersections are perpendicular to the axes, then the Generalized Wu Wen-tsün Formulae are

(1.1) $r_1^2 + h_{10}^2 = r_2^2 + h_{20}^2$,

(1.2) $h_{20}^2 + 2p_1x_0 - p_1h_{10} = h_{10}^2 - r_2^2$,

(1.3) $r_1^2 + 2x_0h_{10} - x_0^2 = r_2^2 + h_{20}^2$, (take $\theta = \frac{\pi}{2}$ in g_1)

(1.4) $h_{10}^2 - r_2^2 = h_{20}^2 - h_{10}^2c_1^2 - x_0^2c_1^2 + 2x_0h_{10}c_1^2$,

(2.2) $h_{10}^2 - 2p_1x_0 + 2p_1h_{10} = h_{20}^2 - 2p_2y_0 + 2p_2h_{20}$,

(2.3) $h_{20}^2 + h_{10}^2 + x_0^2 - r_1^2 - 2x_0h_{10} = h_{10}^2 + 2p_2y_0 - p_2h_{20}$, (take $\theta = \frac{\pi}{2}$ in g_1)

(2.4) $h_{20}^2 - h_{10}^2c_1^2 - x_0^2c_1^2 + 2x_0h_{10}c_1^2 = h_{10}^2 + 2p_2y_0 - p_2h_{20}$,

(3.3) $x_0^2 - r_1^2 - 2x_0h_{10} = y_0^2 - r_2^2 - 2y_0h_{20}$,

(3.4) $2x_0h_{10}c_1^2 - h_{10}^2c_1^2 - x_0^2c_1^2 = h_{10}^2 + y_0^2 - r_2^2 - 2y_0h_{20}$,

(4.4) $h_{20}^2 - h_{10}^2c_1^2 - x_0^2c_1^2 + 2x_0h_{10}c_1^2 = h_{10}^2 - h_{20}^2c_2^2 - y_0^2c_2^2 + 2y_0h_{20}c_2^2$,

where the condition (1.1) is the primary Wu Wen-tsün Formula.

4 Blending of Quadrics with Elliptic Transverse Cross Sections

Now we turn to the discussion on the blending of quadratic surfaces with elliptic transverse cross section. First we study the blending of two elliptic cylinders:

$$\begin{cases} g_1 = \dfrac{(y\sin(\theta)\cos(\alpha) + z\sin(\alpha))^2}{{a_1}^2} + \dfrac{(-y\sin(\theta)\sin(\alpha) + z\cos(\alpha))^2}{{b_1}^2} - 1, \\ g_2 = \dfrac{((-(x + y\cos(\theta))\sin(\theta) + y\sin(\theta)\cos(\theta))\cos(\alpha) + z\sin(\alpha))^2}{{a_2}^2} \\ \quad + \dfrac{(-(-(x + y\cos(\theta))\sin(\theta) + y\sin(\theta)\cos(\theta))\sin(\alpha) + z\cos(\alpha))^2}{{b_2}^2} - 1, \end{cases}$$

and the clipping planes are of the form (8). If $\Delta_{12} \neq 0$, $h_{13} = h_{23} = 0$ and $\sin 2\alpha_1 = \sin 2\alpha_2 = 0$, that is, one of the axes of the elliptic cross section lies in the xy-plane, the corresponding Wu Wen-tsün Formulae hold:

1.If $\alpha_1 = 0, \alpha_2 = 0:$ $\quad a_1^2 b_2^2(\Delta_{20}^2 \sin^2\theta - a_2^2\Delta_{12}^2) = a_2^2 b_1^2(\Delta_{10}^2 \sin^2\theta - a_1^2\Delta_{12}^2);$

2.If $\alpha_1 = \frac{\pi}{2}, \alpha_2 = 0:$ $\quad b_1^2 b_2^2(\Delta_{20}^2 \sin^2\theta - a_2^2\Delta_{12}^2) = a_2^2 a_1^2(\Delta_{10}^2 \sin^2\theta - b_1^2\Delta_{12}^2);$

3.If $\alpha_1 = 0, \alpha_2 = \frac{\pi}{2}:$ $\quad a_1^2 a_2^2(\Delta_{20}^2 \sin^2\theta - b_2^2\Delta_{12}^2) = b_2^2 b_1^2(\Delta_{10}^2 \sin^2\theta - a_1^2\Delta_{12}^2);$

4.If $\alpha_1 = \frac{\pi}{2}, \alpha_2 = \frac{\pi}{2}:$ $\quad b_1^2 a_2^2(\Delta_{20}^2 \sin^2\theta - b_2^2\Delta_{12}^2) = b_2^2 a_1^2(\Delta_{10}^2 \sin^2\theta - b_1^2\Delta_{12}^2).$

In particular, the axes of the two elliptic cylinders meet at a right angle and the intersections are perpendicular to the axes, the corresponding generalized Wu Wen-tsün Formulae are:

1.If $\alpha_1 = 0, \alpha_2 = 0:$ $\quad a_1^2 b_2^2(h_{10}^2 - a_2^2) = a_2^2 b_1^2(h_{20}^2 - a_1^2);$

2.If $\alpha_1 = \frac{\pi}{2}, \alpha_2 = 0:$ $\quad b_1^2 b_2^2(h_{10}^2 - a_2^2) = a_2^2 a_1^2(h_{20}^2 - b_1^2);$

3.If $\alpha_1 = 0, \alpha_2 = \frac{\pi}{2}:$ $\quad a_1^2 a_2^2(h_{10}^2 - b_2^2) = b_2^2 b_1^2(h_{20}^2 - a_1^2);$

4.If $\alpha_1 = \frac{\pi}{2}, \alpha_2 = \frac{\pi}{2}:$ $\quad b_1^2 a_2^2(h_{10}^2 - b_2^2) = b_2^2 a_1^2(h_{20}^2 - b_1^2).$

We also discuss the generalized Wu Wen-tsün Formulae for the blending of two elliptic paraboloids, ellipsoids and elliptic cones, the readers who want to know the details of the above formulae should look up [6].

5 Conclusion

Wu Wen-tsün Formula obtained in [1] is a deep differential geometry formula, which discloses the relationship between the existence condition of the two-way blending and the intrinsic geometric parameters of the primary quadratic surfaces. We generalize the formula to the general case of the two-way blending of the implicit defined quadratic surfaces and derive similar formulae(we also call them Wu Wen-tsün Formulae).

References

1. Wu Wen-tsun and Wang Ding-kang, On surface-fitting problem in CAGD, *Mathematics in Practice and Theory,* No.3, 26–31, 1994.
2. C. Hoffmann and J. Hopcroft, Quadratic blending surfaces, *Computer-Aided Design,* Vol. 18, pp 301–307,1986.
3. J. Warren, Blending algebraic surfaces, *ACM Transactions on Graphics,* Vol.8, No.4, 263–278,1989.

4. Wu Tie-ru, Zhou Yun-shi, *Computer Aided Geometric Design,* Vol.17, No.9, 759-766,2000.
5. Wu Tie-ru, Gao Wei-guo and Feng Guo-chen, Blending of implicit algebraic surfaces, *Proceedings of the ASCM 1995* (Beijing, China, Aug. 18-20), 125–131, Scientists Incorporated,1995.
6. Lei Na, Ph.D. Dissertation, Jilin University, P.R.China,2002.

EPSILON: A LIBRARY OF SOFTWARE TOOLS FOR POLYNOMIAL ELIMINATION

DONGMING WANG

LIP6, UPMC–CNRS, 4 place Jussieu, 75252 Paris Cedex 05, France
E-mail: Dongming.Wang@lip6.fr · http://calfor.lip6.fr/~wang

This article presents a Maple library of functions for decomposing systems of multivariate polynomials into triangular systems of various kinds (regular, simple, or irreducible), with an application package for manipulating and proving geometric theorems.

1 Introduction

It has been widely recognized that zero-decomposition algorithms based on triangular sets are efficient tools for polynomial elimination and computation. They are ideal alternatives to algorithms based on Gröbner bases and resultants for many geometric applications. The popularization of triangular-set-based methods suffers from the lack of available implementations, while in striking contrast Gröbner basis packages may be found in almost all computer algebra systems. A recent effort to improve the situation is made by Aubry and Moreno Maza[1] who have implemented the four methods of Wu, Lazard, Kalkbrener, and the author in Axiom. This comparative implementation has also been integrated into Aldor by Moreno Maza. Another earlier implementation of Wu's method, carried out by Gao and Chou, has been included in the system GEX[3] for automated geometric reasoning.

The author has been advocating the development of triangular set methods for more than a decade. One of his efforts is the maintenance of the CharSets package,[8] which provides a general and complete implementation of Wu–Ritt's method of characteristic sets,[5,14,15] together with a few relevant algorithms, in Maple. It seems that this package still represents one of the most comprehensive and widely available implementations of the characteristic set method. Part of the package was translated by M. W. Messollen into a library for Singular and Macaulay 2 and by T. Shimoyama into Risa/Asir for the purpose of primary decomposition.[6] It is being translated by D. Stanger into Maxima.

In 1993 the author proposed a method for decomposing any system of multivariate polynomials into triangular systems, with or without projection, and into irreducible triangular systems.[7] This method makes use of some elimination strategies from Wu–Ritt's method and Seidenderg's theory and is practically more efficient. In 1998 and 2000, the author proposed another method, based on the computation of regular subresultants, for decomposing any polynomial system into simple systems[11] and regular systems.[12] Meanwhile, we have implemented these methods as two separate packages, named TriSys and SiSys, also in Maple for research experiments. The Epsilon library described in this paper consists of CharSets, TriSys, SiSys, and another application package GEOTHER[10] (for geometric theorem proving). We present the implementation, functionality, performance, and availability of Epsilon, together with some practical information for the user.

Besides the above-mentioned packages, there exist several implementations of triangular set algorithms, in particular, algorithms based on Wu–Ritt's characteristic sets.[2] However, most of these implementations have quite limited availability. The reader is referred to the books[3,13,16] for remarkable applications of triangular set methods and software tools in geometric reasoning.

2 Design and Implementation of Epsilon

Epsilon is a library of functions implemented in Maple mostly and in Java for polynomial elimination with applications. Some of the Java routines for geometric drawing were written previously in C. This library has evolved from the author's continuous effort to put elimination theory and methods into practice. The functions available in the current version of the library are capable of performing a number of computational tasks including:

- Decomposing polynomial systems into triangular systems of various kinds;
- Decomposing algebraic varieties into unmixed or irreducible subvarieties, and polynomial ideals into primary components;
- Factorizing polynomials over successive algebraic extension fields;
- Solving systems of polynomial equations and inequations;
- Manipulating and proving theorems in elementary and differential geometry automatically.

There are also functions which can decompose any system of ordinary differential polynomials into (irreducible) differential triangular systems. Main capabilities and the dependence of the four Epsilon modules are summarized in Fig. 1.

The implementation of Epsilon started with the CharSets module in the fall of 1989. The first version CharSets 1.0 running with Maple 4.3 was ready for public distribution in later 1990. This self-contained package implements a collection of fundamental algorithms, including two little known algorithms for polynomial factorization over algebraic extension fields, based on the computation of Wu–Ritt's characteristic sets.[5,14,15]

Along with the implementation of CharSets, we created an environment GEOTHER for manipulating and proving theorems in geometry. With the powerful algorithms from CharSets and TriSys as its algebraic engine, GEOTHER provides a set of functions for the manipulation, translation, diagram generation, and automated proving of geometric theorems. An early version of GEOTHER was ready for demonstration in 1991, and in 1996 was published a short description of the enhanced version.[10]

CharSets is an independent toolkit, whereas all the other modules need routines from CharSets and thus depend on CharSets. Some of the functions in CharSets, TriSys, and SiSys do the same tasks but with different performances (see Fig. 3). A distinct feature of SiSys lies in its capability of computing regular systems and simple systems, while TriSys' functions for zero decomposition remain the most efficient. The four packages were designed and implemented at different times along with the author's algorithmic research. Most of the algorithms implemented in Epsilon are described formally in the book.[13] Some of them have similar functionalities, but they differ in terms of structure and computational efficiency.

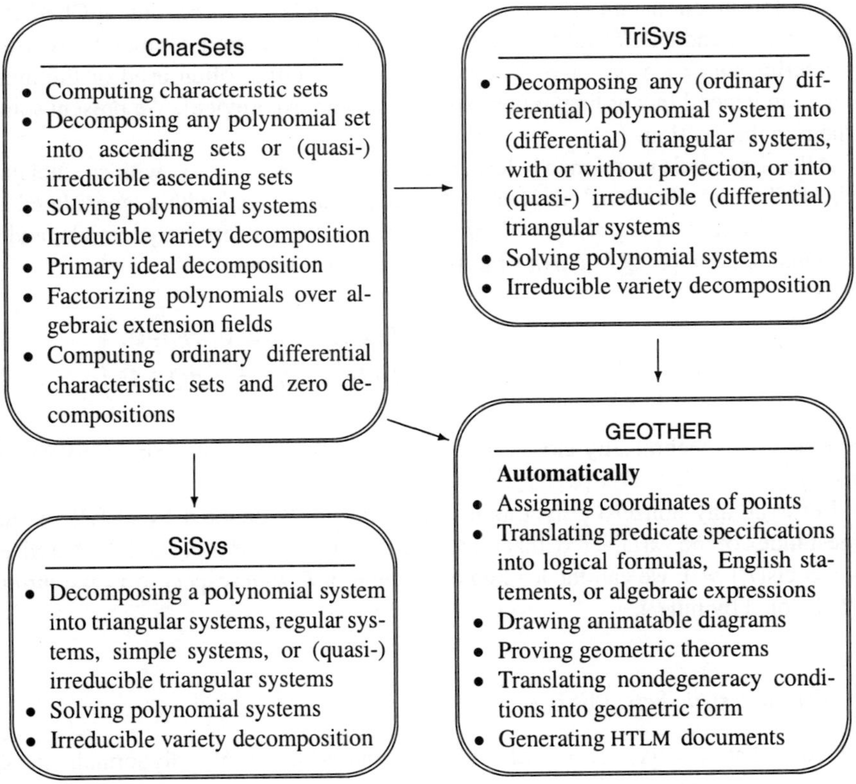

Figure 1. Functionality and relationship of Epsilon packages

Maple was chosen as the basis of our packages for its easy-to-use high-level programming language, its potential and availability, and most importantly, its variety and efficiency of basic functions on polynomial operation and elimination. The flourishing development and increasing popularity of Maple as a general-purpose computer algebra system have proven the soundness of our early choice.

In brief, Epsilon was not designed *a priori*; it has been developed progressively by incorporating up-to-date research accomplishments. It is a unique and rich library for triangular-set-based polynomial elimination. Its functions for zero decomposition are all Gröbner bases free (see relevant remarks by Gräbe[4]). More technical and implementational details for each Epsilon package will be discussed elsewhere.

3 Epsilon Functions

The Epsilon functions are chosen from the large number of functions in CharSets, TriSys, SiSys, and GEOTHER to help the user perform common computations without entering into each package. These functions are either often used or the most efficient in general (according to our experiments and experience). We present some of them in this section.

All polynomials mentioned below are in the variables $x_1,\ldots,x_n$ with coefficients in $\mathcal{K}$, the field Q of rational numbers or the ring $Q[u_1,\ldots,u_d]$ with u_i considered as parametric variables. A *polynomial set* is a finite set of nonzero polynomials, and a *polynomial system* $[\mathbb{P},\mathbb{Q}]$ is a pair of polynomial sets. Define

$$\mathrm{Zero}(\mathbb{P}/\mathbb{Q}) := \left\{ (\bar{x}_1,\ldots,\bar{x}_n) \in \bar{\mathcal{K}}^n \;\middle|\; \begin{array}{l} P(\bar{x}_1,\ldots,\bar{x}_n) = 0, \forall P \in \mathbb{P}, \\ Q(\bar{x}_1,\ldots,\bar{x}_n) \neq 0, \forall Q \in \mathbb{Q} \end{array} \right\},$$

where $\bar{\mathcal{K}}$ is an algebraically closed field containing $\mathcal{K}$. We shall write $\mathrm{Zero}(\mathbb{P})$ for $\mathrm{Zero}(\mathbb{P}/\mathbb{Q})$ when $\mathbb{Q} \subset \mathcal{K} \setminus \{0\}$.

Let P be any nonzero polynomial. The *class* of P, denoted by $\mathrm{cls}(P)$, is the biggest index of the variables x_k that effectively occur in P if $P \notin \mathcal{K}$, and 0 otherwise. Let $p = \mathrm{cls}(P) > 0$; we call the leading coefficient of P with respect to x_p the *initial* of P, denoted by $\mathrm{ini}(P)$.

3.1 Characteristic Sets: `CharSet`

Inputted a list `X` $= [x_1,\ldots,x_n]$ of variables and a set `P` of polynomials in `X`, `CharSet(P,X)` returns an $\mathbb{F}$-*modified quasi-characteristic set* $\mathbb{C}$ of `P` with respect to the variable ordering $x_1 \prec \cdots \prec x_n$. For example,

```
> P := [x^2-3*x*z+2, x*y+z^3, 5*y^2-3*z^3]: X := [z,y,x]:
> CharSet(P,X);
```

$$[-135z^5+25z^6+60z^3+36,\, 15yz+5z^3+6,\, xy+z^3],$$
$$\textit{factors removed} = \{z\}$$

where $\mathbb{C}$ is the list (ordered set) of the three polynomials and $\mathbb{F} = \{z\}$, the set of removed factors.

Definition 1 A finite nonempty ordered set $\mathbb{T} = [T_1, T_2, \ldots, T_r]$ of polynomials is called a *triangular set* if $0 < \mathrm{cls}(T_1) < \mathrm{cls}(T_2) < \cdots < \mathrm{cls}(T_r)$. A *quasi-ascending set* is either a triangular set or a set of a single nonzero polynomial of class 0.

A quasi-ascending set $\mathbb{C}$ is called an $\mathbb{F}$-*modified quasi-characteristic set* of a polynomial set $\mathbb{P}$ if $\mathrm{Zero}(\mathbb{P}/\mathbb{F}) \subset \mathrm{Zero}(\mathbb{C})$ and $\mathtt{remset}(\mathbb{P},\mathbb{C}) = \varnothing$, where $\mathbb{F}$ is another polynomial set.

Here the set of nonzero pseudo-remainders of a polynomial set $\mathbb{P}$ with respect to an arbitrary quasi-ascending set $\mathbb{A} = [A_1, \ldots, A_r]$, with $\mathrm{cls}(A_i) = p_i$, is defined as

$$\mathtt{remset}(\mathbb{P},\mathbb{A}) := \{\mathtt{prem}(P,\mathbb{A}) \mid P \in \mathbb{P}\} \setminus \{0\}, \quad \text{where}$$
$$\mathtt{prem}(P,\mathbb{A}) := \begin{cases} 0 & \text{if } r = 1 \text{ and } p_1 = 0, \\ \mathtt{prem}(\ldots\mathtt{prem}(P, A_r, x_{p_r}), \ldots, A_1, x_{p_1}) & \text{otherwise,} \end{cases}$$

and $\mathtt{prem}(P,A,x)$ denotes the *pseudo-remainder* of P with respect to A in x.

3.2 Triangular Series: `TriSer`

Inputted a list $\mathtt{X} = [x_1, \ldots, x_n]$ of variables and a polynomial system $[\mathtt{P},\mathtt{Q}]$ in $\mathtt{X}$, $\mathtt{TriSer}([\mathtt{P},\mathtt{Q}],\mathtt{X})$ returns a *fine triangular series* $[\mathbb{T}_1,\mathbb{U}_1], \ldots, [\mathbb{T}_e,\mathbb{U}_e]$ of $[\mathtt{P},\mathtt{Q}]$ under the variable ordering $x_1 \prec \cdots \prec x_n$.

Definition 2 A polynomial system $[\mathbb{T},\mathbb{U}]$ is called a *triangular system* if $\mathbb{T}$ is a triangular set and $I(\bar{x}_1, \ldots, \bar{x}_k) \neq 0$ for any $I \in \{\mathrm{ini}(T) \mid T \in \mathbb{T}\}$ of class k and $(\bar{x}_1, \ldots, \bar{x}_n) \in \mathrm{Zero}(\mathbb{T} \cap \mathcal{K}[x_1, \ldots, x_k]/\mathbb{U})$. A triangular system $[\mathbb{T},\mathbb{U}]$ is said to be *fine* if $\mathtt{prem}(U,\mathbb{T}) \neq 0$ for all $U \in \mathbb{U}$.

A sequence of (fine) triangular systems $[\mathbb{T}_1,\mathbb{U}_1], \ldots, [\mathbb{T}_e,\mathbb{U}_e]$ is called a (*fine*) *triangular series* of a polynomial system $[\mathbb{P},\mathbb{Q}]$ if

$$\mathrm{Zero}(\mathbb{P}/\mathbb{Q}) = \bigcup_{i=1}^{e} \mathrm{Zero}(\mathbb{T}_i/\mathbb{U}_i). \tag{1}$$

3.3 Solving Polynomial Systems: `TSolve`

Inputted a set $\mathtt{X} = \{x_1, \ldots, x_n\}$ of variables and a polynomial system $[\mathtt{P},\mathtt{Q}]$ in $\mathtt{X}$, $\mathtt{TSolve}([\mathtt{P},\mathtt{Q}],\mathtt{X})$ returns the isolated solutions of $\mathtt{P} = 0, \mathtt{Q} \neq 0$, and a sequence of fine triangular systems $[\mathbb{T}_1,\mathbb{U}_1],\ldots,[\mathbb{T}_e,\mathbb{U}_e]$ such that $\bigcup_{i=1}^{e}\mathrm{Zero}(\mathbb{T}_i/\mathbb{U}_i)$ represents the set of all the positive-dimensional solutions of $\mathtt{P}=0,\mathtt{Q}\neq 0$.

Note that $\mathbb{P}=0,\mathbb{Q}\neq 0$ mean the system of polynomial equations $P=0$ and inequations $Q\neq 0$ for all $P\in\mathbb{P}$ and $Q\in\mathbb{Q}$. In the books[13,16] one may find different properties of regular, simple, and irreducible triangular systems introduced below.

3.4 *Regular Series:* `RegSer`

Inputted a list $\mathtt{X}=[x_1,\ldots,x_n]$ of variables and a polynomial system $[\mathtt{P},\mathtt{Q}]$ in $\mathtt{X}$, `RegSer([P,Q],X)` returns a *regular series* $[\mathbb{T}_1,\mathbb{U}_1],\ldots,[\mathbb{T}_e,\mathbb{U}_e]$ of $[\mathtt{P},\mathtt{Q}]$ under the variable ordering $x_1\prec\cdots\prec x_n$.

Definition 3 A triangular system $[\mathbb{T},\mathbb{U}]$ is called a *regular system* if (a) $\mathrm{cls}(T)\neq\mathrm{cls}(U)$ for all $T\in\mathbb{T}$ and $U\in\mathbb{U}$; (b) $\mathrm{ini}(U)(\bar{x}_1,\ldots,\bar{x}_{k-1})\neq 0$ for any $1\leq k\leq n$, $U\in\mathbb{U}$ of class k, and

$$(\bar{x}_1,\ldots,\bar{x}_{k-1})\in\mathrm{Zero}(\mathbb{T}\cap\mathcal{K}[x_1,\ldots,x_{k-1}]/\mathbb{U}\cap\mathcal{K}[x_1,\ldots,x_{k-1}]).$$

A sequence of regular systems $[\mathbb{T}_1,\mathbb{U}_1],\ldots,[\mathbb{T}_e,\mathbb{U}_e]$ is called a *regular series* of a polynomial system $[\mathbb{P},\mathbb{Q}]$ if (1) holds.

3.5 *Simple Series:* `SimSer`

Inputted a list $\mathtt{X}=[x_1,\ldots,x_n]$ of variables and a polynomial system $[\mathtt{P},\mathtt{Q}]$ in $\mathtt{X}$, `SimSer([P,Q],X)` returns a *simple series* $[\mathbb{T}_1,\tilde{\mathbb{T}}_1],\ldots,[\mathbb{T}_e,\tilde{\mathbb{T}}_e]$ of $[\mathtt{P},\mathtt{Q}]$ with respect to the ordering $x_1\prec\cdots\prec x_n$.

Definition 4 A pair $[\mathbb{T},\tilde{\mathbb{T}}]$, in which $\mathbb{T}$ and $\tilde{\mathbb{T}}$ are either triangular sets or the empty set, is called a *simple system* if (a) $\mathbb{T}\cap\tilde{\mathbb{T}}=\varnothing$ and $\mathbb{T}\cup\tilde{\mathbb{T}}$ can be reordered as a triangular set; (b) for every $T\in\mathbb{T}\cup\tilde{\mathbb{T}}$ of class k and any

$$(\bar{x}_1,\ldots,\bar{x}_{k-1})\in\mathrm{Zero}(\mathbb{T}\cap\mathcal{K}[x_1,\ldots,x_{k-1}]/\tilde{\mathbb{T}}\cap\mathcal{K}[x_1,\ldots,x_{k-1}]),$$

$\mathrm{ini}(T)(\bar{x}_1,\ldots,\bar{x}_{k-1})\neq 0$ and $T(\bar{x}_1,\ldots,\bar{x}_{k-1},x_k)$ is squarefree.

A sequence of simple systems $[\mathbb{T}_1,\mathbb{U}_1],\ldots,[\mathbb{T}_e,\mathbb{U}_e]$ is called a *simple series* of a polynomial system $[\mathbb{P},\mathbb{Q}]$ if (1) holds.

3.6 *Irreducible Characteristic Series:* `ICS`

Inputted a list $\mathtt{X}=[x_1,\ldots,x_n]$ of variables and a polynomial system $[\mathtt{P},\mathtt{Q}]$ in $\mathtt{X}$, `ICS([P,Q],X)` returns an *irreducible characteristic series* $[\mathbb{T}_1,\mathbb{U}_1],\ldots,[\mathbb{T}_e,\mathbb{U}_e]$ of $[\mathtt{P},\mathtt{Q}]$ with respect to the ordering $x_1\prec\cdots\prec x_n$.

Definition 5 A fine triangular set $\mathbb{T} = [T_1, \ldots, T_r]$ is said to be *irreducible* if there do not exist any integer k $(1 \le k \le n)$ and polynomials D, F, G with $\mathrm{cls}(D) < \mathrm{cls}(F) = \mathrm{cls}(G) = \mathrm{cls}(T_k)$ such that $DT_k = FG$ (when $k = 1$) or $\mathtt{prem}(DT_k - FG, [T_1, \ldots, T_{k-1}]) = 0$.

A fine triangular series $[\mathbb{T}_1, \mathbb{U}_1], \ldots, [\mathbb{T}_e, \mathbb{U}_e]$ of a polynomial system $[\mathbb{P}, \mathbb{Q}]$ is called an *irreducible characteristic series* of $[\mathbb{P}, \mathbb{Q}]$ if $\mathbb{T}_i$ is irreducible for each i.

3.7 *Irreducible Variety Decomposition:* IVD

Inputted a set $\mathtt{X} = \{x_1, \ldots, x_n\}$ of variables and a polynomial set $\mathtt{P}$ in $\mathtt{X}$, $\mathtt{IVD(P,X)}$ returns a sequence of irredundant polynomial sets $\mathbb{P}_1, \ldots, \mathbb{P}_e$ such that

$$\mathrm{Zero}(\mathtt{P}) = \bigcup_{i=1}^{e} \mathrm{Zero}(\mathbb{P}_i)$$

and each $\mathrm{Zero}(\mathbb{P}_i)$ as an algebraic variety is *irreducible*.

We do not define the standard notions of *irreducible algebraic variety* and *primary ideal* (see the references[6,13,16]).

3.8 *Primary Ideal Decomposition:* PID

Inputted a set $\mathtt{X} = \{x_1, \ldots, x_n\}$ of variables and a polynomial set $\mathtt{P}$ in $\mathtt{X}$, $\mathtt{PID(P,X)}$ returns a sequence of irredundant polynomial sets $\mathbb{P}_1, \ldots, \mathbb{P}_e$ such that

$$\mathrm{Ideal}(\mathtt{P}) = \bigcap_{i=1}^{e} \mathrm{Ideal}(\mathbb{P}_i)$$

and each $\mathrm{Ideal}(\mathbb{P}_i)$ is a primary ideal, where $\mathrm{Ideal}(\mathbb{P})$ denotes the ideal generated by all the polynomials in $\mathbb{P}$.

3.9 *Differential Triangular Series:* dTriSer

Inputted a list $\mathtt{X} = [t, x_1, \ldots, x_n]$ of variables and an ordinary differential polynomial system $[\mathtt{P}, \mathtt{Q}]$ in $\mathtt{X}$ and $\mathrm{d}^k x_i / \mathrm{d}t^k$, $\mathtt{dTriSer([P,Q],X)}$ returns a differential triangular series $[\mathbb{T}_1, \mathbb{U}_1], \ldots, [\mathbb{T}_e, \mathbb{U}_e]$ of $[\mathtt{P}, \mathtt{Q}]$ under the variable ordering $x_1 \prec \cdots \prec x_n$.

Definition 6 A sequence of differential triangular systems $[\mathbb{T}_1, \mathbb{U}_1], \ldots, [\mathbb{T}_e, \mathbb{U}_e]$ is called a *differential triangular series* of a differential polynomial system $[\mathbb{P}, \mathbb{Q}]$ if (1) holds, when zeros of polynomials are replaced by differential zeros of differential polynomials.

3.10 Proving Geometric Theorems: `Prove`

Inputted the specification `T` of a geometric theorem, `Prove(T)` proves or disproves `T`, with subsidiary conditions provided.

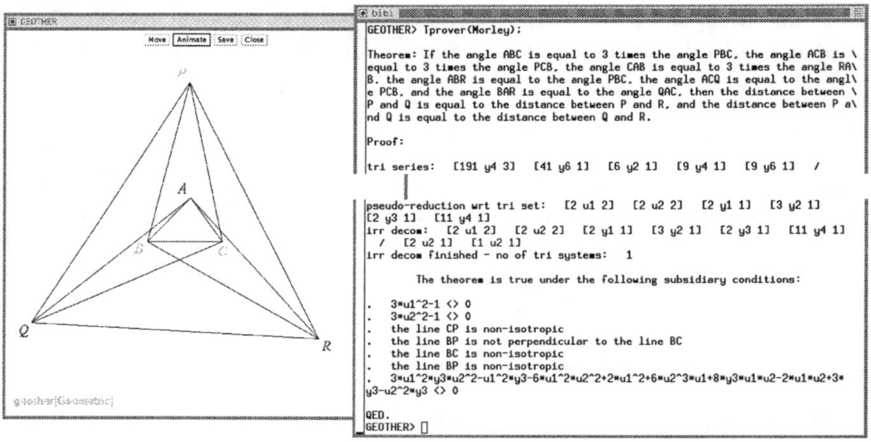

Figure 2. A GEOTHER session

Figure 2 reproduces a proof session and a diagram drawn automatically by GEOTHER for Morley's trisector theorem.

4 Experiments

Experiments[7,9,1] show that the decomposition functions in TriSys are more efficient than those in CharSets in most cases. Table 1 is provided not for comparison but to give the reader an impression about the performance and applicability of Epsilon functions. The computations were done in Maple V.5 on an i686 under Linux 2.2.12 for some of the 50 test examples from the references,[7,9] and ∞ means "> 3600."

Consider, for instance, the decomposition of a polynomial set into irreducible triangular sets, which can be done by the function `trisys[its]=ICS`, `sisys[its]`, or `charsets[ics]` from TriSys, SiSys, or CharSets. Figure 3 illustrates the performances of these functions in terms of computing time for some other examples.

Example	1	2	5	12	14	16	17	19	21	22	24
`TriSer`	1.3	73.9	.8	2.7	123	290	8.3	.6	∞	1.5	1.1
`RegSer`	2.4	28.7	4.8	1.4	∞	8.2	280	7	∞	22.8	7
`SimSer`	6.1	45.2	5.6	2	∞	7.2	∞	8.2	∞	18	28.5
`ICS`	3.4	86	1.2	.6	378	135	19.4	1.2	4.7	3	3.4
`IVD`	7	100	1.2	.8	392	149	19.6	1.2	5.6	3.4	4

26	29	30	32	33	35	39	41	43	44	47	49
.2	.1	.6	.3	6.1	3.4	3	1357	2.7	18.8	.8	6.3
.8	2	88	1	8.9	9.3	10.8	33	966	13.5	13.6	∞
.9	2.6	89	∞	∞	10.5	16.6	44.6	1197	∞	∞	7501
14.6	1.5	1.4	.5	6.5	4.6	4.5	581	17.9	5.7	12.7	31.2
16.4	1.6	1.4	.5	4.8	5	4.6	571	17.5	12.9	∞	35.1

Table 1. Computing times (CPU secs) for five kinds of decompositions

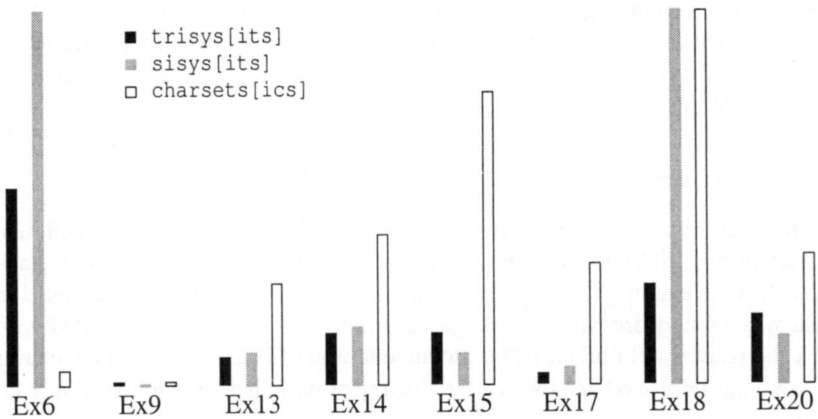

Figure 3. Timings for irreducible decomposition

5 Interface, Availability, and Documentation

5.1 Interface

More than 80% of Epsilon code has been written as Maple programs. This includes the entirety of code for the three modules CharSets, TriSys, SiSys, and the algebraic-computation part of GEOTHER. Actually, Epsilon is nothing else but a user-contributed Maple library and its modules can be used as standard Maple packages. Therefore, Epsilon functions can be defined and called either in command-line Maple or in X-Maple with graphic interface.

However, some of the functions in GEOTHER need external programs written in Java and interact with the operating system. These functions might not work under certain operating systems and Java installations. This concerns in particular the automatic generation of diagrams and HTML documentation for geometric theorems.

The Maple function `system` is simply used for the interaction with the operating system. The pipe facility of Unix and Linux operating systems is used for the implementation of a graphic interface, by means of which the user can call GEOTHER functions with mouse click.

5.2 Availability

The CharSets package has been included in Maple's share library for distribution since 1991 and its latest version may be downloaded from `http://calfor.lip6.fr/~wang/charsets`. The entire Epsilon library version 0.618 will be distributed with a book entitled "Elimination Practice: Software Tools and Applications." The author will continue enhancing the library, and its future versions released will be made available on the author's web page. Since the library is distributed with Maple source code, most of its functions may be used on any platform on which the corresponding version of Maple is installed.

5.3 Documentation

The first part of the above-mentioned book will serve as a general introduction to the Epsilon library. In Maple, the library, its modules, and all the user functions have on-line help with examples. The help facility has been created and may be used in the usual way as standard Maple packages and functions. In the CD included with the book the reader will find an HTML documentation of Epsilon with many examples. Revised and improved versions of the document will also be made available on the author's web page.

Acknowledgments. The development of the Epsilon library has been supported by various funding sources during the author's stay at RISC–Linz and LIFIA/LEIBNIZ–IMAG, and currently by the SPACES project (http://www.spaces-soft.org) and the Chinese national 973 project NKBRSF G19980306.

References

1. Aubry, P., Moreno Maza, M. (1999): Triangular sets for solving polynomial systems: A comparative implementation of four methods. *J. Symb. Comput.* **28**: 125–54.
2. Chou, S.-C., Gao, X.-S., Liu, Z., Wang, D.-K., Wang, D. (2000): Geometric theorem provers and algebraic equation solvers. In: *Mathematics Mechanization and Applications* (X.-S. Gao and D. Wang, eds.), pp. 491–505. Academic Press, London.
3. Gao, X.-S., Zhang, J.-Z., Chou, S.-C. (1998): *Geometry Expert* (in Chinese). Nine Chapters Publ., Taiwan.

4. Gräbe, H.-G. (1999): About the polynomial system solve facility of Axiom, Macsyma, Maple, Mathematica, MuPAD, and Reduce. In: *Computer Algebra Systems: A Practical Guide* (M. Wester, ed.), pp. 121–151. Wiley, Chichester.
5. Ritt, J. F. (1950): *Differential Algebra*. Amer. Math. Soc., New York.
6. Shimoyama, T., Yokoyama, K. (1996): Localization and primary decomposition of polynomial ideals. *J. Symb. Comput.* **22**: 247–277.
7. Wang, D. (1993): An elimination method for polynomial systems. *J. Symb. Comput.* **16**: 83–114.
8. Wang, D. (1995): An implementation of the characteristic set method in Maple. In: *Automated Practical Reasoning: Algebraic Approaches* (J. Pfalzgraf and D. Wang, eds.), pp. 187–201. Springer, Wien New York.
9. Wang, D. (1996): Solving polynomial equations: Characteristic sets and triangular systems. *Math. Comput. Simul.* **42**: 339–351.
10. Wang, D. (1996): GEOTHER: A geometry theorem prover. In: *Automated Deduction* (M. A. McRobbie and J. K. Slaney, eds.), LNAI 1104, pp. 166–170. Springer, Berlin Heidelberg.
11. Wang, D. (1998): Decomposing polynomial systems into simple systems. *J. Symb. Comput.* **25**: 295–314.
12. Wang, D. (2000): Computing triangular systems and regular systems. *J. Symb. Comput.* **30**: 221–236.
13. Wang, D. (2001): *Elimination Methods*. Springer, Wien New York.
14. Wu, W.-t. (1984): Basic principles of mechanical theorem proving in elementary geometries. *J. Syst. Sci. Math. Sci.* **4**: 207–235.
15. Wu, W.-t. (1987): A zero structure theorem for polynomial-equations-solving. *Math. Mech. Res. Preprints* 1: 2–12.
16. Wu, W.-t. (1994): *Mechanical Theorem Proving in Geometries: Basic Principles*. Springer, Wien New York.

MINIMAL SETS OF CRITICAL PAIRS

MASSIMO CABOARA
Dipartimento di Matematica, Università di Genova, Via Dodecaneso 35, I-16146 Genova, Italy
E-mail: caboara@dima.unige.it

MARTIN KREUZER
Fakultät für Mathematik, Universität Regensburg, D-93040 Regensburg, Germany
E-mail: martin.kreuzer@mathematik.uni-regensburg.de

LORENZO ROBBIANO
Dipartimento di Matematica, Università di Genova, Via Dodecaneso 35, I-16146 Genova, Italy
E-mail: robbiano@dima.unige.it

In the computation of a Gröbner basis using Buchberger's algorithm, a key issue for improving the efficiency is to produce good techniques for avoiding as many unnecessary critical pairs as possible. A good solution would be to avoid *all* non-minimal critical pairs, and hence to process only a *minimal set of generators* of the module generated by the critical pairs. In this paper we show how to obtain that desired solution while retaining the same efficiency as with the classical implementation. As a consequence, we get a new Optimized Buchberger Algorithm.

1 Introduction

Ever since practical implementations of Buchberger's famous algorithm for computing Gröbner bases [1] became feasible, it has been clear that, in order to improve the efficiency of this algorithm, one needs to avoid the treatment of as many critical pairs as possible. B. Buchberger [2] was the first to study this, and later (see for instance [3], [5], [6], [9]) his results were substantially improved and expanded. Nevertheless, Gebauer and Möller [6] showed that their method did not always produce a minimal set of generators of the module generated by the critical pairs. However, their method was very efficient and yielded an "almost" minimal set of critical pairs. Since then, many kinds of optimizations of Buchberger's algorithm have been found, in particular by implementators of computer algebra systems. But the problem of efficiently minimalizing the critical pairs has gone largely unnoticed and seems to be overdue for a solution. Indeed, that is the main objective of this paper.

To achieve our goal, we proceed as follows. Foremost, we need a detailed understanding of the entire process of computing Gröbner bases, in particular in the homogeneous case. A fine-tuned algorithm for simultaneously computing a Gröbner basis and a minimal system of generators contained in it leads to a new result on the minimalization of a reduced Gröbner basis. Then this result is applied to critical pairs, using the fact that we show how the old criteria $M(i,j)$ and $F(i,j)$ of Gebauer and Möller [6] yield a reduced Gröbner basis of the module of syzygies of the leading term module. Besides, when applied to this special case, the algorithm admits many subtle optimizations. In the end, we really achieve the goal of minimalizing the critical pairs efficiently.

Now, why do we think that what we have achieved is important?

The first reason is theoretical curiosity. It is common knowledge among the implementators of Buchberger's algorithm that the criteria of Gebauer and Möller produce an *almost* minimal set of critical pairs. We wanted to see whether that "vox populi" is really true. Of course one could use a standard minimalization processes to produce minimal sets of critical pairs, but this method could only handle small examples. Instead, we observed that, after applying the criteria of Gebauer and Möller, a *reduced Gröbner basis* of the module of critical pairs is obtained. Then we were able to see the difference between the reduced Gröbner basis and a minimal set of generators of the critical pairs, and how this difference depends on the size of the example.

Another important reason is that we wanted to be able to compute a minimal set of generators of the module of critical pairs with the *same efficiency* as the usual application of the Gebauer-Möller criteria. And we wanted to do it while computing a Gröbner basis, so that we can replace the Gebauer-Möller criteria by our procedure. As we show in the last section, we have achieved this goal.

A third reason is that our results hold in full generality, namely for Gröbner bases of modules over positively (multi-) graded rings. Other optimizations of Buchberger's algorithm, for instance ideas using trivial syzygies, do not hold in this generality. Moreover, we would like to point out that the pairs we discard are truly useless, whereas, for instance, pairs between elements in a reduced Gröbner basis which reduce to zero can still be useful for the computation of syzygies.

Finally, the readers should know that the basic terminology is taken from the book of the second and third authors [7].

2 Positive Gradings

Since we are interested in optimizing the Buchberger Algorithm in the homogeneous case, we start by saying which gradings we consider. From now on let K always be a field and $P = K[x_1, \ldots, x_n]$ a polynomial ring over K. Moreover, let $W \in \mathrm{Mat}_{m,n}(\mathbb{Z})$ be an $m \times n$-matrix with integer entries.

Definition 2.1. There exists exactly one $\mathbb{Z}^m$-grading on P such that every term $t = x_1^{\alpha_1} \cdots x_n^{\alpha_n}$ is homogeneous of degree $\deg_W(t) = W \cdot (\alpha_1, \ldots, \alpha_n)^{\mathrm{tr}}$. We say that P is **(multi-) graded** by W. The matrix W is called the **weight matrix** and its rows are called the **weight vectors**.

For instance, the grading on P by $W = (1, \ldots, 1)$ is the standard grading. For every $d \in \mathbb{Z}^m$, the homogeneous component of degree d of P is $P_{W,d} = \oplus_{\deg_W(t)=d} K \cdot t$. Given $\delta_1, \ldots, \delta_r \in \mathbb{Z}^m$, the graded free P-module $\oplus_{i=1}^r P(-\delta_i)$ inherits a $\mathbb{Z}^m$-grading from P in the natural way. Again we say that $\oplus_{i=1}^r P(-\delta_i)$ is **graded by** W. In order to be able to use these gradings in our algorithms, we need a positivity assumption.

Definition 2.2. The grading on P by W is called **positive** if $\mathrm{rk}(W) = m$, if no column of W is zero, and if the first non-zero entry in each column of W is positive.

If P is positively graded by W, we have $P_{W,0} = K$ and $\dim_K(M_{W,d}) < \infty$ for every finitely generated graded P-module M and every $d \in \mathbb{Z}^m$. Moreover, the graded version of Nakayama's Lemma holds and we have the following characterization (see Sect. 1.7, 4.1 and 4.2 of Kreuzer – Robbiano[7,8]).

Proposition 2.3 *Let P be graded by W, where W has $\mathbb{Z}$-linearly independent rows and non-zero columns. Then the following conditions are equivalent.*

a) The grading on P given by W is positive.

b) The restriction of `Lex` *to the monoid $\Gamma = \{d \in \mathbb{Z}^m \mid P_{W,d} \neq 0\}$ is a well-ordering, i.e. every non-empty subset of Γ has a minimal element with respect to* `Lex`.

c) The restriction of `Lex` *to the monoid $\Gamma = \{d \in \mathbb{Z}^m \mid P_{W,d} \neq 0\}$ is a term ordering, i.e. every element $d \in \Gamma$ satisfies $d >_{\mathtt{Lex}} 0$.*

d) There exists a term ordering τ on $\mathbb{T}^n$ which is compatible with $\deg_W$.

e) There exists a module term ordering σ on $\mathbb{T}^n\langle e_1, \ldots, e_r\rangle$ which is compatible with the grading given by W.

3 Truncated Gröbner Bases

From now on we shall always assume that $P = K[x_1, \ldots, x_n]$ is a polynomial ring over a field K which is positively graded by a matrix $W \in \mathrm{Mat}_{m,n}(\mathbb{Z})$. The following notation will prove to be convenient. Given a subset S of a graded P-module and a multidegree $d \in \mathbb{Z}^m$, we let $S_{\leq d} = \{v \in S \mid v \text{ homogeneous}, \deg_W(v) \leq_{\mathrm{Lex}} d\}$ and $S_d = \{v \in S \mid v \text{ homogeneous}, \deg_W(v) = d\}$.

Definition 3.1. Assume that $G = \{g_1, \ldots, g_s\}$ is a homogeneous σ-Gröbner basis of a graded P-module M, and let $d \in \mathbb{Z}^m$. Then the set $G_{\leq d}$ is called a d**-truncated Gröbner basis** of M, or a Gröbner basis of M which has been **truncated in degree** d.

Truncated Gröbner bases have a characterization which is analogous to the Buchberger criterion in the usual case. To formulate this precisely, we need to explain what we mean by the **degree of a critical pair**.

Let M be a graded submodule of a graded free P-module $\oplus_{i=1}^r P(-\delta_i)$, and let $g_1, \ldots, g_s \in M \setminus \{0\}$ be homogeneous elements of degree $d_i = \deg_W(g_i)$. Moreover, for $i = 1, \ldots, s$, we write $\mathrm{LM}_\sigma(g_i) = c_i t_i e_{\gamma_i}$, where $c_i \in K \setminus \{0\}$, where $t_i \in \mathbb{T}^n$, and where $\gamma_i \in \{1, \ldots, r\}$. Then we let $\mathbb{B} = \{(i,j) \mid 1 \leq i < j \leq s \text{ and } \gamma_i = \gamma_j\}$ and choose $(i,j) \in \mathbb{B}$. The element $\sigma_{ij} = \frac{\mathrm{lcm}(t_i,t_j)}{c_i t_i} \varepsilon_i - \frac{\mathrm{lcm}(t_i,t_j)}{c_j t_j} \varepsilon_j$ is a syzygy of the pair $(\mathrm{LM}_\sigma(g_i), \mathrm{LM}_\sigma(g_j))$. Clearly, it is a homogeneous element of degree $\deg_W(\sigma_{ij}) = \deg_W(\mathrm{lcm}(t_i,t_j)) + \delta_{\gamma_i}$ of the graded free P-module $\oplus_{k=1}^s P(-d_k)$ whose canonical basis $\{\varepsilon_1, \ldots, \varepsilon_s\}$ satisfies $\deg_W(\varepsilon_k) = d_k$ for $k = 1, \ldots, s$. This degree equals the degree of the corresponding S-vector $S_{ij} = \frac{\mathrm{lcm}(t_i,t_j)}{c_i t_i} g_i - \frac{\mathrm{lcm}(t_i,t_j)}{c_j t_j} g_j$ in the graded P-module $\oplus_{k=1}^r P(-\delta_k)$. The elements of the set $\mathbb{B}$ are called the **critical pairs** of $(g_1, \ldots, g_s)$. For every critical pair $(i,j) \in \mathbb{B}$, we have $\deg_W(\sigma_{ij}) \geq_{\mathrm{Lex}} \max\{d_i, d_j\}$ for all $(i,j) \in \mathbb{B}$.

The following characterization of truncated Gröbner bases can be shown using Props. 2.3.10 and 2.3.12 of Kreuzer – Robbiano [7].

Proposition 3.2 (Characterization of Truncated Gröbner Bases) *Let P be positively graded by $W \in \mathrm{Mat}_{m,n}(\mathbb{Z})$, let $G = \{g_1, \ldots, g_s\}$ be a set of non-zero homogeneous vectors which generates a graded submodule M of $\oplus_{i=1}^r P(-\delta_i)$, and let $d \in \mathbb{Z}^m$. Then the following conditions are equivalent.*

a) The set $G_{\leq d}$ is a d-truncated σ-Gröbner basis of M.

b) For every homogeneous element $v \in M_{\leq d} \setminus \{0\}$, we have $\mathrm{LT}_\sigma(v) \in \langle \mathrm{LT}_\sigma(g) \mid g \in G_{\leq d}\rangle$.

c) For all pairs $(i,j) \in \mathbb{B}_{\leq d}$, we have $\mathrm{NR}_{\sigma,\mathcal{G}_{\leq d}}(S_{ij}) = 0$, where $\mathcal{G}_{\leq d}$ is the *tuple obtained from $\mathcal{G} = (g_1, \ldots, g_s)$ by deleting the elements of degree greater than d.*

As a consequence, we obtain our first important technical tool.

Corollary 3.3 *Let $d \in \mathbb{Z}^m$, let the elements of $\mathcal{G} = (g_1, \ldots, g_s)$ be a d-truncated σ-Gröbner basis of M, and let $g_{s+1} \in \oplus_{i=1}^r P(-\delta_i)$ be a homogeneous element of degree d such that $\mathrm{LT}_\sigma(g_{s+1}) \notin \langle \mathrm{LT}_\sigma(g_1), \ldots, \mathrm{LT}_\sigma(g_s)\rangle$. Then $\{g_1, \ldots, g_{s+1}\}$ is a d-truncated Gröbner basis of $M + \langle g_{s+1}\rangle$.*

Recall that the Lemma of Nakayama guarantees that all minimal systems of generators have the same length in the positively graded situation. Using a clever induction argument based on this lemma, we can also show the following useful result.

Corollary 3.4 *Let $\{g_1, \ldots, g_s\} \subseteq M \setminus \{0\}$ be a minimal homogeneous system of generators of M, and let $g_{s+1} \in \oplus_{i=1}^r P(-\delta_i) \setminus M$ be a homogeneous vector such that $\deg_W(g_{s+1}) \geq_\sigma \max\{\deg_W(g_i) \mid i = 1, \ldots, s\}$. Then $\{g_1, \ldots, g_{s+1}\}$ is a minimal system of generators of the module $M + \langle g_{s+1}\rangle$. In particular, we have $\mu(M + \langle g_{s+1}\rangle) = \mu(M) + 1$.*

4 The Homogeneous Buchberger Algorithm

Now we let $\delta_1, \ldots, \delta_r \in \mathbb{Z}^m$, and we let M be a graded submodule of $\oplus_{i=1}^r P(-\delta_i)$ which is generated by a tuple of non-zero homogeneous vectors $\mathcal{V} = (v_1, \ldots, v_s)$. Furthermore, we choose a module term ordering σ on the monomodule of terms $\mathbb{T}^n\langle e_1, \ldots, e_r\rangle$ in $\oplus_{i=1}^r P(-\delta_i)$.

The homogeneous version of Buchberger's well-known algorithm computes a homogeneous σ-Gröbner basis of M degree-by-degree. To ease the notation, we shall use the following convention: whenever a vector g_i appears, we write $\mathrm{LM}_\sigma(g_i) = c_i t_i e_{\gamma_i}$, where $c_i \in K \setminus \{0\}$, where $t_i \in \mathbb{T}^n$, and where $\gamma_i \in \{1, \ldots, r\}$. For two indices i, j such that $\gamma_i = \gamma_j$, we let $\sigma_{ij} = \frac{\mathrm{lcm}(t_i,t_j)}{c_i t_i} e_i - \frac{\mathrm{lcm}(t_i,t_j)}{c_j t_j} e_j$ and $S_{ij} = \frac{\mathrm{lcm}(t_i,t_j)}{c_i t_i} g_i - \frac{\mathrm{lcm}(t_i,t_j)}{c_j t_j} g_j$.

Theorem 4.1 (The Homogeneous Buchberger Algorithm)
Consider the following sequence of instructions.

1) *Let $B = \emptyset$, $\mathcal{W} = \mathcal{V}$, $\mathcal{G} = \emptyset$, and let s' be the number of elements in $\mathcal{G}$.*

2) *Let d be the smallest degree with respect to* `Lex` *of an element in $\mathcal{W}$ or in B. Form the subtuple $\mathcal{W}_d$ of elements of degree d in $\mathcal{W}$ and the subset $B_d = \{(i,j) \in B \mid \deg_W(\sigma_{ij}) = d\}$, and delete their entries from $\mathcal{W}$ and $\mathcal{B}$, respectively.*

3) *If $B_d = \emptyset$, continue with step 6). Otherwise, choose a pair $(i,j) \in B_d$ and remove it from B_d.*

4) Compute the S-vector S_{ij} and its normal remainder $S'_{ij} = \mathrm{NR}_{\sigma,\mathcal{G}}(S_{ij})$. If $S'_{ij} = 0$, continue with step 3).

5) Increase s' by one, append $g_{s'} = S'_{ij}$ to the tuple $\mathcal{G}$, and append $\{(i,s') \mid 1 \le i < s', \gamma_i = \gamma_{s'}\}$ to the set B. Continue with step 3).

6) If $\mathcal{W}_d = \emptyset$, continue with step 9). Otherwise, choose a vector $v \in \mathcal{W}_d$ and remove it from $\mathcal{W}_d$.

7) Compute $v' = \mathrm{NR}_{\sigma,\mathcal{G}}(v)$. If $v' = 0$, continue with step 6).

8) Increase s' by one, append $g_{s'} = v'$ to the tuple $\mathcal{G}$, and append $\{(i,s') \mid 1 \le i < s', \gamma_i = \gamma_{s'}\}$ to the set B. Continue with step 6).

9) If $\mathcal{W} = \emptyset$ and $B = \emptyset$, return the tuple $\mathcal{G}$ and stop. Otherwise, continue with step 2).

This is an algorithm which returns an ordered tuple $\mathcal{G} = (g_1, \dots, g_{s'})$ whose elements are a homogeneous σ-Gröbner basis of M.

The proof of this theorem is standard Computer Algebra and will for instance be contained in Volume 2 of Kreuzer – Robbiano[8]. The following corollary computes a minimal system of generators of M contained in the given one while computing a Gröbner basis. For the proof we again refer to Kreuzer – Robbiano[8], Sect. 4.3.

Corollary 4.2 (Buchberger Algorithm with Minimalization)
In the situation of the theorem, we replace steps 1), 8), and 9) by the following instructions.

1') Let $B = \emptyset$, $\mathcal{W} = \mathcal{V}$, $\mathcal{G} = \emptyset$, let s' be the number of elements in $\mathcal{G}$, and let $\mathcal{V}_{\min} = \emptyset$.

8') Increase s' by one, append $g_{s'} = v'$ to the tuple $\mathcal{G}$, append v to the tuple $\mathcal{V}_{\min}$, and append $\{(i,s') \mid 1 \le i < s', \gamma_i = \gamma_{s'}\}$ to the set B. Continue with step 6).

9') If $\mathcal{W} = \emptyset$ and $B = \emptyset$, return the pair $(\mathcal{G}, \mathcal{V}_{\min})$ and stop. Otherwise, continue with step 2).

Then the resulting set of instructions defines an algorithm. It returns a pair $(\mathcal{G}, \mathcal{V}_{\min})$ such that $\mathcal{G}$ is an ordered tuple of homogeneous vectors which are a σ-Gröbner basis of M, and $\mathcal{V}_{\min}$ is a subtuple of $\mathcal{V}$ of homogeneous vectors which are a minimal system of generators of M.

In addition, the claim that $\mathcal{V}_{\min}$ contains a minimal system of generators of M holds even if we stop the algorithm after steps 3) to 8) have finished treating elements of degree $d_{\max} = \max_{\mathrm{Lex}}\{\deg_W(v_1), \dots, \deg_W(v_s)\}$.

At this point we are ready to apply the Buchberger Algorithm with Minimalization 4.2 to a reduced Gröbner basis and improve it significantly in that case.

Theorem 4.3 (Minimalization of a Reduced Gröbner Basis)
In the situation of Theorem 4.1, let $\mathcal{V} = (v_1, \ldots, v_s)$ be the reduced σ-Gröbner basis of M. Consider the following instructions.

1) *Let $s' = 0$, $B = \emptyset$, $\mathcal{W} = (v_1, \ldots, v_s)$, $\mathcal{G} = \emptyset$, and $\mathcal{V}_{\min} = \emptyset$.*
2) *Let d be the smallest degree with respect to* `Lex` *of an element of $\mathcal{W}$ or an element of $\{\sigma_{ij} \mid (i,j) \in B\}$. Form $\mathcal{W}_d$ and B_d and delete their entries from $\mathcal{W}$ and B, respectively.*
3) *If $B_d = \emptyset$, continue with step 6). Otherwise, choose a pair $(i,j) \in B_d$ and remove it from B_d.*
4) *Compute $S'_{ij} = \mathrm{NR}_{\sigma,\mathcal{G}}(S_{ij})$. If $S'_{ij} = 0$, continue with step 3).*
5) *Increase s' by one, append $g_{s'} = S'_{ij}$ to $\mathcal{G}$, append $\{(i,s') \mid 1 \le i < s', \gamma_i = \gamma_{s'}\}$ to B, and continue with step 3).*
6) *If $\mathcal{W}_d = \emptyset$, continue with step 9). Otherwise, choose $v \in \mathcal{W}_d$ and remove it from $\mathcal{W}_d$.*
7) *If $\mathrm{LT}_\sigma(v) = \mathrm{LT}_\sigma(g)$ for some $g \in \mathcal{G}$, then replace the element g in $\mathcal{G}$ by v. Continue with step 6).*
8) *Increase s' by one, append $g_{s'} = v$ to $\mathcal{G}$ and $\mathcal{V}_{\min}$, and append $\{(i,s') \mid 1 \le i < s', \gamma_i = \gamma_{s'}\}$ to B. Continue with step 6).*
9) *If $\mathcal{W} = \emptyset$ and $B = \emptyset$, return $\mathcal{V}_{\min}$ and stop. Otherwise, continue with step 2).*

This is an algorithm which computes a subtuple $\mathcal{V}_{\min}$ of $\mathcal{V}$ such that $\mathcal{V}_{\min}$ is a minimal system of generators of M.

5 Minimalizing the Critical Pairs

In this section we continue to use the assumptions and notation of the previous section. In the Homogeneous Buchberger Algorithm 4.1, instead of treating all critical pairs in $\Sigma = \{\sigma_{ij} \mid (i,j) \in \mathbb{Z}\}$, it is enough to treat those pairs corresponding to a subset $\Theta \subseteq \Sigma$ which is a minimal system of generators of $\mathrm{Syz}_P(c_1 t_1 e_{\gamma_1}, \ldots, c_s t_s e_{\gamma_s})$. In order to find Θ, we first show that two of the rules given by Gebauer and Möller [6] produce a minimal Gröbner basis of $\mathrm{Syz}_P(c_1 t_1 e_{\gamma_1}, \ldots, c_s t_s e_{\gamma_s})$ contained in Σ. From this we find Θ by applying Theorem 4.3. We start by modifying Definition 3.1.1. of Kreuzer – Robbiano [7] slightly.

Definition 5.1. On the monomodule of terms $\mathbb{T}^n\langle e_1, \ldots, e_s\rangle$ in $\oplus_{i=1}^s P(-d_i)$

we define a relation τ by letting

$$t\,e_i \geq_\tau t'\,e_j \iff \begin{cases} \mathrm{LT}_\sigma(t\,t_i\,e_{\gamma_i}) >_\sigma \mathrm{LT}_\sigma(t'\,t_j\,e_{\gamma_j}), & \text{or} \\ \mathrm{LT}_\sigma(t\,t_i\,e_{\gamma_i}) = \mathrm{LT}_\sigma(t'\,t_j\,e_{\gamma_j}) & \text{and} \quad i \geq j \end{cases}$$

for $t, t' \in \mathbb{T}^n$ and $i, j \in \{1, \ldots, s\}$. It is easy to check that τ is a module term ordering. It is called the term ordering **induced** by $\mathcal{G} = (t_1 e_{\gamma_1}, \ldots, t_s e_{\gamma_s})$ and σ.

It is well-known that the set Σ is a τ-Gröbner basis of the module $\mathrm{Syz}_P(c_1 t_1 e_{\gamma_1}, \ldots, c_s t_s e_{\gamma_s})$. Using our notation, the main result of Gebauer and Möller [6] reads as follows.

Proposition 5.2 *Consider the following instructions.*

RULE 1. *Delete in Σ all elements σ_{jk} such that there exists an index $i \in \{1, \ldots, j-1\}$ such that t_{ki} divides t_{kj}. Call the resulting set Σ'.*

RULE 2. *Delete in Σ' all elements σ_{ik} such that there exists an index $j \in \{i+1, \ldots, k-1\}$ such that t_{kj} properly divides t_{ki}. Call the resulting set Σ''.*

RULE 3. *Delete in Σ'' all elements σ_{ij} such that there exists an index $k \in \{j+1, \ldots, s\}$ such that t_{ik} properly divides t_{ij} and t_{jk} properly divides t_{ji}. Call the resulting set Σ'''.*

Then the set Σ''' still generates $\mathrm{Syz}_P(c_1 t_1 e_{\gamma_1}, \ldots, c_s t_s e_{\gamma_s})$.

Remark 5.3. Let us interpret the previous proposition in another way. For $1 \leq i < j \leq s$ such that $\gamma_i = \gamma_j$, we have $\mathrm{LT}_\tau(\sigma_{ij}) = t_{ji} e_j$. Hence Rules 1 and 2 can be restated as follows.

RULE 1'. Delete in Σ all elements σ_{ij} such that there exists an element $\sigma_{i'j}$ such that $\mathrm{LT}_\tau(\sigma_{ij})$ is a proper multiple of $\mathrm{LT}_\tau(\sigma_{i'j})$.

RULE 2'. If, among the remaining elements, there are elements σ_{ij}, $\sigma_{i'j}$ such that $\mathrm{LT}_\tau(\sigma_{ij}) = \mathrm{LT}_\tau(\sigma_{i'j})$, then delete the one having the larger index $\max\{i, i'\}$.

From Rules 1' and 2' it follows that the set Σ'' is a minimal τ-Gröbner basis of the module $\mathrm{Syz}_P(c_1 t_1 e_{\gamma_1}, \ldots, c_s t_s e_{\gamma_s})$.

In general, Σ'' is not a minimal system of generators of the P-module $\mathrm{Syz}_P(c_1 t_1 e_{\gamma_1}, \ldots, c_s t_s e_{\gamma_s})$, as our next example shows. (For another example, see Gebauer and Möller [6].)

Example 5.4. Let $P = \mathbb{Q}[x, y, z]$ be standard graded (i.e. graded by $W = (1, \ldots, 1)$), let $r = 1$, $s = 4$ and $t_1 = x^3 z^2$, $t_2 = x^3 y^4$, $t_3 = y^5 z^2$, $t_4 = x^2 y^5 z$. Then we get $\sigma_{12} = y^4 e_1 - z^2 e_2$, $\sigma_{13} = y^5 e_1 - x^3 e_3$, $\sigma_{14} = y^5 e_1 - xz e_4$, $\sigma_{23} = yz^2 e_2 - x^3 e_3$, $\sigma_{24} = yz e_2 - x e_4$, and $\sigma_{34} = x^2 e_3 - z e_4$. By applying Rules 1 and 2, we get the minimal τ-Gröbner basis $\Sigma'' = \{\sigma_{12}, \sigma_{24}, \sigma_{34}, \sigma_{13}\}$

of $\mathrm{Syz}_P(t_1, t_2, t_3, t_4)$, since $\mathrm{LT}_\tau(\sigma_{23}) = \mathrm{LT}_\tau(\sigma_{13})$ and $\mathrm{LT}_\tau(\sigma_{14}) = z \cdot \mathrm{LT}_\tau(\sigma_{24})$. Now we use Rule 3 and find $\Sigma''' = \Sigma''$, but Σ''' is not a minimal system of generators of $\mathrm{Syz}_P(t_1, t_2, t_3, t_4)$, since $\sigma_{13} = y\sigma_{12} + z\sigma_{24} - x\sigma_{34}$.

Before continuing, we mention one important property of the set Σ''.

Proposition 5.5 *The set $\widetilde{\Sigma} = \{-c_j \cdot \sigma_{ij} \mid \sigma_{ij} \in \Sigma''\}$ is the reduced τ-Gröbner basis of the module $\mathrm{Syz}_P(c_1 t_1 e_{\gamma_1}, \ldots, c_s t_s e_{\gamma_s})$.*

Remark 5.6. When we apply the algorithm of Theorem 4.3 to the set $\widetilde{\Sigma}$, we make the following observations.

A *pair of pairs*, i.e. a critical pair between two elements σ_{ij}, $\sigma_{i'j'}$ yields an S-vector $S_{((i,j),(i',j'))} = c\,t\,\sigma_{ij} - c'\,t'\,\sigma_{i'j'}$ such that $c, c' \in K$ and $t, t' \in \mathbb{T}^n$ and $j = j'$, since the two leading terms have to cancel. Without loss of generality, let $i < i'$. The result is of the form $\tilde{c}\,\tilde{t}\,\sigma_{ii'}$ for some $\tilde{c} \in K$ and $\tilde{t} \in \mathbb{T}^n$. The degree of such a pair of pairs is $\deg_W(S_{((i,j),(i',j))}) = \deg_W(t) + \deg_W(\sigma_{ij}) = \deg_W(\frac{\mathrm{lcm}(t_i, t_{i'}, t_j)}{t_j}) + \deg_W(e_j) = \deg_W(\mathrm{lcm}(t_i, t_{i'}, t_j)) + \delta_{\gamma_j}$.

During the course of the algorithm, a new Gröbner basis element can only be obtained if $\tilde{t} = 1$. This is equivalent to $\gcd(t_{ij}, t_{i'j}) = 1$.

Based on these results, we are now ready to optimize the computation of a minimal set of generators of $\mathrm{Syz}_P(\mathrm{LM}_\sigma(\mathcal{G}))$. To ease the notation, we shall minimalize the set Σ'' instead of $\widetilde{\Sigma}$.

Proposition 5.7 (Minimalization of the Critical Pairs)
Let Σ'' be the τ-Gröbner basis of $\mathrm{Syz}_P(\mathrm{LM}_\sigma(\mathcal{G}))$ introduced above. Consider the following instructions.

1) *Let $\mathcal{B}^* = \emptyset$, $\mathcal{W} = \Sigma''$, $\mathcal{A} = \emptyset$, and $\Theta = \emptyset$.*
2) *For all $\sigma_{ij}, \sigma_{i'j} \in \Sigma''$ such that $1 \le i < i' < j \le s$, form the pair of pairs $S_{((i,j),(i',j))} = \tilde{t}\,\sigma_{ii'}$, where $\tilde{t} \in \mathbb{T}^n$. If $\tilde{t} = 1$, append $\sigma_{ii'}$ to $\mathcal{B}^*$.*
3) *Let d be the smallest degree with respect to* `Lex` *of an element of $\mathcal{W}$ or $\mathcal{B}^*$. Form $\mathcal{W}_d$ and $\mathcal{B}^*_d$, and delete their entries from $\mathcal{W}$ and $\mathcal{B}^*$, respectively.*
4) *If $\mathcal{B}^*_d = \emptyset$, continue with step 11). Otherwise, choose an element $\sigma_{ij} \in \mathcal{B}^*_d$ and remove it from $\mathcal{B}^*_d$.*
5) *If $\mathrm{LT}_\tau(\sigma_{ij}) \in \mathrm{LT}_\tau(\mathcal{A}_d)$, then continue with step 4).*
6) *If $\mathrm{LT}_\tau(\sigma_{ij}) = \mathrm{LT}_\tau(\sigma_{i'j})$ for some $\sigma_{i'j} \in \mathcal{W}_d$, then remove the element $\sigma_{i'j}$ from $\mathcal{W}_d$, append it to $\mathcal{A}$, and continue with step 4).*
7) *Find $\sigma_{i'j} \in \mathcal{A}_{<d}$ such that t_{ji} is a multiple of $t_{ji'}$. Then perform the head reduction step $\sigma_{ij} \xrightarrow{\sigma_{i'j}} \tilde{t}\,\sigma_{k\ell}$, where $\tilde{t} \in \mathbb{T}^n$, $k = \min\{i, i'\}$, and $\ell = \max\{i, i'\}$. If $\tilde{t} \ne 1$, continue with step 4).*
8) *If $\mathrm{LT}_\tau(\sigma_{k\ell}) \in \mathrm{LT}_\tau(\mathcal{A}_d)$, then continue with step 4).*

9) If $\mathrm{LT}_\tau(\sigma_{k\ell}) = \mathrm{LT}_\tau(\sigma_{k'\ell})$ *for some element* $\sigma_{k'\ell} \in \mathrm{LT}_\tau(\mathcal{W}_d)$*, then remove the element* $\sigma_{k'\ell}$ *from* $\mathcal{W}_d$*, append it to* $\mathcal{A}$*, and continue with step 4).*

10) If $\sigma_{k\ell} \in \mathcal{B}_d^*$*, then delete* $\sigma_{k\ell}$ *in* $\mathcal{B}_d^*$ *and continue with step 7), applied to this element. Otherwise continue with step 4).*

11) Append $\mathcal{W}_d$ *to* $\mathcal{A}$ *and to* Θ*.*

12) If $\mathcal{W} = \emptyset$ *and* $\mathcal{B}^* = \emptyset$*, return* Θ *and stop. Otherwise, continue with step 3).*

This is an algorithm which computes a subset $\Theta \subseteq \Sigma''$ *such that* Θ *is a minimal system of generators of* $\mathrm{Syz}_P(\mathrm{LM}_\sigma(\mathcal{G}))$*.*

Let us apply this algorithm in the situation of Example 5.4.

Example 5.8. Our task is to minimalize $\mathcal{W} = \Sigma'' = \{\sigma_{12}, \sigma_{13}, \sigma_{24}, \sigma_{34}\}$, where we have $\deg_W(\sigma_{12}) = 9$, $\deg_W(\sigma_{13}) = 10$, and $\deg_W(\sigma_{24}) = \deg_W(\sigma_{34}) = 9$.

In step 2), the algorithm constructs the set $\mathcal{B}^*$. The pair of pairs $((2,4),(3,4))$ yields $S_{((2,4),(3,4))} = z\sigma_{24} - x\sigma_{34} = -yz^2e_2 + x^3e_3 = \sigma_{23}$, and this is the only element of $\mathcal{B}^*$. Notice that it has degree 10.

In step 3), the algorithm starts to operate in degree $d = 9$. Since $\mathcal{B}_9^* = \emptyset$, it appends σ_{12}, σ_{24}, and σ_{34} to $\mathcal{A}$ and Θ in step 11).

Next we process degree 10. In step 4), we choose $\sigma_{23} \in \mathcal{B}_{10}^*$ and set $\mathcal{B}_{10}^* = \emptyset$. Then, in step 6), we find $\mathrm{LT}_\tau(\sigma_{23}) = x^3e_3 = \mathrm{LT}_\tau(\sigma_{13})$, where $\sigma_{13} \in \mathcal{W}_{10}$. Hence σ_{13} is removed from $\mathcal{W}_{10}$ and appended to $\mathcal{A}$ in step 6).

Thus we have $\mathcal{B}^* = \emptyset$ and $\mathcal{W} = \emptyset$ now, and step 12) returns $\Theta = \{\sigma_{12}, \sigma_{24}, \sigma_{34}\}$. We note that this is the correct answer, and there is an improvement over the application of Proposition 5.2 coming from the fact that in step 6) we merely check $\mathrm{LT}_\tau(\sigma_{ij}) \in \mathrm{LT}_\tau(\mathcal{W}_d)$ rather than $\sigma_{ij} \in \mathcal{W}_d$.

The following example provides a case where it is actually necessary to do one head reduction step in 7) in order to find a previously undiscovered non-minimal critical pair.

Example 5.9. Let $P = \mathbb{Q}[x_1, \ldots, x_5]$ be standard graded, let $r = 1$ and $s = 4$. The terms $t_1 = x_2^2x_3^6x_4x_5^2$, $t_2 = x_1^8x_2x_4x_5^4$, $t_3 = x_1^8x_2^2x_3^6$, and $t_4 = x_1^8x_3^6x_5^4$ yield the critical pairs $\sigma_{12} = x_1^8x_5^2e_1 - x_2x_3^6e_2$, $\sigma_{13} = x_1^8e_1 - x_4x_5^2e_3$, $\sigma_{14} = x_1^8x_5^2e_1 - x_2^2x_4e_4$, $\sigma_{23} = x_2x_3^6e_2 - x_4x_5^4e_3$, $\sigma_{24} = x_3^6e_2 - x_2x_4e_4$, and $\sigma_{34} = x_5^4e_3 - x_2^2e_4$. Here steps 1) and 2) of Proposition 5.2 discard σ_{23} and σ_{14}, because $\mathrm{LT}_\tau(\sigma_{23}) = x_4x_5^4e_3 = x_5^2\,\mathrm{LT}_\tau(\sigma_{13})$ and $\mathrm{LT}_\tau(\sigma_{14}) = x_2^2x_4e_4 = x_2\,\mathrm{LT}_\tau(\sigma_{24})$. Thus we have $\Sigma'' = \{\sigma_{12}, \sigma_{13}, \sigma_{24}, \sigma_{34}\}$. We note that $\deg_W(\sigma_{12}) = 21$, $\deg_W(\sigma_{13}) = 19$, and $\deg_W(\sigma_{24}) = \deg_W(\sigma_{34}) = 20$. But Σ'' is not minimal, because we have $\sigma_{12} = x_5^2\sigma_{13} - x_2\sigma_{24} + x_4\sigma_{34}$.

Now we apply our algorithm. In step 2), we compute $S_{((2,4),(3,4))} = x_2\sigma_{24} - x_4\sigma_{34} = x_2x_3^6e_2 - x_4x_5^4e_3 = \sigma_{23}$. Thus σ_{23} is appended to $\mathcal{B}^*$. It has degree $\deg_W(\sigma_{23}) = 21$. No further pairs of pairs are found.

In step 3), the algorithm starts to operate in degree $d = 19$. We have $\mathcal{B}^*_{19} = \emptyset$ and $\mathcal{W}_{19} = (\sigma_{13})$. Thus we append σ_{13} to $\mathcal{A}$ and Θ in step 11).

Next we pass to degree $d = 20$. We still have $\mathcal{B}^*_{20} = \emptyset$, but now we get $\mathcal{W}_{20} = (\sigma_{24}, \sigma_{34})$. In step 11), σ_{24} and σ_{34} are put into $\mathcal{A}$ and Θ.

When we start processing degree $d = 21$, we have to choose $\sigma_{23} \in \mathcal{B}^*_{21}$ and set $\mathcal{B}^*_{21} = \emptyset$ in step 4). The leading term $\mathrm{LT}_\tau(\sigma_{23}) = x_4 x_5^4 e_3$ is not equal to one of the leading terms of the elements of $\mathcal{A}_{21}$ or $\mathcal{W}_{21}$. But we can perform a head reduction step in 7), namely $\sigma_{23} \xrightarrow{\sigma_{13}} -\sigma_{12}$.

Here step 8) does not apply, but in step 9) we have $\mathrm{LT}_\tau(\sigma_{12}) \in \mathrm{LT}_\tau(\mathcal{W}_{21})$. Thus we continue by removing σ_{12} from $\mathcal{W}_{21}$ and appending it to $\mathcal{A}$.

Finally, we get $\mathcal{B}^* = \emptyset$ and $\mathcal{W} = \emptyset$. The algorithm returns $\Theta = \{\sigma_{13}, \sigma_{24}, \sigma_{34}\}$. As mentioned above, the non-minimal critical pair σ_{12} was discovered after one head reduction step in 7).

6 The Optimized Buchberger Algorithm

In this section we combine the results obtained so far. We continue to use the notation and conventions of the previous sections. In particular, we let $P = K[x_1, \ldots, x_n]$ be a polynomial ring over a field K which is positively graded by $W \in \mathrm{Mat}_{m,n}(\mathbb{Z})$, we let M be a graded submodule of a graded free P-module $\oplus_{i=1}^r P(-\delta_i)$ which is generated by a tuple of homogeneous vectors $\mathcal{V} = (v_1, \ldots, v_s)$, and we let σ be a module term ordering on $\mathbb{T}^n\langle e_1, \ldots, e_r\rangle$.

Theorem 6.1 (The Optimized Buchberger Algorithm)
Let $P = K[x_1, \ldots, x_n]$ be positively graded by $W \in \mathrm{Mat}_{m,n}(\mathbb{Z})$, let M be a graded submodule of a graded free P-module $\oplus_{i=1}^r P(-\delta_i)$, let $\mathcal{V} = (v_1, \ldots, v_s)$ be a tuple of non-zero homogeneous vectors which generate M, and let σ be a module term ordering on $\mathbb{T}^n\langle e_1, \ldots, e_r\rangle$. Consider the following sequence of instructions.

1) *Let $\mathcal{W} = \mathcal{V}$, $A = \emptyset$, $B = \emptyset$, $B^* = \emptyset$, $\mathcal{G} = \emptyset$, and let s' be the number of elements in $\mathcal{G}$.*

2) *Let d be the smallest degree with respect to* `Lex` *of an element of $\mathcal{W}$ or $\{\sigma_{ij} \mid (i,j) \in B\}$. Form $\mathcal{W}_d$, B_d, B^*_d and delete their entries from $\mathcal{W}$, B, and B^*, respectively.*

3) *Apply* `MinPairs`(A, B_d, B^*_d).

4) *If $B_d = \emptyset$, then continue with step 7). Otherwise, choose a pair $(i,j) \in B_d$, delete it from B_d, and append it to A.*

5) *Compute S_{ij} and $S'_{ij} = \mathrm{NR}_{\sigma,\mathcal{G}}(S_{ij})$. If $S'_{ij} = 0$, then continue with 4).*

6) *Increase* s' *by one, append* $g_{s'} = S'_{ij}$ *to* $\mathcal{G}$*, perform* `Update`$(B, B^*, g_{s'})$*, and continue with step 4).*

7) *If* $\mathcal{W}_d = \emptyset$ *then continue with 10). Otherwise, choose* $v \in \mathcal{W}_d$ *and delete it in* $\mathcal{W}_d$*.*

8) *Compute* $v' = \mathrm{NR}_{\sigma,\mathcal{G}}(v)$*. If* $v' = 0$*, continue with step 7).*

9) *Increase* s' *by one, append* $g_{s'} = v'$ *to* $\mathcal{G}$*, perform* `Update`$(B, B^*, g_{s'})$*, and then continue with step 7).*

10) *If* $\mathcal{W} = \emptyset$ *and* $B = \emptyset$*, then return* $\mathcal{G}$ *and stop. Otherwise, continue with step 2).*

Here the procedure `Update`$(B, B^*, g_{s'})$ *is defined as follows.*

U1) *Form the set* $C = \{(i, s') \mid 1 \le i < s',\, \gamma_i = \gamma_{s'}\}$*.*

U2) *Delete from* C *all pairs* (j, s') *for which there exists an index* $i \in \{1, \ldots, j-1\}$ *such that* $t_{s'i}$ *divides* $t_{s'j}$*.*

U3) *Delete from* C *all pairs* (i, s') *for which there exists an index* $j \in \{i+1, \ldots, s'-1\}$ *such that* $t_{s'j}$ *properly divides* $t_{s'i}$*.*

U4) *Find in* C *all pairs* (i, s') *and* (j, s') *such that* $1 \le i < j < s'$ *and such that* $\gcd(t_{is'}, t_{js'}) = 1$*. For each of these, check if* (i, j) *is already contained in* B^* *and append it if necessary.*

U5) *Append the elements of* C *to* B *and stop.*

Furthermore, the procedure `MinPairs`(A, B_d, B^*_d) *is defined as follows.*

M1) *If* $B^*_d = \emptyset$*, then stop. Otherwise, choose a pair* (i, j) *in* B^*_d *and remove it from* B^*_d*.*

M2) *If* $t_{ji} = t_{ji'}$ *for some pair* $(i', j) \in A$*, then continue with step M1).*

M3) *If* $t_{ji} = t_{ji'}$ *for some pair* $(i', j) \in B_d$*, then remove this pair from* B_d *and append it to* A*. Continue with step M1).*

M4) *Find* $(i', j) \in A$ *such that* t_{ji} *is a multiple of* $t_{ji'}$*. Let* $(k, \ell) = (\min\{i, i'\}, \max\{i, i'\})$*. If* $\gcd(t_{ij}, t_{i'j}) \ne 1$*, then continue with M1).*

M5) *If* $t_{\ell k} = t_{\ell k'}$ *for some pair* $(k', \ell) \in A$*, then continue with M1).*

M6) *If* $t_{\ell k} = t_{\ell k'}$ *for some pair* $(k', \ell) \in B_d$*, then delete this pair in* B_d*, append it to* A*, and continue with M1).*

M7) *If* $(k, \ell) \in B^*_d$*, then delete* (k, ℓ) *in* B^*_d *and continue with M4), applied to this pair.*

M8) *Continue with step M1).*

Altogether, we obtain an algorithm which computes an ordered tuple $\mathcal{G} = (g_1, \ldots, g_{s'})$ *whose elements form a homogeneous* σ*-Gröbner basis of* M*.*

Moreover, the set of pairs which are treated at some time in steps 4) – 6) of the algorithm corresponds to a minimal system of generators of the module $\mathrm{Syz}_P(c_1 t_1, \ldots, c_{s'} t_{s'})$.

Let us illustrate the performance of this algorithm by a simple example. It shows that cases like Example 5.4 occur naturally during actual Gröbner basis computations.

Example 6.2. Let $P = \mathbb{Q}[x, y, z]$ be standard graded, let $r = 1$, let $\sigma =$ `DegLex`, and let $M \subseteq P$ be the homogeneous ideal generated by $v_1 = x^3z^2 + x^2y^2z$, $v_2 = x^3y^8$, and $v_3 = y^{10}z^2$. Then the leading terms are $t_1 = x^3z^2$, $t_2 = x^3y^8$, and $t_3 = y^{10}z^2$. Let us follow the steps of the Optimized Buchberger Algorithm.

The first degree is $d = 5$. Since $B_5 = \emptyset$, the first actions are to choose $v_1 \in \mathcal{W}_5$ in step 7) and append $g_1 = v_1$ to $\mathcal{G}$ in step 9). Then we continue with $d = 11$ and choose $v_2 \in \mathcal{W}_{11}$ in step 7). Since $v' = \mathrm{NR}_{\sigma,\mathcal{G}}(v_2) = v_2$, we append$g_2 = v_2$ to $\mathcal{G}$ in step 9) and update the set of pairs. The result is $B = \{(1,2)\}$ and $B^* = \emptyset$. Now we have to treat the degree $d = 12$. Notice that the degree of the pair $(1,2)$ is 13. Hence $B_{12} = \emptyset$ and we have to choose $v_3 \in \mathcal{W}_{12}$ in step 7). Since $v' = \mathrm{NR}_{\sigma,\mathcal{G}}(v_3) = v_3$, we append$g_3 = v_3$ to $\mathcal{G}$ in step 9) and update the set of pairs. In step U1), we form $C = \{(1,3),(2,3)\}$. In step U2), we obtain $t_{31} = x^3 = t_{32}$, and therefore $(2,3)$ is deleted in C. The result is $B = \{(1,2),(1,3)\}$ and $B^* = \emptyset$. This completes degree 12, and we continue with degree 13.

We choose the pair $(1,2)$ in step 4) and append it to A. Then we compute $S_{12} = y^8 g_1 - z^2 g_2 = x^2y^{10}z$ and $S'_{12} = \mathrm{NR}_{\sigma,\mathcal{G}}(S_{12}) = x^2y^{10}z$. Thus we have a new Gröbner basis element $g_4 = x^2y^{10}z$ and need to update the pairs again. In step U1), we form $C = \{(1,4),(2,4),(3,4)\}$. Step U2) does not apply, but in step U3) we remove the pair $(1,4)$ from C, since $t_{42} = x$ properly divides $t_{41} = xz$. Now we check that $t_{24} = y^2z$ properly divides $t_{23} = y^2z^2$ and $t_{34} = x^2$ properly divides $t_{32} = x^3$. Hence the pair $(2,3)$ is appended to B^*.

At this point we have finished degree 13, and we have the situation $A = \{(1,2)\}$, $B = \{(1,3),(2,4),(3,4)\}$, $B^* = \{(2,3)\}$, $\mathcal{G} = \{g_1, \ldots, g_4\}$, and $s' = 4$. The next degree is $d = 14$, where we have to deal with the pairs in $B_{14} = \{(2,4),(3,4)\}$. Since $B^*_{14} = \emptyset$, we choose $(2,4)$ in step 4) and append it to A. Then we compute $S_{24} = 0$ and continue by choosing $(3,4)$ in B_{14} and adding it to A. Again $S_{34} = 0$, and degree 14 is finished.

Now we start degree 15 by performing `MinPairs`(A, B_{15}, B^*_{15}), where $A = \{(1,2),(2,4),(3,4)\}$, $B_{15} = \{(1,3)\}$ and $B^*_{15} = \{(2,3)\}$. In step M1), we choose $(2,3)$. In step M3), we discover $t_{32} = x^3 = t_{31}$, where $(3,1) \in B_{15}$. Hence $(1,3)$ is deleted in B_{15} and appended to A. Then the procedure is finished, and the facts that $B_{15} = \emptyset$ as well as $\mathcal{W}_{15} = \emptyset$ allow us to return $\mathcal{G}$ and stop.

As in Example 5.4, we have found one useless pair, namely the pair $(1, 3)$ in degree 15, which would not have been discovered by the Gebauer-Möller Installation, and which we were able to discard by a simple combinatorial check.

Remark 6.3. Let us discuss the efficiency of the algorithm of Theorem 6.1.

a) When we compare this algorithm with the Gebauer-Möller installation, we see that Rules 1) and 2) correspond exactly to steps U2) and U3). However, Rule 3) is not performed by `Update`(...), but by step M3) of `MinPairs`(...). In fact, step M3) gets rid of more pairs, because Rule 3) requires $(i,j) \in B_d^* \cap B_d$, whereas we only need a pair $(i,j) \in B_d^*$ such that $\mathrm{LT}_\tau(\sigma_{ij}) = \mathrm{LT}_\tau(\sigma_{i'j})$ for some $(i',j) \in B_d$.

b) A potential drawback of our approach is that the number of pairs of pairs considered in step U4) is quadratic in the number of elements of C surviving steps U2) and U3). But that number is usually fairly small. Hence the cost of U4) and the cardinality of B^* tend to be rather small. On the other hand, we do not need to check Rule 3 for all elements of the list B which is usually rather long. Our experiments suggest that, on average, the overhead of the two approaches is comparable.

c) Our procedure `MinPairs`(...) is very efficient in treating the elements of B_d^*. Each time we loop through steps M2) – M8), we delete one pair in B_d^*, and B_d^* is never enlarged. In practice, we find that most superfluous pairs are cancelled by steps U2) and U3) and that the lists B_d^* are generally small. Hence our algorithm harnesses the full power and efficiency of the Gebauer-Möller installation, while it simultaneously kills *all* unnecessary pairs at a comparatively small cost.

7 Experimental Data

Experimental data showing the efficiency of the Optimized Buchberger Algorithm will be presented in the full paper.

References

1. B. Buchberger, *Ein Algorithmus zum Auffinden der Basiselemente des Restklassenrings nach einem nulldimensionalen Polynomideal* (Ph.D. Thesis, Universität Innsbruck, 1965).

2. B. Buchberger, A criterion for detecting unnecessary reductions in the construction of Groebner bases, *Proc. EUROSAM 79*, Springer LNCS **72**, 3–21 (1979).
3. B. Buchberger, Groebner Bases: An Algorithmic Method in Polynomial Ideal Theory, in *Multidimensional Systems Theory*, ed. N.K. Bose (D. Reidel Publ. Comp., Dordrecht, 1985)
4. CoCoA, a system for *Co*mputing in *Co*mmutative *A*lgebra, available via anonymous ftp from `cocoa.dima.unige.it`
5. R. Gebauer and H.M. Möller, Buchberger's Algorithm and staggered linear Bases, in *Proceedings of the 1986 Symposium on Symbolic and Algebraic Computation: SYMSAC '86 (Waterloo, Ontario)*, ACM Press, New York, 218–221 (1986)
6. R. Gebauer and H.M. Möller, On an installation of Buchberger's algorithm, *J. Symb. Comput.* **6**, 257–286 (1987).
7. M. Kreuzer and L. Robbiano, *Computational Commutative Algebra 1* (Springer, Heidelberg, 2000).
8. M. Kreuzer and L. Robbiano, *Computational Commutative Algebra 2* (Springer, Heidelberg, in preparation).
9. H.M. Möller, T. Mora and C. Traverso, Computation of Gröbner bases using syzygies, in *Proceedings of ISSAC '92 Intgernational Symposium on Symbolic and Algebraic Computation*, ACM Press, New York, 320–328 (1992)

A STUDY IN THE INTEGRATION OF COMPUTER ALGEBRA SYSTEMS:

MEMORY MANAGEMENT IN A MAPLE–ALDOR ENVIRONMENT

STEPHEN M. WATT
ONTARIO RESEARCH CENTER FOR COMPUTER ALGEBRA
UNIVERSITY OF WESTERN ONTARIO
LONDON CANADA, N6A 5B7

1 Introduction

Many approaches have been followed to allow several computer algebra systems to work together, including MP [1], OpenMath [2], and the IAMC framework [5]. All of these propose mechanisms for loosely coupled systems sharing data by some communications protocol.

For certain types of computation, loosely coupled systems are inappropriate. Examples of this includes situations where

- the problem treated by a foreign system would have its computation costs dominated by communication costs
- the problem treated by a foreign system requires access to some *a priori* undetermined subset of a relatively large data structure or database
- common persistent data structures are maintained and manipulated by operations in each of the computer algebra systems

We have undertaken to study the problems which arise in the tight coupling of computer algebra systems. By this we mean having multiple computer algebra systems share the same address space so that objects created by one system can be passed passed by reference in calls to another system.

As an interesting practical problem, we have explored the case of using Aldor code from within the Maple system, allowing Aldor and Maple programs to share data and call each other. This allows Aldor to be used as a compiled extension language for Maple, and consequently provides Maple an efficient generic programming facility. This required creating an *Enhanced External Function* interface for Maple [7], which has been incorporated as part of the standard release of Maple since Maple 7.

In this paper we describe the low-level memory management issues arising from this integration. The natural interface between Aldor domains and Maple modules shall be reported elsewhere.

Memory management is a well-studied subject (see, e.g. the ISMM proceedings [4 6]), but there has been relatively little work on the integration of existing, mature memory managers. Of interest is the work on CMM [3], the "customizable memory manager," which can be useful in porting multiple C++ applications to a common setting. The work reported here focuses on the integration of mark-and-sweep collectors in a language-independent setting.

2 Systems Issues

The combined Maple-Aldor environment consists of a Maple system with dynamically loaded Aldor code. Maple and Aldor each maintain their own memory heap and use their own native formats for programs, big integers, and other data structures. Pointers to Maple objects may be passed to Aldor functions and *vice versa*. Likewise, data structures in the Maple heap may point to objects in the Aldor heap and *vice versa*.

The main problems in this integration relate to the representation of inter-heap pointers and the behaviour cooperative of garbage collection. The current implementations of both Maple and Aldor use a mark-and-sweep garbage collectors, but problems arise because of different assumptions made by the respective systems:

- Maple assumes that all data objects are self-identifying, and that there are a fixed set of possible object types. The memory manager has intimate knowledge of the fields within each of the possible object types, and has exact knowledge of which fields are pointers and which are data. The structure of the Maple memory manager, however, does not allow a constant time test to determine whether an address points to an object in the heap.

- Aldor assumes that arbitrary data structures may be required, and that data is *not* self identifying. This allows easy inclusion of foreign pointers in data objects, but does not allow the memory manager to distinguish pointer from non-pointers within these objects. The Aldor memory manager allows fast, $O(1)$ time test to determine whether a given address points to the beginning of an object in the heap.

  Because the Aldor garbage collector is more general, it does not have exact knowledge of the internal layout of allocated objects, global data or the program stack. It uses a conservative strategy where the targets of all potential pointers are regarded as live. (A potential pointer is a word that, if interpreted as a pointer, would point to a data object.)

It would be completely impractical to re-work the two systems to use a single memory manager for several reasons. First, it would be prohibitively costly. Second, the memory managers are now mature—any significant changes would risk introducing subtle bugs. Third, this would create a solution applicable only to the Maple-Aldor combination, and would not shed any light on the general problem of system integration.

The approach we have taken is to allow both the Maple and Aldor heaps to co-exist and be for the most part independently managed. The problems are how to represent pointers between heaps, and how to coordinate garbage collection.

3 Pointers between Heaps

Since the Aldor memory manager makes no assumption about the content of heap-allocated objects, pointers from Aldor objects to Maple objects are simply stored as words—*i.e.* in the same representation as pointers to Aldor or C objects.

Life is not so simple for including pointers to foreign objects in Maple data: the Maple memory manager has detailed understanding of all the types of objects in Maple so it was necessary to introduce a new type of basic Maple object. We have called these new objects *external object handles*, or *handles* for short.

Handle objects in Maple allow references to objects in external heaps to be returned from external functions to Maple. The handle objects contain three components:

- one word for data, as immediate data or an external pointer
- one word for type, as immediate data or an external pointer
- a "mark event handler" function

The mark event handler allows the referenced object in the external heap to be aware of garbage collection. If this function pointer is non-zero, it is called with data and type words as arguments when the handle object is marked by the garbage collector (see Section 5).

4 Identification of Objects

While the Aldor memory manager is able to determine whether an address points to the beginning of an Aldor object, the corresponding test is not possible with Maple's memory manager. The only way to determine whether an address points to an object in the Maple heap is to traverse the containing heap segment from the beginning.

To allow garbage collection in the presence of inter-heap pointers, it is necessary to have certain knowledge whether a pointer into the Maple heap is actually pointing at a Maple object. Operations to traverse the Maple heap have therefore been included in the Enhanced External Function interface for this purpose. These operations allow a program to traverse the Maple heap and see the addresses of all objects.

5 Garbage Collection Events

When a reference to a Maple object is passed to an external function, or a Maple object is created by an external function, it is possible that there may be no references to that object from other Maple structures. We call these "captured Maple objects." When such an object is not referenced by other Maple structure, it is susceptible to be garbage collected. Since Aldor functions can create captured Maple objects, it is necessary for a Maple–Aldor environment to ensure that any captured Maple objects are protected from garbage collection. Analogously, captured Aldor objects may arise from calling Maple functions on Aldor objects, and these must also be handled by a Maple–Aldor environment.

Protecting captured objects can be done in any of three ways: through explicit object management, through reference counting, or through proper garbage collection. When there are many objects flowing back and forth across an interface, and user-level library code operates on these objects, explicit management is error-prone. Reference counting does not allow the creation of cyclic data objects, which are not uncommon in a computer algebra setting. We have therefore developed an automatic method to protect captured objects under garbage collection. This method relies on a protocol of "garbage collection events."

There are four garbage collection events which are made known via Maple's new enhanced foreign function interface. These are

1. the start of a garbage collection,
2. the start of the mark phase,
3. the marking of handles to external objects,
4. the end of the garbage collection.

Notification of the start, the mark phase, and end of garbage collection may be requested by registering event listeners. These take functions which should be called when the appropriate event occurs.

All installed start listeners are called at the beginning of a Maple garbage collection. The mark phase listeners are called when the garbage collector is willing to receive mark requests for Maple objects. The end listener is called at the end of a garbage collection.

Notification of the marking of external object handles is handled on a per object basis, and is achieved by placing a mark event handler in the object itself as described in Section 3. An additional Maple kernel operation is made available to allow the external application to request that its captured Maple objects be retained as live during a garbage collection.

6 Garbage Collection Protocol

The garbage collection events are sufficient to allow the Maple garbage collector and the Aldor garbage collector to work together. By registering appropriate Aldor memory management functions for the the Maple GC events, we are able to establish the following protocol:

1. Whichever heap initiates a garbage collection, the first thing it does is to initiate a garbage collection in the other, so that both heaps garbage collect synchronously.
2. Traverse the Maple heap building information allowing exact constant-time identification of Maple objects.
3. The Maple heap is marked starting from the Maple root set.

   Whenever a handle to an Aldor object is encountered, its mark event handler function is called. This will initiate marking within the Aldor heap.
4. The Aldor heap is marked starting from the Aldor root set.

   The targets of all potential pointers are marked. (A potential pointer is a word which, if interpreted as a pointer, would point to an object.) Potential pointers to Maple objects are identified using the procedures describe in Section 4.
5. Discard the information allowing exact constant-time identification of Maple objects.
6. The Maple and Aldor heaps are swept independently.

This protocol ensures that no live object is collected and permits object cycles to traverse heap boundaries.

7 Conclusions

We see that the successful tight coupling of computer algebra systems involves careful attention to interactions between memory managers. To allow the mark-and-sweep garbage collectors of Aldor and Maple to interact properly, it was necessary to

- introduce the concept of garbage collection events
- introduce the concept of a "handle" object in Maple
- expose a limited set of representation-independent operations on the Maple heap

These provide sufficient functionality to allow the Aldor memory manager to work cooperatively with Maple's memory manager. The additional cost associated with this implementation is an additional pass over the Maple heap to gain exact pointer knowledge. This is comparable in cost to the sweep phase of Maple's garbage collector.

References

1. *MP: A protocol for the efficient exchange of mathematical expressions*, S. Gray, N. Kajler and P. Wang, Proc. ISSAC 94 International Symposium on Symbolic and Algebraic Computation, pp. 330-335, ACM Press 1994.
2. *An OpenMath 1.0 Implementation*, Stéphane Dalmas, Marc Gaëtano, Stephen Watt, Proc. ISSAC 97 International Symposium on Symbolic and Algebraic Computation, pp. 241-248, ACM Press 1997.
3. *A Customisable Memory Management Framework for C++*, G. Attardi, T. Flagella and P. Iglio, Software Practice and Experience, 28(11), 1143-1183, 1998.
4. Proc. ISMM'98 International Symposium on Memory Management, ACM Sigplan Notices 34(3), ACM Press 1999.
5. *Design and Protocol for Internet Accessible Mathematical Computation*, Paul S. Wang, Proc. ISSAC 99 International Symposium on Symbolic and Algebraic Computation, pp. 291-298, ACM Press 1999.
6. Proc. ISMM 2000 International Symposium on Memory Management, ACM Press 2000.
7. *An Enhanced External Function Interface for Maple*, S. Watt, Waterloo Maple internal report, 2001.

LOCAL AND REMOTE USER INTERFACE FOR ELIMINO THROUGH OMEI

YONGWEI WU

Institute of Systems Science, Chinese Academy of Sciences, Beijing 100080, China.

WEIDONG LIAO

Institute for Computational Mathematics, Department of Computer Science, Kent State University, Kent, Ohio 44242, U.S.A.

DONGDAI LIN

State Key Laboratory of Information Security, Institute of Software, Chinese Academy of Sciences, Beijing 100080, China.

PAUL S. WANG

Institute for Computational Mathematics, Department of Computer Science, Kent State University, Kent, Ohio 44242, U.S.A.

ELIMINO is a new computer-mathematics research system developed to support Wu's method for computing characteristic sets of polynomials and other related operations. By implementing OMEI as its application programming interface, ELIMINO could make itself IAMC compliant, gain access to the graphical user interface Dragonfly locally and remotely, and provide interoperability with other OMEI compliant tools, such as scientific editors and word processing systems, with minimal efforts.

1 Introduction

ELIMINO[5] is a new symbolic computation system being developed at the Institute of Systems Science of the Chinese Academy of Sciences. Capabilities of ELIMINO include manipulation of multi-precision numbers and polynomials, computation of characteristic set in Wu's method[9], polynomial equation solving, geometric theorem proving, etc..

To make ELIMINO easy to use and widely accessible, a good graphical user interface (GUI) is needed. Preferably, the same GUI will also make ELIMINO accessible remotely, via a LAN and/or the Internet.

Our approach is to apply the *Open Mathematical Engine Interface* (OMEI)[4], an application programming interface (API) specification for mathematical compute engines. OMEI enables tools and applications to use the same uniform interface to access any compliant mathematical engine. The concept of OMEI is similar to that of Microsoft's *Open Database Connectivity*

(ODBC) which has become the standard for PCs and LANs. With OMEI, distributed mathematical components, such as front ends, servers, and tools, can interoperate with different mathematical engines.

OMEI is used in the *Internet Accessible Mathematical Computation* (IAMC) framework[7] as a server-to-external-engine interface. The IAMC framework uses MCP (the mathematical computation protocol)[8] to connect users to remote mathematical engines across the Internet. IAMC already has an OMEI-ready server prototype (Starfish)[1], and a client prototype (Dragonfly)[2] supplying a nice GUI for back-end mathematical systems.

By adopting OMEI as an application interface, ELIMINO achieves the following results with minimal efforts.

- Using Dragonfly as a local GUI
- Using Dragonfly through IAMC as a remote GUI
- Ready to become an IAMC compliant compute engine through Starfish
- Interoperability with other OMEI compliant tools, such as editors and word processing systems.

We begin by briefly overviewing ELIMINO, IAMC, and OMEI. The ANSI C/C++ implementation of OMEI in ELIMINO is described. The way Dragonfly is interfaced to ELIMINO locally and remotely is discussed. Because ELIMINO is the first new symbolic computation system to adopt OMEI, the work tests the feasibility of the OMEI concept and design. A working system will be demonstrated in the conference.

2 ELIMINO

ELIMINO is a new computer-mathematics research system developed at the Mathematics Mechanization Research Center(MMRC), Institute of Systems Science, Chinese Academy of Sciences, as part of the "Mathematics Mechanization and its Applications" project. A long-standing goal at MMRC is to automate Wu's method independent of existing computer algebra systems. In ELIMINO, many different kinds of mathematical objects and data structures are provided. As an interactive system, ELIMINO is designed to focus on the implementation of Wu's method for researchers to perform sophisticated mathematical computations. It has very general capabilities for treating numbers, polynomials and characteristic sets[9].

To facilitate mathematical research, ELIMINO is kept open and flexible. From the user viewpoint, it consists of three parts:

- **Kernel part** is the soul of the system, it contains implementation of number system, polynomial manipulation system, characteristic set method. In fact, it is a powerful algebraic compute engine.
- **Applications** are packages or programs developed using the ELIMINO library. Examples include the polynomial system solver and the geometry theorem prover. A package may be built-in or loaded into ELIMINO on demand.
- **Front-end** is the interface between the system and users, it handles I/O between the user and the system.

From a software architecture viewpoint, ELIMINO has these five layers as shown in Figure 1:

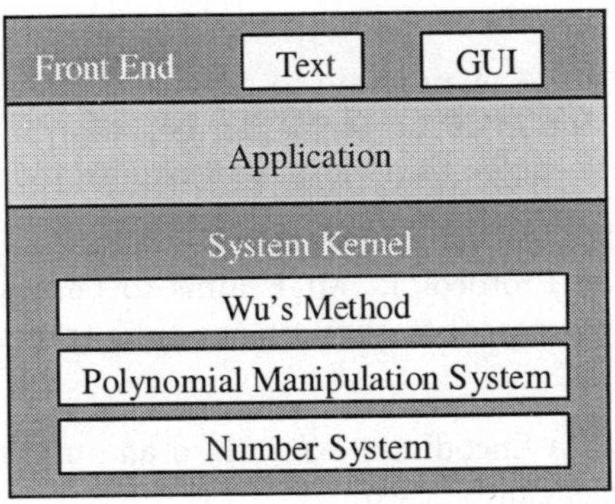

Figure 1. ELIMINO System Architecture

Each layer is an independent module, written in ANSI standard C/C++,providing a collection of closely-related and tightly organized services. A lower layer never calls functions in an upper layer.

The current ELIMINO front-end is text based. A good graphic user interface (GUI) can make ELIMINO easier to use. It would be even better if such a GUI can also make ELIMINO network accessible.

3 IAMC Framework

The IAMC framework is designed to make mathematical computing easily accessible and usable on the Web/Internet. Figure 2 shows the overall IAMC framework architecture.

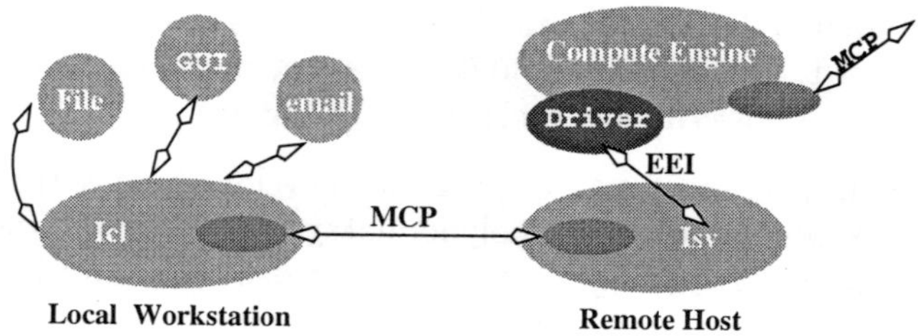

Figure 2. IAMC Architecture

The IAMC framework consists of these components:

1. IAMC client (Icl) — An end-user agent for accessing services provided by any IAMC server.
2. IAMC server (Isv) — A program to provide mathematical computation powers through the MCP protocol. An Isv may or may not employ an external compute engine to perform mathematical computations.
3. Protocol — IAMC clients and servers are connected by the Mathematical Computation Protocol[7]). MCP aims to be a simple and effective request-response protocol to support one-time transactions and interactive sessions.
4. Mathematical Data Encoding — Standard and user-defined mathematical data encodings can be used.
5. External Engine Interface (EEI) — A specification and API implementation for binding existing compute engines to IAMC servers.

4 Open Mathematical Engine Interface (OMEI)

Open Mathematical Engine Interface(OMEI) is an application programming interface(API) specification that aims to be an interface general enough to work for most mathematical compute engines. It specifies a set of function prototypes together with their syntax and semantics to give a unified programming level interface for heterogeneous mathematical engines. These prototypes support all operations that are necessary for server-engine interaction, including connecting to/disconnecting from a compute engine, querying engine capabilities, creating and executing commands, etc..

As an attempt in standardizing programming interface for compute engines, OMEI can achieve several objectives:

- **Make Compute Engine IAMC-Capable**
  OMEI and IAMC framework together can help making compute engine internet accessible. The IAMC framework follows a 3-tier architecture that consists of IAMC client, IAMC server, and back-end compute engine. The OMEI can be used to connect IAMC server and back end engines. Once an OMEI driver of a compute engine has been developed according to OMEI specification, the IAMC server can be configured to dynamically locate and load this OMEI driver, and so will be able to forward the computational requests to the engine and results to remote IAMC clients.

- **Application Portability**
  An application or user interface developed using any OMEI-compatible interface would be portable among different compute engines, as long as those compute engines have OMEI drivers available. That is, the compute engine and its user interface or applications can be developed separately with the help of OMEI specification.

- **Heterogeneous Integrated Compute Engine**
  Since an application can access multiple engines by loading multiple OMEI drivers, an integrated compute engine with combined capabilities can be accomplished under OMEI programming model. A parallel/distributed problem solving environment would be easily achievable over the OMEI programming paradigm.

- **3-tier/multi-tier mathematical system**
  This is very similar to IAMC framework that is MCP oriented. It is proven that the multi-tier architecture can achieve better performance and scalability over traditional client/server architecture. Besides IAMC, OMEI could also be used in other non-MCP distributed mathematical systems. This architecture is also useful for developing Web or other thin client based computational/educational systems.

5 OMEI Implementation in ELIMINO

There are basically two approaches to the implementation of OMEI for ELIMINO compute engine. The most straightforward approach is to use interprocess communication(IPC) to access ELIMINO. ELIMINO reads from *standard input* and writes to *standard output*. The OMEI driver connects to ELIMINO with a two-way pipe, sending computational requests into the pipe and reading computational results from the pipe. The IPC approach is simple and straightforward, but lacks flexibility and user control.

As current ELIMINO comes with an API of a set of ad-hoc C/C++ functions, we have adopted the same-process approach to implement OMEI for ELIMINO (see Figure 3(right)).

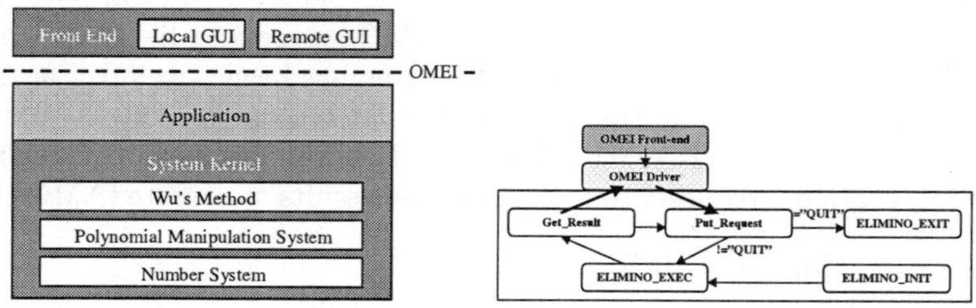

Figure 3. OMEI via Function Calls

This approach is more flexible and allows users to set up the computation environment, input/output mode, command encoding format, and resource allocation. In-process data transfer is also more reliable than IPC.

In our same-process approach, the OMEI is implemented above the application layer of ELIMINO, but below the front-end layer (see Figure 3(left)), providing standard programming interface for front-end layer to access computation powers of the ELIMINO compute engine.

Our implementation of OMEI drivers for ELIMINO is a standard C/C++ library. It contains a total of 18 functions for connecting to/disconnecting from ELIMINO, querying ELIMINO's capabilities, creating computational commands and executing these commands, each of them is programmed as the function prototypes specified in the OMEI specification. The functions `Put_Request`,`Elim_exec` and `Get_Result` are added to ELIMINO for OMEI to send requests, execute commands and obtain results. The three functions access the ELIMINO API to perform computations.

As an example, let's see how to perform the simple computation *gcd((x+y+z)*(7*x-8*z),(7*x-8*z)*(37*z-87*y*x))* in ELIMINO via OMEI.

```
#include "ELIMINO.h"
char *request="gcd((x+y)*(7*x-z),(7*x-z));";    // (1)
char *result;
int main() {
  ELIMINO_INIT();                                // (2)
  Put_Request(char * request);                   // (3)
  ELIMINO_EXEC();                                // (4)
  Get_Result(char * result);                     // (5)
```

```
    ELIMINO_EXIT();                                    // (6)
    return 0;
  }
```

First, two variables, **request** and **result**, are allocated to store the computational request from IAMC servers or other compliant tools and computational result or error information. Second, ELIMINO is initialized as necessary by setting up computation environment and input/ouput modes (line 2). The request is then sent to (line 3) and evaluated by the ELIMINO compute engine (line 4). Finally the result is read into the buffering variable **result** and the ELIMINO engine quits.

6 Applying OMEI to ELIMINO

Dragonfly is an IAMC client prototype and *Starfish* an IAMC server prototype which are implemented in Java. The Dragonfly-Starfish approach makes interactive mathematical computations accessible and interoperable over the Internet. By adopting OMEI as an application interface, ELIMINO can easily use *Dragonfly* as its local interface (see Figure 4 (a)) or remote user interface (see Figure 4 (b)).

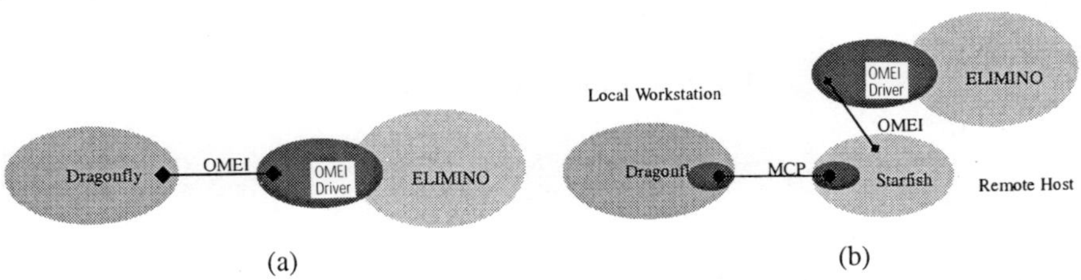

Figure 4. Use Dragonfly as local interface (a) or remote user interface (b)

A graphical user interface like Dragonfly not only helps beginning users to get started in ELIMINO quickly, but also provides potentials for adding into ELIMINO more functionalities, such as mathematical graphing and plotting. To interface Dragonfly to ELIMINO as a local GUI, a Java OMEI driver has to be developed because Dragonfly is developed in Java. In other words, Dragonfly has to load this Java OMEI driver to interact with ELIMINO. The Java OMEI driver could be thought of as a Java Wrapper for OMEI function calls and it is implemented by using Java Native Interface (JNI). Figure 5 shows the user interface of Dragonfly as the local GUI for ELIMINO.

Using Dragonfly through IAMC as a remote GUI is another story. It

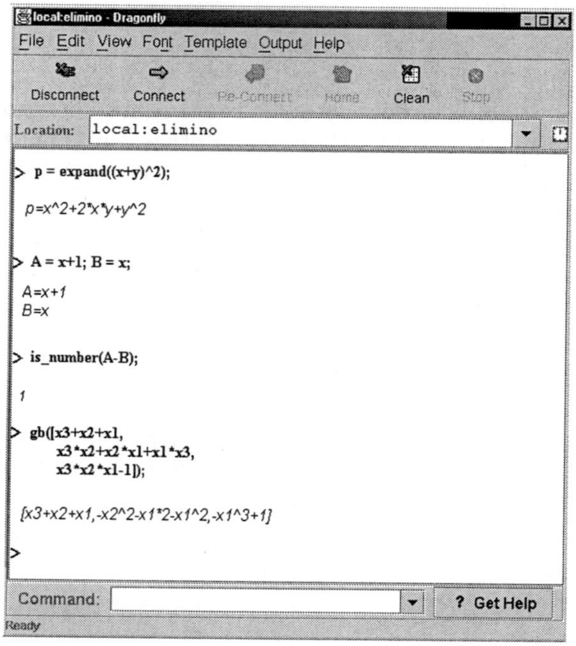

Figure 5. Dragonfly as a local GUI

is possible to turn the ELIMINO compute engine into an IAMC server by integrating MCP library into it. Nevertheless, with the OMEI interface, it is much easier to integrate ELIMINO into an existing IAMC server, such as our IAMC server prototype *Starfish*. *Starfish* is configured to dynamically locate and load the OMEI driver. With the loaded driver, *Starfish* will be able to forward the computational requests to ELIMINO and results to *Dragonfly*, or any other IAMC clients. Figure 6 shows the user interface of *Dragonfly* as the remote GUI for ELIMINO.

Turning ELIMINO into an IAMC server directly (by integrating MCP library) or indirectly (by using an IAMC server like *Starfish*) also has some other potentials. This will make ELIMINO an IAMC compliant engine and it will be able to serve any MCP requests from Internet. For example, a Web-based mathematical education system may send MCP requests to ELIMINO to get answers for some algebra questions, and a distributed crypt algorithm may issue MCP requests to ELIMINO to get some big prime numbers.

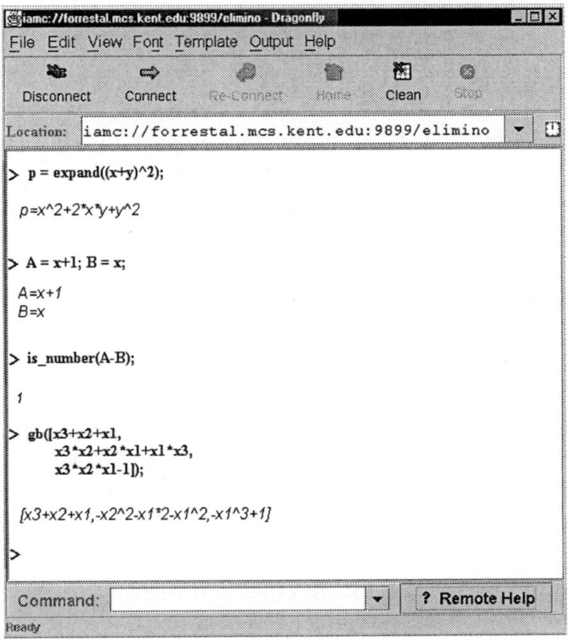

Figure 6. Dragonfly as a remote GUI

7 Conclusion and Future Work

The OMEI interface entitles several new features for ELIMINO, a new computer-mathematics research system, with minimal efforts. An existing graphical user interface, *Dragonfly*, could be used to access ELIMINO locally and remotely. As an IAMC compliant compute engine, ELIMINO can be used as a compute server for Internet based mathematical computational and education systems. Moreover, OMEI also gains interoperability with other OMEI compliant tools, such as scientific editors and word processing systems.

We will integrate XMEC [10], an eXtensible Mathematical Encoding Converter into Java OMEI driver for ELIMINO engine so that the engine could accept input and produce output encoding in standard encoding formats, such as MathML. When this work completes, users will be able to use Dragonfly or other tools to access ELIMINO engine to evaluate MathML code and generate MathML output. The ELIMINO compute engine will gain better interoperability in distributed and Internet based mathematical computation and education environment.

Acknowledgements

Work reported herein has been supported in part by the Chinese National 973 Project NKBRSF G1998030609, Chinese 863 Project 2001AA144030, the USA National Science Foundation under Grant CCR-9721343 and INT-9722919, and the Ohio Board of Regents Computer Science Enhancement Funds.

References

1. LIAO, W. and WANG, P. S. *Building IAMC: A Layered Approach.* Proc. International Conference on Parallel and Distributed Processing Techniques and Applications (PDPTA'00). pp. 1509-1516.
2. LIAO, W. and WANG, P. S. *Dragonfly: A Java-based IAMC Client Prototype.* Lecture Notes on Computing Vol.8. (Proceedings of ASCM 2000.) World Scientific Press, pp. 281-290.
3. LIAO, W. and WANG, P. S. *Specification of OMEI: Open Mathematical Engine Interface.* ICM Technical Report. 2001. http://icm.mcs.kent.edu/reports/index.html.
4. LIAO W., LIN D. and WANG P. S. *OMEI: Open Mathematical Engine Interface.* Proceedings of ASCM'2001, pp 83-91, Matsuyama, Japan, September 26-28, 2001. Lecture Notes Series on Computing Vol. 9, World Scientific.
5. LIN D., LIU J. and LIU Z. *Mathematical Research Software: ELIMINO.* Proceedings of ASCM'98. pp. 107 - 116, Lanzhou Univ., China, 1998.
6. WANG, P. S. *Internet Accessible Mathematical Computation.* In the 3rd Asian Symp. on Computer Mathematics (ASCM'98), pp 1-13, Lanzhou Univ., China, 1998.
7. WANG, P. S., GRAY, S., KAJLER N., LIN D., LIAO W. etc. *IAMC Architecture and Prototyping: A Progress Report.* Proceedings of ACM ISSAC'01, University of Western Ontario, London, Ontario, Canada, July 22-25, 2001.
8. WANG, P. S. *Design and Protocol for Internet Accessible Mathematical Computation.* In Proc. ISSAC'99 (1999), ACM Press, pp. 291-298.
9. WU, W. T. *Basic Principle of Mechanical Theorem Proving in Elementary Geometries,* J. Syst. Sci. Math. Sci. 4, 207-235 (1984).
10. ZOU, X. *XMEC: An Extensible Mathematical Encoding Converter.* ICM Technical Report. 2001. http://icm.mcs.kent.edu/reports/index.html.

PROGRAMMING THE IBM MATHML EXPRESSION EDITOR FOR INTERACTIVE MATHEMATICAL APPLICATIONS

SAMUEL S. DOOLEY
IBM T. J. WATSON RESEARCH CENTER
P.O. BOX 714, HAWTHORNE, NY 10532 USA
dooley@watson.ibm.com

The IBM MathML Expression Editor is a two-dimensional mathematical editor for expressions encoded in content and presentation MathML. It allows a user to interact with the visual presentation of an expression, while simultaneously creating the underlying content structure of the expression. The component edition of the editor provides programming interfaces and customization facilities that support rapid development of mathematical software targeted for deployment in a web browser or other container application. This paper describes the component architecture of the editor, the programming interfaces provided for interactive mathematical application development, and the configuration framework that allows each application built using the editor to provide custom mathematical notations, input templates, menus, palettes, and other editing conventions. Using these features, the editor can supply natural visual editing for symbolic expressions to a variety of mathematical applications.

1 Introduction

User interface development is typically one of the most expensive and least portable aspects of any large software system. The problem is somewhat more vivid in the development of mathematical software due to the rich structure found in common mathematical notation. Historically, the most powerful computer algebra systems have relied on equally powerful one-dimensional plain-text input formats closely tied to the mathematical programming languages they support. These input formats are commonly augmented by programs that generate markup in widely-used two-dimensional mathematical presentation languages such as LaTeX or MathML[1] that can be used as export formats for display by other programs. Two-dimensional mathematical editing software that hopes to achieve the same success for mathematical input as has been achieved for two-dimensional mathematical presentation is often difficult to develop and relies heavily on the particular notational conventions of the systems with which it communicates. Relying solely on presentational markup as the standard for mathematical communication makes it more difficult to extract the computational expression structure necessary for true round-trip communication between the user interface and software applications. The goal of content markup for mathematics, as exemplified in MathML, is to capture a representation of the compositional structure of an

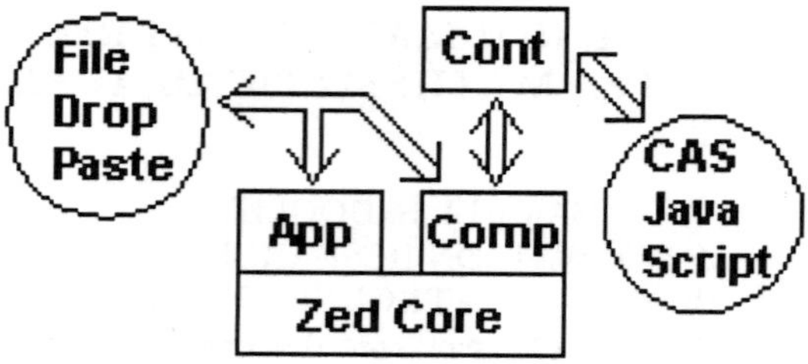

Figure 1. The Zed Component Architecture

expression that allows for more robust communication between software components for mathematics. Coordinating this content representation with the presentational markup within the user interface allows support for an extensible model for the human user to interact with presentational structures in ways that communicate with other software components using unambiguous, portable content representations.[2] [3]

The IBM MathML Expression Editor,[4] known informally as Zed, is a two-dimensional mathematical editor for expressions encoded in content and presentation MathML. It allows a user to interact with the visual presentation of an expression, while simultaneously creating the underlying content structure of the expression. The architecture of the editor as a software component that supports two-dimensional editing of both content and presentation markup allows Zed to be included as a reusable resource in the development of a wide variety of mathematical software applications. The programming interfaces provided by the editor allow access to both the presentation and content markup, and provide the high degree of extensibility and customization that is often a design requirement for mathematical software. As a result, many different flavors of mathematical software applications can be rapidly prototyped, tested, and deployed to an extent not possible with previous generations of mathematical authoring software and tools.

2 Architecture

Figure 1 illustrates the component architecture for Zed, that is, the functional relationships detailing how the component edition of Zed operates as an ActiveX control hosted in an ActiveX container. The component edition inherits a great deal of its editing support from a common core framework for editing

MathML elements shared with the application edition of the editor.[5] This core editing framework is itself built on rendering support for MathML presentation elements provided by techexplorer.[6] The component edition of the editor is implemented by subclassing generic Zed document, view, and application classes to support the interfaces expected of an ActiveX control. Incorporating the core technology into another component framework could easily be accomplished by duplicating this thin subclassing layer to target a different model for component extensions, whether IE behaviors, Netscape-style plugins, Adobe Acrobat extensions, or other types of addins for multimedia content authoring environments.

The component edition of the editor defines a collection of custom properties, methods, and events for use from the container. Collectively these are known as the Zed Custom Application Programming Interface (API), that is described in the next section. The properties, methods and events in the Zed Custom API operate within a browser, word processor, presentation viewer, or other container to provide computationally interesting user interfaces for mathematical applications. There are custom properties for accessing the MathML markup for the current expression, either to get the markup as a string or to replace the expression by parsing a new string. Subexpressions of the current expression can be accessed using the `id` attribute to identify the desired expression. Analogous properties allow access to the internal Lisp-like expression format used by the editor, for those who prefer parentheses to angle brackets. Both the content and presentation markup can be accessed in these ways, allowing the editor to parse both content and presentation as input, and to convert from content to presentation. The custom events provided by Zed allow script writers to synchronize actions with the completion of loading the component, with any modification to the current expression, or with a user-initiated action to submit the current expression for some further computation. These events provide a foundation for incorporating symbolic expressions into more interesting dynamic mathematical effects.

The custom methods provided in the API support a high degree of configuration for many aspects of the behavior of the mathematical elements and operators Zed supports. Six general categories of operator configuration information can be customized: layout formats, named templates, dynamic menus, palettes, traversal order, and other properties. The component edition provides custom methods for getting, setting, or adding to any of these six configuration sets and for loading and saving configuration files to the filesystem. These methods can be used by application developers to supply custom layout formats for mathematical operators, to define custom menus and palettes for their applications, to alter the default bindings for many of the keyboard accelerators used by the editor, or to supply editing behavior for anonymous function symbols. These methods are described in the next section along with the rest of the Zed API, and their role in supporting an extensible framework for mathematical software is detailed in section 4.

3 Interfaces

The properties, methods, and events contained in the Zed Custom API are listed in Table 1. The Zed Custom API can be conceptually divided into a number of smaller APIs, each relating to a particular aspect of the behavior of the editor.

3.1 Initialization

Zed provides the boolean property `AutoSize` to allow the container to specify that it supports the ability of the component to automatically resize itself as the expression changes. If this property is set to true, the component will resize itself and notify the container with appropriate events. Typically this property is used by the container to flow text or other objects around the component.

Zed provides the boolean property `GrabFocus` to allow the container to arrange for a particular instance of the editor to have the focus when the container document is first loaded. This property is useful if there are several instances of the editor on a single page, and the intent is that one particular instance should get the attention of the user first.

3.2 Synchronization

In addition to the stock event `ReadyStateChange` and the stock property `ReadyState`, Zed also provides three custom events that control synchronization with external scripts and other components. The `Onload` event is fired once Zed has been completely loaded and all of its data structures have been initialized. This event allows script writers a means of ensuring that Zed is ready to accept other method calls before the script initiates other actions. The `Update` event is fired each time the expression being edited is changed. This event can signal any number of different conditions, including user input, script actions, or file or configuration loading. The `Submit` event is fired upon request by the user, typically in response to a keyboard accelerator as specified by the configuration. This event is typically bound to a key such as the `Enter` key, but may be rebound as desired by the script writer.

Properties		
Type	Name	Parameters
Boolean	AutoSize	
Boolean	GrabFocus	
Integer	ReadyState	
String	MmlSource	
String	MmlContent	
String	MmlPresent	
String	MmlCurrent	
String	ZedSource	
String	ZedContent	
String	ZedPresent	
String	ZedCurrent	
String	IdMmlContent	(String id)
String	IdMmlPresent	(String id)
String	IdZedContent	(String id)
String	IdZedPresent	(String id)

Methods		
Type	Name	Parameters
Boolean	LoadConfig	(String name)
Boolean	SaveConfig	(String name)
String	GetConfig	(String key)
Boolean	SetConfig	(String key, String value)
Boolean	AddConfig	(String key, String value)
Boolean	ReadConfig	(String key, String name)
Boolean	WriteConfig	(String key, String name)
void	ShowPalette	(String name)
void	HidePalette	(String name)

Events		
Type	Name	Parameters
void	ReadyStateChange	()
void	Onload	()
void	Update	()
void	Submit	()

Table 1. The Zed Custom Application Programming Interface

3.3 Configuration

Zed provides the `LoadConfig` and `SaveConfig` methods for loading and saving the entire collection of configuration information to a named file in a compressed format. The `GetConfig` and `SetConfig` methods allow access to a particular section of the configuration information using an expanded string format, and the `AddConfig` method can be used to augment that string with additional declarations. The `ReadConfig` and `WriteConfig` methods allow these section-specific strings to be read from or written to a particular file, aiding the creation of the compressed configuration files. Many of these methods take as their first argument the key name of the configuration section being accessed, identified using one of the six built-in strings used to name each section, as described in section 4.

3.4 Document Markup

Zed provides properties for accessing the document markup string for the expression being edited. This markup is somewhat human-readable, but is typically more suited for parsing by another application or component in cooperation with the editor. The `MmlContent` and `MmlPresent` properties give read/write access to the MathML content and presentation markup for the document, and the `MmlCurrent` property gives access to the MathML content markup for the current expression selected by the expression cursor. The `MmlSource` property gives access to the filename used to initialize the document. Each of these methods has a corresponding version for accessing the internal Zed markup for each expression.

3.5 Document Labels

In addition to giving access to the current expression, Zed provides methods for accessing the content or presentation markup (in either MathML or Zed format) for elements in the document identified with an `id` attribute. These methods would ultimately be superceded by providing more complete support for the Document Object Model (DOM)[7] in addition to the custom API.

3.6 User Interface

This first version of the editor provides only minimal programmatic control over the user interface for the editor, other than via the configuration customization interfaces to be detailed in the next section. The `ShowPalette` and `HidePalette` methods can be used to show or hide particular palettes available to the user. Ultimately, one would like to have additional methods for controlling the state of the expression cursor, and for programmatic invocation of named templates.

4 Customization

Mathematical applications often require a high degree of extensibility in the collection of operators they support. Zed satisfies this requirement by providing a framework for customizing operator editing behavior. Customization information is associated with property values attached to the internal symbols representing MathML content elements and user-defined operators appearing in `<apply>` elements. Using this framework, the default editing support for MathML elements can be configured to meet the requirements of a particular application, and editing support for operators not provided by MathML can be defined. Six categories of properties can be customized: layout formats (`layouts`), named templates (`templates`), dynamic menus (`menus`), palettes (`palettes`), traversal order (`orders`), and other properties (`bindings`).

The layout formats provide layout transformation rules for converting content to presentation.[3] The named templates define the collection of editing transformations that can be used to create expression structure for both content and presentation.[3] Aspects of the Math context menu Zed provides for invoking named templates can be customized by the dynamic menu information, including its submenus, menu item labels, and template bindings. Two-dimensional palettes can also be specified, where each palette button is associated with a named template, and the template provides an expression to be displayed on the palette button. The default traversal order for the subexpressions of an operator can be overridden by associating a permutation with the operator. Finally, other property/value pairs specify operator precedence values for parenthesization, declarations of operators as associative, nary, or array operators, and other bindings.

Zed uses its generic expression data type for representing the customization properties associated with each operator, both for convenience in the internal implementation and because it provides an easily implementable and uniform linear syntax for configuring the customization properties. Using this linear syntax to modify the default behavior of the editor is a reasonably convenient way for application developers, using the Zed Custom API, to meet the needs of their applications. The Configuration API described in section 3.3 supports the dynamic modification of the customization properties, as well as the creation of precompiled configuration files that can be distributed with the application.

$$\frac{-b+\sqrt{b^{2}-4\cdot a\cdot c}}{2\cdot a} \qquad \frac{-b+\sqrt{b^{2}-4\times a\times c}}{2\times a}$$

Figure 2. Layout Customization Example

$$\frac{-b\pm\sqrt{b^{2}-4ac}}{2a}$$

```
<apply><ci>pm</ci>
  <apply><minus/><ci>b</ci>
  </apply>
  <apply><root/> ...
```

Figure 3. Custom Operator Example

The size of the customization information for an operator language as extensive as MathML could easily create a performance issue. While the information could be compiled into the program for speed, storing the configuration properties as external files provides greater flexibility. Zed stores the configuration expressions in a compressed form in an external binary-format configuration file. The file contains a header with a table of section offsets, a symbol table listing the names of the symbols used, and a binary encoding of each of six configuration expressions, one for each section listed above. In practice, the size of a configuration file is quite small, and it can be expanded in memory as needed.

Two brief examples suffice to suggest the extensibility of the configuration framework. In Figure 2, instances of multiplication in the quadratic formula are shown using two alternative layout transformations for the `<times>` element; the same content markup is used in each instance, and the transition from one layout to any other can be scripted dynamically using the API. In Figure 3, the quadratic equation is presented again, this time by introducing a custom content operator for the plus/minus infix operator; both layout and editing transformations for the new operator can be defined using the API.

5 Applications

The component architecture for the editor and the custom API exposed for manipulation within the container environment combine to provide an extensible, reusable software module that can be incorporated into a wide variety of mathematical applications. The Zed component edition provides conve-

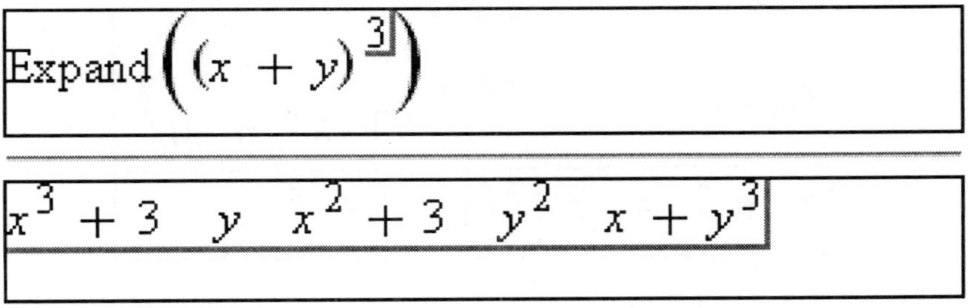

Figure 4. Computer Algebra Interface

nient two-dimensional input and output of mathematical expressions using open standards for representing both the content and presentation markup for an expression, and for communicating that markup with the rest of the application. By handling the complexity of two-dimensional expression input, Zed can greatly simplify the design and implementation of the user interface for a mathematical application, allowing the developer to focus on the application itself while delegating expression input to the Zed component. A wide range of potential applications could be developed with such a component; two examples are presented here to suggest the possibilities.

One way the Zed component can be used is as a generic front-end interface for computer algebra systems. In the traditional user interface model for these systems, a command is read by the system, evaluated by means of highly sophisticated computational algorithms and symbolic simplifiers, and the result is output back to the user. For those computer algebra systems that are capable of reading and writing MathML, creating a web page that imitates this behavior can be accomplished quite easily by embedding instances of the editor for the input and output expressions, and using JavaScript to collect the MathML from the editor and send it to the back-end system. Figure 4 illustrates this arrangement. The upper instance of the editor is used to type an expression that represents a command to be executed by the system, and the lower instance is used to display the result. Using content MathML for both input and output ensures that the communication is unambiguous.

More elaborate symbolic spreadsheets can be supported using the editor in much the same way. Provided with a computing engine capable of maintaining computational dependencies between variable assignments, the editor can be used to modify any one of several interdependent assignments, then update events fired by the editor can be used to provoke updates to the state

$x := 2$	$M := \begin{pmatrix} x & 1 \\ 1 & x^2 + 1 \end{pmatrix}$
$x + 1 = 3$	$M_{1,1} = 4$

Figure 5. Symbolic Spreadsheet Example

of the symbolic spreadsheet, which are then communicated back to the editor instances in the document. Figure 5 demonstrates a very simple such spreadsheet, where the dynamic effects are suggested by the fact that the state of the spreadsheet is captured after adding +1 to the second cell in the first row, and just before firing the update event that would lead to the recalculation of the second cell in the second row upon updating the document.

Rapid prototyping of various interface models for computer algebra systems is only one example of how the Zed component could be used for more elaborate mathematical applications. While authoring interactive mathematical courseware for the internet has steadily been made easier, one remaining obstacle Zed addresses is the ability of the courseware author to incorporate symbolic information in an essential way in the exposition. With the ability to include symbolic expressions in dynamic presentational effects, either in a web browser or in more standard slide presentation software, a much wider range of expositional techniques can be employed. With appropriate support from instant messaging containers, math chat rooms could be implemented allowing for collaborative symbolic problem solving. The ultimate goal of such developments is to open up access to what are now primarily closed systems, to increase the range of applications that can be built using available tools.

6 Conclusions

The component edition of the IBM MathML Expression Editor provides an open architecture for developing user interfaces for mathematical applications that incorporate two-dimensional mathematical input. The open architecture includes full support for MathML content and presentation elements in the interface that can be communicated to other components of the application. The editing support provided in Zed can be customized and extended to meet the notational and representational needs of the application, and the behavior of the editor can be scripted by external tools to bring forward a wide range of expositional effects. Packaging these facilities for mathematical editing as a reusable component with well-defined programming interfaces provides a valuable tool for reducing the barriers for developing mathematical software for the internet or for other application environments.

7 Acknowledgements

The work described here was implemented by the author as part of the IBM MathML Expression Editor, and would not have been possible without the support of other members of the techexplorer group: Luis Alvarez, Stephen Bachop, Angel Díaz, Bill Jecusco, Richard Jenks, Mimi Jett, Thomas Ling, and Michael Reiter. In particular, work toward an early prototype for an expression editor written in Java using techexplorer by Richard Jenks and Luis Alvarez provided a great deal of insight on the issues involved in two-dimensional editing for mathematical expressions.

References

1. David Carlisle, Patrick Ion, Robert Miner, and Nico Poppelier. Mathematical Markup Language (MathML) Version 2.0. W3C Recommendation REC-MathML2-20010221, W3C, 21 February 2001. `http://www.w3.org/TR/2001/REC-MathML2-20010221`.
2. Samuel S. Dooley. Coordinating Mathematical Content and Presentation Markup in Interactive Mathematical Documents. In [8], pages 54–61, 1998.
3. Samuel S. Dooley. Editing Mathematical Content and Presentation Markup in Interactive Mathematical Documents. To appear, ISSAC 02.
4. Samuel S. Dooley. *Users Guide for the IBM MathML Expression Editor, Version 1.1.0.* IBM, 25 March 2002.
5. Samuel S. Dooley. Bringing MathML Content and Presentation Markup to the Web with the IBM MathML Expression Editor (extended abstract). To appear, MathML 02.
6. IBM Software. techexplorer Hypermedia Browser, January, 2002. `http://www.software.ibm.com/techexplorer`.
7. Lauren Wood et al. Document Object Model (DOM) Level 1 Specification. W3C Recommendation REC-DOM-Level-1-19981001, W3C, 1 October 1998. `http://www.w3.org/TR/1998/REC-DOM-Level-1-19981001`.
8. Oliver Gloor, editor. *Proceedings of the 1998 International Symposium on Symbolic and Algebraic Computation*, New York, 13–15 August 1998. ACM SIGSAM, ACM Press. Rostock, Germany.

FROM MATHEMATICAL SERVERS TO MATHEMATICAL SERVICES

MIKE DEWAR & DAVID CARLISLE
NAG LTD.
WILKINSON HOUSE
JORDAN HILL RD.
OXFORD
UNITED KINGDOM
MIKED@NAG.CO.UK & DAVIDC@NAG.CO.UK

1 Introduction

Most mathematical software in use today is either packaged as components (numerical libraries, add-ins for Excel etc.) or as interactive packages (Maple, Mathematica, Matlab). From a user's perspective, the software is stored and administered locally, and in the (still surprisingly rare) case when computation is distributed across several machines, this distribution is set up and controlled locally as well. The purpose of this paper is to describe work which could ultimately break down this centralised approach to software use and development, and replace it with a more flexible, distributed, networked solution.

Before being seduced by the latest trends in software engineering and the worldwide web however, it is worth noting the advantages of the current approach. A user knows exactly what version of a piece of software he or she is using, and can choose whether or not to upgrade to new versions. Different software packages can give different but equally valid answers to the same problem (due to the selection of branch cuts, for example[1]), and of course will contain different bugs, and while backwards compatibility is usually promised in software releases it is not always realised in practice. Networks are still not 100% reliable and while standards such as IEEE arithmetic[4] have reduced problems caused by different behaviour at a hardware level, it is still an issue, particularly where parallel computation is concerned. Finally security and data integrity are often important and are much easier to guarantee in a closed environment than across the whole internet.

From a developer's point of view the conventional approach offers many advantages too. At NAG, for example, each release of a software package is subjected to rigorous quality assurance procedures, and support queries from our users are much easier to address because we know exactly what environment they are using. From a commercial point of view our existing business model is well understood and has a long and successful track record.

A more distributed arrangement, on the other hand, would place the burden of maintenance and upgrading on the software developer. Users would be guaranteed access to the latest versions of software and, if the protocols for communicating between the different software components were open and published, could choose from amongst a number of different suppliers. Increasingly science and engineering is a collaborative activity, even in industry (for example the European Jet Fighter is being developed by a Consortium of seven organisations in four countries), and so distributed software systems might model more accurately the way research and development is carried out.

The first step away from the traditional method of software delivery is for applications to become more component-based. As an internal software architecture this is increasingly the case, but it is still comparatively rare for this to be accessible to a user. However many packages now offer a "foreign function interface" of some kind which allows foreign code (usually packages as shared libraries/DLLs) to be linked in. This is not a generic mechanism by any means and often requires some knowledge of the underlying data structures being used, but it does provide an increased degree of flexibility. Later in this paper we shall discuss how we can take advantage of these interfaces.

Recently the Worldwide Web Consortium (W3C) has started to develop technology to support the *semantic web*[11], a vision of a global network where information has semantically-rich representations defined by well-defined *ontologies* and where *web services*, software agents which understand a particular set of ontologies, are available to handle a wide range of tasks. While open standards to support this are still under development, software tools to support many of the individual tasks are already becoming available, the most ambitious of which is probably Microsoft's .Net framework. While the semantic web vision is attractive, a more practical application of the emerging tools may well be to allow scalable, flexible application integration, an idea which we will address later in this paper.

Finally, another activity within the scientific computation community has been the emergence of *grid computing*. The idea is that a high-speed international network links together computational resources around the globe, and that when a user needs to use a resource it is supplied regardless of where it is physically located, by analogy with the grids which supply energy, water etc. in many countries. Some proponents of the grid believe that it needs to adopt similar technologies to the semantic web, to form a *semantic grid* [2].

2 OpenMath & MathML

OpenMath[8] is a standard for representing mathematical data in as unambiguous a way as possible. The original motivation for OpenMath came from the Computer Algebra community. Computer Algebra packages were getting bigger and more unwieldy, and it seemed reasonable to adopt a generic "plug and play" architecture to allow specialised programs to be used from general purpose environments. There were plenty of mechanisms for connecting software components together, but no common format for representing the underlying data objects. It quickly became clear that any standard had to be vendor-neutral and that objects encoded in OpenMath should not be too verbose. It can be used to exchange mathematical objects between software packages or via email, or as a persistent data format in a database. It is tightly focussed on representing semantic information and is not intended to be used directly for presentation, although tools exist to facilitate this. It is extensible, and its most common representation is as XML.

The heart of OpenMath is a collection of definitions of mathematical concepts, organised into *Content Dictionaries* or CDs, each of which defines a small collection of closely related symbols. Collections of related CDs are formed into *CD Groups.* For example the *linalg1* CD defines simple operations such as *transpose* and *determinant*, *linalg2* and *linalg3* supply symbols for creating objects such as matrices and vectors from a row- and column-oriented perspective, *linalg4* defines more complicated operators such as *characteristic equation* and *eigenvalue*, and *linalg5* defines special matrix structures such as *Hermitian*, *banded* etc. Together these CDs form the linalg CD Group. In addition *linalg1* and *linalg2* are part of the MathML CD Group, since they contain symbols which have exact counterparts in content MathML (see below).

Symbols from CDs can be combined according to the OpenMath language to form OpenMath objects. For example sin is an OpenMath symbol described in an OpenMath CD, while $\sin(x)$ is an OpenMath object (the application of the symbol sin defined in CD *transc1* to the variable x). A CD can be regarded as a *micro-ontology*, while any collection of CDs plus the OpenMath language might be thought to form an ontology.

In 1998, the Worldwide Web Consortium (W3C) produced its first recommendation for the Extensible Markup Language (XML) [12], intended to be a universal format for representing structured information on the worldwide web. It was swiftly followed by the first MathML recommendation[10] which is an XML application oriented mainly towards the presentation of mathematical expressions. MathML also supports a fixed set of content tags describing a very limited subset of high-school mathematics, and a more flexible mechanism (via on the `csymbol` element) for annotating symbols with arbitrary semantic information.

From the point of view of representing semantics, the main differences between OpenMath and MathML are the following:

- MathML uses XML exclusively, whereas OpenMath can have alternative encodings (although in what follows we will assume that the XML encoding is being used).
- The syntax of the XML encoding of OpenMath is more regular than that of MathML, but MathML is better integrated with other XML recommendations.
- OpenMath offers a language for defining the semantics of symbols, whereas MathML does not. A user can use the syntax of MathML and the semantics of OpenMath if he or she wishes, via the `csymbol` element. However there is no guarantee that all the `csymbols` in an expression use the same mechanism for defining their semantics, or even a consistent one.

There are widely-available and well-understood technologies for carrying out XML⟶XML transformations (e.g. the eXtensible Stylesheet Language (XSL)[9]) which make mapping between well-defined subsets of OpenMath and MathML relatively straightforward. Equally this can be used to map OpenMath to presentation MathML (or TEX or any other rendering language).

We believe that it makes little difference whether a user prefers the OpenMath or MathML syntax for an object, and that the important distinction between the two languages is that:

OpenMath provides a mechanism for describing the semantics of mathematical symbols;

MathML provides an XML-based mechanism for presenting mathematical expressions.

This is clearly a highly-complementary arrangement.

3 Mathematical Servers

In the OpenMath ESPRIT project, we developed a number of mathematical servers which were capable of being deployed on an intranet or the internet.

The servers used OpenMath as a communication protocol and their interfaces were defined in terms of OpenMath objects. The client simply created an object of the appropriate type, transmitted it to the server via RPC, and waited for the result. Translation from OpenMath to the native representation was handled independently by *phrase books* on both the client and server side. One example we constructed was a numerical optimisation server which could be used from a computer algebra system or a Java applet running in a web page. Another example was an applet which could use either AXIOM[5] or Reduce [3] as a "calculus server" to integrate or differentiate expressions. This demonstrated that on the one hand such a server could be used by a wide variety of clients, while on the other hand this approach offered a user a genuine opportunity to "plug-and-play".

There are two major problems with this approach however. First of all there is the overhead of encoding, transmitting and decoding the mathematical objects being manipulated. Whether this is an issue depends on both the client application and the computation time. In the examples we investigated the computation time was often non-trivial so the extra overhead introduced by the client-server architecture was negligible. Moreover the native data structures in the client were not necessarily appropriate for the computation in the server so translation was actually desirable (for example until recently most computer algebra systems used arrays of pointers to floats rather than packed arrays of floats which are preferable for numerical computation). Nevertheless in practice such an architecture should be scalable so that the server can be downloaded and run locally using shared memory to minimise the overhead.

The second problem concerns the integration of the server into what we might loosely term the "programming infrastructure" of the client. For example any interactive mathematics package probably contains debugging and tracing facilities as well as mechanisms for issuing errors and exceptions. Even if the server returns an error message it may not be understandable to the user for a number of reasons: it might use different terminology, different names for parameters, or even be in a different language to that used by the client. Solving these problems on a case-by-case basis is tedious and often not very cost-effective.

3.1 Bespoke Interface Generation

The NAG Fortran Library[7] contains over 1000 user-callable subroutines for performing a wide range of mathematical operations. Embedding the Library in other software environments, whether general packages like Excel or the

Maple Computer Algebra System, or more specialised systems such as Telegnomic's Enterprise Asset Optimisation System or PeopleSoft 8, is increasingly becoming an important part of NAG's business. To simplify this process NAG has developed interface specifications in XML for every routine in the Library, combined with language-neutral documentation. XSL stylesheets are used to produce the "glue code" necessary to interface the Library code to the foreign function interface of the host system, and to generate natural documentation.

Component-based frameworks such as CORBA and Active-X use the Interface Description Language (IDL) to describe software interfaces. These descriptions are used to generate intermediary code to allow components to interact with one another. For our purposes, however, we found that IDL was too low-level, and that we needed to invent a more specialised mechanism. In our scheme, for instance, the interface descriptions include the following information for each parameter:

- its concrete data type (e.g. array of double precision floating point numbers);
- an abstract type describing its rôle (e.g. *data*, *workspace*, *principal dimension*, ...)
- its intent (*input*, *output*)
- for an array, its dimension and storage structure;
- any constraints on permitted input values;
- its definition, if this can be described in terms of other parameters;
- any default or suggested values, if appropriate.
- ...

As mentioned, the routine documentation has been processed to remove language-specific features (e.g. array indices which may be 0- or 1-based, statements such as "`A.LT.0`" etc.). Critically, error conditions and explanatory messages have also been abstracted so that they can be used and understood in different environments.

It should be noted that software reuse is a difficult problem and that as part of this exercise the Fortran Library was substantially re-structured so that it could be called more easily from different languages and embedded more naturally in different environments. The details of that work are beyond the scope of this paper.

As indicated earlier, we can now generate interface (or "glue") code automatically to integrate NAG software into many different environments. This glue code is often quite complicated and might, for example, do constraint checking, perform memory allocation, package and unpackage data, and return error information to the host system in the most natural manner available. Although we are currently only doing interface generation statically, it could be done dynamically to adapt to conditions on the client system[a].

3.2 Mathematical Components

The system as described so far has perhaps one major flaw. To generate the glue code linking NAG and a client piece of software requires information to be encoded explicitly in the interface description about both the NAG software and the client. The main reason for this is to map higher-level data structures in the client into the more primitive structures which the NAG Library uses for efficiency and portability reasons. This requirement can, however, be eliminated if the interfaces are described using a common abstraction such as OpenMath, when the tailoring of an interface becomes purely a transformation issue. Note that none of this implies that OpenMath will be used to encode data, although this is certainly a possibility.

4 Mathematical Services

The infrastructure for mathematical servers described in the previous section makes it possible, with a little effort, to combine two particular pieces of software. The concept of a *service* however, is rather different.

The basic notion of a *web service* is of an autonomous piece of software which runs somewhere on the internet and is able to describe what it does and how it can be invoked according to some general protocol or language. At the time of writing (April 2002) no consensus has emerged as to what that protocol or language should be, but there are a number of contenders and W3C has started an activity to find a common standard. It is likely that the more mundane aspects of service description, such as who is authorised to use the service, how much it costs to access, and a very broad categorisation of what it does will be handled by this mechanism. However the details of what specific tasks a service performs and the techniques which it uses to do them will most likely be handled by domain-specific descriptions expressed in an appropriate ontology. In this model, the most critical aspect of the

[a]this would, however, raise a number of challenging questions related to quality assurance

description of a web service's functionality is therefore the statement of which ontologies it "understands", since without that no client can progress to the point where it learns how to invoke the service. For the purposes of this paper we will ignore the generic issues, which we assume will either be solved by W3C or by market forces, and focus on the issues which are specific to mathematical services.

An example of a mathematical web service might be an "integration service" based on the quadrature routines in NAG's Fortran Library. How would it express the fact that it performed integration? If it were an OpenMath application then it might say that it understood the OpenMath CDs necessary to describe definite integrals. However OpenMath is only a data representation, it does not define what an application does with an OpenMath object when it receives it. Thus we might also want to indicate the fact that our server attempts to *evaluate* the objects which it receives (as opposed to looking them up in a database, rendering them etc.). We also would like to express the fact that the results returned are exclusively numerical if the evaluation is successful, which can be done by describing the result in terms of OpenMath CDs as well. We note that the granularity provided by OpenMath CDs makes it possible to define the language "understood" by a service quite tightly. While it would be possible to us MathML as the language for representing the objects received by the service, there is no facility in MathML for describing the subset of mathematics which the service can deal with.

In this kind of environment, services can invoke other services and the actual path taken to arrive at a result can be unexpectedly complicated. We would like, therefore, for services to (optionally) support some form of *mathematical explanation ontology* which would describe the choices made inside the service and list any other services invoked. For example if our integration service was asked to compute an integral over the region $[0, \infty]$ it might split it into $[0, 1]$ and $[1, \infty]$ and then transform the second integrand so that it is also over the region $[0, 1]$. This transformation might be accomplished through the use of a computer algebra system. A weaker form of the mathematical explanation ontology would be an audit trail which would simply describe the sequence of services invoked, the data they received and the data they returned. Services which did not support the former could still support the latter. Of course an important practical application of "explanation" is actually debugging, i.e. understanding what went wrong!

In many ways the concept of supporting a hierarchy of optional ontologies is similar to the concept of supporting an *interface* in Java or C^{++}. A first use of this technology might therefore be to redesign the internal structure of existing large systems to use this architecture, making integration of other components and later deployment of parts of the system as web services easier. We note that Microsoft has announced that it is taking precisely this approach with respect to its Office product.

The MONET (Mathematics On The Net)[6] project has just begun to investigate how to implement all these ideas in practice. As well as looking at web service description and invocation, the project is also investigating how to define ontologies to encapsulate the whole problem solving process and report errors, facilitate debugging etc.

5 Conclusions

We do not believe that mathematical services will completely displace conventional mathematical software, in the same way as we do not believe that the semantic web will completely replace the existing web. However we believe that the infrastructure of the semantic web is relatively lightweight, and may well scale down successfully to allow the same architecture to be used inside applications as between applications.

Although OpenMath was designed as a data exchange format, its most important rôle in web services would appear to be in defining ontologies for mathematics. While this does not mean that the language used for transmitting data between services has to be OpenMath, in practice we believe that services will use either OpenMath or MathML, with native formats (such as the Mathematica or Maple languages) being accepted in addition if appropriate.

Although it will take time for mathematical services to evolve and to be accepted by the scientific community, the availability of the computational grid and the need to distribute applications across it may provide a natural niche for this technology.

Acknowledgements

The authors gratefully acknowledge the support of the European Union via the OpenMath Project (ESPRIT 24969), the OpenMath Thematic Network (IST–2000–28719), and the MONET Project (IST–2001–34145).

References

1. R.M. Corless, J.H. Davenport, D.J. Jeffrey and S.M. Watt. "According to Abramowitz and Stegun" or arccoth needn't be uncouth. In *ACM SIGSAM Bulletin*, volume 34, number 2, pages 58-65, 2000.

2. David De Roure, Nicholas Jennings, Nigel Shadbolt and Mark Baker. *Research Agenda for the Semantic Grid: A Future e-Science Infrastructure* The UK National e-Science Centre, Technical report UKeS-2002-02, 2001.
3. A. H. Hearn. *The REDUCE User's Manual*, The RAND Corporation, 1987.
4. *IEEE standard for binary Floating-point arithmetic.* ANSI/IEEE Standard 754, 1985.
5. Richard D. Jenks and Robert S. Sutor, *Axiom — The Scientific Computation System*, Springer-Verlag, 1992.
6. *The MONET Project.* `http://www.nag.co.uk/Projects/MONET`.
7. *The NAG Fortran Library, Mark 20*, NAG Ltd, Oxford, 2001.
8. The OpenMath Consortium. *The OpenMath Standard*, February 2000. O. Caprotti, D. P. Carlisle and A. M. Cohen Eds.
9. World-Wide Web Consortium, *The Extensible Stylesheet Language (XSL) Version 1.0.* Recommendation, October 2001. `http://www.w3.org/TR/2001/REC-xsl-20011015/`.
10. World-Wide Web Consortium, *Mathematical Markup Language (MathML) 1.0 Specification.* Recommendation, April 1998. `http://www.w3.org/TR/1998/REC-MathML-19980407`.
11. World-Wide Web Consortium, *Semantic Web Activity.* `http://www.w3.org/2001/sw`
12. World-Wide Web Consortium, *Extensible Markup Language (XML) 1.0.* Recommendation, February 1998. `http://www.w3.org/TR/1998/REC-xml-19980210`.

TOWARDS A MATHEMATICAL SERVICES DESCRIPTION LANGUAGE

OLGA CAPROTTI
WOLFGANG SCHREINER
Research Institute for Symbolic Computation
Johannes Kepler University, A-4040 Linz, Austria
{ocaprott,schreine}@risc.uni-linz.ac.at

As a foundation for a framework for brokering mathematical services distributed among networked servers, we propose a language that describes the mathematical problems solved by the services. This mathematical service description language extends WSDL, the web service description language. Servers register their problem solving capabilities expressed in this language with a "semantic broker" to which clients submit corresponding task descriptions. The broker (possibly in cooperation with a deduction system) determines the suitable services and returns them to the client for invocation. This mechanism thus hides from the client the actual implementation of mathematical services and focuses on the semantical aspects.

1 Introduction

In the recent years, mathematical software systems, i.e., computer algebra systems like Maple, Mathematica, or GAP, automated theorem provers like Otter, PVS, or Theorema, or systems for numerical scientific computing like Matlab are slowly migrating from being standalone desktop tools to be used by a human to computational engines used as "mathematical services" by other programs. There are numerous examples in the literature of scientific software needing access to computer algebra methods, of computer algebra systems requiring advanced reasoning capabilities vice-versa of automated theorem provers exploiting computer algebra systems for their system solving facilities, see e.g. [1]. Some systems provide built-in communication protocols for interfacing to third-party software, whereas others have to be accessed via the standard input/standard output interface. Frequently however these approaches suffer from the inadequate specification of the functionality of a system. The result may take surprising forms, a challenge for the interpretation capabilities of the client software sometimes forced to mimic a human user.

Furthermore the Internet and the World Wide Web have further extended the range of possibilities by enabling networked access to mathematical services hosted in distant locations. Clients may use these services via remote access protocols [2,3,4]. The more networked mathematical services become available, the less it is desirable that each speaks its own language, therefore common mathematical communication protocols are required. Since more and more clients are programs rather than humans, mechanisms to convey some understanding on the meaning of the offered services need to be worked out.

The technologies for accessing mathematical services over the network focus mainly on the problem of portable data representation (OPENMATH [5], MathML [6], MathBus [7], IAMC [2], MathWeb [8], JAVAMath [3]). IAMC does provide a protocol for the exchange of meta-data about the names of services offered and about accepted data formats, but it does not yet explicitly address the problem how a client can learn about the service semantics. Only the MathWeb activities deal with the semantic level focusing on the integration of automated theorem provers.

Thus, while the basic communication protocols for mathematical data have emerged, there has been only little work on the higher protocol levels that make mathematical services actually usable for automated tools without human insight. In particular, additional protocol levels have to be designed in order to support programming of mathematical applications that employ mathematical services distributed over the Internet. The final goal of a mathematical service description language is to be able to write such programs from an abstract point of view, without knowing about the actually available implementations. In this paper we advocate a mathematical service description format that abstracts from the details of an implementation and focuses on the actual *problem* being solved. The paper begins with a section on some of the considerations that contributed to our design of a language for mathematical services. Section 3 explains how the Web Service Description Language can be extended by additional meta-data to become more useful for our purposes. The following section describes ongoing work on what we believe is a promising way of describing the implementation meta-data associated with a service. The paper ends by some final remarks.

2 Description of Mathematical Services

From a pragmatic point of view, the description of a mathematical service is intended to facilitate the discovery of a certain computational facility, either abstractly from a specific mathematical question or more informally by a name of an algorithm or via any other item that might identify a piece of software. Keeping this in mind, the overall design of the description language is structured in layers which proceed from an abstract view, namely the problem addressed by the mathematical service, all the way down to the hostname-port address of the service as depicted in Figure 1.

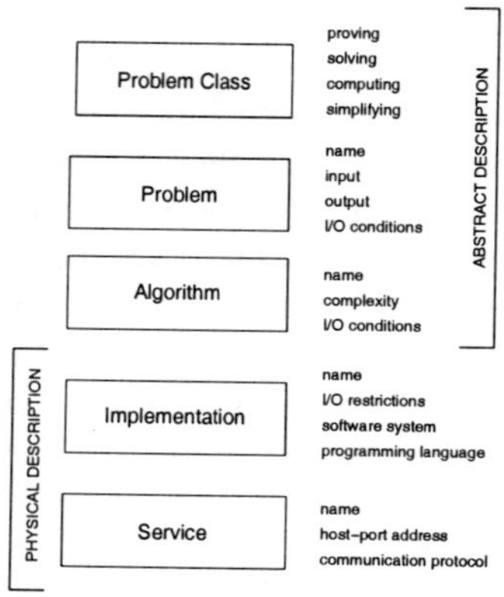

Figure 1. Mathematical Service Description Language Layers

2.1 Problem Class

We begin by attempting a classification/taxonomy of problem classes. This first coarse distinction is intended as a guideline on how to further specify a specific problem instance. The underlying idea is that a *computing* problem might have a different description from a *proving* problem. For instance, stating that a problem belongs to the class of *simplification* problems, amounts to choosing an equivalence relation $\equiv$ and a *reduction* relation since these are key components of the problem class definition.

A *problem class* description defines the structure of classes of problems (parametrized by a logical system L and a theory T of L) including for instance

- *Computing*, namely the problem of carrying out operations in a domain, in a theory T i.e., given a term f with a free variable x and a value a in a properly chosen domain, return the value of f with a substituted for x;

- *Simplifying*, namely the problem of obtaining simpler yet equivalent objects in a domain, i.e., given a term t and a set of simplification rules R, return a term $t' = t$ that satisfies some "simplicity" condition,

- *Solving*, i.e., given a set of formulas $F[\vec{x}]$ in a particular logic, find (some or all) substitutions $\vec{a}$ in a chosen domain that make $F[\vec{a}]$ true in a certain theory;

- *Proving*, i.e., given a closed formula F in a particular logic, determine whether the formula is true in a particular domain or not. In particular, for some logic, this is equivalent to returning a proof.

In the logic broker architecture [9], a uniform way of specifying abstractly a service employs the notion of consequence relation [10]. Because this notion of consequence can be used for characterizing several known logics (including Linear Logic and non-monotonic logics) and in general, for semantics-independent classification of standard connectives (via equations on consequence relations), it is enough for formally specifying services provided by reasoning systems which are the focus of the logic broker architecture. Our approach aims at describing not only reasoning services (hence it must accommodate a description of consequence relations) but also computational ones.

2.2 Problem

A *problem* is a concrete instance of a problem class. The specific format of the problem instance is defined by the problem class it belongs to, eventually with additional (optional) fields that enhance its formalization. The name of a problem description may be used by a task to refer to a particular problem in a direct way; this allows a broker to identify corresponding services easily.

Additionally, if we consider that mathematical services as intended in this paper are computational services provided by mathematical software, it is possible to formally describe the problem that is solved as inspired by formal methods, that is using of techniques from logic and discrete mathematics supported by mechanized proof methods.

The motivation for including a formal definition is one of usage. The formal definition of a problem description can be used by a broker (possibly in collaboration with a deduction system) to identify services that satisfy a description given in terms of the content of a problem. A specification of a problem expresses constraints on what is to be computed in a style à la Floyd-Hoare by describing input and output parameters $\vec{x}$ and $\vec{y}$, their domains, and eventually a precondition $p(\vec{x})$ and a postcondition $q(\vec{x}, \vec{y})$. Notice that in order to formally define a problem, we will require a knowledge base of mathematics to use as reference for the mostly used concepts. Our preferred choice in this regard involves OPENMATH.

For instance, a task requesting an integer GCD computation can be satisfied by the broker using a service advertising extended GCD computation. Although the names of the problems do not match, formal manipulation of the specification of the service allows to satisfy the request in an alternative way by reasoning on the preconditions and postconditions of the problem.

2.3 Algorithm

An *algorithm* is a solution strategy for a problem (possibly, by strengthening the precondition or weakening the postcondition, of a more special one). Apart from referring to the solved problem, an algorithm description may contain information about its asymptotic time complexity and memory requirements, i.e., about its general dynamic behavior. Potentially, also a description of the algorithm in a formal algorithmic language may be added for reasoning about additional properties. Apparently, there may be multiple algorithms for a given problem.

This layer captures the abstract computational essence of a mathematical service, which often is the interface of an algorithm. Requiring a certain algorithm for solving a problem is the natural mathematical activity it aims to support.

2.4 Implementation

An *implementation* is a particular realization of an algorithm as executable software, possibly with additional constraints on the input and additional possibilities for the output (error cases, implementation restrictions, etc). Apart from referring to the algorithm applied (or at least to the problem solved), an implementation description may contain additional information about the underlying software system, about its efficiency (relative to a fixed architecture and a fixed performance factor). Sometimes it is interesting to know the author of a certain implementation, and the date or version. This kind of meta-data is the object of this description layer.

2.5 Service

A *service* is a set of implementations running on a particular machine; its description contains information about the underlying machine (contact address, computational performance, communication performance, memory equipment, ...), about the actual communication protocol for invoking and controlling the service (including the choice of synchronous or asynchronous communication, the possibilities for querying the status of a computation, the potential of aborting a computation, etc), for querying the status of the machine, and about the way in which service requests are handled (one at a time, in parallel up to some limit, etc). This description is mostly achieved using WSDL [11].

In summary, every description has a unique resource identifier (URI) and may embed a formal definition. Less sophisticated applications may associate a fixed meaning with a URI; this is analogous to the standard set of OPEN-MATH CDs which informally agree on the meaning of every symbol. More advanced applications may refer to the formal definition to learn about unknown entities; this is analogous to KQML-based approaches for describing mathematical services. While the first strategy makes the description framework quickly usable, the formal ultimately unleashes its power; combinations of both strategies (formal descriptions relying on informally defined entities) are also possible.

3 Service Layer: Exploiting WSDL

In order to have a running service it would suffice to limit the description to the service layer, and in fact most available client-server architectures work at this level and hide the location details and the objects' encoding to the client. This is the level of WSDL, the Web Services Description Language (WSDL) [11], currently only a W3C note, but already supported by several tools and programming interfaces for quickly developing services based on it.

A WSDL document is a collection of definitions aiming at specifying *services* as collections of *ports*, i.e. network endpoints. Several elements are used in the description of services: **<Types>** is a container for data types given according to a certain type system. In particular, WSDL supports the XML Schemas specification as preferred language for type definition but it also allows using other languages via extensibility. **<Message>** is an abstract, typed definition of the data being communicated in the input and output. **<Operation>** is an abstract description of an action supported by the service in terms of input and output. **<Port Type>** is an abstract set of operations supported by one or more endpoints. Here below is the abstract portion of a WSDL specification consisting of a service of type **Integrator** providing two operations. For reason of space, we have abbreviated some of the details.

```
<?xml version="1.0" encoding="UTF-8"?>
<wsdl:definitions
          targetNamespace="http://.../services/SymbolicIntegrator"
          xmlns="http://schemas.xmlsoap.org/wsdl/"
          xmlns:SOAP-ENC="http://schemas.xmlsoap.org/soap/encoding/"
          xmlns:impl="http://.../services/SymbolicIntegrator-impl"
          xmlns:intf="http://.../services/SymbolicIntegrator"
          xmlns:wsdl="http://schemas.xmlsoap.org/wsdl/"
          xmlns:wsdlsoap="http://schemas.xmlsoap.org/wsdl/soap/"
          xmlns:xsd="http://www.w3.org/2001/XMLSchema">

  <import namespace="some-URI" location="some-URI"/>*
  <wsdl:documentation .... />?

  <wsdl:types>?
```

```
    <wsdl:documentation .... />? <-- extensibility element -->*
    <xsd:schema .... />*
 </wsdl:types>

 <wsdl:message name="DefIntegralResponse" .../>
      <wsdl:documentation .... />?
      <wsdl:part name="return" type="xsd:string"/>
      <wsdl:part name="nmtoken" element="qname"? type="qname"?/>*
 </wsdl:message>

 <wsdl:message name="DefIntegralRequest" .../>
 <wsdl:message name="nmtoken" .../>*

 <wsdl:portType name="Integrator">
   <wsdl:operation name="DefIntegral" parameterOrder="in0">
     <wsdl:input message="intf:DefIntegralRequest"/>
     <wsdl:output message="intf:DefIntegralResponse"/>
   </wsdl:operation>
   <wsdl:operation name="IndefIntegral" parameterOrder="in0" .../>
 </wsdl:portType>
```

The concrete set of definitions is given using two more elements. `<Binding>` is the association of a concrete protocol and data format specification to a particular abstract port type. For instance, the information that a service sends OpenMath data using the XML encoding of OpenMath is given in the binding. Finally, `<Service>` is a collection of related endpoints, (`<Port>`s), each defined as a combination of a binding and a network address. In order to know the functionalities of the service `IntegratorService` we have to check its `portType`, referenced by its binding. This describes the abstract syntactical interface in terms of operations with their input and output types.

```
 <wsdl:binding name="SymbolicIntegratorSoapBinding" type="intf:Integrator" .../>
       <wsdl:documentation .... />? <-- extensibility element -->*
       <wsdl:operation name="nmtoken">*
          <wsdl:documentation .... />? <-- extensibility element -->*
          <wsdl:input|output>?
              <wsdl:documentation .... />? <-- extensibility element -->
          </wsdl:input|output>
          <wsdl:fault name="nmtoken">*
              <wsdl:documentation .... />? <-- extensibility element -->*
          </wsdl:fault>
          </wsdl:operation>
   </wsdl:binding>

 <wsdl:service name="IntegratorService">
   <wsdl:port binding="intf:SymbolicIntegratorSoapBinding" name="SymbolicIntegrator">
     <wsdlsoap:address location="http://.../services/SymbolicIntegrator"/>
   </wsdl:port>
 </wsdl:service>

</wsdl:definitions>
```

Our experiments with WSDL[a] have shown that although it is very convenient to setup a service using WSDL, and to generate automatically clients for a service described by a WSDL specification, in order to reason about the service, more information is needed, for instance it would be desirable to know the abstract type of the arguments (that is, beyond the primitive types provided by XML schemas). Meta-data concerning the hardware type, its performance and details on the computational package used in the underlying implementation of the service is often useful for choosing a certain mathematical service instead of another one.

A key design feature of WSDL are the extensibility elements, used to include details on the transport protocol adopted by the service. Extensibility allows to modify the rule for the `<Binding>` in order to support a number of concrete communication protocols like SOAP, HTTP_GET_POST, SMTP_MIME. The extensibility fields of WSDL are the natural place where to convey additional meta-data of the kind we are interested in. We have decided to exploit extensibility within the **port** element in order to include in the WSDL meta-data concerning the hardware of the physical machine connected to the endpoint. This meta-data describes e.g. the processor type, the amount of memory, and a measure of performance using the Standard Performance Evaluation Corporation reference [b]. We have produced an XSD schema to express software and hardware characteristics to be included in a WSDL.

```
<wsdl:port binding="intf:SymbolicintegratorSoapBinding" name="SymbolicIntegrator">
  <wsdlsoap:address location="http://.../services/SymbolicIntegrator"/>
  <sdl:machine xmlns="http://perseus.risc.uni-linz.ac.at:8080/msdl/service">
     <processor>Asus A7V Athlon</processor>
     <memory>1.2GHz</memory>
     <performance SPECtype="SPEC CINT2000" metric="speed">
        <CPU value="1"/> <base value="409"/> <peak value="458"/>
     </performance>
  </sdl:machine>
</wsdl:port>
```

When enhanced by this kind of meta-data, it is possible to use the WSDL specification to answer queries such as "is the service **SymbolicIntegrator** running on a machine with performance SPEC CINT2000 at least equal to or better than peak value 450?". To be able to answer queries such as "is there an implementation of the service **SymbolicIntegrator** using the package Mathematica?" we need to add an implementation description layer on top of the service layer.

[a]See `http://perseus.risc.uni-linz.ac.at:8080/openmath/index.html`
[b]`http://www.specbench.org/`

4 Implementation Layer

The next layer of description of a mathematical service collects the meta-data specific to the implementation. In a way similar to the service description, which has an abstract part and a concrete part, an implementation can be viewed as a collection of concrete realizations of an algorithm (with a problem specification) by a specific service. Each realization binds an algorithm or a problem to a service description. At the time of writing, we are drafting schemas for a mathematical implementation and problem definition language that will allow to specify implementation and problem descriptions such as the one here below.

```
<midl:definitions  xmlns:wsdl="http://schemas.xmlsoap.org/wsdl/"
                   xmlns:midl="http://perseus.risc.uni-linz.ac.at:8080/msdl/midl">
                   xmlns:mpdl="http://perseus.risc.uni-linz.ac.at:8080/msdl/mpdl" ...>

  <import namespace="some-URI" location="some-URI"/>*
  <wsdl:documentation .... /> ?

  <midl:algorithm name="BubbleSort" >
      <mpdl:problem>
         <input name="nmtoken" .../>*
         <inputcondition .../>*
         <output name="nmtoken" .../>*
         <outputcondition .../>*
      </mpdl:problem>
  </midl:algorithm>

  <midl:servicebinding  name="BSortRiscBinding" type="BubbleSort">
    <wsdl:documentation .... />?  <-- extensibility element -->*

    <wsdlbinding:service portType="sort:BubbleSortPortType"
                         xmlns:sort=".../services/SortingService">
      <!-- address the exact operation for BubbleSort -->
      <invocation name="BubbleSortExec">*
          <wsdlbinding:operation name="sort:BSortOperation" .../>
      </invocation>
      <!-- concrete messages are related to inputs and outputs  -->
      <parameter name="nmtoken">*
          <wsdlbinding:message name="sort:some_messageName" .../>
      </parameter>
    </wsdlbinding:service>
  </midl:service>

  <midl:implementation name="BSort">
    <midl:realization name="riscBSort" servicebinding="BSortRiscBinding">*
      <midl:serviceaddress name=".../services/SortingService?wsdl" />
      <midl:bibliography xmlns:dc="dublin_core" .../>
      <midl:efficiencyfactor  .../>
      <midl:software .../>
    </midl:realization>
   </midl:implementation>

</midl:definitions>
```

As WSDL teaches, a specific XML language defined by a collection of XML schemata [12] is a suitable language for describing the functionality of mathematical services. Each problem class can be associated with a specific namespace defined by an XML schema, so for instance there is a schema for a computing problem. This schema acts as a specification of the basic skeleton structure of the problem instance. Problem descriptions will also be defined by XML schemas, as to associate a namespace to each specific problem, for instance a real root isolation problem could be described using the schema identified by the URI `http://www.mathbroker.net/computing/algebra/realrootisolation`. Schemas for problem instances include the schema of the problem class they belong to so that the class requirements on the descriptions are satisfied.

5 Conclusion and Work Plan

We have described the design of a language for the description of mathematical services in a distributed networked environment. The language is structured in layers, each one capturing a particular aspect of a mathematical problem solution, implementation and service details. This structured approach has the advantage of allowing a modular development of the language by starting from the bottom layer, which is in part covered by existing standard languages, and then adding onto it.

The final goal of the description language is to support the electronic processing of mathematical services in the scenario where the Net is a global problem solving framework where each computation can be accessed/found by a multitude of entry points.

Our current work is the exact definition of the XML schemas that will correspond to the layers, the identification of some case studies that should serve as a guideline, and the study of available on-line knowledge bases of mathematics to be used as reference. We have begun with analyzing available web services standards such as WSDL and related ones. Our experiments have convinced us that WSDL alone does not suffice if we want to eventually be able to reason about mathematical services, to streamline them and to do mathematics on-line.

References

1. Alessandro Armando and Tudor Jebelean, editors. *Special Issue on Calculemus-99*, volume 32 of *Journal of Symbolic Computation*. Academic Press, September 2001.

2. Paul Wang. Design and Protocol for Internet Accessible Mathematical Computation. In Sam Dooley, editor, *ISSAC'99 International Symposium on Symbolic and Algebraic Computation*, pages 291–298, Vancouver, Canada, July 28–31, 1999. ACM, New York.
3. JavaMath — Internet Accessible Mathematical Services, February 2001. `http://javamath.sourceforge.net`.
4. O. Caprotti, A. M. Cohen, H. Cuypers, M. Riem, and H. Sterk. Using OpenMath Servers for Distributing Mathematical Computations. In Wei Chi Yang, Sung-Chi Chu, and Jen-Chung Chuan, editors, *ATCM 2000: Proceedings of the Fifth Asian Technology Conference in Mathematics*, pages 325–336, Chiang-Mai, Thailand, December 17–21 2000. ATCM Inc., USA.
5. Stephane Dalmas, Marc Gaetano, and Stephen Watt. An OpenMath 1.0 Implementation. In Wolfgang Küchlin, editor, *ISSAC'97 International Symposium on Symbolic and Algebraic Computation*, pages 241–248, Maui, Hawaii, July 21–23, 1997. ACM Press, New York.
6. Patrick Ion and Robert Miner. Mathematical Markup Language (MathML) Version 2.0. W3C proposed recommendation, World Wide Web Consortium, January 2001. `http://www.w3.org/TR/2001/PR-MathML2-20010108/`.
7. Richard Zippel. The MathBus. In *Workshop on Internet Accessible Mathematical Computation at ISSAC'99*, Vancouver, Canada, July 28–31, 1999. `http://icm.mcs.kent.edu/research/iamc99proceedings.html`.
8. Andreas Franke and Michael Kohlhase. System Description: MathWeb, an Agent-Based Communication Layer for Distributed Automated Theorem Proving. In Harald Ganzinger, editor, *16th Conference on Automated Deduction*, number 1632 in Lecture Notes in Artificial Intelligence, pages 217–221. Springer, 1999.
9. A. Armando and D. Zini. Interfacing Computer Algebra and Deduction Systems via the Logic Broker Architecture. In *Proceedings of the XXth Calculemus symphosium*, St. Andrews, Scotland, 6-7 August 2000.
10. Arnon Avron. Simple consequence relations. *Information and Computation*, 92(1):105–140, 1991.
11. E. Christensen, F. Curbera, G. Meredith, and S. Weerawarana. Web Services Description Language (WSDL) 1.1. W3C Note, 15 March 2001.
12. Paul V. Biron and Ashok Malhotra. XML Schema Part 2: Datatypes. W3C candidate recommendation, World Wide Web Consortium, October 2000. `http://www.w3.org/TR/2000/CR-XMLschema-2-20001024`.

AN ALGORITHM TO COMPUTE THE MINIMAL TELESCOPERS FOR RATIONAL FUNCTIONS (DIFFERENTIAL – INTEGRAL CASE)

K. O. GEDDES
Symbolic Computation Group
University of Waterloo
Waterloo, N2L 3G1, Canada
E-mail: kogeddes@scg.math.uwaterloo.ca

H. Q. LE
Symbolic Computation Group
University of Waterloo
Waterloo, N2L 3G1, Canada
E-mail: hqle@scg.math.uwaterloo.ca

We present an algorithm to compute the minimal telescopers for rational functions for the differential–integral case. We also describe a Maple implementation of the algorithm and show timing comparisons between this algorithm and other related algorithms.

1 Introduction

Let K be a field of characteristic zero, D_x and D_y be the differential operators w.r.t. x and y, respectively. For a given *hyperexponential function* $F(x,y)$, i.e., the logarithmic derivatives D_xF/F and D_yF/F are rational functions of x and y, the differential–integral analogue of Zeilberger's algorithm [1], named hereafter as $\mathcal{Z}$, constructs a non-trivial linear differential operator L of order ρ with polynomial coefficients of the form

$$L = a_\rho(x)D_x^\rho + \cdots + a_0(x)D_x^0,\ a_\rho(x) \neq 0,\ a_i(x) \in K[x],\ 0 \leq i \leq \rho, \quad (1)$$

and a hyperexponential function $G(x,y)$ such that

$$LF(x,y) = \sum_{i=0}^{\rho} a_i(x)D_x^i F(x,y) = D_y G(x,y). \quad (2)$$

The computed *telescoper* L is of minimal order. It is unique up to a left-hand factor $P(x) \in K[x]$, and is named the *minimal telescoper*. We call a Z-pair (L,G) where L is the minimal telescoper the *minimal Z-pair*.

The construction of a Z-pair (L,G) is the central part of the unified theory of differentiation under the integral sign. It is also useful in proving theorems in the theory of special functions [1]. It is shown in [1] that $\mathcal{Z}$ always succeeds in constructing the minimal Z-pair for a given hyperexponential function $F(x, y)$. Let $F(x, y)$ be a rational function of x and y. Since the class of rational functions is a proper subset of the class of hyperexponential functions, there always exists a Z-pair for $F(x, y)$.

In this paper we present an algorithm to construct the minimal Z-pair when the input $F(x, y)$ is a rational function of x and y. The algorithm is based on a special form of representation of $F(x, y)$ (this helps to break down the problem to be solved into smaller ones), along with an efficient construction of the minimal Z-pair for each member of this representation. Namely, we show that when one applies $\mathcal{Z}$ to each member of this representation, many expensive steps in $\mathcal{Z}$ can be safely ignored. The minimal Z-pair for $F(x, y)$ is then obtained by using Least Common Left Multiple (LCLM) computation. We also describe a Maple implementation of the algorithm and show timing comparisons between this algorithm, the original $\mathcal{Z}$ [1], and a modified version of $\mathcal{Z}$ [4].

The algorithm in this paper is an improvement of the modified version of $\mathcal{Z}$ as presented in [4].

2 Gosper's and Zeilberger's Algorithms

In this section, we summarize Gosper's and Zeilberger's algorithms. This is needed in the analysis of our proposed algorithm.

2.1 Gosper's Algorithm

For a given hyperexponential function $F(y)$, Gosper's algorithm, named hereafter as $\mathcal{G}$, provides a decision procedure to determine if there exists a hyperexponential function $S(y)$ such that

$$D_y S(y) = F(y). \tag{3}$$

It also simultaneously constructs $S(y)$ provided that such a function exists. The algorithm consists of two main steps:

1. Construct $p(y), q(y), r(y) \in K[y]$ such that

$$\frac{D_y(F(y)/p(y))}{F(y)/p(y)} = \frac{q(y)}{r(y)}, \tag{4}$$

   where $q(y)$ and $r(y)$ satisfy the condition

$$\gcd(r(y), q(y) - j\, D_y r(y)) = 1 \text{ for every non-negative integer } j. \tag{5}$$

Note that if there exists an $j \in \mathbb{N}$ such that

$$\gcd(r(y), q(y) - j\, D_y r(y)) = g(y),\ \deg\, g(y) \geq 1, \tag{6}$$

then the value of j can be computed as the non-negative integer roots of

$$\text{res}_y(r(y), q(y) - j\, D_y r(y)). \tag{7}$$

2. Find a polynomial solution $f(y) \in K[y]$ of the first-order linear differential equation with polynomial coefficients

$$p(y) = (q(y) + D_y r(y)) f(y) + r(y) D_y f(y). \tag{8}$$

A polynomial solution $f(y)$ of (8), if it exists, can be obtained by using the method of undetermined coefficients. One first needs to compute an upper bound k for the degree of the polynomial $f(y)$. This can be achieved by considering the following two cases:

Case 1. $\deg\, r(y) \leq \deg\, (q(y) + D_y r(y)) = l$. It follows from (8) that

$$k = \deg\, f(y) = \deg\, p(y) - l.$$

Case 2. $\deg\, (q(y) + D_y r(y)) < \deg\, r(y) = l$. Set

$$f(y) = c_k\, y^k + c_{k-1}\, y^{k-1} + O\left(y^{k-2}\right).$$

It also follows from (8) that $p(y) = L(k)\, c_k\, y^{k+l-1} + O\left(y^{k+l-2}\right)$, where $L(k)$ is linear in k and free of c_k. Let k_0 be the root of $L(k)$. Then

$$k = \begin{cases} \max\{k_0, \deg\, p(y) - l + 1\}, & \text{if } k_0 \text{ is an integer,} \\ \deg\, p(y) - l + 1, & \text{otherwise.} \end{cases}$$

2.2 Zeilberger's Algorithm

$\mathcal{Z}$ uses an *item-by-item* examination on the order ρ of the operator L in (1). It starts with the value of 0 for ρ and increases ρ until it is successful in finding a Z-pair (L, G) for F. For a given guessed order ρ of the operator L where the coefficients $a_i(x), 0 \leq i \leq \rho$ are unknown, $\mathcal{Z}$ applies L to $F(x, y)$. This results in another hyperexponential function $H(x, y)$. The function $H(x, y)$, considered as a hyperexponential function of y, is then used as input to $\mathcal{G}$. Since $\mathcal{Z}$ is guaranteed to be successful, there exist a value of $\rho \in \mathbb{N}$ and a hyperexponential function $G(x, y)$ such that $H(x, y) = D_y\, G(x, y)$, i.e., the system of linear equations, constructed as a consequence of solving (8) via the method of undetermined coefficients, is consistent and we get the desired minimal Z-pair.

3 On the Minimal Telescoper of a Sum of Rational Functions

The following theorem is the differential analogue of Theorem 1 in [5] (which is for the difference case).

Theorem 1 *Let $L_1, \ldots, L_s$ be the minimal telescopers for the rational functions $F_1, \ldots, F_s$, respectively. Then*

(i) $\mathrm{lclm}(L_1, \ldots, L_s)$ *is a telescoper for $F = F_1 + \cdots + F_s$.*

(ii) *For any telescoper L^* for F, if L^* is also a telescoper for every F_i, $i = 1, \ldots, s$, then* $\mathrm{lclm}(L_1, \ldots, L_s)$ *is the minimal telescoper for F.*

Set $L = \mathrm{lclm}(L_1, \ldots, L_s)$, i.e., there exist $L'_1, \ldots, L'_s \in K(x)[D_x]$ such that $L = L'_1 \circ L_1 = \cdots = L'_s \circ L_s$. Then $(L, L'_1 G_1 + \cdots + L'_s G_s)$ is the minimal Z-pair for F.

4 The Decomposition Problem and the Existence of the Minimal Z-pair

For a given rational function $F(y) \in K(y)$, an algorithm to solve the decomposition problem (also known as a reduction algorithm) for indefinite integrals of rational functions [2,3,6] constructs two rational functions $S(y) \in K(y)$, $T(y) = a(y)/b(y)$, $a(y), b(y) \in K[y]$ such that

$$F(y) = D_y\, S(y) + T(y), \tag{9}$$

and the denominator $b(y)$ of $T(y)$ is of minimal degree. Moreover, $T(y)$ is *proper*, i.e., $\deg a(y) < \deg b(y)$, and $b(y)$ is monic and square-free. Note that the algorithm allows one to write $T(y)$ in the form $T(y) = \sum_i \frac{u_i(y)}{v_i(y)}$ such that

$$\deg u_i < \deg v_i,\ \ v_i(y) \text{ is square-free },\ \gcd(v_i, v_j) = 1 \text{ for } i \neq j. \tag{10}$$

If $T(y)$ in (9) vanishes, then $F(y)$ is said to be *rational integrable.*

For $F(x, y) \in K(x, y)$, denote $F(x; y)$ as an element of $K(x)(y)$ (sometimes, when fitting, also as an element of the ring $K(x)[y]$). By applying to $F(x; y)$ any of the algorithms to solve the decomposition problem, we can represent $F(x; y)$ in the form

$$F(x; y) = D_y\, S(x; y) + T(x; y),\ \ S,\, T \in K(x)(y). \tag{11}$$

The following proposition can be easily derived from (11).

Proposition 1 *Let (L, G) be the minimal Z-pair for $T(x, y)$. Then the minimal Z-pair for $F(x, y)$ is $(L, L\,S(x, y) + G(x, y))$.*

Consider the rational function $T(x;y)$ in (11) which is written in the form

$$T(x;y) = \sum_i \frac{u_i(x;y)}{v_i(x;y)}, \quad u_i(x;y),\ v_i(x;y) \in K(x)[y] \qquad (12)$$

such that the three conditions in (10) hold.

Theorem 2 *If L is a telescoper for $T(x,y)$, then L is also a telescoper for every $T_i = u_i(x,y)/v_i(x,y)$.*

Proof: Suppose that $L \in K[x, D_x]$ is not a telescoper for $T_i = u_i(x,y)/v_i(x,y)$. This implies LT_i is not rational integrable w.r.t. y. It can be shown (see [4], for instance) that

$$LT_i = a_\rho(x)\sigma_\rho(x)\frac{y^{\rho d+s} + O\left(y^{\rho d+s-1}\right)}{v_i^{\rho+1}(x,y)} + \cdots + a_0(x)\frac{u_i(x,y)}{v_i(x,y)}, \quad \sigma_\rho \in K(x). \qquad (13)$$

Therefore, the application of a reduction algorithm w.r.t. y to LT_i results in

$$LT_i = L\frac{u_i}{v_i} = D_y R_i + \frac{u_i'}{v_i}, \quad R_i \in K(x,y),\ u_i' \in K(x)[y],\ \deg u_i' < \deg v_i.$$

Since LT is rational integrable w.r.t. y, $L(T - T_i)$ is *not* rational integrable w.r.t. y, i.e.,

$$\frac{u_i'}{v_i} + \frac{u'}{\prod_{j\neq i} v_j} = 0, \quad u' \in K(x)[y].$$

Consequently, $v_i | (u_i' \prod_{j\neq i} v_j)$. Since $\gcd(v_i, \prod_{j\neq i} v_j) = 1$, $v_i | u_i'$. This is impossible since $\deg u_i' < \deg v_i$. ∎

5 The Basic Case

In this section, we analyze $\mathcal{Z}$ when it is applied to each simple fraction in (12); namely to

$$T_i = \frac{u_i}{v_i}, \quad u_i,\ v_i \in K(x)[y],\ v_i \text{ is square-free, } \deg u_i < \deg v_i, \qquad (14)$$

and show that by applying $\mathcal{Z}$ to T_i instead of to the input rational function F, one can safely ignore a number of expensive steps in the algorithm. It follows from the description of $\mathcal{Z}$ in subsection 2.2 that $\mathcal{Z}$ simply does some initial set-up and calls $\mathcal{G}$. Hence, our focus will be on the two main steps of $\mathcal{G}$.

The following lemma is proven in [4].

Lemma 1 *In step 1 of $\mathcal{G}$, the* gcd *computation (6), the resultant computation and the computation of non-negative integer roots of the polynomial resultant (7) are redundant, and hence can be safely ignored.*

Set deg $u_i = s$, deg $v_i = d$. Let ρ be the guessed order of the minimal telescoper L.

Lemma 2 *In step 2 of $\mathcal{G}$, the computed upper bound k for the degree of the polynomial solution $f(x;y) \in K(x)[y]$ in (8) is $k = \rho\, d$.*

Proof: It follows from (8) that

$$\begin{aligned} p(x;y) &= (q(x;y) + D_y r(x;y)) f(x;y) + r(x;y) D_y f(x;y) \\ &= -\rho\, D_y v_i(x;y)\, f(x;y) + v_i(x;y) D_y f(x;y). \end{aligned} \tag{15}$$

We now look into the two cases that help determine an upper bound k as described in subsection 2.1. Since

$$\begin{aligned} &\deg\, r(x;y) = \deg\, v_i(x;y) = d, \text{ and} \\ &\deg\,(q(x;y) + D_y r(x;y)) = \deg\,(-\rho\, D_y v_i(x;y)) = d-1, \end{aligned} \tag{16}$$

the condition for Case 1 is not satisfied while the one for Case 2 is. Set

$$f(x;y) = c_k\, y^k + \cdots + c_1\, y + c_0,\ \ v_i(x;y) = b_d\, y^d + \cdots + b_1\, y + b_0,\ \ c_i,\, b_j \in K(x).$$

Then $p(x;y) = (-\rho\, d + k)\, b_d\, c_k\, y^{k+d-1} + O\left(y^{k+d-2}\right)$. Therefore, $L(k) = (-\rho\, d + k) b_d$, and

$$k_0 = \rho\, d \in \mathbb{N}. \tag{17}$$

On the other hand, it follows from (13) that

$$\deg\, p(x;y) - \deg\, r(x;y) + 1 \leq \rho\, d + s - d + 1. \tag{18}$$

Since $s < d$, a comparison between (17) and (18) results in deg $p(x;y) -$ deg $r(x;y) + 1 \leq k_0$. ∎

The following algorithm, called *ZpairT*$_i$, computes the minimal *Z*-pair for $T_i \in K(x,y)$ in (14).

algorithm *ZpairT*$_i$;
input: $T_i = u_i(x,y)/v_i(x,y)$ in (14);
output: the minimal *Z*-pair (L_i, G_i) for T_i;
  $\rho := 1$;
  do
    $L := \sum_{i=0}^{\rho} a_i(x) D_x^i$;
    $R(x,y) := L T_i$;
    $p(x,y) :=$ numerator$(R(x,y))$;
    $r(x,y) := v_i(x,y)$; $q(x,y) := -(\rho+1)\, D_y\, v_i(x,y)$;
    use the method of undetermined coefficients to find a
      non-zero polynomial solution $f(x;y)$ of
      $p(x;y) = -\,\rho\, D_y\, v_i(x;y)\, f(x;y) + v_i(x;y)\, D_y\, f(x;y)$
      of degree (ρ deg$_y$ $v_i(x;y)$) or less;

if there exists a polynomial solution **then**
  return $(L, r(x,y)f(x,y)T_i(x,y)/p(x,y))$;
else
  $\rho := \rho + 1$;
fi;
od;

The correctness of the algorithm follows from Lemmas 1 and 2. Note that since the input rational function T_i is not rational integrable w.r.t. y, the value of the guessed order ρ starts at 1, instead of at 0.

6 The General Approach and an Algorithm

In [4], we proposed a slightly modified version of $\mathcal{Z}$ for rational functions. For a given $F(x,y) \in K(x)(y)$, one first applies a reduction algorithm to construct $S, T \in K(x)(y)$ such that relation (11) holds, and then applies $\mathcal{Z}$ to T. It is shown that the resultant computation and non-negative integer root finding of the polynomial resultant (7) can be safely ignored. The minimal Z-pair for $F(x,y)$ is then constructed using Proposition 1. In this paper, we further decompose T and compute the minimal telescopers for each simple fraction in (12). In addition to being able to avoid the expensive computations in step 1 of $\mathcal{G}$, we show in Section 5 that one can also determine an upper bound k for the polynomial solution $f(x;y)$ in (15) very quickly. By decomposing T (without bearing any additional cost), we also break the problem to be solved into smaller problems. This is useful since $\mathcal{Z}$ is most expensive when the value of the guessed order ρ of L in (1) is high. Theorems 1 and 2 guarantee that the lclm of the minimal telescopers for these simple fractions is indeed the minimal telescoper for T. Finally, one applies Proposition 1 to construct the minimal Z-pair for $F(x,y)$. Note that the algorithm proposed in this paper requires some “extra” lclm computation.

We also adopt the heuristic as proposed in [4] which is used to determine whether one should first apply a reduction algorithm to the input $F(x,y) \in K(x;y)$. Let T in (11) be written as $T = u/v$, $u, v \in K(x)[y]$. The heuristic takes into account the fact that the numerator u can have very large coefficients in $K(x)$. This case is not unusual when one applies a reduction algorithm to $F(x;y)$. Another important factor is the structure of $F(x,y)$. In step 1 of $\mathcal{G}$, we first compute $D_yF(x,y)/F(x,y)$ in order to construct the polynomials $p(x;y), q(x;y), r(x;y) \in K(x)[y]$ in (4). Hence, we need to pay attention to the case when there are common factors between the numerator and the denominator of $D_y F(x,y)/F(x,y)$, and they cancel out nicely. Let

$$H(x,y) = \frac{\sigma_1(x,y)}{\sigma_2(x,y)}, \ \sigma_1, \sigma_2 \in K(x)[y], \ \gcd(\sigma_1(x,y), \sigma_2(x,y)) = 1. \qquad (19)$$

Define $deg_y\, H(x,y) = deg_y\, \sigma_1(x,y) + deg_y\, \sigma_2(x,y)$. Set

$$d_1 = \deg_y \frac{D_y\, T(x,y)}{T(x,y)}, \ d_2 = \deg_y \frac{D_y\, F(x,y)}{F(x,y)}.$$

If $d_1 \leq d_2$, then first apply a reduction algorithm; otherwise, invoke the original $\mathcal{Z}$ directly. A justification for this simple heuristic is given in [4].

We conclude this section with a description of the algorithm $Modified\mathcal{Z}$.

algorithm $Modified\mathcal{Z}$;
input: a rational function $F(x,y) \in K(x,y)$;
output: the minimal *Z*-pair (L,G) for F;
  apply a reduction algorithm w.r.t. y to construct
    $S(x,y)$, $T(x,y) \in K(x,y)$ in (11). Note that $T(x,y)$ can be written as
    $T(x,y) = \sum_{i=1}^{m} T_i(x,y)$ where each T_i is of the form (14);
  if $T(x,y) = 0$ then
    return $(1, S(x,y))$;
  fi;
  for $i = 1, 2, \ldots, m$ **do**
    $(L_i, G_i) := ZpairT_i(T_i(x,y))$;
  od;
  apply *lclm* computation to construct $L'_1, \ldots, L'_m \in K(x)[D_x]$ such that
    $L = \mathrm{lclm}(L_1, \ldots, L_m) = L'_1 \circ L_1 = \cdots = L'_m \circ L_m$;
  return $(L, L'_1 G_1 + \cdots + L'_m G_m)$.

7 Implementation

The algorithm is implemented and incorporated into the Maple module `dZeilberger` [4]. The functionalities include

```
> eval(dZeilberger);
```

module dZeilberger ()
export IsHyperexponential, HermiteReduce, Gosper, Zeilberger;
option package;
description "Implementation of Gosper's algorithm and Zeilberger's algorithm for the differential case";
end module

The exported variables indicate the functions that are available. The main function of the package is **Zeilberger** which has the calling sequence

$$\text{Zeilberger}(F, x, y, D_x); \qquad (20)$$

where F is a hyperexponential function of x and y, D_x is the differential operator w.r.t. x. The output is a list of two elements $[L, G]$ specifying the minimal Z-pair (L, G) for F, i.e., $LF = D_y G$.

For the case when the input F is a rational function of x and y, there also exist three variants of **Zeilberger** with the calling sequences

$$\text{Zeilberger}(F, x, y, D_x, 'reduction' = false); \qquad (21)$$

$$\text{Zeilberger}(F, x, y, D_x, 'reduction' = true, 'decomp' = false); \qquad (22)$$

$$\text{Zeilberger}(F, x, y, D_x, 'reduction' = true, 'decomp' = true); \qquad (23)$$

These variants provide users with the flexibility to invoke different versions of $\mathcal{Z}$. The calling sequence (21) is for the original $\mathcal{Z}$ (initialized as O) where no reduction is done to $F(x, y)$. For the next two (22) and (23), which are initialized as M_1 and M_2, respectively, a reduction algorithm is first applied to $F(x; y)$ to obtain $S(x; y), T(x; y)$ in (11). If (22) is used, then $\mathcal{Z}$ is applied directly to $T(x, y)$; otherwise, $T(x, y)$ is further decomposed and $\mathcal{Z}$ is applied to each simple fraction of $T(x, y)$. Note that when (20) is used, the heuristic (H) is invoked to decide whether a reduction algorithm should be applied first to the input rational function. If the answer is positive, then (23) is used.

Various experiments have been done to compare the time (in seconds) and space (in kilobytes) requirements for different versions of $\mathcal{Z}$. We now show a few examples of our experimentations[a].

Example 1 Consider the rational function

$$F(x, y) = \frac{6\,y^2 + 2\,x^2}{(2\,x^2 - 2\,x^2 y - 2\,y^3)^2} + \frac{1}{(2\,y^3 - 2\,xy - 2\,x^3)^2} + \frac{1}{x^2 y - xy - x^2}.$$

Compute the minimal Z-pair for $F(x, y)$ using H, O, M_1, and M_2, and record the time and space requirements. The figures we obtained are as follows:

$$\begin{aligned} H &: \text{timing} = \ \ 1.68, \text{memory} = \ \ 7,729 \\ O &: \text{timing} = 18.85, \text{memory} = 68,364 \\ M_1 &: \text{timing} = \ \ 7.06, \text{memory} = 23,057 \\ M_2 &: \text{timing} = \ \ 1.67, \text{memory} = \ \ 7,336 \end{aligned}$$

Example 2 For the rational function

$$F(x, y) = \frac{7y + 5x - 2}{(7 - 7x + 4y - 8xy + 2x^2 + 8y^2)^2 (7 + 8x - 2y + 3xy + 6x^2 - y^2)^2},$$

[a] All reported timings were obtained on a 400Mhz SUN SPARC SOLARIS with 1Gb RAM.

we obtained the following figures:

$$\begin{aligned} H &: \text{timing} = 3.04, \text{ memory} = 14,625 \\ O &: \text{timing} = 2.30, \text{ memory} = 11,819 \\ M_1 &: \text{timing} = 252.38, \text{ memory} = 205,316 \\ M_2 &: \text{timing} = 53.30, \text{ memory} = 111,539 \end{aligned}$$

Observe that the heuristic (H) chose the right version in both Examples 1 and 2. In the next two examples, our experiments were performed on three sets of samples each of which consists of 20 rational functions.

Example 3 Each element of the first two sets is of the form

$$F = \frac{\partial}{\partial y} R(x,y) + \sum_{i=1}^{n} T_i(x,y), \; R = \frac{1}{q(x,y)^j}, \; T_i = \frac{1}{p_i(x,y)^j}, \qquad (24)$$

q, p_i are randomly-generated polynomials of x and y with small integer coefficients, and both the numerator and denominator of F are in expanded forms. For elements in the first set S_1, $n = 2$, $j \in \{1,2\}$. As for the second set S_2, $n = 3$, $j = 1$. Note that the polynomials p_i's are not necessarily square-free w.r.t. y.

Tables 1 and 2 show the time and space requirements to complete the samples in the sets using different versions of $\mathcal{Z}$. It is understandable that if the number of rational functions T_i's in (24) is increased, the difference in time and space requirements between M_2 and M_1, and between M_1 and O would be much more significant.

		Timing (seconds)			
		H	O	M_1	M_2
S_1	Total	26.85	178.21	32.91	23.29
	Average	1.34	8.91	1.64	1.16
S_2	Total	22.39	1021.84	142.64	21.16
	Average	1.11	51.09	7.13	1.05

Table 1. Comparison of time requirement among different versions of $\mathcal{Z}$.

		Memory (kilobytes)			
		H	O	M_1	M_2
S_1	Total	136,061	658,578	153,950	114,679
	Average	6,803	32,928	7,697	5,733
S_2	Total	104,278	2,927,157	630,308	100,593
	Average	5,213	146,357	31,515	5,029

Table 2. Comparison of memory requirement among different versions of $\mathcal{Z}$.

Example 4 Each element of the set is of the form

$$F = \frac{F_{11}(x,y)}{F_{21}(x,y)^2 F_{22}(x,y)^2} \text{ where}$$

$$\deg F_{11}(x,y) = 1, \ \deg F_{21}(x,y) = 2, \ \deg F_{22}(x,y) = 2.$$

The figures in Table 3 show that the original $\mathcal{Z}$ (O) outperformed the two modified versions. This is due to the "nice" structure of F. It is also worth mentioning that M_2 gave a much better performance than M_1.

	Timing (seconds)				Memory (kilobytes)			
	H	O	M_1	M_2	H	O	M_1	M_2
Total	62.71	50.81	2488.09	688.09	276,328	225,748	4,042,810	2,263,657
Average	3.13	2.54	124.40	34.40	13,816	11,287	202,140	113,182

Table 3. Comparison of time and memory requirements among different versions of $\mathcal{Z}$.

Acknowledgements

This work is partially supported by Natural Sciences and Engineering Research Council of Canada Grant No. RGPIN8967-01, No. CRD215442-98.

References

1. G. Almkvist, D. Zeilberger, The method of differentiating under the integral sign. *J. Symb. Comput.* **10**, 1990, 571–591.
2. Ch. Hermite, Sur l'integration des fractions rationelles, *Nouvelles annales de mathematiques (2 serie)* **11**, 1872, 145–148.
3. E. Horowitz, Algorithms for Partial Fraction Decomposition and Rational Integrations, *Proc. SYMSAM'71*, ACM Press, 1971, 441–457.
4. H.Q. Le, On the differential-integral analogue of Zeilberger's algorithm to rational functions, *Proc. ASCM '2000, Lecture Notes Series on Computing*, World Scientific Publ., Singapore/River Edge, 2000, 204–213.
5. H.Q. Le, A Direct Algorithm to Construct Zeilberger's Recurrences for Rational Functions, *Proc. FPSAC'2001*, 2001, 303–312.
6. M. Ostrogradsky, De l'integration des fractions rationelles, *Bull. de la classe physico-mathématique de l'Académie Imperiale des Sciences de Saint-Petersburg* **IV**, 1845, 147–168, 286–300.
7. D. Zeilberger, The method of creative telescoping, *J. Symb. Comput.* **11**, 1991, 195–204.

AN AUTOMATED ALGEBRAIC METHOD FOR FINDING EXACT SOLUTIONS TO NONLINEAR EVOLUTION EQUATIONS

ZHI-BIN LI, YIN-PING LIU, RUO-XIA YAO

Department of Computer Science, East China Normal University, Shanghai, 200062, China

E-mail: lizb@cs.ecnu.edu.cn

Based on three types of Riccati equations, an algebraic and straightforward method for finding exact travelling wave solutions to nonlinear evolution equations is proposed, and a *Maple* package REAM which can entirely automatically output the exact solitary wave solutions as well as periodic wave solutions is presented. Using REAM the solutions for more than one hundred nonlinear evolution equations have been obtained. Not only are previously known solutions recovered but, in some cases, new solutions or more general form of solutions are obtained.

1 Introduction

Nonlinear evolution equations are widely used to describe complex phenomena in various fields of science, especially in physical science. A large variety of methods including the Bäcklund transform[1], inverse scattering theory[2] and Hirota's bilinear methods[3] has been developed to obtain analytic solutions to nonlinear evolution equations. But in some cases, they can not do very well.

Recently, with the development of computer symbolic systems such as *Macsyma*, *Mathematica*, *Maple*, which allow us to perform some complicated and tedious algebraic calculations on a computer, some algebraic and straightforward methods such as the hyperbolic function method[4-8], the Jacobi elliptic function method[9,10] and the trigonometric function method[11] have become more and more attractive. Through carefully observation we find that these methods are all based on such an idea that one can use the solutions of a given differential equation or equations with lower order to construct the solutions of a desired differential equation(s) with higher order, which can be assumed as the polynomial in terms of the solutions of the given equation(s). Armed with this novel view of point, in this paper, choosing three types of Riccati equation(s) as the given differential equation(s), a unified algebraic method is proposed, and a *Maple* package REAM to carry out the idea of the method is given, which can output directly the exact solitary wave solutions as well as periodic wave solutions automatically for a class of nonlinear evolution equations.

2 Method for finding travelling wave solutions

Given a nonlinear evolution equation, say, in two independent variables,

$$H(u, u_x, u_t, u_{xx}, u_{tt}, \ldots) = 0, \tag{1}$$

where $u = u(x,t)$, H is a polynomial about u and its derivatives. We want to construct, if exist, such exact solutions that can be expressed as a polynomial of solutions of the differential equation(s) with lower order, whose solution should be easily got, such as Riccati equation(s). In this paper, three types of Riccati equation(s) were considered.

1) Single Riccati equation

$$f' = 1 - f^2, \tag{2}$$

where $f = f(\xi)$. This equation admits the hyperbolic tangent function solution

$$f = \tanh(\xi). \tag{3}$$

2) Coupled Riccati equations

$$f' = g^2, \;\; g' = -fg, \tag{4}$$

where $f = f(\xi)$, $g = g(\xi)$. This set of equations admits a pair of hyperbolic function solutions

$$f = \tanh(\xi), \;\; g = \mathrm{sech}(\xi), \tag{5}$$

which satisfy

$$g^2 = 1 - f^2. \tag{6}$$

3) Tripled Riccati equations

$$f' = gh, \;\; g' = -fh, \;\; h' = -r^2 fg, \tag{7}$$

with the modulo $r\,(0 < r < 1)$, and $f = f(\xi)$, $g = g(\xi)$, $h = h(\xi)$. This set of equations admits a set of Jacobi elliptic function solutions

$$f = \mathrm{sn}(\xi, r), \;\; g = \mathrm{cn}(\xi, r), \;\; h = \mathrm{dn}(\xi, r), \tag{8}$$

which satisfy

$$g^2 = 1 - f^2, \;\; h^2 = 1 - r^2 f^2. \tag{9}$$

Obviously, when $r \to 1$, then equation (7) is reduced to equation (4), and when $r \to 0$ (setting $h = 1$), then (7) is reduced to

$$f' = g, \ \ g' = -f, \tag{10}$$

which admits a pair of trigonometric function solutions

$$f = \sin(\xi), \ \ g = \cos(\xi), \tag{11}$$

where f and g also satisfy the constraint (6).

The method for finding the exact travelling wave solutions of equation (1) proceeds in the following steps:

1) Convert the original PDE into an ODE

Taking a simple travelling wave transformation $\xi = k(x - ct)$, ξ is the travelling wave variable, k and c (wave speed) are constants to be determined, (1) is converted into an ordinary differential equation

$$H(u, u', u'', \ldots) = 0, \tag{12}$$

where $' := d/d\xi$.

2) Make assumption to the form of the solution

This step is crucial for the approach, for it determines the types of the solutions we are seeking for. In this paper, the required solutions of equation (12) can be assumed as one of the following three forms:

$$u(\xi) = \sum_{i=0}^{m} a_i f(\xi)^i, \tag{13}$$

where $f(\xi)$ satisfies the single Riccati equation (2).

$$u(\xi) = \sum_{i=0}^{m} a_i f(\xi)^i + \sum_{i=1}^{m} b_i g(\xi) f(\xi)^{i-1}, \tag{14}$$

where $f(\xi), g(\xi)$ satisfy the coupled Riccati equations (4).

$$u(\xi) = \sum_{i=0}^{m} a_i f(\xi)^i + \sum_{i=1}^{m} [A_i g(\xi) + b_i h(\xi)] f(\xi)^{i-1} + \sum_{i=2}^{m} B_i g(\xi) h(\xi) f(\xi)^{i-2}, \tag{15}$$

where $f(\xi), g(\xi)$ and $h(\xi)$ satisfy the tripled Riccati equations (7).

In (13), (14) and (15), m is the polynomial degree of the solution in f, (f, g) and (f, g, h), respectively, which can be determined in the next step.

3) Determine the polynomial degree m

In view of the equations (2), (4) and (7), we can see easily, if the polynomial degree of $u(\xi)$ is m, that

$$O(u(\xi)) = m,$$

then the degree of $d^p u/d\xi^p$ is

$$O(\frac{d^p u}{d\xi^p}) = m + p, \quad p = 1, 2, 3, \cdots,$$

and

$$O(u^q \frac{d^p u}{d\xi^p}) = (q+1)m + p, \quad q = 0, 1, 2, \cdots, \quad p = 1, 2, 3, \cdots.$$

Substituting (13), (14) and (15) into (12) and balancing the derivative term of the highest order with the nonlinear term in (12), we can determine the polynomial degree m.

4) Form the nonlinear algebraic systems

Once m is determined, three sets of nonlinear algebraic systems can be obtained:

(a) Substituting (13) into (12), using (2) repeatedly, yields an algebraic equation in powers of $f(\xi)$. Collecting the coefficients of the same powers and then setting them to zero will form a system of nonlinear algebraic equations for unknowns $a_i (i = 0, \cdots, m), k, c$.

(b) Substituting (14) into (12), and eliminating any derivative of (f, g) and any power of g higher than one with (4) and (6), yields an algebraic equation in powers of $f(\xi)$ and $g(\xi)$. Likewise, a system of nonlinear algebraic equations for unknowns $a_i(i = 0, \cdots, m)$, $b_j(j = 1, \cdots, m)$, k, c can be got.

(c) Substituting (15) into (12), and eliminating any derivative of (f, g, h) and any power of g and h higher than one with (7) and (9), yields an algebraic equation in powers of $f(\xi), g(\xi)$ and $h(\xi)$. Same as before, we also can get a system of nonlinear algebraic equations for unknowns $a_i(i = 0, \cdots, m)$, $A_i, b_i(i = 1, \cdots, m)$, $B_i(i = 2, \cdots, m)$, k, c. It is noticeable that if $m = 1$ then B_i must be zero.

5) Solve the nonlinear algebraic equations

The Ritt-Wu characteristic sets method[1213] is a powerful tool for solving nonlinear algebraic equations, which is an algorithmic method and has been implemented and contributed by Dongming Wang[14] to the *Maple* share library with the package Charsets. We use the function `Csolve` involved in Charsets directly to solve the nonlinear algebraic equations, and then, if possible, find all the symbolic solutions satisfying some physical requirements, e.g. $k \neq 0$ and $c \neq 0$.

6) Obtain the exact travelling wave solutions

All that remains to be done is to return to the original variables to obtain the desired travelling wave solutions of (1). In general, the exact solitary wave solutions (when we use the form (13) or (14)) and periodic wave solutions (when we use the form (15)) can be constructed.

Usually m is a positive integer. However, once in a while, the value of m may be a negative or a fraction. In these cases, we can introduce a transformation $u = v^{\mathrm{sign}(m)/d}$, where d is the denominator of m, and change (1) into another equation for v, whose balancing number will be a positive integer. Then it can be dealt with by the above method. Therefore, the final solutions u may be a rational or radical function in solutions of a given Riccati equation.

As a simple example to demonstrate the method outlined above, we have chosen the modified Kawachara equation[15]

$$u_t + u_x + u^2 u_x + p u_{xxx} + q u_{xxxxx} = 0, \tag{16}$$

where p, q are constants.

By introducing the new independent variable $\xi = k(x - ct)$, (16) can be transformed into an ODE for $u(\xi)$:

$$-cu' + ku' + ku^2 u' + pk^3 u''' + qk^5 u''''' = 0. \tag{17}$$

To look for exact travelling waves, the parameter m should first be determined. If u has a global degree m in f or (f, g) or (f, g, h), then u''''' has degree $m+5$, and u^2u' has degree $3m+1$; the only possibility for m is $m = 2$.

Now we turn to steps 4) and 5), there are three cases:

Case I: For the assumption (13), step 4) gives the following nonlinear system of equations:

$$\begin{aligned}
&-720qk^4 - 2a_2^2 = 0,\\
&-5a_1(24qk^4 + a_2^2) = 0,\\
&2a_2^2 - 4a_2a_0 + 1680qk^4 - 24pk^2 - 4a_1^2 = 0,\\
&-a_1(-5a_2^2 + a_1^2 - 240qk^4 + 6a_2a_0 + 6pk^2) = 0,\\
&4a_2^2a_0 - 1232qk^4a_2 - 2a_2a_0^2 + 4a_1^2a_2 + 2ca_2 - 2a_0a_1^2 + 40pk^2a_2 - 2a_2 = 0,\\
&a_1(a_1^2 - 1 - 136qk^4 + 6a_2a_0 + c - a_0^2 + 8pk^2) = 0,\\
&2a_2 - 16pk^2a_2 + 272qk^4a_2 + 2a_0a_1^2 + 2a_2a_0^2 - 2ca_2 = 0,\\
&-a_1(-1 + c - 16qk^4 + 2pk^2 - a_0^2) = 0,
\end{aligned} \tag{18}$$

from which it is determined

$$\{a_0 = \frac{40qk^2 - p}{\sqrt{-10q}},\ a_1 = 0,\ a_2 = -6\sqrt{-10q}k^2,\ c = 1 - 24qk^4 - \frac{p^2}{10q}\},$$

$$\{a_0 = -\frac{40qk^2 - p}{\sqrt{-10q}},\ a_1 = 0,\ a_2 = 6\sqrt{-10q}k^2,\ c = 1 - 24qk^4 - \frac{p^2}{10q}\}.$$

Thus two solitary wave solutions of (16) can be obtained with the parameters constraints $q < 0$

$$u(x,t) = \pm\frac{p - 40k^2q}{\sqrt{-10q}} \pm \frac{60k^2q}{\sqrt{-10q}}\tanh^2[\frac{k(10qx + 240q^2k^4t - 10qt + p^2t)}{10q}], \tag{19}$$

where k is an arbitrary constant. When choosing $k = \frac{\sqrt{-5qp}}{10q}$, (19) becomes

$$u(x,t) = \pm\frac{3p}{\sqrt{-10q}}\mathrm{sech}^2[\sqrt{\frac{p}{-20q}}(x - (1 - \frac{8p^2}{50q})t)], \tag{20}$$

which is the same as that given in Ref.[15].

Case II: For the assumption (14), step 4) gives the following nonlinear system of equations:

$$\begin{aligned}
&b_2 - cb_2 - 5pk^2b_2 + 61qk^4b_2 + b_2b_1^2 + b_2a_0^2 + 2a_1a_0b_1 = 0,\\
&b_2^3 - 2b_2 + 28pk^2b_2 - 2b_2a_0^2 + 6a_1a_2b_1 - 662qk^4b_2 - 4a_1a_0b_1 + 6a_2a_0b_2 + 2cb_2\\
&+3a_1^2b_2 - 5b_2b_1^2 = 0,\\
&-b_1^3 - b_1 + 5pk^2b_1 + 4a_2a_0b_1 + 4a_1a_0b_2 + cb_1 - 61qk^4b_1 - b_1a_0^2 + 2a_1^2b_1\\
&+2b_2^2b_1 = 0,\\
&8a_1a_2b_1 - 4b_2b_1^2 - 5a_2^2b_2 - 1320qk^4b_2 + 8a_2a_0b_2 + 24pk^2b_2 + 3b_2^3 + 4a_1^2b_2 = 0,\\
&2b_2(-360k^4q - 3a_2^2 + b_2^2) = 0,\\
&-10a_1a_2b_2 + 5b_2^2b_1 - 120qk^4b_1 - 5a_2^2b_1 = 0,\\
&b_1^3 - 7b_2^2b_1 - 3a_1^2b_1 - 6a_1a_0b_2 + 180qk^4b_1 - 6pk^2b_1 - 6a_2a_0b_1 + 4a_2^2b_1\\
&+8a_1a_2b_2 = 0,\\
&a_1 + 16qk^4a_1 + a_1b_1^2 + 2b_2a_0b_1 + a_1a_0^2 - ca_1 - 2pk^2a_1 = 0,\\
&2a_2(-a_2^2 - 360k^4q + 3b_2^2) = 0,\\
&a_1^3 - 8b_2a_0b_1 + a_1(c - 1 - a_0^2 + 8pk^2) + 3a_1b_2^2 - 136qk^4a_1 + 6a_2(b_1b_2 + a_0a_1)\\
&-4a_1b_1^2 = 0,\\
&272qk^4a_2 + 4a_1b_1b_2 - 16pk^2a_2 - 2b_1^2a_0 + 2a_1^2a_0 + 2a_2(1 - c + a_0^2 + b_1^2)\\
&+2b_2^2a_0 = 0,\\
&240qk^4a_1 + 3a_1b_1^2 - 6a_1a_0a_2 - a_1^3 - 6pk^2a_1 - 8a_1b_2^2 + 5a_1a_2^2 + 6b_2a_0b_1\\
&-16a_2b_1b_2 = 0,\\
&-120qk^4a_1 - 5a_1a_2^2 + 10a_2b_1b_2 + 5a_1b_2^2 = 0,\\
&-24pk^2a_2 - 4a_2^2a_0 - 10a_2b_2^2 + 2a_2^3 - 4a_1^2a_2 + 4a_2b_1^2 + 4b_2^2a_0 + 1680qk^4a_2\\
&+8a_1b_1b_2 = 0,\\
&-2a_1^2a_0 - 6a_2b_1^2 - 2a_2 - 12a_1b_1b_2 + 2ca_2 + 2b_1^2a_0 + 40pk^2a_2 - 6b_2^2a_0 + 4a_2b_2^2\\
&+4a_1^2a_2 - 2a_2a_0^2 - 1232qk^4a_2 + 4a_2^2a_0 = 0,
\end{aligned} \tag{21}$$

from which three solitary wave solutions are obtained. Two of them are listed in case 1. In addition, a new type of solitary wave solution with the parameters constraints $q > 0$ and $p > 0$ is listed as follows:

$$u(x,t) = -\frac{3\sqrt{10}p}{5\sqrt{q}} \tanh[w(100q(x-t) - 11p^2t)]\mathrm{sech}[w(100q(x-t) - 11p^2t)], \tag{22}$$

where $w = \sqrt{10pq}/1000q^2$. Notice that in (22) all parameters are fixed.

Case III: For the assumption (15), step 4) gives the nonlinear system consisting of 28 equations, from which two periodic wave solutions in terms of Jacobi elliptic functions are obtained

a) For $q > 0$ and $p > 0$,

$$u(x,t) = \pm\frac{3\sqrt{10}r^2 p}{5\sqrt{q}(r^2-2)}\mathrm{cn}[k(x-ct), r]\mathrm{sn}[k(x-ct), r], \tag{23}$$

where $k = \mp\frac{\sqrt{p}}{\sqrt{-10q(r^2-2)}}$, $c = 1 + \frac{p^2(64r^2 + 11r^4 - 64)}{100q(r^2-2)^2}$.

b) For $q < 0$,

$$u(x,t) = \pm\frac{\sqrt{-q}}{\sqrt{10q}}\left(p - 20\,qk^2(r^2+1) + 60r^2k^2q\mathrm{sn}^2[k(x-ct), r]\right) \tag{24}$$

where $c = \frac{240\,q^2k^4(r^2 - r^4 - 1) + 10\,q - p^2}{10\,q}$, and k is an arbitrary constant.

When $r \to 1$, (23) is reduced to (22), (24) is reduced to (19). Similarly, when $r \to 0$, then (23) and (24) are reduced to constant solutions to be rejected. This equation does not admit solution in terms of trigonometric functions, which is deduced from the Riccati equation (10).

¿From the above example, it can be seen that some solitary wave solutions can be obtained from the feature of (15). However, we have found some equations, which have no periodic wave solution in terms of Jacobi elliptic functions, but indeed have solitary wave solutions.

3 The package REAM

REAM consists of four subprograms denoted as RATH, ERATH, AJFM, TRFM. As for the method mentioned in the last section, RATH implements the branch based on the features of (13) and (2). ERATH implements the branch based on the features of (14) and (4). These two programs can output exact solitary wave solutions in terms of hyperbolic functions automatically. AJFM implements the branch based on the features of (15) and (7), which can entirely output exact periodic wave solutions in terms of Jacobi elliptic functions. TRFM implements the branch based on the features of (14) and (10), which can entirely output exact periodic solutions in terms of trigonometric functions. For simplicity, here we just give a brief description of RATH, the others are similar.

RATH contains five sub-procedures denoted as: twvar(), findm(), coeff(), sol() and prin().

twvar(): In this procedure we first extract the independent variable from the function u; then make the travelling wave variable ξ, at the same time, we also convert the original partial differential equation into an ordinary differential equation; finally, we extract arbitrary coefficients of the original equation and form the parameters' list PL. For example, for the modified Kawachara equation (16), $\xi = k(x - ct)$ and $PL = [p, q]$.

findm(): In this procedure the starting point is the ansatz that the solution $u(\xi)$ is (the assumption is just for deducing the value of m):

$$u(\xi) = T^m \tag{25}$$

with $T' = 1 - T^2$. Substituting (25) into the ordinary differential equation, and using the relation $T' = 1 - T^2$ repeatedly, one can changes the equation to a polynomial equation in terms of T. We extract the power of T in each term of the polynomial and form a list LM. Setting

$$y = \max(\mathrm{op}(LM)), \tag{26}$$

where op operator extracts components of an expression. Apparently, (26) represents a flexuous line in (m, y) plane, finding the sharp points (usually, only one sharp point), we can determine the value of m. For example, for equation (16), $y = \max(m + 1, 3m + 1, m + 3, m + 5)$, there is only one sharp point (2 , 7), hence $m = 2$.

coeff(): If $m = \mathrm{sign}(m)n/d$, where n, d are both positive integer, we make the following transformation: $u = v^{\mathrm{sign}(m)/d}$ (for a positive integer m, $d = 1, n = m$, and $\mathrm{sign}(m)$ is equal to 1), and change the ordinary differential equation in u into an equation in v.

Then we can make the assumption:

$$v(\xi) = \sum_{i=0}^{m} a_i f(\xi)^i. \tag{27}$$

Substituting (27) into the ordinary differential equation in v and also using (2), repeatedly. It will yield a polynomial equation in $f(\xi)$. We set the polynomial in descending order and then extract the coefficients one by one. The coefficients set, denoted as PS, is a set of polynomials in $a_i(i = 0, \cdots, m), k, c$, and parameters appearing in the initial partial differential equation. It is not difficult to prove that the former $m+1$ polynomials form a "triangular" set for $a_i(i = 0, \cdots, m)$, which is denoted as $PS1$. The procedure also simplifies PS and $PS1$ and gets rid of all unnecessary factors in each polynomial.

sol(): In this procedure we proceed 6 steps as follows:

(1) Solve $PS1$ using `Csolve` and determine some parameters as well as $a_i(i = 0, \cdots, m)$. These values are delivered as a list of lists denoted as $SOL1$.

(2) Eliminate the trivial solutions from $SOL1$, and also denote the list consisting of the remaining solutions as $SOL1$. For every list $SOL1_i$ in $SOL1$ do the following steps (3) to (6) repeatedly.

(3) Substitute a list $SOL1_i$ of $SOL1$ into PS and get PPS. PPS is a list, in which the first $m+1$ entries are zeros. The remaining entries denoted as $PS2$ is a set of polynomials for the rest of parameters.

(4) For $PS2$, choose `CSolve` or *Maple* command `Solve` to determine the rest of parameters according to whether there is radical expression contained in PPS or not. The solutions of $PS2$ are returned as a list of lists denoted as $SOL2$.

(5) For each element $SOL2_j$ of $SOL2$, integrate it with $SOL1_i$ to construct a solution of PS denoted as SOL_k.

(6) Merge the solution SOL_k into SOL, which is a solutions' list of PS. Namely, when the procedure finished, SOL contains all solutions of PS.

prin(): In this procedure we first eliminate all particular cases and trivial solutions from the whole solutions' list, and eliminate, if do not need, all complex solutions; then, output the eligible solutions as desired format.

4 Summarize

We have described an unified automated algebraic method based on the idea that one can use the solutions of a given differential equation(s) with lower order to construct the solutions of a desired differential equation(s) with higher order. A *Maple* package REAM that can entirely automatically output exact solitary wave solutions as well as periodic wave solutions to nonlinear evolution equations is given. Using REAM, we have computed the solutions of more than one hundred nonlinear evolution equations on a PC Pentium III. The running time for most of the equations is less than 30 seconds. Furthermore, all results have been verified automatically before output. It successfully recovered all previously known solutions that had been found by other more sophisticated methods. More importantly, for some equations, with no extra effort we also pick up other new and more general types of solutions at the same time. REAM can also be used as a tool to test a nonlinear evolution equation whether admitting exact solitary wave solution or periodic wave solution.

Acknowledgments

This work is supported by W.T. Wu Foundation on Mathematics Mechanization, and the "Shu-Guang" Project of Shanghai, China.

References

1. R. K. Dodd, Solitons and Nonlinear Wave Equations. London: Academic Press, Inc, Ltd, 1982.
2. M. J. Ablowitz, P. A. Clarkson, Solitons, Nonlinear Evolution Equations and Inverse Scattering. Cambridge University Press, 1991.
3. R. Hirota, J. Satsuma, Soliton solutions of a coupled Korteweg-de Vries equation. *Phys. Lett. A* 85 (1981) 407.
4. W. Malfliet, Solitary wave solutions of nonlinear wave equations. *Am. J. Phys.* 60 (1992) 650.
5. E. J. Parkes, B. R. Duffy, An automated tanh-function method for finding solitary wave solutions to nonlinear evolution equations. *Comp. Phys.*

Comm. 98 (1996) 288.
6. E. G. Fan, H. Q. Zhang, New exact solutions to a system of coupled KdV equations. *Phy. Lett. A* 245 (1998) 389.
7. Z. B. Li, Exact solitary wave solutions of non-linear evolution equations, in *Mathematics Mechanization and Applications*, 389-408, Academic Press, SanDiego, CA, 2000.
8. E. G. Fan, Extended tanh-function method and its applications to nonlinear equations. *Phys. Lett. A* 277 (2000) 212.
9. S. K. Liu, Z. T. Fu, S. D. Liu, Q. Zhao, Jacobi elliptic function expansion method and periodic wave solutions of nonlinear wave equations. *Appl. Math. Mech.* 22 (2001) 326.
10. Z. T. Fu, S. K. Liu, S. D. Liu, Q. Zhao, New Jacobi elliptic function expansion and new periodic solutions of nonlinear wave equations. *Phys. Lett. A* 290 (2001) 72.
11. C. Yan, New explicit and exact travelling wave solutions for a system of variant Boussinesq equations in mathematical physics. *Phys. Lett. A.* 224 (1996) 77.
12. W. T. Wu, Mechanical Theorem Proving in Geometries: Basic Principles (Springer-Verlag, New York, 1994. English translation by Xiaofan Jin and Dongming Wang. Originally published as " Basic Principles of Mechanical Theorem Proving in Geometry" in Chinese language by Science Press, Beijing, 1984).
13. B. Mishra, Algorithmic Algebra (Springer-Verlag, New York, 1993).
14. D. M. Wang, An Implementation of the Characteristic Set Method in Maple, *RISC-Linz* Series 91-25, 1991.
15. J. K. Hunter, J. Scheurle, Existence of perturbed solitary wave solutions to a model equation for water waves. *Phys. D* 32 (1988) 253.

PARAMETRIC ANALYSIS OF SYSTEMS OF ALGEBRAIC INEQUALITIES *

ANDREY V. BANSHCHIKOV

Institute of Systems Dynamics and Control Theory
Siberian Branch, Russian Academy of Sciences
134, Lermontov St., Irkutsk, 664033, Russia
E-mail : bav@icc.ru

Unique software elaborated on the basis of the computer algebra software package "Mathematica" has been employed in investigations of stability of a relative equilibrium position for an uncontrolled satellite with a gravitation stabilizer on a circular orbit. Domains of various degrees of Poincare instability have been revealed in the space of introduced parameters. Under the assumption that the potential system is unstable (the degree of instability being even) the problem of possibility of its gyroscopic stabilization is considered. Parametric analysis of stability conditions has been conducted and two-parameter sections of the system's domains of gyroscopic stabilization have been constructed with the aid of a software system intended for graphic solving of systems of algebraic inequalities.

1 Introduction

Investigation of problems of stability, stabilization of nonlinear or linearized models of mechanical systems often leads to the problem of "parametric analysis" of the conditions (inequalities) obtained. In case of parametric analysis it is important to have the possibility to estimate the domain of values of the parameters under which a desired system's state is provided. Naturally, it is hard to hope to obtain any readable analytical results for the models which have high dimensions and contain many parameters. On this stage one can efficiently use software packages of computer algebra (SPCA) as well as the corresponding software elaborated on the basis of these software packages.

The author has elaborated the software PASI [1] intended for obtaining the solutions for systems of multiparametric algebraic inequalities. The software PASI represents a complex of interactive user programs which are executed in interpretation mode in the environment of SPCA "Mathematica" [2]. If the system of inequalities under scrutiny has more than two parameters then employing the abovementioned software the user construct some cuts (sections) of the space of all the parameters by the planes of the two parameters x_1 , x_2 chosen by him (the rest of the parameters being fixed). As a result, an initial

*THE WORK HAS BEEN PARTIALLY SUPPORTED BY THE RUSSIAN FOUNDATION FOR FUNDAMENTAL RESEARCH (GRANT NO. 00-01-00730).

system of algebraic inequalities of several variables – after the corresponding automatic transformations – is reduced to a system of one-parameter polynomials with numerical coefficients

$$F_i(x_j) = a_n^{(i)} x_j^n + a_{n-1}^{(i)} x_j^{n-1} + \ldots + a_1^{(i)} x_j + a_0^i \geq 0 \ (\leq 0), \quad (i = 1, \ldots, k_1;\ j = 1, 2) \tag{1}$$

and to a system of two-parameter inequalities with numerical coefficients of the form

$$F_i(x_1, x_2) \geq 0 \ (\leq 0), \quad (i = k_1 + 1, \ldots, k), \tag{2}$$

which are nonresolvable explicitly with respect to one of the parameters. To the end of solving the system of inequalities (1) we have developed a unique software program **SolveSystemInequality**, which searches for the solution in symbolic-numeric form as a union of nonoverlapping intervals of the numerical axis. For the purpose of solving the system of inequalities (2) on account of solutions of inequalities (1) a unique software program **Region** has been elaborated which finds and performs graphical constructing a 2D domains in the plane of parameters x_1, x_2. A list of symbolic-numeric inequalities, names (identifiers) of parameters and intervals of variation of their values with respect to abscissa and ordinate axes are assigned as the input for this software program. Now consider briefly the main steps of the algorithm.
1) For the functions $F_i(x_1, x_2)$ "singular" points (their real values at the abscissa-axis) are found. Singular points define, in particular, vertical asymptotes or points where the sign of the coefficient at the highest power of parameter x_2 changes.
2) The given interval on abscissa-axis is split into subintervals on account of the number of singular points. On each subinterval, in accordance with the step of variation of values along the abscissa-axis, the following operations are performed:

2.1) numerical solutions of the equation $F_i(x_1, x_2) = 0$ are obtained;

2.2) all the branches of the boundary of the desired domain are found;

2.3) a small neighbourhood of each branch of the boundary is investigated and verified, whether these points satisfy above inequalities or not. Information about the neighbourhood is automatically analyzed, and the found domain is painted to be more explicit.
3) The resulting 2D image of the domain on a given interval at the output is represented as a union of domains for each of the subinterval on a unified scale.

The algorithms for graphic representation of solutions for the systems of algebraic inequalities (1), (2) in its complete form, examples of loading a software programs, its usage as well as specific traits of the programs **SolveSystemInequality**, **Region** are described in detail in [3]. These two programs form a kernel of the software PASI.

Note, in the last version of SPCA "Mathematica" there appeared the user package **InequalityGraphics** (its author is Roger Germundsson, 2000), which is also intended for finding graphic 2D and 3D-solutions of systems of algebraic inequalities. Maple V is capable of finding solutions only for systems of linear inequalities.

It is known that stability of rotation of conservative satellite systems of rigid bodies, as well as secular stability, can be bound up also with gyroscopic stabilization [4,5]. The objective of this work was to employ the elaborated software, conduct parametric analysis of obtained stability conditions and construct domains of gyroscopic stabilization for a well-known model [6].

2 Problem Statement

Consider a problem of stability of a position of relative equilibrium in the orbital coordinate system for an uncontrolled satellite with a gravitational stabilizer. The stabilizer represents a rigid rod having the point mass at its free end. The rod is connected to the satellite with the aid of a 2-degree-of-freedom suspension. The rotation axes of the rod coincide with the direction of the axes of pitch and bank. The hinges are supposed to be frictionless. The system's mass center performs some passive motion along the Kepler circular orbit. The system is influenced by some gravitation moment (other moments being neglectibly small). In undisturbed motion the system's principal central axes of inertia coincide with orbital axes, and the rod is oriented along the radius of the orbit.

Linearized equations of motion are decomposed into two subsystems. Respectively, in the pitch channel (θ) and in the yaw-and-bank channel (ψ, φ):

$$M_1\,\ddot{q}_1 + K_1\,q_1 = 0\,; \qquad M_2\,\ddot{q}_2 + G\,\dot{q}_2 + K_2\,q_2 = 0 \tag{3}$$

$$K_1 = 3\omega^2\begin{pmatrix} b-a & f \\ f & f \end{pmatrix}; \qquad M_1 = \begin{pmatrix} c & f \\ f & d \end{pmatrix} > 0\,; \qquad q_1 = \begin{pmatrix} \theta \\ \delta_1 \end{pmatrix};$$

$$K_2 = \omega^2\begin{pmatrix} c-b & 0 & 0 \\ 0 & 4(c-a) & 4f \\ 0 & 4f & 3f+d \end{pmatrix}; \qquad M_2 = \begin{pmatrix} a & 0 & 0 \\ 0 & b & f \\ 0 & f & d \end{pmatrix} > 0\,;$$

$$G = \omega \begin{pmatrix} 0 & \mathrm{g} & 0 \\ -\mathrm{g} & 0 & 0 \\ 0 & 0 & 0 \end{pmatrix} ; \qquad q_2 = \begin{pmatrix} \psi \\ \varphi \\ \delta_2 \end{pmatrix} ,$$

where ω is the modulus of the orbital angular velocity; δ_1 , δ_2 are rotation angles of the rod with respect to the satellite's body;

$$a = J_y \ ; \quad b = J_x + m\,r\,(r+l) + \frac{1}{3} m l^2 + m_0\,(r+l)^2 \ ; \quad c = b + J_z - J_x \ ;$$

$$d = (\frac{1}{3} m + m_0)\,l^2 \ ; \quad f = (\frac{1}{2} m + m_0)\,r\,l + d \ ; \quad \mathrm{g} = J_z - J_x - J_y \ ;$$

m, m_0 – masses of the rod and the point load at the end, resp.; $l > 0$ – the rod's length; $r > 0$ – the distance from the system's mass center to the point of attachment of the rod; J_x , J_y , J_z – principal inertia moments of the satellite.

Equations (3) may be interpreted as equations of oscillations of a mechanical system influenced by potential (with the matrices K_1 , K_2) and gyroscopic (with the matrices G) forces. These forces are determined by the forces of gravitation as well as by orbital motion. M_1 , M_2 play the role of diagonal blocks of the kinetic energy matrix.

Introduce the four dimensionless parameters:

$$\alpha = \frac{c-b}{a} = \frac{J_z - J_x}{J_y} \ ; \quad \gamma = \frac{b-a}{c} \ ; \quad p_1 = \frac{d}{f} \ ; \quad p_2 = \frac{f}{c} \ . \tag{4}$$

The physically obtainable values of the parameters lie within the intervals:

$$|\alpha| < 1 \ ; \quad |\gamma| < 1 \ ; \quad 0 < p_1 < 1 \ ; \quad 0 < p_2 < 1 \ . \tag{5}$$

The secular equation of the system (3) : $\Delta(\lambda^2) = \Delta_1(\lambda^2) * \Delta_2(\lambda^2) = 0$ contains λ only in even powers. Here the characteristic determinants for the channel of pitch and the channel of yaw-and-bank are, respectively :

$$\Delta_1 = \begin{vmatrix} \lambda^2 + 3\omega^2\gamma & p_2(\lambda^2 + 3\omega^2) \\ \lambda^2 + 3\omega^2 & \lambda^2 p_1 + 3\omega^2 \end{vmatrix} =$$

$$= \lambda^4\,(p_1 - p_2) + 3\omega^2\lambda^2\,(1 + \gamma\,p_1 - 2\,p_2) + 9\omega^4\,(\gamma - p_2) \ ;$$

$$\Delta_2 = \begin{vmatrix} \lambda^2 + \omega^2\alpha & \lambda\omega(\alpha-1) & 0 \\ \lambda\omega(\gamma-1)(\alpha-1) & \lambda^2(1+\gamma\alpha) + 4\omega^2(\alpha+\gamma) & p_2(\alpha+1)(\lambda^2+4\omega^2) \\ 0 & \lambda^2 + 4\omega^2 & \lambda^2 p_1 + (3+p_1)\omega^2 \end{vmatrix} =$$

$$= v_6 \lambda^6 + v_4 \lambda^4 + v_2 \lambda^2 + v_0 = \lambda^6 (p_1 - p_2 + \alpha (\gamma p_1 - p_2))$$

$$+ \lambda^4 (3+3\alpha\gamma+2p_1+3\alpha p_1+\alpha^2 p_1+3\gamma p_1+3\alpha\gamma p_1-8p_2-9\alpha p_2-\alpha^2 p_2)\omega^2$$

$$+ \lambda^2 (3+9\alpha+3\alpha^2+9\gamma+6\alpha\gamma+p_1+3\alpha p_1+5\alpha^2 p_1+3\gamma p_1+6\alpha\gamma p_1$$

$$-16p_2-24\alpha p_2-8\alpha^2 p_2)\omega^4 + 4\omega^6 \alpha ((\gamma+\alpha)(3+p_1)-4p_2(\alpha+1)) .$$

It is known that stability in such systems is possible only in the case when all the roots of the polynomial Δ with respect to λ^2 are negative and real. The corresponding criterion [7] of negativity and reality of the roots for the $n-$ th degree polynomial is realized by the author in [8] with the aid of SPCA "Mathematica". Algebraic conditions, which provide the desired properties of the roots, respectively, for the channels of pitch and yaw-and-bank,

$$\begin{cases} p_1 - p_2 > 0 , \quad 1 + \gamma p_1 - 2p_2 > 0 , \quad \gamma - p_2 > 0 , \\ p_1^2 \gamma^2 + (4p_2(1-p_1) - 2p_1)\gamma + (1 - 4p_2(1-p_1)) > 0 \end{cases} \tag{6}$$

$$\begin{cases} v_6 > 0 , \quad v_4 > 0 , \quad v_2 > 0 , \quad v_0 > 0 , \\ Dis = v_4^2 v_2^2 - 4v_2^3 v_6 - 4v_4^3 v_0 + 18 v_6 v_4 v_2 v_0 - 27 v_0^2 v_6^2 > 0 \end{cases} \tag{7}$$

will define the relations between the system's parameters, the satisfaction of which makes stable the motion of a satellite with a stabilizer. The latter inequality in (7) have also been obtained in analytical form with the aid of a special program and written down in the form of an explicit dependence on the parameters (4). But it is too awkward to be given here. To emphasize the complexity of the analyzed expression it is sufficient to note only that maximum powers of the parameters α, γ, p_1, p_2 in Dis are, respectively, 8, 4, 4, 3.

3 Stability in the Pitch Channel

Note, the first of the conditions (6) is satisfied due to definite positiveness of the matrix of kinetic energy M_1 . It can be easily shown that the last inequality in (6) also holds. To this end let us write down the discriminant of this polynomial with respect to γ : $16(p_1-1)^2 p_2 (p_2 - p_1)$, which is always negative for $p_1 > p_2$. Due to both above assumptions and positivity of the coefficient for γ^2 it is possible to conclude that the polynomial under scrutiny assumes only positive values. Graphically it is easy to show (see Fig. 1) that

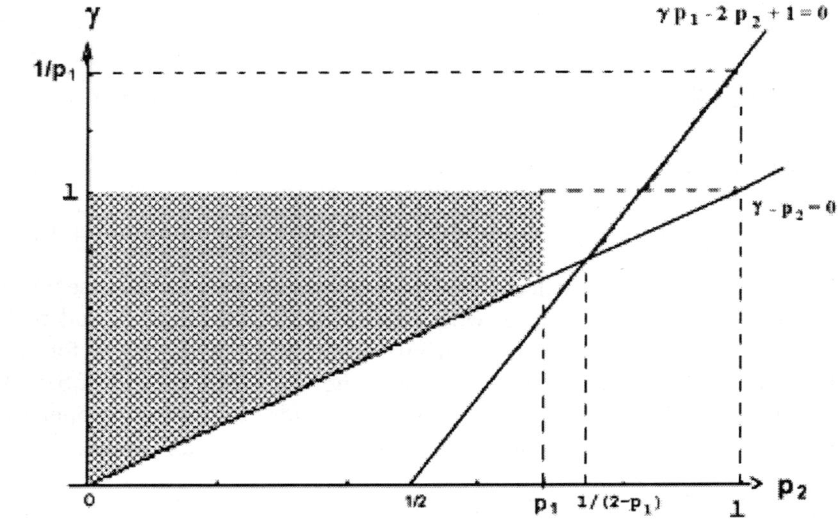

Figure 1. Domain of stability in the pitch channel.

for two inequalities in (6), which have not yet been considered, satisfaction of the condition $1+\gamma p_1 - 2p_2 > 0$ is provided by the inequality $\gamma - p_2 > 0$.

So, the solution in the pitch channel is stable if and only if $\gamma - p_2 > 0$.

Before proceeding to the parametric analysis of stability conditions in the yaw-and-bank channel, let us define the domain of real values of parameters α , γ on account of conditions (5) and $\gamma - p_2 > 0$. Denote $k = a/b$ and after several simple transformations the following relation can be obtained :

$$\gamma = \frac{1-k}{1+k\,\alpha}\,, \quad \text{where} \quad 0 < k < 1 - \frac{f}{b}\,. \tag{8}$$

4 Gyroscopic stabilization

Kelvin-Chetayev theorems [4] allow one to start investigation of stability of the trivial solution for the yaw-and-bank channel from analysis of the matrix of potential forces (in denotations of (4)) :

$$K_2 = 4\omega^2 \begin{pmatrix} \alpha & 0 & 0 \\ 0 & \alpha+\gamma & 4\,p_2\,(\alpha+1) \\ 0 & 1 & 3+p_1 \end{pmatrix}.$$

Even (odd) character of the degree of Poincare instability is determined

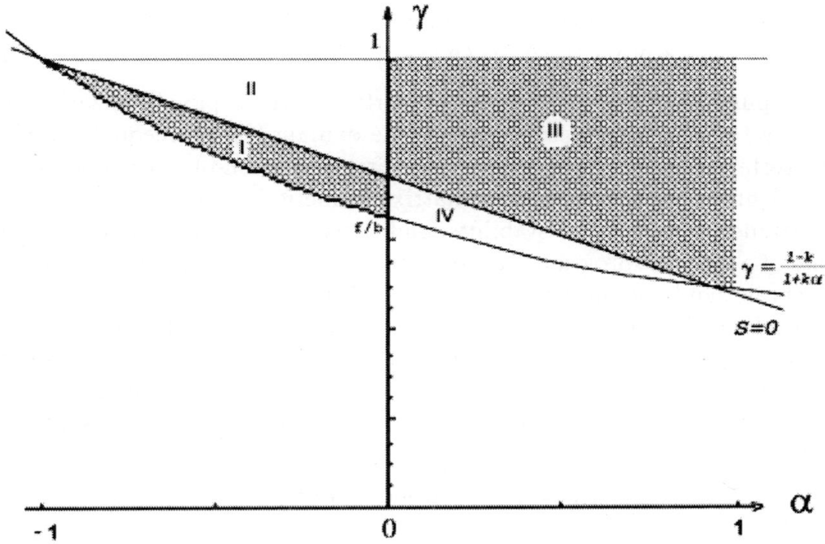

Figure 2. Domains having different degrees of instability.

by positivity (negativity) of the determinant of the matrix K_2. Note that $det K_2 = v_0$. The relations $\alpha = 0$ and $S = (\gamma+\alpha)(3+p_1)-4p_2(\alpha+1) = 0$ assign in the space of parameters the surfaces which separate the domains having different degrees of instability. For example, Fig. 2 qualitatively shows these domains for fixed values of parameters p_1, p_2. Here, resp., instability domains have : I – an even degree ; II, IV – odd degree ; III – zero degree. The shaded area is bounded with the lines obtained from (8) by substituting the ultimate possible values of $k = 0$ and $k = 1 - f/b$. In domain I the subsystem of yaw-and-bank is potentially unstable, but gyroscopic stabilization is possible in it. In domain III addition of gyroscopic forces to potential ones helps to retain the property of stability of the motion under scrutiny. Gyroscopic stabilization is impossible in domains II and IV.

Consider now the question of possibility of gyroscopic stabilization of the system when the condition $J_y > J_x > J_z$ for the satellite's inertia moments holds. Note that in this case $\alpha < 0$, and hence, the subsystem of yaw-and-bank is potentially unstable (i.e. K_2 is not a positive definite matrix). Obviously, the domain of gyroscopic stabilization, which has an even degree of instability, can be defined as a solution of the system of inequalities:

$$\alpha < 0 \, ; \qquad (\gamma + \alpha)(3 + p_1) - 4 p_2 (\alpha + 1) < 0 \, . \tag{9}$$

For the purpose of finding out the property of gyroscopic stabilization it is necessary to determine in what part of the domain (9) the inequalities (7) is valid. Note that $v_6 = det M_2$, hence, the first of the conditions (7) is satisfied due to definite positiveness of the matrix of kinetic energy M_2 .

Parametric analysis of stability conditions (7) has been conducted with the aid of a software program **Region** intended for graphic solution of systems of algebraic symbolic-numeric inequalities. Within the frames of this analysis the initial 4-dimensional space of parameters was partitioned by the planes of two chosen parameters (other parameters being fixed). As a result, several 2-parameter sections (cuts), which are of interest for us, have been constructed.

For example, let us fix the parameters of the stabilizer by the values of : $p_1 = 0.8$, $p_2 = 0.7$. In this case, the expressions from (7) will write:

$$v_0 = 4\alpha (-2.8 + \alpha + 3.8 \gamma) ;$$

$$v_2 = -7.4 + 1.4 \alpha^2 + 11.4 \gamma + \alpha (-5.4 + 10.8 \gamma) ;$$

$$v_4 = -1. + 0.1 \alpha^2 + 2.4 \gamma + \alpha (-3.9 + 5.4 \gamma) ;$$

$$\begin{aligned} Dis = {} & 0.0004 \alpha^8 + \alpha^7 (0.096 - 0.106 \gamma) - 1.14 \alpha^6 (-1.622 + \gamma)(-0.904 + \gamma) \\ & - 51.341 \alpha^5 (-1.221 + \gamma) (-1.057 + \gamma) (-0.889 + \gamma) \\ & - 158.995 \alpha^4 (-1.683 + \gamma) (-1.275 + \gamma) (-0.916 + \gamma) (-0.787 + \gamma) \\ & + 424.57 \alpha^3 (-1.311 + \gamma) (-1.031 + \gamma) (0.572 - 1.496 \gamma + \gamma^2) \\ & - 257.227 \alpha^2 (-1.214 + \gamma) (-0.765 + \gamma) (0.465 - 1.298 \gamma + \gamma^2) \\ & - 88.282 \alpha (-1.827 + \gamma) (-0.573 + \gamma) (0.561 - 1.493 \gamma + \gamma^2) \\ & + 83.174 (0.686 - 1.625 \gamma + \gamma^2) (0.422 - 1.298 \gamma + \gamma^2) . \end{aligned}$$

The result of operation of the software program

Region$[\{ v_4 > 0 , \ v_2 > 0 , \ v_0 > 0 , \ Dis > 0 \} , \{ \alpha , -1 , 0 \} , \{ \gamma , 0.7 , 1 \}]$

is shown in Fig. 3. As obvious from Fig. 3 gyroscopic stabilization is possible not for the whole domain I (Fig. 2) but only for its shaded "needle-shape" part.

5 Conclusion

It is necessary to emphasize that problems of reliability and precision of computations, as well as problems of explicitness and speeding-up of the process

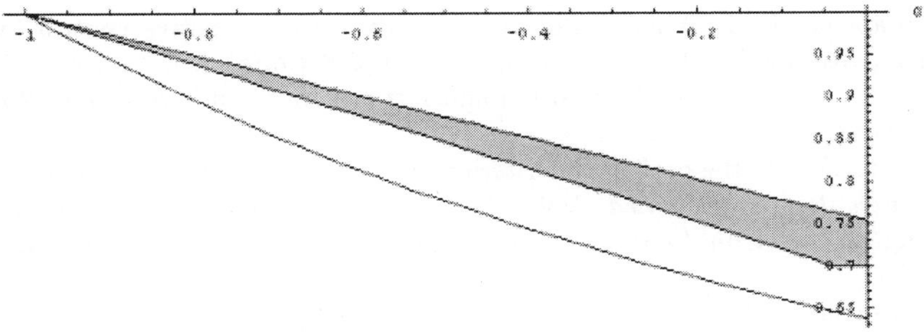

Figure 3. Domain of gyroscopic stabilization.

of investigations can be partially solved when SPCA is chosen in the capacity of the software tool. Side by side with employment of the SPCA (in the capacity of "a calculator") for solving a definite problem the approach, which implies elaboration of some software for solving a definite class of problems on the basis of the internal programming language of the SPCA (in our case – "Mathematica"), is quite important. Practically, all the above analysis has been conducted with the use of this software.

References

1. Banshchikov A.V. A Software System for Parametric Analysis of Systems of Algebraic Inequalities (Software PASI). *Patent of Russian Federation for Software No. 2000611004.* Rospatent, 5 Oct., 2000.
2. Wolfram S. Mathematica: A System for Doing Mathematics by Computer. (Addison-Wesley Publ. Co., Redwood City, 1988).
3. Banshchikov A.V., Burlakova L.A. Algorithms of Symbolic Computation Used in Stability Analysis. *Programmirovanie.* No. 3, (1997), pp. 72 - 80 (Russian). English transl. in *Programming and Computer Software.* **23**, 3 (1997) 173 - 179.
4. Chetayev N.G. Stability of Motion. Works on Analytical Mechanics. (Izd. AN SSSR, Moscow, 1962) (Russian).
5. Davyskib A., Samsonov V.A. On the possibility of gyroscopic stabilization of rotation of a system of rigid bodies. *Prikladnaya Matematika i Mekhanika.* Vol. 59, No. 3, (1995), pp. 385 - 390 (Russian). English transl. in *Journal of Applied Mathematics and Mechanics.* **59**, 3 (1995).

6. Potapenko E.M. Dynamics of a spacecraft with line active control by a gravitational stabilizer. *Kosmicheskie issledovania.* Vol. 26, No. 5, (1988), pp. 699 - 708 (Russian). English transl. in *Cosmic Research.* **26**, 5 (1988).
7. Katz A.M. On the problem of criterion of aperiodic stability. *Prikladnaya Matematika i Mekhanika.* Vol. 15, No. 1, (1951), pp. 120. (Russian). English transl. in *Journal of Applied Mathematics and Mechanics.* **15**, 1 (1951).
8. Banshchikov A.V., Burlakova, L.A. Information and Research System "Stability". *Izvestia RAN. Teoria i sistemi upravlenia.* No. 2, (1996), pp. 13 - 20 (Russian). English transl. in *Journal of Computer and Systems Sciences International.* **35**, 2 (1996) 177-184.

Σ^{it} – AN ALDOR LIBRARY FOR LINEAR DIFFERENTIAL AND DIFFERENCE EQUATIONS (ABSTRACT)

MANUEL BRONSTEIN
INRIA – Projet CAFÉ
2004, Route des Lucioles, B.P. 93
F-06902 Sophia Antipolis Cedex, France
E-mail: `Manuel.Bronstein@sophia.inria.fr`

We describe how the computer algebra library Σ^{it} has evolved from a special–purpose embeddable library[2] to a general purpose ALDOR[1] computer algebra library with special emphasis on efficient algorithms for solving linear ordinary differential and (q)–difference equations. After a short tour of the algorithms recently implemented in Σ^{it}, we outline in particular its following design aspects:

- Its use of advanced features of the ALDOR programming language (post–facto extensions and callbacks) to combine abstract data types and generic algorithms with efficient special algorithms for specific types.
- Its implementations of the algorithms of differential Galois theory over non-algebraically closed constant fields, allowing for efficient future generalisations to linear partial differential equations.
- Its use of abstract generic parsers and interpreters to allow easy communications between Σ^{it} clients and interactive programs (e.g. MAPLE, graphical user interfaces, web services, etc).

References

1. The ALDOR compiler, available from `http://www.aldor.org`.
2. M. Bronstein. Σ^{it}– a strongly-typed embeddable computer algebra library. In J. Calmet, editor, *Proceedings of DISCO'96*, LNCS 1128, pages 22–33. Springer, 1996.

ALLTYPES: AN ALGEBRAIC LANGUAGE AND TYPE SYSTEM

FRITZ SCHWARZ

FhG, Institut SCAI, 53754 Sankt Augustin, Germany
Email: fritz.schwarz@gmd.de

The software system ALLTYPES provides an environment that is particularly designed for developing software in differential algebra. Its most important features may be described as follows: A set of about thirty parametrized algebraic types is defined. Data objects represented by these types may be manipulated by more than one hundred polymorphic functions. Reusability of code is achieved by genericity and multiple inheritance. The user may extend the system by defining new types and polymorphic functions. A language comprising seven basic language constructs is defined for implementing mathematical algorithms. The easy manipulation of types is particularly supported due to a special portion of the language dedicated to manipulating typed objects, i. e. for performing user-defined or automatic type coercions. Type inquiries are also included in the language.

1 Organization of Computer Algebra Software

The software described in this article originated from the desire to provide an environment for implementing high quality computer algebra software. Areas of application are for example the symmetry analysis of ordinary and partial differential equations, finding closed form solutions of ordinary differential equations and Janet base algorithms.

ALLTYPES provides a collection of types especially designed for modeling mathematical data objects occuring in differential algebra. Its fine-structured type system requires special tools for manipulating objects of these various types easily, they are made available in terms of a specialized portion of the language. Furthermore this language defines a small number of powerful control constructs that are especially well suited for implementing computer algebra code. It may be written such that each line may be executed individually, its action may be specified in mathematical terms, and the result may be checked against this specification.

It turns out that many aspects that arise during the design and the implementation of computer algebra software may be better understood if they are considered in the more general context of software engineering. Good references for these aspects are the books by Coad and Yourdon [1], [3] and Meyer [2]. A recent review by Taivalsaari [4] is also useful.

A little consideration leads to the conclusion that the ultimate reason for most of the problems which occur during the development and the maintenance of large pieces of software is the fact that too many lines of code have to be considered at a single time. As a consequence, the two most important principles of software design follow rather naturally.

→ Decompose large pieces of software into smaller ones through *modularity*;

→ Limit the growth of software through *reusability*.

The principles according to which modularity is achieved is the most important part of the design process. In a computer algebra context appropriate modules are represented by *algebraic types*. Each type should represent a concept from the underlying mathematical field. In order to achieve a high degree of reusability, the flexibility of types is increased by *parametrization*, and the system design should support *inheritance*. The basic idea of this latter concept is to allow new types to be based on already existing ones. Since most types in the system are not defined from scratch but inherit from one or more ancestors, inheritance decreases the total size of the system considerably.

The distinction between an *computer algebra language* and a *computer algebra type system* is especially emphazised. The former provides above all *efficiency*, the latter *structure*. The current version of ALLTYPES is implemented on top of REDUCE [6].

2 Design of ALLTYPES

There are basically two different constituent parts of ALLTYPES. On the one hand there is the type system comprising about thirty parametrized algebraic types for modeling mathematical objects supplemented by about one hundred polymorphic functions for working with them. Their action is achieved by a multiple dispatching mechanism that selects the proper methods for the data objects at hand, either at compile time or at run time. A good comparison of single dispatching vs. multiple dispatching may be found in the article by Taivalsaari [4], Section 3.9. Secondly there is a language provided that is especially designed for implementing mathematical algorithms efficiently and safely on top of this type system. These components are described in this section. Theoretical questions on the design of an algebraic type system are discussed in [7], see also [5].

Informally a type describes the features that are common among a collection of mathematical objects. The user interface of any type is made up of its name, its parameters and perhaps certain attributes that are collectively called the *category* of the type. They may be considered as the *type of a type* or a *second order type*. The totality of all types decomposes naturally into the following three kinds.

→ *Basic types* are data objects supported by the underlying language, they are essentially various kinds of numbers, variables and collections of any kind of objects represented as lists. The representation of basic types is chosen for efficiency reasons.

→ *Algebraic types* are the main components of the type system, they serve to represent structured objects like for example univariate polynomials, partial fractions or linear differential equations. Most of them have one or more parameters and belong to one or more categories. Internally they are represented by records.

→ *Set types* are applied for representing homogeneous collections of objects of the same type like for example polynomial ideals, rational function fields or Lie algebras. In the object-oriented terminology they correspond to *container classes*. They are also represented by records.

The most important types that are currently supported by the system are listed below. It shows that the design of ALLTYPES is strongly influenced by its application for differential equations. For each type a full name, the abbreviation used in the system, its parameter list and categories are given.

Basic Types. For efficiency reasons, the representation of numbers and variables is essentially taken over from the underlying computer algebra language. *Number: NUM.* Subtypes *Integer: Z*, *Rational Number: Q*, *Algebraic Number: AN(k)*, *Modular Number: MN(k)* are the number types currently supported by the system. The single integer argument k of MN defines the modulus. The integer argument of AN has the following meaning. Algebraic numbers are considered as members of an appropriate algebraic number field that is defined by a primitive element. This number field determines the type. The totality of algebraic number fields required in a problem is organized globally, the individual number fields are enumerated by the integer k. Any change of these number types is handled by user defined or automatic type coercisions. The general principle is to keep this type information as local as possible and thereby to avoid the introduction of unnecessary field extensions.
Variable: VAR, *Derivative: DER*. Next to the numbers, these are the most elementary building blocks of all symbolic expressions. A derivative may have a number of dependencies attached to it so that it behaves differently within different types belonging to category *DIFFERENTIAL*.

List: LIST. At a low level of the type system *LIST*'s are used to store collections of objects of any kind temporarily.

Algebraic Types. They form the backbone of the type system of ALLTYPES. Each individual type is characterized by its parameters and the categories which are attached to it. There are four categories known to the system at the moment, they are

RING, *FIELD*, *DIFFERENTIAL* and *CONTAINER*.

Membership in any category implies that certain operations are available at compile time. They are required for controling e. g. how to divide out the content of a polynomial, or how to apply derivatives.

A type parameter denoted by an identifier of capitals means a type followed by a colon and the required category, or a data object denoted by an identifier, followed by a colon and its type. Examples of this notation are $C : RING$, $v : VAR$, $\{v_1, v_2 \ldots\} : LIST(VAR)$ or $n : Z$. The *Ordering* parameter O is especially handled. It is either one of the reserved words *LEX*, *GRLEX* or *REVLEX* with the obvious meaning, a matrix of cotes, or a predicate with two arguments deciding the correct order. Subsequently a selection of types is described that form the core of the type system. The category of each type is given after the colon.

Polynomial: $POLY(NUM) : RING$. The main purpose of this type is to act as a mediator between the representation of polynomials in the underlying computer algebra language. Within the type system they often represent those parts of a more structured object for which at a certain point additional structure is not available or not relevant, for example the coefficients of a univariate polynomial that depend on some additional parameters. The coefficient type of any $POLY$ is some number type NUM. It is carried along with the individual objects.

Rational Function: $RATF(NUM) : FIELD$. The same remarks apply as for the preceding type.

Univariate Polynomial: $UPOLY(C : RING, v : VAR) : RING$. The first parameter determines the coefficient type, the second parameter is a variable name with respect to which the data object must be a polynomial.

Univariate Rational Function: $URATF(C : RING, v : VAR) : FIELD$. The parameters have the same meaning as for $UPOLY$.

Partial Fraction: $PARF(C : RING, v : VAR) : FIELD$. The parameters have the same meaning as for $UPOLY$. Univariate partial fractions may be considered as the natural extension of univariate polynomials.

Distributive Polynomial: $DPOLY(C : RING, \{v_1, v_2, \ldots\} : LIST\ VAR, O) : RING$. The first argument determines the coefficient type, objects of this type must be polynomial in the variables $v_1, v_2, \ldots$.

Distributive Rational Function: $DRATF(C : RING, \{v_1, v_2, \ldots\} : LIST\ VAR) : FIELD$. Objects of this type must be rational in the variables $v_1, v_2, \ldots$. Numerator and denominator are $DPOLY$'s with coefficients determined by the first parameter.
Function Field: $FFIELD(C : RING, \{e_1, e_2, \ldots\} : LIST\ VAR) : FIELD$. Objects of this type must be rational in the variables $e_1, e_2, \ldots$ that will be elementary function kernels. Their special handling is taken from the underlying computer algebra language.
Linear Form: $LFORM(C : RING, \{v_1, v_2, \ldots\} : LIST\ VAR, O)$. The first parameter determines the coefficient type. Objects of this type must be linear and homogeneous w.r.t. the variables on the list $\{v_1, v_2, \ldots\}$ or w.r.t. to those variables for which the predicate returns true. If the *Ordering* parameter is provided, it is applied for establishing the respective term ordering, if not, a system dependent default ordering is established.
Quadratic Form: $QFORM(C : RING, \{v_1, v_2, \ldots\} : LIST\ VAR, O)$. The meaning of the parameters is analogous to those in $LFORM$.
Differential Polynomial: $DFPOLY(NUM, d, v, O) : \{DIFFERENTIAL, RING\}$ where NUM is the coefficient type, $d = \{d_1, d_2, \ldots\} : LIST\ DER$ are the dependent variables and $v = \{v_1, v_2, \ldots\} : LIST\ VAR$ are the independent ones.
Linear Differential Form: $LDF(C : RING, d, v, O)$ where C is the coefficient type, $d = \{d_1, d_2, \ldots\} : LIST\ DER$ are the dependent variables and $v = \{v_1, v_2, \ldots\} : LIST\ VAR$ are the independent ones. Objects of this type must be linear and homogeneous in the derivatives of a set of functions $d_1, d_2, \ldots$. With respect to these derivatives they are $LFORM$'s with coefficient type determined by the first parameter.
Linear Differential Operator: $LDO(C : RING, \{v_1, v_2, \ldots\} : LIST\ VAR)$. The first argument determines the coefficient type, the first order partial derivatives are taken w.r.t. to the variables $v_1, v_2, \ldots$.
Linear Ordinary Differential Equation: $LODE(C : RING, y : DER, x : VAR)$. Objects of this type are linear and homogeneous in y and its derivatives w.r.t. x, with coefficient type determined by the first parameter. Subtypes are LODE1, LODE2 and LODE3 with the same parameters.
Ordinary Differential Equation: $ODE(C : RING, y : DER, x : VAR))$. Objects of this type must be polynomial in y and its derivatives w.r.t. x with coefficient type determined by the first parameter. Subtypes are ODE1, ODE2, ODE3, ABEL and RICCATI with the same parameters.
Partial Differential Equation: $PDE(C : RING, u, x)$ where C is the coefficient type, $u = \{d_1, d_2, \ldots\} : LIST\ DER$ are the dependent variables and $x = \{v_1, v_2, \ldots\} : LIST\ VAR$ are the independent ones.

Vector: $VECTOR(E : TYPE, n : Z)$. The first argument determines the element type, the second argument the number of components, they are enumerated from 1 to n.
Matrix: $MATRIX(E : TYPE, m, n : Z)$. The first argument determines the element type, the second and third argument the number of rows and columns, they are enumerated from 1 to m and 1 to n respectively.

Set Types. Many mathematical objects are a homogeneous collection of some other objects in the first place. They are homogeneous in the sense that each object has the same type. Additional features make them into new mathematical entities deserving a new representation within the type system. All types in this subesction belong to categeroy $CONTAINER$.
Set: SET(TYPE):CONTAINER. The main purpose of this type is to provide the operations typical for a set from which all types of category $CONTAINER$ inherit them.
Polynomial Ideal: $POLID(NUM)$. Elements generate ideal of polynomials of type $POLY$ with like number type.
Differential Polynomial Ideal: $DFPOLID(NUM, d, v, O)$ describes a differential ideal generated by $DFPOLY$'s of the same type arguments.
Linear Form Module: $LFMOD(C : RING, \{v_1, v_2, \ldots\} : LIST\ VAR, O)$. Elements generate a module of $LFORM$'s with like parameters.
Linear Differential Form Module: $LDFMOD(C : RING, d, v, O)$ describes a module generated by LDF's with the same type arguments.
Lie Algebra of Vector Fields: $LIEVEC(C : RING, \{v_1, v_2, \ldots\} : LIST\ VAR)$. Elements are LDO's of the type determined by the parameters, they must generate a Lie algebra.
Lie Algebra: $LIEALG(n : Z)$. The single argument determines the dimension of an abstract Lie algebra. It is specified by its structure constants. If $e_1, e_2, \ldots, e_n$ is the basis w.r.t. which they are defined, they are given in the arrangement

$$(e_1, e_2), \ldots (e_1, e_n), (e_2, e_3), \ldots, (e_2, e_n), \ldots, (e_{n-1}, e_n).$$

Polymorphic Functions and Methods. The operations that may be performed on the various data objects are specified in mathematical terminology for the *polymorphic functions*. They are applied by the user for writing the code that realizes the desired algorithms. A polymorphic function cannot come into action by itself, only after it is replaced by the appropriate *method* the code may be executed. In general many methods corresponding to different types, possibly with varying number of arguments, may be defined for a single polymorphic function name. The proper method is determined by the types of the data objects to be operated upon. It is the responsibility of the user to assure that the required method is available at execution time. A method may either be defined in the code of the respective type, or it may be inherited from some other type.

There are several advantages of organizing software in this manner. On the one hand, the degree of reusablitiy is increased by writing code in terms of polymorphic functions instead of the methods itself. Secondly introducing a new type means going through the collection of polymorphic functions and implementing those methods for the newly defined type that are not inherited or have to be redefined, and possibly additional methods that are valid exclusively for this newly defined type and its descendants. This is a unique and well-defined procedure including the guarantee that *all* necessary changes have been performed *without* explicit changes of already existing code. Finally for the user it is much more efficient to become acquainted with a single set of polymorphic functions providing essentially the same functionality for the complete type system, instead of repeating this effort for the individual types. In other words, the *learning curve* increases faster.

The mechanism applied in ALLTYPES that gives rise to a polymorphic function call being replaced by the proper method will be explained now in more detail. The methods corresponding to a polymorphic function are denoted by extending the polymorphic function name by the type names, separated by underscores; they are called *type extensions*. There are as many type extensions as there are parameters. The significance of the various type extensions is not symmetric however. The first type extension of a method determines the modulet where it is defined. For a *type homogeneous* polymorphic function the first extension of any of its methods determines also the type of the returned result. As an example the type homogeneous polymorphic function $Multiply$ may be called as $Multiply(u, v)$ by a user. Valid examples of method calls that may be generated from it by ALLTYPES are

$$Multiply\_POLY\_POLY(u : POLY, v : POLY) : POLY,$$
$$Multiply\_POLY\_BASIC(u : POLY, v : BASIC) : POLY,$$
$$Multiply\_LFORM\_TYPE(u : LFORM, v : TYPE) : LFORM.$$

The first two methods define the multiplication of an object of type $POLY$ by another $POLY$, or by an object of type $BASIC$ which may be some number or a variable. The last method defines the multiplication of an object of type $LFORM$ by an object of any type that is consistent with the type homogeneity of $Multiply$. Essentially this means that the coefficients of the first argument u are multiplied by v. In accordance with the type homogeneity of the polymorphic function $Multiply$ the type of the returned object is as indicated after the closing bracket of the method definition. A method

$$Multiply\_LFORM\_LFORM(u : LFORM, v : LFORM)$$

is not allowed by type homogeneity because the result cannot be coerced to an object of type $LFORM$. If u and v for example both have the same type $LFORM(POLY\ NUM, \{a, b, c\})$, an appropriate call for multiplying them would be

$$Multiply(u|QFORM(POLY\ NUM, \{a, b, c\})|, v)$$

from which the method call

$$Multiply\_QFORM\_LFORM(u|QFORM(POLY\ NUM, \{a, b, c\})|, v)$$

is generated and the result of type $QFORM(POLY\ NUM, \{a, b, c\})$ is returned. Type homogeneity entails also the different answer for the two polymorphic function calls

$$Multiply(u : LFORM, v : RATF) : LFORM.$$

and

$$Multiply(u : RATF, v : LFORM) : RATF$$

Another example showing the flexibility gained by this software organization is the polymorphic function $Coefficient$. Some of its method calls are

$Coefficient\_POLY\_POLY(u : POLY, v : POLY)$,
$Coefficient\_POLY\_BASIC(u : POLY, v : VAR)$,
$Coefficient\_UPOLY\_BASIC(u : UPOLY, n : Z)$,
$Coefficient\_LODE\_BASIC(u : LODE, n : Z)$,
$Coefficient\_LFORM\_BASIC(u : LFORM, v : VAR)$.

In the first call, the second argument v must be a monomial, its coefficient in u is returned. In the second call, u must be linear in the variable v, otherwise a runtime error occurs. In the two subsequent method calls, n must be a positive integer, the coefficient of the $n-th$ power of the univariate polynomial in the distinguished variable, or the coefficient of the $n-th$ derivative in the linear ordinary differential equation respectively are returned. In the last call, v must be a variable w.r.t. which u is linear and homogeneous.

The complete set of methods cannot be discussed in this article, there are almost one thousand of them defined in the system. All the information necessary for working with them may be obtained by applying a few utility functions provided for this purpose.

The Language Constructs. The complete language is composed of two fundamentally different parts. On the one hand, there is the *algebraic language* which is applied for manipulating algebraic objects according to the mathematical rules that are valid for them, for example arithmetic, differentiation, integration and many others. Secondly, there is the *type manipulation part* that is applied for manipulating the types of the mathematical objects such that they satisfy the constraints imposed by the type system. This latter part of the language is syntactically separated from the former by the occurrence of a pair of bars |..|, i. e. any piece of input code enclosed between two bars specifies a type manipulation. These two parts of the language will be described now one after the other.

The fundamental syntactic units of the ALLTYPES language are the *FunctionConstructor* and the *IterationConstructor*. They are partially described by the following productions where keywords are given in boldface letters. Here a *Set* has the meaning of a *LIST*.

$$FunctionConstructor = FunctionName(Expression|$$
$$Predicate, Iterator\{, Iterator\})$$
$$FunctionName =$$
$$\textbf{applies}|\textbf{select}|\textbf{exists}|\textbf{satisfies}|\textbf{separate}$$
$$Iterator = Variable\ \textbf{from}\{..\textbf{from}\}Set|Variable$$
$$\textbf{from}\ Integer, .., [\textbf{by}\ Integer, ..,]Integer$$
$$IterationConstructor = \textbf{iterate}(\textbf{true}?Predicate,$$
$$Expression\{, Expression\})|\textbf{iterate}($$
$$Expression\{, Expression\}, \textbf{true}?\ Predicate)$$
$$OrderingConstructor = \textbf{orders}(BooleanExpression,$$
$$Variable\ \textbf{and}\ Variable\ \textbf{from}\ Set[, Set])$$

The remaining part of the syntax and the semantics will be explained informally. An *Expression* is any piece of code that is syntactically and semantically accepted by the interpreter or compiler and returns a result. In *Axiom* it is essentially an expression, in a LISP system it is very similar to a *form*. The free variables occurring in an *Expression* or a *Predicate* are evaluated in the environment in which the *FunctionConstructor* is called. All local variables are attached to certain sets by the subsequent *Iterator*.

The *Variable* in the first production alternative for the *Iterator* determines a variable name which may occur in the *Expression*, it specifies an element of the *Set*. Each occurrence of the keyword **from** determines a level of nesting of *Set*. If there is a single occurrence, the iteration occurs over elements of a set, if there are two, elements of a set of sets are taken, and so forth. If there is more than a single iterator, the iteration takes place over several sets, set of sets etc. in parallel. Both the number of iterators and the depth of nestings of sets are arbitrary up to the following constraints. The level of nestings in each iterator must be the same, the same is true for the cardinality of sets at each level. The second production for the *Iterator* allows it to iterate over intervals of integers, the stepsize being determined by the optional *Integer* following the keyword **by**.

The **applies** constructor causes the *Expression* to be evaluated for each selection of variables as determined by the *Iterator*. It returns an object with the same structure as it occurs in the latter with the result of the evaluations taken at the respective positions. The objects may be sets, sets of sets etc.

In the **select** and the **exists** constructors the *Expression* must be a predicate. The returned object of the **select** constructor has the same structure as those occurring in the iterators, it contains those elements of the *Set* occurring in the last *Iterator* for which the *Expression* evaluates to true. The **exists** constructor returns only the first element occurring in this *Set* for which the *Expression* is true. If the *Expression* begins with the keywords **maximal** or **minimal**, the remaining part of it must evaluate to an integer. In this case the **select** constructor returns all elements from the *Set* occurring in the last *Iterator* for which the *Expression* returns the maximal or minimal value respectively. The same applies to the **exists** constructor except that only the first element with this property is returned. A meaningful application will presume the proper number of occurrences in either case.

The **satisfies** constructor is actually a predicate. It returns true if all evaluations of its first argument which must be a predicate also evaluate to true.

The **separate** constructor generates new sets of objects from the *Set* which occurs in the last *Iterator*. The newly created sets are formed such that the value of the *Expression* is the same for all elements of its member sets.

The two versions of the *IterationConstructor* are the basic syntactic unit for an unspecified number of repetitions of an action. In the first version, the *Predicate* is evaluated before any other action is taken. If the result is true, the *Expression*'s are evaluated from left to right one after the other. This proceeding is repeated until the *Predicate* evaluates to something that is not true. In this case the result of the last evaluation in the *Expression* is returned. If the *Predicate* is not true at the first evaluation, the value of the complete *IterationConstructor* is false. In the second version, a similar evaluation scheme is applied except that the *Predicate* is evaluated at the end of each round. The returned value of the constructor is always that of the last evaluation in the *Expression*.

The *BooleanExpression* in the **orders** constructor contains the two arguments *Variable* of the subsequent **and** clause. It causes the *Set* to be rearranged such that the evaluation of the *BooleanExpression* with any pair of elements such that the first member is located to the left of the second member evaluates to *true*. If the last optional argument of type *Set* is provided, its elements are rearranged in the same way as the first argument and this latter set is returned.

3 Defining Methods and Local Functions

Methods and local functions are the pieces of code that actually execute the algorithms, they are implemented in terms of *Reduce* procedures. Whereas the methods are supposed to be used in as many modules as possible via polymorphic function calls and the inheritance scheme, the latter concept is local and should protect the user against unintentional reuse of function names. This is achieved essentially by introducing a name space for each module. However in order to increase the flexibility of the system, a mechanism is provided for evading this constraint in a clear and well defined manner by means of the *Export* and *Import* constructs that are also explained there.

Standard Method Definition. Any method definition consists of two major parts.

→ The headline. The general structure for defining the outside view of a new method for a polymorphic function *Foo* is

Method Foo(u:$TYPE_1$*,v:*$TYPE_2$*,...):*$TYPE_3$*;*

Any method name must begin with a capital letter.

→ The body. It is a *BEGIN - END* block that starts with a declaration part which is followed by the implemenation part.

There is some interrelation between these two entries that will become clear in the detailed description below.

Any argument u of the newly defined method for *Foo* has to be accompanied by a type declaration, separated by a colon. In general it has the form

$u{:}TYPE_i$ or $u,v,\ldots :TYPE_i$

if several arguments of the same type occur.

$TYPE_i$ is the name of any type defined in the ALLTYPES type system including *BASIC*. If the type of a method argument is declared to be $TYPE_i$, the object occuring in that position at compiletime or at runtime must be equal to that type or one of its descendants.

The declaration *TYPE* has a special meaning. Because it is the ancestor of any other type, this method will be attached to the polymorphic function call only if no other method is found when traversing the inheritance tree of the type of the actual argument.

The significance of method arguments is not completely symmetric. The type of the first argument determines the module of the algebraic type to which it belongs. Therefore it cannot be *BASIC* or *TYPE* because no methods do exist for them.

To some extent it is possible to enforce type parameter checking for the arguments of a method definition. If the headline of a method definition is

Method Foo(u:$TYPE_1$*(A,B),v:*$TYPE_2$*(A,C,D))TYPE;*

the first parameters of $TYPE_1$ and $TYPE_2$ must be identical to each other, whereas there is no relation between the remaining parameters. In general, the same identifier in the same parameter position means that the corresponding parameter values must be the same. If this is violated at compiletime or at runtime, an error message is obtained.

The declaration $TYPE_i$ following the closing bracket in the headline specifies the type of the result returned by the method. In the current version of ALLTYPES there is made no use of it except in the special case described below.

The $BEGIN - END$ block of any method definition starts with the type declaration part of the local variables. Valid type declarations are

- → $TYPE_i$ *a,b,... ;* declares a, b etc. to be of the algebraic type $TYPE_i$, but *not* one of its descendants.

- → *SET* $TYPE_i$ *p,q,... ;* declares p, q etc. to be SET types, i. e. homogeneous collections of the respective element type.

- → *scalar x,y,... ;* is applied if there is no type information available for this variable at compile time.

The main purpose of the type declarations both in the headline and the body are its use for replacing polymorphic function calls by the proper methods. Whenever an argument declared as `scalar` occurs, the method cannot be determined at compile time. In these cases a piece of code is inserted that performs this substitution at runtime. Therefore it should be avoided whenever possible.

The type of a method argument may not be redefined in the declaration part of the body. The only exception from this rule is the declaration *BASIC*, an argument declared as *BASIC* may be redefined in the method body as *SET* $TYPE_i$.

Defining Local Functions. The basic construct for defining local functions is identical to that for defining methods. That means, the headline for defining a local function *Foo* is

$LocalFunction\ Foo(u{:}TYPE_1, v{:}TYPE_2){:}TYPE_3;$

The effect of this definition however is completely different from a method definition. The function with the name *Foo* is known *only within the module* in which it is defined. Local functions with the same name *Foo* may be defined in any other module as well with different specification and body. The purpose of this construct is to allow defining functions that are local to a single module without interfering with the name space of any other module.

It turns out that this concept becomes much more useful if this locality principle is slightly softened according to the following rules. If a local function *Foo* is defined in a certain module $MODULE_1$ and it ought to be called also in $MODULE_2$ without redefining it there, the declaration

Export Foo $\{MODULE_2\}$

must be included in the code of $MODULE_1$ where *Foo* is defined, and the declaration

Import $MODULE_1$ $\{Foo\}$

must be included in the code of $MODULE_2$.

4 Extending the Type System

Defining new types or packages in ALLTYPES requires a deeper knowledge of the system architecture than just defining new functions. This is particularly true for the type definition because the complete inheritance scheme is involved. Introducing a new type into the system essentially involves three major steps.

→ Define the new type on a separate file.

→ Provide the proper type coercion functions.

→ Recompile the extended ALLTYPES system.

Creating New Types. The language construct *DefineAlgebraicType* is provided for extending the type system. The syntax for the headline of a newly defined algebraic type is formally defined by the following production rules.

DefineAlgebraicType *Type*[**:** *Category*] [**Is_a** *Parent*];

Type::= *TypeName*(*ParameterName* [,*ParameterName*]*)

Category::=CategoryName | {*CategoryName* [,*CategoryName*]*}
Parent::=ParentName | {*ParentName* [,*ParentName*]*}

Any *TypeName* must be an identifier of capitals in order to avoid name clashes with already existing identifiers in the *Reduce* system.

The optional clause *Is_a* of a *DefineAlgebraicType* statement allows it to define a new type as a descendant of one or more already existing types. Any type occuring as a *ParentName* in the *Is_a* clause must be defined before, otherwise an error message is obtained at compile time. If there is more than a single parent type on the *Parent* list, the order of types on this list determines the traversal of the inheritance tree when methods are attached to polymorphic function names. The rightmost type has the highest priority for this selection. This completes the discussion of the headline of a new type.

The optional argument *Category* defines the type of a type, or a second order type [7]. It is defined by the statement

DefineCategory *CategoryName* [*Is_a CategoryName*|
{*CategoryName**}] {*FunctionName* [,*FunctionName*]*}

Each *FunctionName* is the name of a polymorphic function. The list of functionnames is optional if the new category inherits from two or more categories. For any type belonging to this category a corresponding method must be available, either by definition in the respective type module or by inheritance. Otherwise an error message is obtained at compile time. The optional *Is_a* clause allows it to inherit from already existing categories. All methods of the argument category, or the list of arguments, are required for the newly defined category.

The inside of a type comprises its *internal representation* and the *local functions.* The latter have been discussed in the preceding section. For any new type there is the optional first statement

InternalRepresentation(*FieldName:=nil[,FieldName:=nil]**)

Each *FieldName* must be an identifier of capitals. This statement is necessary if in a new type the internal representation requires fields which do not occur already as parameters of this type, and are not inherited from the parameters or an interal representation of a parent.

If a new type is defined from scratch, *DefineAlgebraicType* is applied without the *Is_a* clause. On the other hand, if it is based on types already existing in the system, the option with the *Is_a* clause is applied. Any type A may inherit from one or more types B, $C, \ldots$ in the system. Valid type definitions are

$$\begin{array}{l} DefineAlgebraicType\ \ A\ \ Is\_a\ \ B\quad or \\ DefineAlgebraicType\ \ A\ \ Is\_a\ \ \{B, C, D, \ldots\} \end{array} \tag{1}$$

defining a single or multiple inheritance relation respectively. In detail, any inheritance relation implies the following features for the descendant.

→ All methods of the ancestors are inherited to the descendant. The details of how repeated inheritance and multiple inheritance are handled are explained subsequently;

→ All categories of the ancestors are inherited to the descendant;

→ In the internal representation, the fields of the descendant are the union of the fields of its anchestors.

Inheritance relations create a hierarchical structure between the individual types of the system. If new types are defined by the user, these hierarchies are extended appropriately by the system.

5 Summary and Outlook

The system ALLTYPES described in this article is very well suited for implementing sophisticated software for working with differential equations. Its organization by means of a fine structured type system of well adjusted types has turned out to be a fundamental prerequisite for developing high quality software efficiently. A detailed description of ALLTYPES including a manual for working with it will be available soon [8].

References

1. P. Coad, E. Yourdon, *Object-Oriented Analysis* and *Object-Oriented Design*, both published by Yourdon Press, Englewood Cliffs, 1991.
2. B. Meyer, *Object-oriented Software Construction*, Prentice Hall, 1997.
3. I. Graham, *Object-Oriented Methods*, Addison-Wesley, 2001.
4. A. Taivalsaari, *On the Notion of Inheritance*, ACM Computing Surveys **28**, 438-479 (1996).
5. J. F. Fedoraro, *The Design of a Language for Algberaic Computation Systems*, Thesis, Berkeley, 1983.
6. A. C. Hearn, *Reduce User's Manual*, part of Reduce.
7. A. Weber, *Structuring the Type System of a Computer Algebra System*, Dissertation, Universität Tübingen, 1992.
8. D. Scheller and F. Schwarz, *ALLTYPES: An ALgebraic Language and TYPE System*, system manual, to appear.

SECURE INTERNET ACCESSIBLE MATHEMATICAL COMPUTATION FRAMEWORK*

DONGDAI LIN[†] ZHIMIN SONG[†] PAUL S. WANG[‡]

1 Introduction

The Mathematical computing is inevitably becoming distributed over the Internet – for easy distribution of mathematical materials; to make specialized computations widely accessible; to allow easy interoperability; and to aggregate functionalities from different systems. Indeed, making mathematical communication easy over the Internet has *many* potential applications.

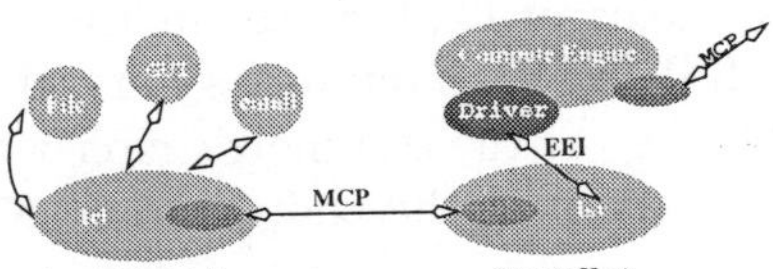

Figure 1. IAMC Architecture

Internet Accessible Mathematical Computing(IAMC) Framework developed at Kent State University is an infrastructure for bringing mathematical computational and educational services over the internet, it includes an IAMC client, an IAMC server, and the *Mathematical Computation Protocol* for connecting IAMC clients and servers effectively and efficiently over the internet, see Figure 1. The computation powers of an existing mathematical compute engine can be made available on the Web/Internet by an IAMC server.

Currently, IAMC depends on the standard TCP/IP protocol suite for security. Due to the vulnerabilities of present IP-based network, in current IAMC framework, trusted hosts can be impersonated by exploiting existing vulnerabilities in IP-based networks, data and messages delivered between server and client are vulnerable to eavesdropping and interference from unauthorized intruders. Depending on the sensibility of the information and the type of tampering, the potential damage can be significant. To reduce these risks, we have incorporated SSL/TLS into IAMC. SSL/TLS protocol is the most popular cryptographic security protocol, which provides communication privacy over the Internet. The protocol allows client/server applications to communicate in a way that is designed to prevent eavesdropping, tampering, or message forgery. The new framework can then provide cryptographic authentication, data privacy and integrity.

---

*see ICM Technical Report ICM-200202-0002, http://icm.mcs.kent.edu/reports/index.html, Kent State University.

†Institute of Software, Chinese Academy of Sciences, Beijing 100080, China.
‡ICM, Department of Computer Science, Kent State University, Kent, Ohio 44242, U.S.A.

2 Secure Mathematical Computation Protocol (S+MCP)

IAMC clients and servers are connected by the Mathematical Computation Protocol(MCP). MCP handles one-time computation requests and persistent computation sessions. In order to make IAMC framework secure, we need to extend MCP to be able to invoke SSL/TLS security service. The extension is done by adding a method, called **`security-negotiation`**, to the class of **`Initialization`** of MCP protocol. Method **`security-negotiation`** is used to trigger a SSL/TLS negotiation between the client and server, it has three arguments: **`Authentication`**, **`Ciphersuite Priority`** and **`Key Length`**.

Ciphersuite Priority in the S+MCP client's security-negotiation method offers a Ciphersuite list which will be used by SSL/TLS client in its ClientHello message. Ciphersuite Priority in the S+MCP server's security-negotiation method indicates the SSL/TLS server's choice of Ciphersuite in the order of its preference which will be used by SSL/TLS server in its ServerHello message. The final key length of the public key or the symmetric cipher that will be used by SSL/TLS will be the maximum of the choices of S+MCP client and S+MCP server.

When security-negotiation starts, a dialogue box is displayed for the end-user to choose among available Ciphersuites, the corresponding key lengths and authentication flags. The user input is passed to the client-side security-negotiation method. On the server side default arguments for S+MCP server's security-negotiation method are preset. A successful SSL/TLS negotiation sets up a secure channel, over which IAMC client and server can communicate securely.

During the initialization stage, the IAMC client and server begin the security negotiation immediately after the client-server connection. After the security-negotiation methods, regular MCP messages can be sent over a secure channel.

S+MCP supports three authentication modes: server and client both authenticated, server authenticated, and neither authenticated (total anonymity). Whenever the server is authenticated, the channel is secure against man-in-the-middle attacks, but total anonymous sessions are inherently vulnerable to such attacks. Anonymous servers cannot authenticate clients. Notice that the client authentication is important in practice, it is useful for servers that either charge a fee or provide highly secure computations.

In order to use S+MCP with IAMC, IAMC clients and servers must first obtain their certificates from a Certificate Authority (CA). These certificates will be used for signing, for key exchange or for both.

SOLUTION SPACE SMOOTHING FOR VLSI MODULE PLACEMENT: A COMPUTATIONAL STUDY

SHEQIN DONG, XIANLONG HONG, SONG CHEN AND SHUO ZHOU

Department of Computer Science and Technology, Tsinghua University, Beijing, P.R.C,100084, Tel:+86-010-62785564 Fax:+86-010-62781489

E-mail: dongsq@tsinghua.edu.cn

Over last eight years or so, solution space smoothing has been used to improve the performance of heuristics that solve traveling salesman problem. Solution space smoothing allows a local search heuristic to escape from a poor, local minimum. In recent years, many floorplanning representations, such as Sequence-Pair (SP), O-tree, Bound-Sliceline-Grid (BSG), Corner Block List(CBL) , have been proposed. To make the coding scheme based placement algorithm more efficient and to reduce the effect of local minimum point, we propose a solution space smoothing method for VLSI placement in this paper. We design five smoothing functions and test them with placement with pre-placed modules. Experiment results demonstrated that the proposed method is very efficient for VLSI module placement.

1 VLSI Placement Using Solution Space Smoothing Method

Given a set $M=\{M_1, M_2, \ldots, M_n\}$ of circuit building blocks lie parallel to the coordinate axes in the first quadrant of the plane. A set of nets specifying the interconnections between pins of blocks and a set of pads are also given. A placement $P = \{M_i(x_i, y_i)|1 \leq i \leq n\}$ is an assignment of coordinates to the lower left corners of n rectangular blocks such that there is no two rectangular blocks overlapping. Each M_i is defined by a tuple (h_i, w_i) , where h_i andw_i are the height and the width of block M_i, respectively . The objective of the placement is to find an assignment so that the chip area and interconnection wire-length between blocks are minimized while satisfying the given constraints, if any.

A simple placement instance, without considering the nets among blocks, is that all the circuit building blocks to be placed on a chip have the same size. We can use it as an initial smoothed solution space for a placement instance that has the same number of circuit blocks. In practice, we set the initial placement instance with all circuit blocks have the same size as (1) and (2). Every pin on the blocks will have its new position. Assume that, on block M_i, the jth original pin position is $(PinX^j, PinY^j)$, after the smooth operation that all blocks were reduced to the same size, the new pin position will be as (3).

$$\begin{array}{l}\overline{w} = \min\{w_1, \ldots, w_i, \ldots, w_n\} \\ \overline{h} = \min\{h_1, \ldots, h_i, \ldots, h_n\}\end{array} \quad (1) \qquad (\overline{w}, \overline{h}) \quad (2) \qquad \begin{array}{l} PinX^j * (\overline{w}/w_i) \\ PinY^j * (\overline{h}/h_i)\end{array} \quad (3)$$

A series of simplified placement instances can be created by a specified smoothing factor, α. To keep the similarity between instances in the series, we require the sizes of blocks changed in a slightly, gradually and monotonously increasing mode. After apply a smoothing function $S(\alpha)$, The size of ith block can be derived by applying (4), and the new pin position will be derived by applying (5).

$$S(\alpha) : \begin{cases} w_i(\alpha) = \overline{w} + (w_i - \overline{w})^\alpha \\ h_i(\alpha) = \overline{h} + (h_i - \overline{h})^\alpha \end{cases} \quad (4) \qquad \begin{array}{l} PinX^j * (w_i(\alpha)/w_i) \\ PinY^j * (h_i(\alpha)/h_i) \end{array} \quad (5)$$

To guarantee the sizes of blocks changed in a slightly, gradually and monotonously increasing mode, we normalize items in formula $S(\alpha)$ to the range (0, 1).

When α is decreased from a large number to 1, a series of simplified placement instances are generated according to formula $S(\alpha)$. A solution space generated from a larger α exhibits a smoother terrain surface, and a solution space generated from a smaller α exhibits a more rugged terrain surface. When $\alpha >> 1$, each block size will be reduced to the initial one; when $\alpha = 1$, each block will be of its original size.

Except for the smoothing function $S(\alpha)$, we also design another five smoothing functions similar to the functions that were used to solve traveling salesman problem.

We use BSG coding scheme and its neighborhood solution producing strategies to search an optimal or nearly optimal solution under certain α value using the cost function $(area + \lambda * total\_wirelength)$. Where $area$ is the area of the smallest rectangle enclosing the placement, $total_w irelength$ is the sum of all the net length using the half perimeter model, while λ is the weight. Start from the initial placement instance that has a fairly "flat" solution space, and find the solution to this simplified problem using local search algorithm. The solution of the placement instance is then taken as the initial placement to the next placement instance that has a slightly more complicated solution space. The placement instance is again solved using the same local search algorithm. The above process is repeated until the final placement instance having the original solution space is solved.

For placement with pre-placed modules, in object function, modules whose coordinates are more than pre-placed coordinates receive penalty without fail, otherwise, the modules will be forced to have the pre-defined coordinates.

SIMATH — RECENT DEVELOPMENTS IN TMU

MATSUI T. KOBAYASHI D. ABE M. NAKAMULA K.
Department of Mathematics, Tokyo Metropolitan University
*E-mail: simath@tnt.math.metro-u.ac.jp**

1 New Centre of SIMATH

We announce that **SIMATH**, which is a computer algebra system for algebraic number theory developed at Universität des Saarlandes in Saarbrücken [5], Germany, is now maintained and developed at TMU (Tokyo Metropolitan University) in Tokyo, Japan. It is significant that **SIMATH** has various functions for elliptic curves [12]. The development was, however, almost stopping for a few recent years. We decided to take over the project at TMU for inheritance of those packages. Two new features are in the scope of the next release:

(i) supporting the system for 64-bit platforms,
(ii) adding a package by the CEM (Cyclo-Elliptic Method).

The former is a common problem for many projects. The latter is an algorithm for a certain type of number fields.

2 Supporting 64-bit Platforms

For supporting 64-bit platforms, we decided to use `int` as the basic data type. It is the simplest way to make the system work on both 32-bit and 64-bit computers. We can assume that `int` is always 32 bit wide, see [7], [8].

It is a desirable way to separate the software into a lower level multiprecision library and an upper level mathematical component. Actually **SIMATH** uses third parties' multiprecision packages, but it uses them as locally modified source files. The use of the packages is not fully documented, no one knows whether they are still maintained. This has made maintenance hard. We plan to accomplish the separation in the next major release.

3 Installing the CEM

Starting from cyclotomic or elliptic units, the CEM simultaneously computes the class number and fundamental units for a subfield of an abelian extension

*Research in part supported by the JSPS and the Siemens AG.

of an imaginary quadratic field. The simplest one is a real quadratic field. In this case, we have already translated the code in [6] to a program on **SIMATH**.

The next targets are a cyclic cubic field [4] and a cubic field with negative discriminant [6]. Several problems must be solved in advance: computation in the ray class group of the rational number field or an imaginary quadratic field, computation of special functions such as the Dedekind eta function or other elliptic modular functions, some arithmetic of cyclotomic fields.

4 Future Development

The **SIMATH** system contains not so many functions for number fields. It is our future important task to install algorithms as is described in [1], [2], [9], [10].

The system has been developed by a rather small group. For continuous and stable development, it is better to make the group independent of one person or one small organization. We start to use the CVS (Concurrent Versions System) [3] for open source development of the system.

Adding the CEM as a package of **SIMATH** and fixing a known bug, we find that we can get useful information neither from the manual [11] nor the sources of other packages. There is no manual for developers. For open source development by the CVS, it is strongly required to make a manual for developers open to the public.

References

1. Cohen H., GTM 138, Springer-Verlag, 1993.
2. Cohen H., GTM 193, Springer-Verlag, 2000.
3. Fogel, K., *Open Source Development with CVS*, The Coriolis Group, 2000.
4. Gras, M.-N., J. reine angew. Math., **277** (1975), 89–116.
5. Hollinger, C. and Serf, P., in *Computational Number Theory*, Walter de Gruyter, 1991, 331–342.
6. Nakamula, K., in *Computational Number Theory*, Walter de Gruyter, 1991, 139–162.
7. Open Group, `http://www.opengroup.org/public/tech/aspen/lp64_wp.htm`.
8. Open Group, *Go Solo 2 — The Authorized Guide to Version 2 of the Single UNIX Specification*, Prentice Hall, 1997.
9. Pohst, M. and Zassenhaus, H., *Algorithmic Algebraic Number Theory*, Cambridge University Press, 1989, Rev. 1997.
10. Pohst M., DMV Seminar 21, Birkhäuser, 1993.
11. SIMATH Gruppe, `http://simath.info/html_manual/index.html`.
12. Zimmer H.G., in *Number Theory*, Walter de Gruyter, 1998, 541–595.

STRUCTURES OF THE POLYNOMIAL TYPES IN PROBLEMS OF THE INVARIANT ANALYSIS AND SYNTHESIS OF DIFFERENTIAL EQUATIONS

A. V. FLEGONTOV

Institute for Informatics and Automation of the Russian Academy of Science, 14-th Line, 39, 199178, St.-Petersburg, Russia

E-mail: afleg@spiiras.nw.ru

Consider new methods of structural synthesis of dynamical systems admitting the basic structure of analytical representation of the solutions in a class of polynomial functions with a priori symmetry. Examples of systems of differential equations and dynamical systems with control are given.

1 Invariant Synthesis of Differential Equations

The methods of invariant synthesis of ordinary differential equations are based on engineering of representation of integrated manifold P_i above a differential ring of polynomials $P[\tau]$ (the initial items of information are indicated in work [1], [2]). The problems of synthesis and of analysis in doing so are formulated as follows: 1. The invariant synthesis. Construct a system ODE with a priori symmetry properties on a given pair of functions $(\phi(\tau), \psi(\tau))$ of a determined structure; 2. The invariant analysis. The problem of analysis is put as return to 1. On a system ODE to restore manifold in a given class of functions with by symmetry properties.

2 D - complex

Shaping such models by the essential image rests in symmetred and structured analysis and syntheses, so knowledge of symmetry of equations brings about using the group operators and the whole multitude methods of modern theory of Lie-Baklund groups. Appear, herewith, generators of groups of transformations (particularly discrete) and generate families of commensurable dynamic models. But structured resemblance, in turn, allows to select a class of models in the forms of certain D - complex.

For given manifolds M_{ki}, M_{kj}, full differentiation operator D and discrete symmetry operator $X_i \in G_a$ (generator of group) D - complex defines for $\forall q$ l - parametric family of solutions for l - classes of differential equations F_k^{q-l}, general solutions a class of equations F_k^l and particular solutions for their differential prolongations $(n > q)$.

Consequent using the full differentiation operators and symmetry for problems of syntheses of dynamic models [3] brings about the algorithm of autogenerations of models, where are saved all a priori symmetries for getting models and are generated new prolongue for each k - prolongation. Using the discrete symmetries brings about changing a structure integral manifolds got models, definied by same source differential field of algebraic functions.

3 Examples

1. At the study of behaviour dynamics 2 order in theories of dynamic systems are often meet multivariable nonlinear first order equation belong to nonwhich family of polynomial type models, which factors are multinomials comparatively hang and independent variable [3]. Such models are presented in the forms $y' = P/Q$, where number of free parameters in P and Q is defined by the structure and corresponding parameters integral manifold. Shall synthesize such models from holomorphic subset in parametric type $x = f_1/g$, $y = f_2/g$, where

$$f_i = \alpha_{i1}C_1\tau + \sum_{k=2}^{n} \alpha_{ik}\tau^k, \quad g = \gamma_0 C_1 + \sum_{k=2}^{n} \gamma_k \tau^k. \tag{1}$$

Using an algorithm to autogenerations will bring to 10-parametric nonlinear polynomial models not only, partly provide, but, under $n \geq 4$, and to 18-parametric type. 2. Consider dynamic models with control of a kind: $y_i' = f_i(x, y_i, u), \quad y_i \in R^n, \quad u \in U \subset R^m$. The condition of convertibility of systems simplifies the analysis symmetry, enables to use engineering of integrated manifold (after the manipulation a structure of control is defined and it is required to decide a problem of restoration of control on a given system with by symmetry properties), but also to modify some equations to a more simple kind. We shall cause an example of such updating, arising at the group analysis of one problems of automatic control [1].

References

1. A. Flegontov, *Informatics and Control.* Proceedings of the Int. Conf. ICI&C'97. V.2, St.Petersburg: SPIIRAS, 1997.
2. A. Flegontov, *Computer Algebra in Scientific Computing.* Int. Conf. CASC-98. St.Petersburg: Euler IMI, 1998.
3. A. Flegontov, *6th International IMACS Conference on Applications of Computer Algebra* IMACS ACA 2000, St.Petersburg: Euler IMI, 2000.

Author Index

Index of Mathematical Software Systems